of the Elements

☐ Metals
☐ Semi-metals
☐ Non-metals

Y0-BYI-777

			3a 13	4a 14	5a 15	6a 16	7a 17	8a 18
								2 He 4.003
			5 B 10.81	6 C 12.01	7 N 14.01	8 O 16.00	9 F 19.00	10 Ne 20.18
8b 10	1b 11	2b 12	13 Al 26.98	14 Si 28.09	15 P 30.97	16 S 32.06	17 Cl 35.45	18 Ar 39.95
28 Ni 58.70	29 Cu 63.55	30 Zn 65.38	31 Ga 69.72	32 Ge 72.59	33 As 74.92	34 Se 78.96	35 Br 79.90	36 Kr 83.80
46 Pd 106.4	47 Ag 107.9	48 Cd 112.4	49 In 114.8	50 Sn 118.7	51 Sb 121.8	52 Te 127.6	53 I 126.9	54 Xe 131.3
78 Pt 195.1	79 Au 197.0	80 Hg 200.6	81 Tl 204.4	82 Pb 207.2	83 Bi 209.0	84 Po (209)	85 At (210)	86 Rn (222)

63 Eu 152.0	64 Gd 157.3	65 Tb 158.9	66 Dy 162.5	67<:br>Ho 164.9	68 Er 167.3	69 Tm 168.9	70 Yb 173.0	71 Lu 175.0
95 Am (243)	96 Cm (247)	97 Bk (247)	98 Cf (251)	99 Es (252)	100 Fm (257)	101 Md (258)	102 No (259)	103 Lr (260)

* The main groups are numbered 1a to 8a, while the transition elements are numbered 1b to 8b. Recently, the American Chemical Society and the International Union of Pure and Applied Chemistry recommended a revised system in which the groups are numbered consecutively from 1 to 18 as shown here. Because it may be some time before this revised numbering system becomes widely accepted we have not adopted it in the main body of the text.

Basic Chemistry

Basic Chemistry

Paul S. Cohen
Trenton State College

Milton A. Rothman

Allyn and Bacon, Inc.
Boston   London   Sydney   Toronto

Chapter-opening photo credits: Introduction, Paul Conklin; Chapter 1, David Conklin; Chapter 2, Paul Conklin; Chapter 3, E. R. Degginger; Chapter 4, Jerry Berndt/Stock Boston; Chapter 5, Jon Rawle/Stock Boston; Chapter 6, Fredrik D. Bodin/Stock Boston; Chapter 7, Allyn and Bacon; Chapter 8, Dave Schaefer; Chapter 9, The Bettmann Archive; Chapter 10, Courtesy The Bethlehem Steel Corporation; Chapter 11, W. B. Finch/Stock Boston; Chapter 12, Peter Menzel/Stock Boston; Chapter 13, Steven Baratz/The Picture Cube; Chapter 14, Dave Schaefer; Chapter 15, NASA; Chapter 16, Frank Siteman/The Picture Cube; Chapter 17, Sol Mednick; Chapter 18, Peter Menzel/Stock Boston.

Composition Buyer: Linda Cox
Manufacturing Buyer: Andy Rosenau
Editorial-Production Service: Lifland et al., Bookmakers
Text Designer: Eileen Katin
Photo Researcher: Susan Van Etten
Cover Coordinator and Designer: Linda Dickinson

Copyright © 1986 by Allyn and Bacon, Inc., 7 Wells Avenue, Newton, Massachusetts 02159. All rights reserved. No part of the material protected by this copyright notice may be reproduced or utilized in any form or by any means, electronic or mechanical, including photocopying, recording, or by any information storage and retrieval system, without written permission from the copyright owner.

Library of Congress Cataloging-in-Publication Data

Cohen, Paul S., 1938–
    Basic chemistry.

    Includes index.
    1. Chemistry.   I. Rothman, Milton A.   II. Title.
QD31.2.C62 1986   540   85-23038
ISBN 0-205-08516-4

Printed in the United States of America.

10 9 8 7 6 5 4 3 2 1   90 89 88 87 86

To our wives, Brenda and Miriam,
in appreciation of their patience

Contents

Preface xi

Introduction 1

I.1 Chemistry and Its Uses 1
I.2 Chemistry as a Science 3
I.3 On Studying Chemistry 5

1 Mathematical Preliminaries 11

1.1 Large and Small Numbers: Scientific Notation 12
1.2 Multiplication and Division 14
1.3 Addition and Subtraction 19
1.4 Powers and Roots 20
1.5 Solving Algebraic Equations 24

2 Measurements and Operations 37

2.1 Problem-solving Techniques 38
2.2 Measurements in the Metric System 45
2.3 English-Metric Conversions 51
2.4 Density and Specific Gravity 53
2.5 Analysis of Graphs 58
2.6 Temperature Scales 61
2.7 Accuracy and Precision 66
2.8 Significant Figures 68

3 Matter and Energy 83

3.1 The Nature of Matter 84
3.2 States of Matter 88

3.3 Types of Substances 92
3.4 Energy 95
3.5 Measuring Quantities of Heat 102
3.6 Specific Heat 105
3.7 Heat Changes in Chemical Reactions 110

4 Elements and Compounds 121

4.1 The Elements and their Symbols 122
4.2 The Law of Definite Proportions 125
4.3 Atomic Theory and the Law of Multiple Proportions 127

5 Atomic Structure 137

5.1 Atomic Theory 138
5.2 Atomic Number and Atomic Mass 152
5.3 Ions and Isotopes 155
5.4 Electromagnetic Radiation and Photons 159
5.5 Electron Orbits 161
5.6 Multielectron Atoms and Shell Structure 167

6 The Periodic Table 179

6.1 Historical Background 180
6.2 The Periodic Law and Atomic Numbers 186
6.3 Groups and Their Properties 188
6.4 Trends in Properties 201

7 Bonding 211

7.1 The Nature of the Chemical Bond 212
7.2 Ionic and Covalent Bonding 215
7.3 Polar Covalent Bonds 221
7.4 Oxidation Numbers 224
7.5 The Lewis Electron-Dot Model 226
7.6 The Geometry of Molecules 228

8 Nomenclature 234

8.1 Names and Formulas 236
8.2 Writing Formulas 238
8.3 Naming Compounds 246

9 Avogadro and the Mole 260

- 9.1 Historical Review 262
- 9.2 Avogadro's Number 266
- 9.3 Molar Mass and the Mole 272

10 Chemical Equations 285

- 10.1 The Meaning of a Chemical Equation 286
- 10.2 Four Types of Equations 290
- 10.3 Balancing Equations 297

11 Stoichiometry 307

- 11.1 Equations and Mole Ratios 308
- 11.2 Mass Problems 309
- 11.3 Percent Composition 321

12 The Gas Laws 331

- 12.1 Measuring Gas Pressure 332
- 12.2 Boyle's Law 344
- 12.3 Charles's Law 349
- 12.4 Varying Both Temperature and Pressure 358
- 12.5 Dalton's Law of Partial Pressures 364
- 12.6 Kinetic Theory and the Ideal Gas Law 366

13 Solutions 381

- 13.1 Properties of Solutions 382
- 13.2 Water and the Formation of Solutions 386
- 13.3 Molarity and Normality 390
- 13.4 Colligative Properties of Solutions 397

14 Acids, Bases, and Salts 407

- 14.1 Models of Acids, Bases, and Salts 408
- 14.2 Acid-Base Strength 414
- 14.3 Ionization Concentrations—pH 422

15 Reaction Dynamics 435

- 15.1 Reactions and Molecular Collisions 436
- 15.2 Reaction Rates 441

15.3 Equilibrium 448
15.4 Le Chatelier's Principle 455
15.5 Heat, Temperature, and Equilibrium 460

16 Redox Reactions 469

16.1 The Concept of Redox 470
16.2 Balancing Redox Equations 472

17 Carbon Chemistry 479

17.1 Historical Introduction 480
17.2 Properties of Carbon 482
17.3 Some Simple Carbon Compounds 486
17.4 Nomenclature of Organic Compounds 499
17.5 Other Common Classes of Compounds 508

18 Radioactivity 527

18.1 Radioactive Isotopes 528
18.2 The Properties of Nuclear Radiation 537
18.3 The Detection and Measurement of Radiation 540
18.4 Half-Life 545
18.5 Effects and Uses of Radiation 546

Answers to Problems and Questions 557

Answers to Self Tests 561

Index 573

Preface

Most full year college chemistry courses assume a minimum background of one year of high school algebra and one year of high school chemistry. Many college students who wish to study college chemistry either have a nonexistent or poor secondary school mathematics or chemistry background, or have not studied chemistry in many years. They require an introduction to or review of the principles and concepts necessary for successful learning in an intensive one year survey course. We thus present a text designed for a one semester course preparatory to a full year course.

We begin with an introduction to the mathematical skills necessary for success in first year chemistry courses. Many texts place the mathematics in an appendix. To do so is to imply that this section is not of primary importance, and that if the student needs to use this section it is a need to be hidden away from sight. Students frequently have gaps in their mathematical training. These are not signs of personal failure; they indicate areas of learning that should be conquered to become areas of academic strength.

An important feature of this book is the extensive use of historical and philosophical references. If the discipline of chemistry is to become a vital subject, the student needs to see it as a body of knowledge that is studied and thought about by people who seek answers to many questions. We have endeavored to explain not only the problems faced by scientists but also the milieu in which they work and the many attempts required to reach a successful solution.

Another feature is the extensive use of sample solutions to mathematical application problems. These are followed at the end of each chapter with questions and problems for student review. The problems and questions are divided into sections related to the text so that the student may quickly reinforce learning. Where possible, problems are related to everyday chemicals found in the kitchen or medicine cabinet. Finally, the student is presented with an end-of-chapter self test designed to determine if sufficient learning has taken place to proceed into the next chapter. Answers to selected mathematical problems are provided.

The book contains many other learning tools designed to help students focus on the important points to be learned. These include a pre-chapter list of objectives and an end-of-chapter glossary. Students are often bewildered by the seemingly endless list of ideas and new words presented in a chapter. Here the student is guided to ensure proper use of study time and effort. Frustrations are reduced, and active, positive learning enhanced. To be sure that the text is suitable for the intended audience, a preliminary copy of the manuscript was tested by the authors in a one semester "Preparatory Chemistry" course, and the results incorporated in the text.

Our aim is to provide a basic chemistry background so that students will be able to succeed in a full year college survey course. We thus omitted many of the enrichment chapters often found in the larger full year text. It is not necessary to cover every thought mentioned in a year course to prepare for the course. Rather, one only needs the core of information. The selection of topics is designed to provide this core background to the student.

We wish to acknowledge the support of our wives, Brenda and Miriam. A project of this size could not have been completed without their understanding. In addition, we want to thank Stephen M. Cohen and Mara W. Cohen, both college students, for their reviews of segments of the text and their helpful suggestions. We also wish to thank two people who helped us with the production and development of this text: James Smith, science editor at Allyn and Bacon, without whose ability to see our potential this project would never have begun, and Sally Lifland, who improved the text significantly with her careful editorial commentary and suggestions. Finally, we wish to thank those who reviewed the various parts of the text and who offered many helpful suggestions:

David M. Piatak, Northern Illinois University
Mary C. Cavallaro, Salem State College
O. Jerry Parker, Eastern Washington University
Mark G. Rockley, Oklahoma State University
Larry M. Nicholson, Fort Hays State University
Caroline L. Ayers, East Carolina University
Philip A. Kint, University of Toledo
Hugh J. Bronaugh, Georgia State University
Robert J. Munn, University of Maryland
James A. Campbell, El Camino Community College
Gordon J. Ewing, New Mexico State University
Judith Grobe Sachs, University of Illinois at Chicago
Jeffrey A. Hurlbut, Metropolitan State College
Forrest C. Hentz, Jr., North Carolina State University
Ruth J. Bowen, California State Polytechnic University
Patricia A. O'Neill, Monroe Community College
Jack E. Powell, Iowa State University
Charles Corwin, American River College
William K. Plucknett, University of Kentucky
Frank Hoggard, Southwest Missouri State University
Donald Peterson, California State University, Hayward

Basic Chemistry

Introduction

I.1 Chemistry and Its Uses

Chemistry is one of the most practical of the sciences. The products of chemistry are so common in the home that users rarely wonder how they are made. The ink in your pen, the gasoline in your tank, the drugs in your medicine chest, the dyes that give your clothes their many hues—all of these were invented in chemical laboratories and manufactured in chemical processing plants.

Chemists—those engaged in chemistry professionally—come in many varieties, among which are the following:

1. The research chemist studies the basic properties of matter, learns the rules governing the way different materials react with each other, and learns to use these reactions to form new materials. **Research** itself is the process of creating new knowledge and solving problems. One current problem is finding a way to manufacture insulin in the laboratory. Insulin is a substance required by the body in order to carry on certain biological processes. Those people whose bodies cannot make their own insulin suffer from diabetes, a sometimes fatal illness. The ability of the chemist to extract this material from animal organs saves many lives each year. Learning to make it artificially would reduce costs. However, before insulin can be manufactured, chemists must determine what it consists of. They must take it apart (analysis) before they can put it together (synthesis).

2. The analytical chemist knows how to tell what substances exist within samples of unknown materials and knows how to measure the amounts of these substances. This type of chemist tests water for pollution, tests human tissue for drugs, and follows the course of manufacturing processes to make sure that all of the necessary ingredients are present in the right quantities.

3. The industrial chemist, or chemical engineer, designs the large, complex masses of tanks, pipes, and controls used in the manufacturing of chemical products on a large scale. This type of chemist must understand the organization of large systems, the use of measuring devices, the functions of pumps, valves, and other mechanisms, and how to put these together so they will operate most efficiently and economically.

Other disciplines use chemistry as a tool. The geologist (the scientist who studies rock and minerals) analyzes the composition of the minerals in the earth. The paleontologist (the scientist who studies prehistoric life) measures tiny quantities of radioactive materials to determine the age of fossils found in rock strata. The art historian analyzes paints to determine the age and origin of works of art. The physician analyzes blood, urine, and other body materials as an aid to diagnosing illness.

Chemistry is a very large discipline, with many branches and specialties. In this book we shall give only a brief introduction to the subject. You should, after completing this text, be able to go on to the more detailed and advanced books. You will learn the basic language of chemistry—the language by which the chemist identifies and discusses processes and materials. This language is not a secret jargon designed merely to be difficult. It is a logical terminology designed to help chemists communicate.

For this reason, the study of chemistry requires learning the names of many substances (chemical compounds) and learning how the compounds are assembled from simpler substances (elements). In addition, it is important to learn the properties of these substances—what they look like, how they behave, how they react with other substances, and so on.

When we want to manufacture a particular material, we must know *how much* of each ingredient to use, both to avoid waste and to make sure that we end up with a satisfactory product. The process is not unlike baking a cake. To determine these quantities, we must know how to calculate the amount of each ingredient that should go into the reaction. Therefore, a large part of chemistry is devoted to the mathematics of such calculations.

A most important part of chemistry is the theoretical foundation of the subject—the understanding of the structure of matter and the relationship between matter and energy. Much of this book will deal with one very fundamental concept: the idea that all matter is made up of extremely tiny particles (atoms), which, in turn, are made up of even smaller particles (electrons, protons, and neutrons). You will see that the properties of these small particles determine the behavior of all the different substances in the world. Even the processes inside living organisms are determined by the behavior of the materials out of which we are all made. These processes make up the subject matter of biochemistry and molecular biology.

As you learn the theory of chemistry and learn to write down the chemical equations describing the reactions that take place between a variety of substances, you will encounter some of the ways in which chemistry is related to your own life experiences and to the development of society as a whole.

1.2 Chemistry as a Science

What is it that makes chemistry a science, and how does it differ from other sciences? Science is more than a fixed body of knowledge; it is an ongoing search for new knowledge, continually taking place in university and industrial laboratories. It is also a way of thinking based on knowledge of the laws of nature.

Science is based on the observation of things that exist and events that happen in nature. Scientists talk only about things that can be observed, either directly or indirectly. Although we often talk about things too small to be seen with the naked eye, photographs of these objects can be taken with microscopes and other instruments, so there is no doubt as to their existence. Observations by instruments are generally more dependable than observations by unaided human senses. Therefore, research in science—especially in chemistry and physics—is almost always done with the aid of recording instruments.

The actions of things in nature are described by generalized rules called **laws of nature**, which are powerful statements describing the behavior of everything from tiny particles to huge planets and stars. Laws of nature are arrived at by observing the behavior of some things acting in a given situation and then generalizing from these observations by making a hypothesis about what happens to everything in that same kind of situation. (A **hypothesis** is an educated guess based on observation.) Scientists then test the hypothesis by further experiment and observation to make sure that it is correct.

For example, from the Middle Ages to the present many inventors have tried to build a machine that would turn wheels and do work without burning any kind of fuel. Such a gadget is called a "perpetual motion machine" because once you start it going it is supposed to keep on going forever. Unfortunately, nobody has ever been able to make such a machine work. In the nineteenth century a general law of nature was formulated which explained this failure. This law—the **law of conservation of energy**—states that in order to operate any kind of machine, something called energy (usually obtained by burning fuel) must be put into the machine, and the machine can never produce more energy than is put into it. The term "conservation" refers to a quantity that *never changes*—in this case, a quantity of energy (more about this in Chapter 3).

Once a law of nature has been well tested, it can be used to predict how things in nature will behave. It is this predictability that allows chemists to know what will happen when they mix a known pair of chemicals under a given set of conditions. Using the law of conservation of energy, chemists can predict how much heat will be produced when a given amount of fuel is burned. They can even predict whether or not two chemical compounds will react when mixed together and what the products of the reaction will be.

There are two major types of natural laws. One type is a set of laws that tells us what things will do under a given set of conditions. For example, Newton's laws of motion tell us how objects move when they are acted on by a known set of forces. With these laws, we can predict the orbit of a space satellite as well as the operation of a piece of machinery. Such laws might be called "laws of determination" or "laws of permission." A second type of law tells us what actions *cannot* take place. These are "laws of denial." One such law is the law of conservation of energy mentioned above, which informs us that no action can take place unless there is a way of supplying the energy needed. Laws of denial are extremely powerful, for many of them have been tested to a very high degree of accuracy and so allow us to say very specifically that certain events are impossible. (For example, no scientist would hesitate to say that perpetual motion machines are impossible. For the same reason, scientists are always highly skeptical when they read claims by hopeful inventors that a small chemical pill dissolved in a gallon of water can drive a car for a hundred miles.)

An experiment is not considered to be a good scientific experiment unless it can be repeated. One of the purposes of publishing the results of scientific research in journals is so that people around the world can know exactly how a given experiment was done so that they can repeat it if they wish. It is this possibility of replication that gives scientists confidence in the results of an experiment. There have been many cases of experiments that failed to give the original results when repeated by others. The results of those experiments invariably turn out to be incorrect.

The most important feature of science is that it is based on observation of what actually happens in nature. Sometimes when people say, "That's only a theory," they imply that the theory referred to is just a speculation. But nothing could be further from the truth. In science, when a theory has been verified by experiment and observation, it is much more than a speculation; it is a generalization that explains and predicts events that have been or will be observed.

Thus, the atomic theory, which underlies all chemistry, is no longer a speculation, although at one time it was. During the past century, an enormous number of facts have been accumulated about the structure of matter, so today the term "atomic model" is more appropriate than the term "atomic theory." The same can be said about the theory of evolution, the theory of relativity, and other theories that used to be very speculative but have become widely accepted during the twentieth century because of the great number of facts supporting them.

The science of chemistry takes as its domain of operation the structure of material substances and the behavior of these substances as they interact with each other. It also deals with the creation of new substances and learning methods of manufacturing them for human use. There is great overlapping between chemistry and other sciences. Both chemists and physicists study the atoms that make up all matter within the discipline of chemical physics. The study of energy relationships in chemical reactions falls under the heading of physical chemistry, a major branch of chemistry. The role of elemental substances in living organisms is studied by biochemists and by molecular biologists. The study of chemicals as drugs and the reactions of these drugs in the body is pharmacology.

From this very partial list of examples, we see that chemistry is a many-tentacled science, with branches extending into a variety of important subjects.

1.3 On Studying Chemistry

What Has To Be Learned

There are a great many things to learn in chemistry. If chemistry were nothing but a list of facts to be memorized, the sheer number of such facts would make the study of chemistry a hopeless task. Fortunately, much of chemistry consists of general rules that tie the facts together. If you learn the general rules, you don't have to learn so many separate facts.

To illustrate what we mean by general rules, take a simple example. Consider two facts: (1) A substance called hydrochloric acid has a sour taste, and (2) a chemical called acetic acid (found in vinegar) also has a sour taste. We notice that both substances have the word "acid" in their names, and when we investigate further by testing many acids, we find that substances called acids tend to have sour tastes. This, then, is a general rule: *Acids are sour-tasting substances.* (The evidence does not necessarily prove the converse rule—that all sour-tasting substances are acids.)

Rules that are extremely broad and cover a great many facts are called *general principles*, and those that are most general are the *fundamental laws of nature*. For example, the universal gas laws are general principles because they apply to all kinds of gases; *you don't have to learn a different kind of rule for each kind of gas.* Knowing the universal gas laws enables you to predict what will happen to any container of gas when you heat it or when you compress it.

As you learn the rules of chemistry, you will also absorb a certain frame of mind. The idea that all matter consists of small particles that obey certain general laws of nature leads to the attitude that nature is regular and predictable. This attitude, fundamental to all the sciences, has powerful consequences, for it allows us to predict what is going to happen or what cannot happen when particles get together under a given set of conditions.

To summarize, we learn three major kinds of things in chemistry:

1. Facts about chemical substances: their names, their appearances, the way they are manufactured, what their uses are, the way they react with each other, and how they are identified.

2. General rules about the behavior of chemical substances and how to use these rules to predict what is going to happen in a given situation or to understand why certain reactions take place.

3. An attitude about nature: knowing the laws of nature makes it possible for us to understand nature and to use it to our benefit. (It is a misconception to think that human beings can "control" nature. Nature does whatever it has to do, and if we can arrange things so that the results of nature's actions are what we want, then we are satisfied.) Complete understanding allows us to be aware of the hazards and side-effects of whatever we do and to make judgments as to whether the long-term costs are worth the benefits obtained. (For example, is the pollution caused by the burning of coal worth the energy obtained from it? If not, can this pollution be reduced enough to make the burning of coal worthwhile?)

In the study of chemistry, all three of these items are important. None of them can be left out. Some people believe that it is possible to get by without memorizing facts about individual substances and reactions. They think that learning the rules is enough. However, rules by themselves are very abstract and hard to understand. Learning the properties of acids in general is much easier if you know something about a number of particular acids. Once you know some specific facts, then you can generalize and learn abstract principles. For this reason, it is important to commit to memory as much of the material in this book as you can. On the other hand, memorizing alone is not enough; memorization is not a substitute for understanding what the rules mean and how they work. Learning science requires both memorizing and understanding.

Another important feature of the study of chemistry is learning to deal with quantities of matter. A problem commonly encountered is this: If we mix an amount of substance A with an amount of substance B, how much of substance C will we get as a result? We must use mathematics to answer this question. Therefore, a large part of this book is devoted to solving chemical problems using mathematical methods. The mathematical tools most commonly used in solving chemical problems are simple algebraic equations, scientific notation (powers of ten) for dealing with very large and very small numbers, and conversion of units.

The aim of the first two chapters of this book is to get you so familiar with these topics that you will be able to use these tools without hesitation. There are two reasons for putting all of the mathematical tools in one place instead of scattering them throughout the book. First, you will learn these techniques more thoroughly if there is systematic drill. Second, it is easier to refer back to these techniques if they are all in one place.

The present time is a period of transition between the use of the English system of units and the use of the metric system; some books use one system, and others use the other. In this book we attempt to consistently use the *International System*, which is the latest version of the metric system. At the same time, since work situations often require that a large amount of time be spent converting one set of units to another, we include a section on conversion.

How to Learn Chemistry

A great many students approach the study of chemistry with fear and trepidation, yet there is no need to fear this subject. Any student with average intelligence and a well-organized set of study habits can learn the material in this text. These study habits are essential:

1. Read the text carefully.

2. Outline the highlights. Note words encountered for the first time, and make sure that you understand them. Also note definitions and special information required for complete undertstanding.

3. Carefully study sample problems found throughout the text. If necessary, write down the steps of a problem to make sure you understand the reasoning.

4. Do the problems, and answer the questions at the end of each section.

5. Test yourself with the questions and problems at the end of the chapter.

6. Consult your instructor on any area or specific point that confuses you.

Be sure to allow enough time for the information to "percolate." Cramming at the last minute before an exam usually results in failure. When the solution to a problem is not immediately apparent, you need more time to think the matter through. Often it helps to let a particularly sticky problem sit until the next day. It is amazing what thoughts will come to mind if you give them the chance.

Some of the material in this text is hard enough that you will not be able to understand it immediately, and some of the problems are difficult enough that you will not be able to solve them right away. Do not be dismayed. Some of the material must be difficult; otherwise, it would not serve the purpose of education.

Writing down an outline of the material after you read a chapter or section will help you to learn it. An outline of this section might go as follows:

1. Many students fear chemistry needlessly.

2. What is needed to study chemistry?
    a. Average intelligence
    b. Good study habits

3. Good habits include:
    a. Reading carefully (and more than once)
    b. Highlighting info and making an outline
    c. Studying sample problems
    d. Doing problems at end of sections
    e. Doing self tests at end of chapters
    f. Consulting instructor

    4. Consider time factors in studying.
        a. Info needs time to percolate
        b. Problems need time to be solved

Glossary

conservation of energy, law of Principle stating that in a closed system the total amount of energy is constant—i.e., energy can never be created or destroyed.

hypothesis A conjecture that accounts for a set of facts and that, with further investigation, can serve as the basis for a theory.

laws of nature A set of statements describing, in a very general way, how events take place in nature. Some of these laws describe how objects move and are set into motion (for example, Newton's three laws of motion). Others describe important regularities (symmetries) found in nature (for example, the law of conservation of energy).

research A set of activities performed with the purpose of obtaining new knowledge. Some research is theoretical—theories are created to explain what is observed. Other research is experimental—observations are made to verify existing theories or as a basis for creating new theories. Applied research aims at using known facts to develop processes and products for everyday use.

Problems and Questions

Note: Both word questions and numerical problems are found at the ends of chapters. Questions are labeled Q, and problems are labeled P.

I.1. Chemistry and Its Uses

I.1.Q Describe the functions of
    a. research chemists
    b. analytical chemists
    c. chemical engineers

I.2.Q Describe what a chemist might do
    a. in a mystery story
    b. on an archeological expedition
    c. in an oil refinery
    d. in a hospital
    e. in an art gallery

I.3.Q **a.** Describe three applications of chemistry in your own experience.
    b. Describe three activities of chemists with which you have become acquainted either from reading or from watching TV.

I.4.Q What is the most fundamental and central concept of modern chemistry?

I.5.Q Why is a pharmacist called a "chemist" in England?

I.6.Q Why was World War I called the "war of chemistry," and World War II the "war of physics"? (You may have to do some research to find the answer.)

I.2 Chemistry as a Science

I.7.Q What is the difference between a speculation and a scientific theory?

I.8.Q Look at the person sitting next to you.
    a. Make five statements about that person based on observation.
    b. Make two statements about that person based on speculation.
    c. Do any of your five observations have any theories or speculations hidden in them?

I.9.Q You have heard the "theory" that flying saucers are vehicles from another world exploring the earth. Discuss what would have to be done to convert this speculation into a scientific theory.

I.10.Q **a.** State three facts about each of the following substances: water, orange juice, milk, and lubricating oil.

    b. What do all of these substances have in common?
    c. State a generalization that applies to all of them.

I.11.Q It is often said that "anything is possible if you only try hard enough." Discuss this statement in view of our observation that creating a perpetual motion machine is impossible.

I.12.Q What are the two major types of laws of nature?

I.13.Q Why is it important that experiments be repeatable?

I.3 On Studying Chemistry

I.14.Q What are the three major kinds of things we learn in chemistry?

I.15.Q Write down three facts about several different substances with which you are familiar. How many of these facts do you know from your own observations? Try to recall where you learned each of the facts.

I.16.Q Write down the names of four substances that have similar properties. State a general rule about these substances that describes what they have in common.

I.17.Q How can you tell when you understand a rule or a general principle? (That is, how would you demonstrate to somebody else that you understood the rule?)

I.18.Q Why is it necessary to learn mathematics in order to study chemistry?

I.19.Q Name six good study habits.

Self Test

1. Speculate on what interest a chemist might have in studying about
    a. geology   **b.** biology
    c. medicine   **d.** computers
    e. history   **f.** law
    g. accounting   **h.** art
    i. music   **j.** theater
    k. philosophy

2. Explain why the following people would have to know something about chemistry:
    a. a physician
    b. a paleontologist
    c. a biologist
    d. an electrical engineer
    e. an architect
    f. a lawyer
    g. a journalist
    h. a farmer

3. Name three things that you consider to be impossible, and state your reasons. Are there any laws of nature among your reasons?

4. Define the following terms:
    a. analysis
    b. synthesis
    c. research
    d. particle
    e. science
    f. laws of nature
    g. hypothesis
    h. perpetual motion machine
    i. law of permission
    j. law of denial

5. What is the difference between a speculation and a scientific hypothesis?

6. What is the function of molecular biology?

7. Look at the person nearest you and write down at least five observations about that person. Eliminate all speculations about that person.

8. State three generalizations about the members of your class.

9. Proper study of chemistry requires good study habits. List six study habits important for success in completing this text.

10. What is chemistry?

1
Mathematical Preliminaries

Objectives After completing this chapter, you should be able to:
1. Convert from decimal numbers to scientific notation (powers of 10) and vice versa.
2. Multiply and divide numbers expressed in scientific notation.
3. Add and subtract numbers expressed in scientific notation.
4. Raise a number expressed in scientific notation to any integer power.
5. Take the square root of a number expressed in scientific notation.
6. Solve simple algebraic equations.
7. Solve simple problems involving linear equations.
8. Solve problems involving direct proportionality.

1.1 Large and Small Numbers: Scientific Notation

> The walrus and the carpenter were walking close at hand;
> They wept like anything to see such quantities of sand:
>
> —Lewis Carroll, *Through the Looking Glass*

Large Numbers

What kind of number system would enable you to count the grains of sand on a beach? What kind of number system would permit you to express the size of the smallest particle of matter—a particle billions of times smaller than the dot under this question mark? One thing is certain: You would not find it easy to do the necessary arithmetic with numbers such as 19840000000 or 0.00000000135.

However, very big and very small numbers can easily be handled by making use of the *power-of-ten system*, often called **scientific notation**. The power-of-ten system is based on the following definitions:

$$10^1 = 10$$
$$10^2 = 10 \times 10 = 100$$
$$10^3 = 10 \times 10 \times 10 = 100 \times 10 = 1000$$

The number 10^2 (read as "ten squared" or "ten to the second power") is ten times ten, or two tens multiplied together. The number 10^3 (read as "ten cubed" or "ten to the third power") is three tens multiplied together. The **exponent** (the number above and to the right of the 10) tells you the number of tens that are multiplied together. Therefore, 10^1 represents just a single ten.

Continuing in the same vein, we write

$$10^4 = 10^3 \times 10^1 = 10{,}000$$
$$10^5 = 10^4 \times 10^1 = 100{,}000$$

Notice that 10^5 (read as "ten to the fifth power") consists of a one followed by five zeros. *The number in the exponent always equals the number of zeros following the first digit.* Using this rule, we can write 10^6 immediately: it is a one followed by six zeros. (Be sure not to write a 10 followed by six zeros.)

$$10^6 = 1{,}000{,}000$$

You can see that 10^6 is one million. The rule of exponents helps you write any number, no matter how big. The number 10^{52} is a one followed by 52 zeros. The power-of-ten system lets you use the biggest and the smallest numbers in any calculation. It provides an easy way to find the right answers to seemingly difficult arithmetic problems.

How is a number such as 1,560,000 handled? Making use of the fact that

$$1{,}560{,}000 = 1.56 \times 1{,}000{,}000$$

we write the number as 1.56×10^6.

In scientific notation it is customary to put only one digit to the left of the decimal point. Therefore, if the solution to a problem is a number such as 123×10^7, change it to 1.23×10^9, which is the same number written in the standard form of scientific notation. (We will see how to make this decimal point change in the next section.)

It should be clear that in scientific notation the number 1×10^3 is the same as 10^3.

Those familiar with computers will recognize that the scientific notation we are discussing is exactly the same as the exponential notation used in computer languages. The number 1.23×10^9 is the same as 1.23E09. In typing manuscripts and computer programs, it is much easier to use 1.23E09 than to type 1.23×10^9.

Small Numbers

To handle small numbers, we use negative powers. By definition, any number raised to a negative power is the reciprocal of that number raised to the same positive power. In other words,

$$10^{-n} = \frac{1}{10^n}$$

(Here n stands for any number.)

Then we can say

$$10^{-1} = \frac{1}{10^1} = \frac{1}{10} = 0.1$$

$$= \text{one-tenth}$$

$$10^{-2} = \frac{1}{10^2} = \frac{1}{10 \times 10} = \frac{1}{100} = 0.01$$

$$= \text{one-hundredth}$$

$$10^{-3} = \frac{1}{10^3} = \frac{1}{10 \times 10 \times 10} = \frac{1}{1000} = 0.001$$

$$= \text{one-thousandth}$$

And so on. *When the exponent is* -3, *you move three places to the right of the decimal point before writing the first digit other than zero.*

To write the number 0.0000234 using this system, we first see that it is the same as 2.34×0.00001. Then we see that the one is five places to the right of the decimal point. Therefore, in scientific notation the number is 2.34×10^{-5}. To convert a decimal number into scientific notation, *move the decimal point to the right until it is just to the right of the first nonzero digit. The number that goes in the exponent is equal to the number of places that you moved the decimal point.*

In the above example, the decimal point is moved five places to the right in order to get 2.34, so the entire number is written as 2.34×10^{-5}.

Example 1.1   Write the number 421,000,000 in scientific notation.

Solution   Break the problem down into the following steps. First, see that the given number is 4.21 multiplied by 100,000,000. Then, see that 100,000,000 is a one followed by 8 zeros; therefore, it equals 10^8. Thus the answer is 4.21×10^8.

Example 1.2   Write the number 0.0000357 in scientific notation.

Solution   The given number is 3.57×0.00001. Since the one is 5 places to the right of the decimal point, the exponent must be -5. Accordingly, the answer is 3.57×10^{-5}.

Example 1.3   Write the number 9.46×10^4 in decimal notation.

Solution   The number 10^4 is a one with 4 zeros after it. Therefore, the answer to the problem is $9.46 \times 10,000 = 94,600$.

Example 1.4   Write the number 8.2×10^{-4} in decimal notation.

Solution   The number 10^{-4} is a decimal that has a one located 4 places to the right of the decimal point: 0.0001. The answer to the problem is therefore $8.2 \times 0.0001 = 0.00082$.

1.2 Multiplication and Division

We will look first at multiplication and division, because these operations are easier to perform in scientific notation than are addition and subtraction.

Multiplication

Suppose we multiply two numbers by the decimal system:

$$10 \times 100 = 1000$$

Translating this equation into powers of ten, we have

$$10^1 \times 10^2 = 10^3$$

Notice that the exponent on the right equals the sum of the exponents on the left. That is,

$$1 + 2 = 3$$

Let's try another example:

$$100 \times 10{,}000 = 1{,}000{,}000$$

Putting this equation into powers of ten, we have

$$10^2 + 10^4 = 10^6$$

The exponents are related by the equation

$$2 + 4 = 6$$

From these examples we discover the general rule for multiplying powers of ten: *The exponent of the product equals the sum of the exponents of the numbers being multiplied.* That is, if m and n are any numbers, then

$$10^m \times 10^n = 10^{m+n}$$

Using this rule, we can immediately say that

$$10^5 \times 10^7 = 10^{12}$$

or that

$$10^2 \times 10^6 \times 10^3 = 10^{11}$$

You can multiply as many factors as you like in this manner—simply add up all the exponents. It makes no difference if one or more of the exponents are negative. You just add them, taking the algebraic sign into account. For example,

$$10^6 \times 10^{-4} = 10^{6+(-4)} = 10^{6-4} = 10^2$$

It is always a good idea to check the result of a problem by looking to see if it makes sense. In the above example, we have multiplied a very large number (10^6) by a number less than one (10^{-4}). (Any number with a negative exponent is less than one.) We expect the result to be smaller than the original large number, and it is.

Example 1.5

Solution

$$10^6 \times 10^3 \times 10^4 = ?$$

$$10^{(6)+(+3)+(+4)} = 10^{13}$$

Just add the numbers in the exponents.

Example 1.6

$$10^{-5} \times 10^{-2} \times 10^{-7} = ?$$

Solution

$$10^{(-5)+(-2)+(-7)} = 10^{-14}$$

Add the numbers in the exponents, taking their signs into account.

Example 1.7

$$10^4 \times 10^{-9} \times 10^7 = ?$$

Solution

$$10^{(4)+(-9)+(+7)} = 10^2$$

Some of the numbers in the exponent are positive and some are negative; we add the positives and subtract the negatives.

Up to now we have dealt with problems that have nothing but powers of ten in them. How do we treat a problem such as

$$500 \times 93{,}000{,}000$$

The method is straightforward. First, write the numbers in scientific notation:

$$5.00 \times 10^2 \times 9.3 \times 10^7$$

Next, rearrange the factors so that the two decimal numbers and the two powers of ten are together:

$$5.00 \times 9.3 \times 10^2 \times 10^7$$

Then multiply the numbers to obtain

$$46.5 \times 10^9$$

Although this number is correct, it is customary to convert it into a form in which there is only one digit to the left of the decimal point. That is, the number multiplying the power of ten should be between 1 and 10.

To do this, we divide 46.5 by 10, moving its decimal to the left one place. To keep the entire number unchanged, we must multiply the factor 10^9 by 10, which means adding 1 to the exponent. The result is then written as 4.65×10^{10}.

The general rule is that *if we move the decimal point to the left by n places, then the exponent has to be increased by n.* In the same manner, *if we move the decimal point to the right by n places, then the exponent has to be decreased by n.*

Example 1.8

$$(3 \times 10^4) \times (2 \times 10^6) = ?$$

Solution

$$3 \times 2 \times 10^4 \times 10^6 = 6 \times 10^{10}$$

Example 1.9

$$(2 \times 10^2) \times (6 \times 10^{-5}) = ?$$

Solution

$$2 \times 6 \times 10^2 \times 10^{-5} = 12 \times 10^{(2)+(-5)} = 12 \times 10^{-3}$$
$$= 1.2 \times 10^{1+(-3)} = 1.2 \times 10^{-2}$$

Here we moved the decimal point one place to the left and added 1 to the exponent, changing the -3 to a -2.

Example 1.10

$$(3.14 \times 10^6) \times (6.00 \times 10^{-12}) = ?$$

Solution

$$3.14 \times 6.00 \times 10^6 \times 10^{-12} = 18.8 \times 10^{-6}$$
$$= 1.88 \times 10^{-5}$$

When you multiply 3.14 by 6.00, you get the answer 18.84. However, for reasons that we will give in the next chapter, you must drop the final 4 to "round off" the answer.

Division

Division is just as easy to do as multiplication. We simply make use of the fact that the reciprocal of a number is written with a negative exponent; for example,

$$\frac{1}{10^3} = 10^{-3}$$

Now dividing by a number is the same as multiplying by the reciprocal of that number; that is,

$$\frac{10^5}{10^3} = 10^5 \times \frac{1}{10^3} = 10^5 \times 10^{-3} = 10^{5+(-3)} = 10^{5-3} = 10^2$$

These examples demonstrate the general rule for dividing powers of ten: *The exponent of the denominator is subtracted from the exponent of the numerator.* That is,

$$\frac{10^n}{10^m} = 10^{n-m}$$

You have to be on your guard if either the numerator or the denominator has a negative exponent. This situation may be handled in the same way we normally deal with the addition or subtraction of negative numbers. Consider the following example:

$$\frac{10^6}{10^{-3}} = 10^{6-(-3)} = 10^{6+3} = 10^9$$

Here we have subtracted -3 from 6, which is the same as adding $+3$. Since the problem involves dividing a large number by a very small number, we expect to end up with an even larger number, and we do.

In the following example, we subtract a negative number from another negative number:

$$\frac{10^{-4}}{10^{-6}} = 10^{-4-(-6)} = 10^{-4+6} = 10^2$$

We have divided a small number by an even smaller number, so we end up with a large number.

Example 1.11

$$\frac{10^{23}}{10^{10}} = ?$$

Solution

$$10^{(+23)-(+10)} = 10^{23-10} = 10^{13}$$

Example 1.12

$$\frac{10^5}{10^{-9}} = ?$$

Solution

$$10^{(+5)-(-9)} = 10^{5+9} = 10^{14}$$

Notice that dividing by 10^{-9} is the same as multiplying by 10^{+9}.

Example 1.13

$$\frac{(10^2 \times 10^8)}{(10^4 \times 10^{10})} = ?$$

Solution

$$\frac{(10^{2+8})}{(10^{4+10})} = \frac{10^{10}}{10^{14}} = 10^{10+(-14)} = 10^{-4}$$

Here we have chosen to perform the multiplications first and then do the divisions, although it actually makes no difference in which order you do the various operations.

Example 1.14

$$\frac{2 \times 10^4}{8 \times 10^3} = ?$$

Solution

$$\tfrac{2}{8} \times 10^{4-3} = 0.25 \times 10^1 = 0.25 \times 10 = 2.5$$

The Zeroth Power

Very often we meet a problem like this:

$$\frac{4 \times 10^6}{2 \times 10^6} = 2 \times 10^0$$

What are we to make of 10^0 (read as "ten to the zeroth power")? This problem in ordinary decimals becomes

$$\frac{4{,}000{,}000}{2{,}000{,}000} = 2$$

Since the two answers above represent the same number, it must be true that $2 \times 10^0 = 2$.

This result can be true only if $10^0 = 1$. In general, any number divided by itself equals 1:

$$\frac{10^n}{10^n} = 10^{n-n} = 10^0 = 1$$

So, the first problem above can be answered directly as follows:

$$\frac{4 \times 10^6}{2 \times 10^6} = \tfrac{4}{2} \times 1 = 2 \times 1 = 2$$

We have seen that all numbers with positive exponents are greater than 1 and all numbers with negative exponents are less than 1. Now we see that the number with a zero exponent is equal to 1.

1.3 Addition and Subtraction

When adding decimals, we must first line up the decimal points so that they appear in the same column. For example, if we are given the problem $1.23 + 0.04 + 23.22 = ?$, we write it in the form

$$\begin{array}{r} 1.23 \\ 0.04 \\ 23.22 \\ \hline 24.49 \end{array}$$

Similarly, when adding or subtracting powers of ten, we must make sure that the same number appears in all the exponents. For example, if we want to add 5,000 to 621,000, we do not write

$$5 \times 10^3 + 6.21 \times 10^5 = ?$$

Instead, we change one of the numbers in such a way that it has the same exponent as the other. The simplest way to do this is to add a 2 to the exponent of the first number, which is the same as multiplying it by 100. To keep the entire quantity unchanged, we must divide the number 5 by 100, which means shifting the decimal point two places to the left. The example then becomes

$$0.05 + 10^5 + 6.21 \times 10^5 = (0.05 + 6.21) \times 10^5$$
$$= 6.26 \times 10^5$$

What we are doing is making use of the rules for factoring an expression of the form

$$a \times 10^n + b \times 10^n = (a + b) \times 10^n$$

We can also write the example in column form:

$$0.05 \times 10^5$$
$$+ 6.21 \times 10^5$$
$$\overline{6.26 \times 10^5}$$

Subtraction is treated in the same manner as addition. For example,

$$5.781 \times 10^7 - 4.31 \times 10^6 = ?$$

would be rewritten as

$$5.781 \times 10^7 - 0.431 \times 10^7 = ?$$

or, in column form,

$$5.781 \times 10^7$$
$$-0.431 \times 10^7$$
$$\overline{5.350 \times 10^7}$$

Example 1.15

$$6.23 \times 10^{-3} + 2.30 \times 10^{-4} = ?$$

Solution

$$6.23 \times 10^{-3} + 0.23 \times 10^{-3} = 6.46 \times 10^{-3}$$

We multiply the 10^{-4} by 10, so we divide the 2.30 by 10 to keep the number unchanged.

Example 1.16

$$8.182 \times 10^6 + 3.52 \times 10^5 + 4.22 \times 10^5 = ?$$

Solution

$$8.182 \times 10^6 + 0.352 \times 10^6 + 0.422 \times 10^6 = 8.956 \times 10^6$$

1.4 Powers and Roots

The rules for multiplying and dividing powers of ten can be used to find the rules for other operations, such as finding powers and roots. Suppose, for example, we want to find the area of a square 10^{-6} in on a side (see Figure 1-1). If s represents the length of a side of the square, the area (A) is given by:

$$A = s \times s = s^2$$
$$= (10^{-6} \text{ in})^2$$

To square a number, multiply the number by itself; thus,

$$A = (10^{-6} \text{ in}) \times (10^{-6} \text{ in})$$

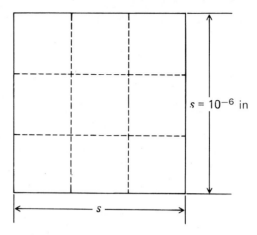

FIGURE 1-1 The area of a square is equal to s × s, where s is the length of a side. If s = 3 units, then the area is 9 square units.

Next, separate the numbers from the units (inches) and treat the unit symbols as though they were quantities to be multiplied:

$$A = 10^{-6} \times 10^{-6} \text{ in} \times \text{in}$$
$$= 10^{(-6)+(-6)} \text{ in}^2$$
$$= 10^{-6-6} \text{ in}^2$$
$$= 10^{-12} \text{ in}^2$$

Notice that 12 is the same as 2 × 6. This means that instead of going through all the steps above, we could have immediately written

$$A = (10^{-6} \text{ in})^2 = 10^{2 \times (-6)} \text{ in}^2$$
$$= 10^{-12} \text{ in}^2$$

(Remember that in² is the same as square inches.)

The rule is that *to square a power of ten we multiply the number in the exponent by* 2. By going through a similar process, you can verify for yourself that to cube a power of ten (to raise it to the third power), you multiply the exponent by 3.

In general, *when 10^m is raised to the nth power, the exponent m is multiplied by the number n.*

$$(10^m)^n = 10^{m \times n}$$

This rule is valid for any value of *m* or *n*—positive, negative, decimal, or fractional.

To find the meaning of a fractional power such as $10^{1/2}$, let's go back to the previous example and take the one-half power of both sides of the equation:

$$A = s^2$$
$$A^{1/2} = (s^2)^{1/2}$$
$$= s^{2 \times 1/2}$$
$$= s^1 = s$$

The above equation states that $s = A^{1/2}$. But now let us see what happens when we take the square root of both sides of the starting equation. (The **square root** of A is defined as that number which, when multiplied by itself, will result in A.) The square root of s^2 is simply s, by definition, because $s \times s = s^2$. Therefore, if we start with

$$A = s^2$$

and take the square root of both sides, we end up with

$$\sqrt{A} = \sqrt{s^2} = s$$

Since $s = \sqrt{A}$ and $s = A^{1/2}$, it must be true that

$$A^{1/2} = \sqrt{A}$$

Thus, *finding the square root of a number is the same as raising that number to the one-half power.* This is true for any kind of number, not just for powers of ten.

Suppose we want to find the length of the side of a square whose area is 10^6 square feet. We have just shown that if the area A is known, the length of a side s is given by

$$s = A^{1/2}$$
$$= (10^6 \text{ ft}^2)^{1/2}$$
$$= 10^{6 \times 1/2} \times (\text{ft}^2)^{1/2}$$
$$= 10^3 \text{ ft}$$

Notice that the unit symbol—feet (or ft)—is handled just like any algebraic quantity. This is a very powerful principle for dealing with any kind of problem involving dimensioned quantities.

Now we proceed to a more difficult question. Suppose we have a square whose area is 10^5 square feet. How do we find the length of one side? Since 5 is not evenly divisible by 2, how do we handle this situation? We go right ahead, using the rules just learned:

$$s = A^{1/2}$$
$$= (10^5)^{1/2}$$
$$= 10^{5 \times (1/2)}$$
$$= 10^{5/2}$$
$$= 10^{(2 + 1/2)}$$
$$= 10^2 \times 10^{1/2}$$

1.4 POWERS AND ROOTS

$$= 100 \times \sqrt{10} = 100 \times 3.16$$
$$= 316 \text{ ft}$$

You can see that in going from line 4 to line 5 in the above solution, we broke the 5/2 in the exponent into a whole number and a fraction. This process is the reverse of what we do when we multiply two exponentials together. The result is the product of two numbers: a whole number and a square root whose value is easily found in a table of square roots or with a calculator.

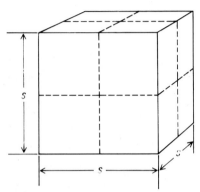

FIGURE 1-2 The volume of a cube is equal to s × s × s, where s is the length of a side. If s = 2 units, then the area is 8 cubic units.

Example 1.17   A grain of salt is a cube 0.01 in on a side (see Figure 1-2). What is the volume of this salt grain?

Solution   If s stands for the length of a side, then the volume of a cubic grain is given by

$$V = s^3$$

In this problem, $s = 0.01$ in $= 10^{-2}$ in. Thus we have

$$V = s^3 = (10^{-2} \text{ in})^3$$
$$= 10^{-6} \text{ in}^3$$

which is one-millionth of a cubic inch. Note that the symbol for the unit (in) is treated just like a number and is cubed (raised to the third power). The symbol in^3 is sometimes read as "inches cubed" and sometimes "cubic inches." Both names mean the same thing.

Example 1.18   A cubic storage bin has a volume of 1000 ft³. What is the length of each side of this bin?

Solution   The volume V is related to the length of the side s by the equation

$$V = s^3$$

Raising both sides of the equation to the one-third power, we have

$$s = V^{1/3}$$

so

$$s = (1000 \text{ ft}^3)^{1/3} = (10^3 \text{ ft}^3)^{1/3}$$
$$= 10^{3 \times 1/3} \text{ ft} = 10^1 \text{ ft} = 10 \text{ ft}$$

1.5 Solving Algebraic Equations

Simple Mathematical Relationships

An **algebraic equation** is a shorthand way of writing down relationships among mathematical quantities. For example, to calculate the volume of a rectangular solid, we use the word formula

$$\text{volume} = \text{length} \times \text{width} \times \text{height}$$

(See Figure 1-3.) This word formula is turned into an algebraic equation by representing each of its four quantities as a single letter:

$$L = \text{length}$$
$$W = \text{width}$$
$$H = \text{height}$$
$$V = \text{volume}$$

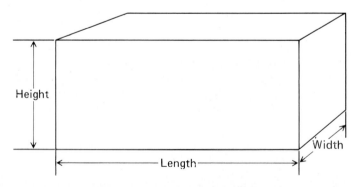

FIGURE 1-3 The volume of a rectangular solid is equal to $L \times W \times H$, where L is the length of the solid, W is its width, and H is its height.

1.5 SOLVING ALGEBRAIC EQUATIONS

With these meanings in mind, we write the equation as follows:

$$V = L \times W \times H$$

Going further, we can eliminate the multiplication signs ($\times$) by using the rule that two letters written next to each other are understood to be multiplied by each other:

$$V = LWH$$

The equation in this form is useful if we know the length, width, and height and want to calculate the volume. We merely substitute numbers into the right-hand side of the equation and perform the required multiplications; the result is the volume.

Example 1.19 A rectangular tank has a length of 4 ft, a width of 2 ft, and a depth of 1.5 ft. What is the volume of the tank?

Solution
$$V = LWH = 4 \text{ ft} \times 2 \text{ ft} \times 1.5 \text{ ft} = 12 \text{ ft}^3$$

Note that ft $\times$ ft $\times$ ft = ft^3, which is read as either "cubic feet" or "feet cubed."

Another problem arises if we know the volume, the length, and the width of an object and want to find its height. The unknown quantity is now H. In order to calculate H, we must have an equation in which H equals some combination of the other quantities, V, L, and W. To obtain such an equation, we solve the original equation for the quantity H. One universal rule is used in solving all equations of this type: *The equation is unchanged if both sides are multiplied or divided by the same quantity or if the same quantity is added to or subtracted from both sides of the equation.*

In the example above, we want to get the quantity H off by itself on one side of the equation. The given equation is

$$V = LWH$$

in which H is multiplied by L and W. We make use of the fact that since any quantity divided by itself equals 1, dividing LW by LW will eliminate LW from that side of the equation, leaving a 1 in its place. Our strategy, then, is to divide both sides of the equation by LW:

$$V = LWH$$

$$\frac{V}{LW} = \frac{LWH}{LW}$$

$$\frac{V}{LW} = 1 \times H = H$$

We then reverse the two sides of the equation to put the quantity H on the left, as is customary:

$$H = \frac{V}{LW}$$

We are now ready to substitute numbers into the right-hand side of the equation to find the value of H.

Example 1.20 A rectangular box has a volume of 2400 in³. The box is 24 in long and 12 in high. How wide is the box?

Solution
$$V = LWH$$

Divide both sides of the equation by LH ($L \times H$):

$$\frac{V}{LH} = \frac{LWH}{LH}$$

$$\frac{V}{LH} = W$$

$$W = \frac{V}{LH} = \frac{2400 \text{ in}^3}{24 \text{ in} \times 12 \text{ in}}$$

$$= 8.33 \text{ in}$$

Note that in³ divided by in × in (in²) gives in¹, or in.

Here is a slightly more complicated equation:

$$\frac{PV}{T} = K$$

(Let us not concern ourselves about the meanings of the various quantities at this point. We will meet this equation later, in Chapter 12.) As you see, we have written this equation without the × signs indicating multiplication, using the previously described convention that two quantities next to each other are understood to be multiplied by each other.

The equation as it stands allows us to find K by substituting numbers for P, V, and T. However, suppose we know K, P, and V and our goal is to find T. In order to isolate T, two things must be done. First, we must move T up into the numerator, and second, we must move all the other quantities to the side of the equation opposite the T. We can get T into the numerator very simply by multiplying both sides of the equation by T. On the left, $T/T = 1$.

$$\frac{PV}{T} = K$$

$$\frac{PVT}{T} = KT$$

$$PV = KT$$

We now have T in the numerator, but it is multiplied by K, which we must move to the other side of the equation. We do this by dividing both sides of the equation by K, so that it is eliminated from the right and ends up in the denominator on the left:

$$\frac{PV}{K} = \frac{KT}{K}$$

$$\frac{PV}{K} = T$$

$$T = \frac{PV}{K}$$

We could have gotten the same result in one step by multiplying the original equation by the fraction T/K, which is the same as multiplying by T and dividing by K.

Linear Equations

A very common kind of equation—one encountered frequently in this book—has the form

$$y = mx + b$$

In this equation, m and b are constants, and x and y are variables. (A **constant** is a quantity that maintains a single value in this equation, while a **variable** may take on a number of different values.) If x takes on various values, y must assume the values forced on it by the relationship. In this situation, x is called the **independent variable** and y is called the **dependent variable** because the value of y *depends* on the value of x.

For example, consider the situation where $m = 2$ and $b = 5$. Give x a number of different numerical values, and then find the corresponding values of y. To do this, simply substitute the given value of x in the equation and then perform the indicated mathematical operations to find the value of y:

	$y = 2 \times x + 5$
$x = 0$	$y = 2 \times 0 + 5 = 5$
$x = 1$	$y = 2 \times 1 + 5 = 7$
$x = 2$	$y = 2 \times 2 + 5 = 9$
$x = 3$	$y = 2 \times 3 + 5 = 11$

and so on.

One way to find the kind of relationship between x and y is by plotting a **graph**, which is a diagram that represents the relationship between two variables. The value of y corresponding to each value of x is plotted in Figure 1-4, and we see that the shape of the graph is a straight line. For this reason, an equation of the general form $y = mx + b$ is called a **linear equation** (or a linear relationship between x and y). Of course, any other set of symbols can be used to represent a linear equation—for example, $z = aw + c$. It is customary to use letters at the end of the alphabet for the variables and letters at the beginning of the alphabet for constants.

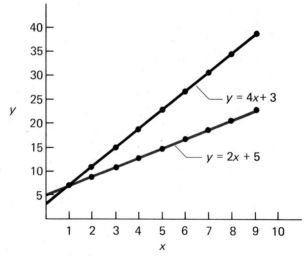

FIGURE 1-4 This graph shows values of y corresponding to various values of x for two different linear relationships.

Even the constants may change, however. The equation $y = 4x + 3$ is a linear relationship between y and x that uses a different set of values for the constants m and b. In Figure 1-4 we see that the graph of $y = 4x + 3$ is a straight line with a slope that is steeper than that of $y = 2x + 5$. (That is, the line makes a greater angle with the horizontal line.) The constants m and b, which may vary from one equation to the next, are called *parameters*.

An important special relationship occurs when the constant b has the value of zero. Then the equation has the simple form

1.5 SOLVING ALGEBRAIC EQUATIONS

$$y = mx$$

See what happens when we put different values of x into this equation, using the value $m = 5$ as an example:

$$x = 0 \quad y = 5 \times 0 = 0$$
$$x = 1 \quad y = 5 \times 1 = 5$$
$$x = 2 \quad y = 5 \times 2 = 10$$
$$x = 3 \quad y = 5 \times 3 = 15$$
$$x = 4 \quad y = 5 \times 4 = 20$$

and so on.

Two important facts about this relationship must be noted:

1. When x equals 0, y also equals 0.

2. When x doubles, y also doubles. That is, when x goes from 1 to 2, y goes from 5 to 10. When x goes from 2 to 4, y goes from 10 to 20.

Such a relationship is called a *direct proportion*; y is directly proportional to x when any change in the value of x causes a proportional change in the value of y. In other words, when one quantity doubles, the other quantity also doubles. Direct proportions are very important in all aspects of science. We find, for example, that the amount of heat released by the burning of a particular kind of fuel is directly proportional to the weight of the fuel. We also find that the distance a car travels is directly proportional to the amount of fuel burned (other factors being equal). Therefore, we know that the distance a car travels must be related to the heat released by the burning fuel.

Very often we are given a linear relationship of the form

$$y = mx + b$$

and are required to find the value of x that goes with a given value of y. In other words, the equation is to be solved for x. To solve this equation, we use the general rule given in the last section: The equation is unchanged if both sides are multiplied or divided by the same quantity or if the same quantity is added to or subtracted from both sides of the equation. Begin by subtracting b from both sides of the equation to obtain

$$y - b = mx + b - b$$

or

$$y - b = mx$$

Now divide both sides of the equation by m:

$$\frac{y-b}{m} = \frac{mx}{m} = x$$

Then exchange the two sides of the equation to put x by itself on the left:

$$x = \frac{y-b}{m}$$

Example 1.21 A car travels 27 mi on a gallon of gas (that is, the mileage is 27 mi/gal). How far does the car go on 5 gal of gas? What kind of relationship is there between the distance traveled and the amount of fuel?

Solution If the car travels a distance of 27 mi for each gallon of gas, then the distance traveled on 5 gal of gas is

$$27 \text{ mi/gal} \times 5 \text{ gal} = 135 \text{ mi}$$

We see that the general relationship can be written

$$\text{mi} = \text{mi/gal} \times \text{no. of gal}$$

or

$$\text{distance} = \text{mileage} \times \text{no. of gal}$$

Letting y = distance, m = mileage, and x = number of gallons, we see that the relationship has the form

$$y = mx$$

and is a direct proportion.

Example 1.22 Find the value of m needed to satisfy the equation

$$y = mx + b$$

if $y = 100$, $x = 5$, and $b = 20$.

Solution To get m by itself, subtract b from both sides of the equation:

$$y - b = mx + b - b$$
$$y - b = mx$$

Divide both sides of the equation by x, and interchange the left and right sides:

$$m = \frac{y-b}{x}$$

> Substitute the given numbers into the equation:
>
> $$m = \frac{100 - 20}{5}$$
>
> $$m = 16$$
>
> *Note*: We can solve an equation for any one of the quantities in it if we know all the others. Any of the terms in the equation may be considered to be the unknown quantity under the right conditions.

Glossary

algebraic equation A symbolic expression in which two or more quantities are related to each other. This equation may contain numbers, letters, and algebraic operations such as addition, subtraction, multiplication, division, exponentiation, or root extraction.

constant A numerical quantity that does not change within the context of a given problem or equation. The constant may be either a number or a letter representing a number.

dependent variable A variable in an equation that changes in response to the change of another (independent) variable; commonly known as the "unknown quantity."

exponent A number or expression placed above and to the right of another number, representing the power to which the second number is raised. For example, in the expression 10^4, the number 4 is the exponent.

graph A diagram that represents a relationship between two variables.

independent variable A variable in an equation that can take on different values, causing the dependent variable to change in response.

linear equation An equation containing only the first powers of both the dependent and independent variables, so that when the relationship between the two variables is plotted on a graph, the result is a straight line. Linear equations in two variables can always be reduced to the form $y = mx + b$, where x is the independent variable and y is the dependent variable.

power Exponent. The expression 10^3 is read "ten to the third power," and is the same as $10 \times 10 \times 10$.

scientific notation A system of writing numbers in which each number is written in the form $C \times 10^N$, where C is a decimal number between 1 and 10 and N is an integer. For example, $3140 = 3.14 \times 10^3$.

square root For a given number X, that number Y which, when multiplied by itself, equals X. That is, $Y \times Y = Y^2 = X$; therefore $Y = \sqrt{X}$. If $X = 9$, then $Y = 3$; or $3 = \sqrt{9}$.

Problems and Questions

1.1 Large and Small Numbers

1.1.P Write these numbers in scientific notation:
- **a.** 23
- **b.** 589,000
- **c.** 93,000,000
- **d.** 389,200,000,000
- **e.** 0.034
- **f.** 0.00000678
- **g.** 1/10,000
- **h.** 1/1,000,000

1.2.P Write these numbers in decimal notation:
- **a.** 5.6×10^3
- **b.** 3.87×10^6
- **c.** 2.783×10^{-3}
- **d.** 9.9×10^{-8}
- **e.** 1.5932×10^{-12}
- **f.** 7.4510×10^{12}
- **g.** 6.022×10^{23} (This number happens to be very important in chemistry. Can you imagine what it would be like if you had to deal with it in decimal notation?)

1.2 Multiplication and Division

1.3.P Multiply these numbers:
- **a.** $10^5 \times 10^7 = ?$
- **b.** $10^2 \times 10^5 \times 10^4 = ?$
- **c.** $10^9 \times 10 = ?$
- **d.** $10^3 \times 10^6 \times 10^9 = ?$
- **e.** $10^{10} \times 10^{15} \times 10^{20} = ?$

1.4.P Multiply these numbers:
- **a.** $10^{-2} \times 10^{-3} = ?$
- **b.** $10^{-4} \times 10^{-5} \times 10^{-6} = ?$
- **c.** $10^{-8} \times 10^{-1} \times 10^{-9} = ?$
- **d.** $10^{-5} \times 10^{-3} \times 10^{-8} = ?$
- **e.** $10^{-20} \times 10^{-8} \times 10^{-85} = ?$

1.5.P Multiply these numbers:
- **a.** $10^5 \times 10^{-7} = ?$
- **b.** $10^{-4} \times 10^5 \times 10^{-6} = ?$
- **c.** $1000 \times \frac{1}{10} \times 10^{-4} = ?$
- **d.** $10^{-2} \times 10^3 \times 10^{-7} = ?$
- **e.** $10^{-5} \times \frac{1}{10^{-6}} \times 10^{-12} = ?$

1.6.P Multiply these numbers, and make sure the answers are in scientific notation:
- **a.** $(2 \times 10^3) \times (6 \times 10^5) = ?$
- **b.** $(3.5 \times 10^2) \times (7.2 \times 10^4) = ?$
- **c.** $(5.4 \times 10^{-7}) \times (9.8 \times 10^{-9}) = ?$
- **d.** $(1.23 \times 10^{52}) \times (6.1 \times 10^{-18}) = ?$
- **e.** $(7.4 \times 10^{15}) \times (1.5 \times 10^{-8}) \times (9.8 \times 10^{-6}) = ?$

1.7.P Divide these numbers:
- **a.** $\dfrac{10^8}{10^4} = ?$
- **b.** $\dfrac{10^3}{10^9} = ?$
- **c.** $\dfrac{10^6}{10^3} = ?$
- **d.** $\dfrac{(10^{15} \times 10^3)}{(10^2 \times 10^4)} = ?$
- **e.** $\dfrac{(10^9 \times 10^3)}{(10^8 \times 10^6)} = ?$

1.8.P Divide these numbers:
- **a.** $\dfrac{10^{-5}}{10^{-7}} = ?$
- **b.** $\dfrac{10^{-8}}{10^{-10}} = ?$
- **c.** $\dfrac{10^{-12}}{10^{-4}} = ?$
- **d.** $\dfrac{(10^{-16} \times 10^{-8})}{(10^{-5} \times 10^{-9})} = ?$
- **e.** $\dfrac{(10^{-6} \times 10^{-2})}{(10^{-5} \times 10^{-11})} = ?$

1.9.P Divide these numbers:
- **a.** $\dfrac{10^8}{10^{-13}} = ?$
- **b.** $\dfrac{10^{-12}}{10^5} = ?$
- **c.** $\dfrac{(10^2 \times 10^5)}{(10^{-7} \times 10^6)} = ?$
- **d.** $\dfrac{(10^{-8} \times 10^5)}{(10^{-7} \times 10^6)} = ?$
- **e.** $\dfrac{(10^6 \times 10^{-2})}{(10^8 \times 10^{-8})} = ?$

1.10.P Divide these numbers, and make sure that the answer is in correct scientific notation:

a. $\dfrac{2.5 \times 10^6}{5.0 \times 10^{-6}} = ?$

b. $\dfrac{7.0 \times 10^{-5}}{1.4 \times 10^{-8}} = ?$

c. $\dfrac{3.5 \times 10^9 \times 4.2 \times 10^7}{7.5 \times 10^5} = ?$

d. $\dfrac{1.23 \times 10^{-6} \times 4.36 \times 10^8}{6.78 \times 10^{-9}} = ?$

e. $\dfrac{4.20 \times 10^{12} \times 8.36 \times 10^{-4}}{5.67 \times 10^{-6} \times 3.14 \times 10^{-9}} = ?$

1.11.P Show that $x^0 = 1$, where x is any number.

1.12.P Multiply these numbers:
a. $(2 \times 10^5) \times (3 \times 10^{-5}) = ?$
b. $(9.36 \times 10^0) \times (7.20 \times 10^7) = ?$
c. $10^0 \times 10^0 \times 6.02 \times 10^{23} = ?$

1.13.P Divide these numbers:

a. $\dfrac{(6 \times 10^4)}{(2 \times 10^4)} = ?$

b. $\dfrac{(10^5 \times 10^0)}{(10^5 \times 10^0)} = ?$

c. $\dfrac{(10^7 \times 10^0)}{(10^{-7} \times 10^0)} = ?$

1.14.P Solve these problems:

a. $\dfrac{(5 \times 10^{15}) \times (10^0)}{5} = ?$

b. $\dfrac{6 \times 10^{12}}{12} = ?$

c. $\dfrac{6.02 \times 10^{23}}{(3.01 \times 10^5) \times (10^0)} = ?$

1.3 Addition and Subtraction

1.15.P Solve these addition problems:
a. $3.4 \times 10^6 + 1.45 \times 10^7 = ?$
b. $5.78 \times 10^{-4} + 6.9 \times 10^{-5} = ?$
c. $3.40 \times 10^3 + 8.64 \times 10^4 + 7.21 \times 10^4 = ?$
d. $2.876 \times 10^6 + 4.67 \times 10^5 + 3.2 \times 10^4 = ?$
e. $1.897 \times 10^{-9} + 6.24 \times 10^{-8} + 9.41 \times 10^{-9} = ?$

1.16.P Solve these subtraction problems:
a. $6.15 \times 10^3 - 5.61 \times 10^2 = ?$
b. $7.21 \times 10^{-7} - 8.32 \times 10^{-8} = ?$
c. $9.52 \times 10^{20} - 6.18 \times 10^{21} = ?$
d. $2.11 \times 10^{-8} - 5.43 \times 10^{-7} = ?$
e. $3.45 \times 10^2 - 7.11 \times 10^3 = ?$

1.17.P Solve these problems:
a. $1.00 \times 10^9 + 7.43 \times 10^{10} = ?$
b. $7.98 \times 10^5 + 1.15 \times 10^4 - 6.50 \times 10^{-3} = ?$
c. $6.71 \times 10^{-5} + 4.5 \times 10^5 = ?$
d. $2.46 \times 10^{-7} + 4.89 \times 10^{-7} + 7.5 \times 10^{-8} = ?$
e. $1.72 \times 10^0 + 4.51 \times 10^1 + 6.51 \times 10^{-1} = ?$

1.4 Powers and Roots

1.18.P Perform these operations:
a. $(10^4)^2 = ?$ b. $(10^5)^3 = ?$
c. $(10^{-6})^2 = ?$ d. $(10^6)^{1/3} = ?$
e. $(10^{-6})^{1/2} = ?$

1.19.P Perform these operations:
a. $(2 \times 10^4)^2 = ?$ b. $(9 \times 10^{12})^{1/2} = ?$
c. $(8 \times 10^{-12})^{1/3} = ?$ d. $(5 \times 10^{-6})^3 = ?$
e. $(6 \times 10^{-3})^2 = ?$

1.20.P a. What is the area of a square that is 10^{-3} in (0.001 in) on a side?
b. What is the volume of a cube that is 10^{-3} in on a side?

1.21.P A square solar collector has an area of 100 ft². What is the length of each side?

1.22.P A cubic chamber is 30 ft on a side. What is the volume of this chamber?

1.23.P A square photographic developing tray has an area of 400 in². What is the length of each side of this tray?

1.5 Solving Algebraic Equations

1.24.P Solve the equations for the variable indicated:
a. $c = a + b$ Solve for b.
b. $15 = a + 7d$ Solve for d.
c. $x = b + (c - 1)d$ Solve for d.
d. $y = 25x + 15$ Solve for x.
e. $xy = 6$ Solve for x.
f. $2x + 7y = 16$ Solve for y.
g. $ma + nb = 24c$ Solve for b.

1.25.P $PV = nRT$
 a. Solve for R.
 b. Given that $P = 760$, $V = 22.4$, $n = 1$, and $T = 273$, find the numerical value of R.

1.26.P $y = mx + b$
 a. Solve for x.
 b. Given that $y = 256$, $m = 3.5$, and $b = -14$, find the numerical value of x.

1.27.P $z = (a - 21)p + (b - 14)q$
 a. Solve for q.
 b. Given that $z = 23.5$, $a = 18$, $b = 25$, and $p = 78$, find the numerical value of q.

Self Test

1. Convert these numbers into scientific notation:
   a. 623,000,000,000
   b. 0.000 000 000 236
   c. $\dfrac{1}{2,000,000}$
   d. 123,456,789
   e. 0.00540

2. Convert these numbers into decimal notation:
   a. 7.91×10^6
   b. 8.5×10^{-4}
   c. 6.02×10^{10}
   d. 1.98×10^{-5}
   e. 9.064×10^{15}

3. Perform these multiplication operations, and check that each result is in correct scientific notation:
   a. $(2.3 \times 10^6) \times (8.9 \times 10^3) = ?$
   b. $(6.9 \times 10^5) \times (1.5 \times 10^{-4}) \times (7.2 \times 10^{-3}) = ?$
   c. $(8.0 \times 10^0) \times (1.4 \times 10^{60}) \times (1.0 \times 10^1) \times 5.0 = ?$
   d. $(3.5 \times 10^{-2}) \times (3.5 \times 10^2) = ?$
   e. $(6.0 \times 10^5)^2 \times (8.5 \times 10^{-3})^3 \times (4.0 \times 10^{-4})^{1/2} = ?$

4. Perform these division operations, and give each answer in scientific notation:
   a. $\dfrac{2.3 \times 10^6}{8.9 \times 10^3} = ?$
   b. $\dfrac{6.9 \times 10^5}{(1.5 \times 10^{-4}) \times (7.2 \times 10^{-3})} = ?$
   c. $\dfrac{(8.0 \times 10^0) \times (1.4 \times 10^{60})}{(1.0 \times 10^1) \times 5.0} = ?$
   d. $\dfrac{3.5 \times 10^{-2}}{3.5 \times 10^2} = ?$
   e. $\dfrac{(6.0 \times 10^5)^2}{(8.5 \times 10^{-3})} = ?$

5. Add or subtract these numbers:
   a. $6.2 \times 10^{-3} + 3.22 \times 10^{-2} = ?$
   b. $4.6 \times 10^2 - 8.00 \times 10^3 = ?$
   c. $2.83 \times 10^0 + 5.625 \times 10^2 - 7.6 \times 10^{-1} = ?$
   d. $3.57 \times 10^{-8} + 6.01 \times 10^{-8} - 7.5 \times 10^{-8} = ?$
   e. $3.22 \times 10^{-15} + 8.99 \times 10^{-15} - 6.02 \times 10^{23} = ?$

In Problems 6-9, be sure to keep track of the units.

6. The volume of a cylinder is given by the formula $V = \pi r^2 h$, where π is the number 3.14. A gas tank has a radius of 500 in and a height of 400 in. What is the volume of the tank in cubic inches?

7. A cubic storage bin is 65 in along each side. What volume of sugar can be stored in the bin?

8. A cylindrical tower has a volume of 1.41×10^6 ft^3. If its height is 500 ft, what is its radius?

9. A spherical gas storage tank has a volume of 4.18×10^9 in^3. What is the diameter of this tank, in inches? (The volume is given by $V = \frac{4}{3}\pi r^3$, and the diameter is twice the radius, or $d = 2r$.)

10. Given the equation $y = mx + b$:
    a. Let $m = 21.5$ and $b = 10.3$. If $x = 5.5$, what is the value of y?
    b. What value must x have in order for y to equal 150?
    c. If the value of x doubles, does y double?
    d. Is y directly proportional to x?
    e. How would this equation have to be changed in order to make y directly proportional to x?

11. Given the equation $w = (c - 2)x + (d + 12)y$, what value must d have to make $w = 123$ if $c = 13.2$, $x = -2.5$, and $y = 15.7$?

12. Given the equation

$$\frac{y}{a} = \frac{x}{b} + c$$

what is the value of x if $y = 235.6$, $a = 23$, $b = 3.4$, and $c = -26.5$? Is there a linear relationship between y and x?

2
Measurements and Operations

Objectives After completing this chapter, you should be able to:
1. Analyze the essential parts of a word problem.
2. Change a word problem into an equation problem.
3. Use the unit factor method of solving problems.
4. State conversion factors for the common sets of units.
5. Solve rate problems.
6. Make measurements of length, volume, and mass in the metric system.
7. Use the metric prefixes.
8. Convert measurements from metric to English units and vice versa.
9. Solve problems involving the concepts of density and specific gravity.
10. Read and draw graphs.
11. Determine a linear relationship and a direct proportion from a graph.
12. Use the Fahrenheit and Celsius temperature scales and convert between the two scales.
13. Estimate the accuracy and precision of a measurement or calculation.
14. Round off numbers to an appropriate number of significant figures.

2.1 Problem-Solving Techniques

Changing Words into Equations

Problems come to us in many forms. In school we are given straightforward arithmetic problems, such as "Find the product of 34 times 68." Using the rules of multiplication (or a calculator), we go step by step through the procedures that give the answer to the problem. No creative thought is needed.

The real world does not spell out problems so simply. In the real world, problems arise from life and work situations, and you must learn methods for restating these problems in a form that permits a mathematical solution. For example, suppose you are working in a chemistry laboratory and have been given the job of filling 12 small bottles with alcohol from a one-gallon container. Each bottle can hold 16 fl oz (fluid ounces), and you wonder if there is enough alcohol in the large container to do the job. You can answer the question by trial and error—just filling bottles until you either fill them all or run out of alcohol. Or you can work out the answer in advance to see if you need to order additional alcohol.

This is a simple problem, but it illustrates how we can break any problem into three basic parts:

1. *The input.* This is information needed to solve the problem. In the present case the input is the number of small bottles, the amount of liquid each bottle can hold, and the amount of liquid in the large container. Notice that the *kind* of liquid does not make any difference to the solution of the problem. Therefore, the fact that the liquid in this problem is alcohol does not matter. In the language of problem solving, we call this fact a "red herring" because it is not needed to solve the problem and may distract your attention from the important facts. In solving problems, you must pick out just those facts that are necessary and ignore red herrings.

2. *The output.* This is the solution we want. In the present case we want to know if a gallon of liquid is enough to fill the 12 bottles.

3. *The method of solution.* The actual solution of the problem requires two ingredients: (a) one or more relationships between the input numbers and the output numbers and (b) a logically ordered set of operations that can be used to manipulate the relationships between the input and the output in order to find the answer to the problem. (In computer terminology, this set of operations is called an algorithm. In the theory of games, these operations make up a *game plan*.)

In the present problem we have two questions to answer:

1. What is the total volume of liquid needed?

2. Is this volume greater than or less than a gallon?

To answer the first question we use a relationship between the input and the output quantities, which we first state in words: *The total amount of liquid needed equals the quantity of liquid in each bottle multiplied by the number of bottles.* Often (especially if there are several mathematical steps) it helps to change the word equation into a symbol equation by letting one letter stand for each quantity in the equation. Therefore, we make the following definitions:

$$V = \text{total volume of liquid}$$

$$v = \text{volume of liquid in each bottle (volume/bottle)}$$

$$N = \text{number of bottles}$$

The word equation then becomes this mathematical equation:

$$\text{total volume} = \text{volume per bottle} \times \text{no. of bottles}$$

or

$$V = v \times N$$

Now we are ready to put numbers into the equation:

$$V = \frac{16 \text{ fl oz}}{1 \text{ bottle}} \times 12 \text{ bottles}$$

$$= 192 \text{ fl oz}$$

We now know the number of fluid ounces needed. In order to compare this quantity with the gallon in the large container, we must know that a U.S. liquid gallon contains 128 fl oz. We are now able to answer the second question: Since 192 fl oz (the volume needed) is greater than 128 fl oz (the amount available), we must order another container of alcohol.

Note that units of measurement are treated just like algebraic quantities. Thus in the above problem the term "bottles" in the numerator canceled the "bottle" in the denominator:

$$\frac{\text{fl oz}}{\text{bottle}} \times \text{bottles} = \text{fl oz}$$

In the next section the unit factor method of setting up problems will be used to make calculations involving units of various kinds. This method helps to ensure that the result of a calculation has the correct set of units as well as the correct numerical value.

The Unit Factor Method

The **unit factor method** is a practical approach to problem solving in chemistry. The key concept in this method is the use of units (feet, pounds, gallons, etc.) to direct the plan of solution. To use this method, first look at the units connected with the quantities given in the problem (the input). Then look at the units expected in the answer (the output). Finally, ask how the first set of units can be converted into the second set.

In this technique conversion factors are important. A **conversion factor** is a number that converts quantities of one unit into quantities to another. To illustrate how conversion factors are used, let us convert pounds into ounces. We know that

$$16 \text{ oz (ounces)} = 1 \text{ lb (pound)}$$

You might ask how we can have 16 on one side of the equation and 1 on the other side. The answer is that, strictly speaking, this is not an equation; it is a statement telling us that 16 oz of anything is equivalent to 1 lb.

If we divide both sides of this expression by 1 lb, we obtain this interesting statement:

$$\frac{16 \text{ oz}}{1 \text{ lb}} = 1$$

The quantity on the left is read as "16 ounces per pound" and may be written as 16 oz/lb. It tells us how many ounces there are in each pound of any substance. The above expression indicates that 16 oz/lb is equivalent to the number 1 (unity). This means that if we multiply a number of pounds by this quantity, the amount of material is unchanged, but *the number of pounds is converted into the equivalent number of ounces.*

The quantity 16 oz/lb is a *unit conversion factor*. It has the ability to change numbers of pounds into numbers of ounces. For example, to find how many ounces there are in 5 pounds, we multiply the number of pounds by the number of ounces per pound:

$$\text{no. of oz} = \text{conversion factor} \times \text{no. of lb}$$

$$\text{no. of oz} = \text{no. of oz/lb} \times \text{no. of lb}$$

$$= \frac{16 \text{ oz}}{1 \text{ lb}} \times 5 \text{ lb} = 80 \text{ oz}$$

Notice that the "lb" in the numerator cancels the "lb" in the denominator, so only ounces remain in the numerator. *The cancellation of units is the key feature of the unit factor method.*

All problems involving conversion factors are set up according to the following general scheme:

$$\text{no. of output units} = \text{conversion factor} \times \text{no. of input units}$$

The Handbook of Chemistry and Physics (CRC Press, Inc., published annually) and many other technical handbooks and encyclopedias present comprehensive tables of conversion factors for all sorts of units.

In problems involving addition or subtraction of dimensional quantities, the units are simply carried along from the left to the right side of the equation:

$$75 \text{ gal} + 25 \text{ gal} = 100 \text{ gal}$$

In general, *the units on the left side of any equation must be the same as the units on the right side of the equation after all multiplications and divisions have been carried out.* Use of this rule helps us to set up equations correctly. The following examples illustrate other features of the unit factor method.

Example 2.1 The price of graphite (carbon) is $3.20 per lb. How much does 5.4 lb of graphite cost?

Solution The conversion factor between dollars and pounds is 3.20 dollars/lb. What number of dollars corresponds to 5.40 lb? We must use the conversion factor in such a way that the pounds cancel and the dollars remain:

$$\text{no. of dollars} = \text{conversion factor} \times \text{no. of lb}$$

$$= \frac{3.20 \text{ dollars}}{1 \text{ lb}} \times 5.40 \text{ lb} = 17.3 \text{ dollars}$$

Example 2.2 Using the same price as in Example 2.1, how many pounds of graphite can we buy for $20?

Solution The answer must be expressed in units of pounds, so the dollars must cancel. Therefore we write

$$\text{no. of lb} = \text{conversion factor} \times \text{no. of dollars}$$

$$= \frac{1 \text{ lb}}{3.2 \text{ dollars}} \times 20 \text{ dollars} = 6.25 \text{ lb}$$

Example 2.3 A tank contains 55.0 ft³ of a chemical liquid. How many 5-gal containers can we fill from this tank?

Solution *The Handbook of Chemistry and Physics* informs us that the conversion factor between cubic feet (ft³) and gallons (U.S. liquid) is 7.48 gal/ft³. To find the number of gallons in 55.0 cu ft, we say

$$\text{no. of gal} = \frac{7.48 \text{ gal}}{1 \text{ ft}^3} \times \text{no. of ft}^3$$

$$= \frac{7.48 \text{ gal}}{1 \text{ ft}^3} \times 55.0 \text{ ft}^3$$

$$= 411 \text{ gal}$$

We have rounded off the answer to three digits because there are no more than three digits in the numbers we started with (more about rounding off later in this chapter).

The above solution is not complete, for we are asked to find the number of containers, each one containing 5.00 gal. We now have another conversion factor to use: 1 container/5 gal. (We don't say 5 gal/1 container because in this problem we want the gallons to be in the denominator.)

$$\text{no. of containers} = \text{no. of containers/gal} \times \text{no. of gal}$$

$$= \frac{1 \text{ container}}{5.00 \text{ gal}} \times 411 \text{ gal}$$

$$= 82.2 \text{ containers}$$

Example 2.4 A chemistry class is doing an experiment in which each student needs to use 3 beakers of distilled water. Each beaker holds 4.00 fl oz of water, and there are 24 students in the class. How many gallons of water are required for the entire class?

Solution Several steps can be combined into one by making sure that the units in the conversion factors cancel so as to give gallons in the final answer. The conversion factors needed are 1 gal/128 oz, 4 oz/beaker, and 3 beakers/student. Then the following calculation can be set up:

$$\text{no. of gal} = \frac{1 \text{ gal}}{128 \text{ oz}} \times \frac{4 \text{ oz}}{1 \text{ beaker}} \times \frac{3 \text{ beakers}}{1 \text{ student}} \times 24 \text{ students}$$

$$= 2.25 \text{ gal}$$

We see that the units "oz," "beakers," and "students" cancel, leaving gallons.

Rate Problems

A common type of problem involves the concept of **rate**, which tells how fast something happens or how many things happen per unit of time. For example, if a laboratory uses 2 gal of distilled water each day, 2 gal per day is the rate at which water is used. The number of gallons used in a given number of days is found by multiplying the number of gallons per day by the number of days:

$$\text{no. of gal} = \frac{\text{no. of gal}}{1 \text{ day}} \times \text{no. of days}$$

or

$$\text{amount} = \text{rate} \times \text{time}$$

Speed is a rate. The distance traveled by a car in a given amount of time is found by figuring

$$\text{no. of mi} = \frac{\text{no. of mi}}{1 \ \cancel{\text{hr}}} \times \text{no. of } \cancel{\text{hr}}$$

or

$$\text{distance} = \text{rate} \times \text{time}$$

We see that the hours units cancel out, and we are left with units of distance. *A rate always consists of some unit of measurement divided by a unit of time.* Therefore, when the rate is multiplied by an amount of time, the units of time cancel. A rate is a conversion factor between a unit of time and another unit of measurement.

Example 2.5 How far will a car travel if it goes 55.0 mi/hr for 10.4 hr?

Solution
$$\text{distance} = \text{rate} \times \text{time}$$
$$= \frac{55.0 \ \text{mi}}{1 \ \cancel{\text{hr}}} \times 10.4 \ \cancel{\text{hr}}$$
$$= 572 \ \text{mi}$$

Example 2.6 A storage tank in a chemical manufacturing plant holds 10,000 gal of acid. Each day an average of 1500 gal of acid is used. The next shipment is 7 days away. Will the present supply of acid be enough?

Solution Find the number of gallons used in 7 days:

$$\text{no. of gal} = \text{rate} \times \text{time}$$
$$= \frac{\text{no. of gal}}{1 \ \text{day}} \times \text{no. of days}$$
$$= \frac{1500 \ \text{gal}}{1 \ \cancel{\text{day}}} \times 7 \ \cancel{\text{days}}$$
$$= 10{,}500 \ \text{gal}$$

Since the tank holds 10,000 gallons, the plant will run out of acid on the last day.
How do we know the tank is empty on the last day? We know that the plant uses 1500 gal per day, and if it uses 10,500 gal by the end of the seventh day, this means it has used 10,500 gal − 1500 gal = 9000 gal by the end of the sixth day. So somewhere during the seventh day it will have used up all 10,000 gal. For a more exact calculation, see the next example.

Example 2.7 How many days will it take for the tank in Example 2.6 to be emptied?

Solution Solve the rate equation for time by multiplying both sides of the equation by 1/rate:

$$\text{no. of gal} = \text{rate} \times \text{time}$$

$$\text{no. of gal} \times \frac{1}{\text{rate}} = \frac{1}{\cancel{\text{rate}}} \times \cancel{\text{rate}} \times \text{time}$$

Therefore

$$\text{time} = \text{no. of gal} \times \frac{1}{\text{rate}}$$

$$= 10{,}000 \text{ gal} \times \frac{1}{1500 \frac{\text{gal}}{\text{day}}}$$

$$= 10{,}000 \cancel{\text{ gal}} \times \frac{1 \text{ day}}{1500 \cancel{\text{ gal}}}$$

$$= 6.7 \text{ days}$$

Note:

$$\frac{1}{\frac{\text{gal}}{\text{day}}} = \frac{\text{day}}{\text{gal}}$$

Example 2.8 A certain chemical is manufactured at the rate of 250 lb per day. How many days does it take to make 45,000 lb of this material?

Solution

$$\text{no. of lb} = \text{rate} \times \text{time}$$

Multiply both sides of this equation by 1/rate:

$$\text{no. of lb} \times \frac{1}{\text{rate}} = \frac{1}{\cancel{\text{rate}}} \times \cancel{\text{rate}} \times \text{time}$$

$$\text{time} = \text{no. of lb} \times \frac{1}{\text{rate}}$$

$$= 45{,}000 \text{ lb} \times \frac{1}{250 \frac{\text{lb}}{\text{day}}}$$

$$= 180 \text{ days}$$

2.2 Measurements in the Metric System

Scientific measurements are generally made using the **metric system**, a rational system of units that originated in France late in the eighteenth century and has now been adopted by most of the countries of the world. (A **unit** is a precisely defined quantity in terms of which physical properties are measured.) The English name for the present version of the metric system is the **International System**. (It is abbreviated SI after the French name *Système International*.)

The advantage of the metric system is that all the units of measurement are divided into ten subunits, so there is no need to remember bothersome conversions such as 12 in per ft, 5280 ft per mi, 16 oz per lb, etc.

In the International System, the four fundamental units with which we will be concerned are those of length, mass, time, and quantity of electric current. We will shortly see that other units, such as those for measuring velocity, energy, and electric charge, can be defined in terms of the four fundamental units.

Units of Length and Area

The fundamental unit of length in the metric system is the *meter*. Originally, the meter was defined to be 10^{-7} of the distance from the North Pole to the equator. Later, a platinum-iridium bar was constructed whose length was as close as possible to the original definition. This bar, which became known as the *standard meter*, is kept at the Bureau of Weights and Measures near Paris. A similar bar was installed at the U.S. Bureau of Standards in Washington, D.C. All rulers, meter sticks, and other length-measuring devices were in the past calibrated by comparison with the standard meter bar. An even more precise standard is currently being used. The meter is now defined to be a length equal to the distance light travels in a vacuum in 1/299,792,458 of a second. The number of digits in this figure shows the great precision with which such measurements can be made.

For comparison with the English units commonly used in the United States, note that a meter is equal to about 3.28 ft, or 39.37 in; it is just a little longer than a yard (36 in).

Many times it is convenient to use units that are either greater or smaller than a meter. A set of prefixes has been devised for this purpose. For example, the prefix *kilo* stands for a thousand, so a thousand meters is called a kilometer. (The word "kilometer" is often pronounced kil*o*meter, with the emphasis on the second syllable, although strictly speaking the pronunciation *kilo*meter is more correct.) One advantage of the prefix system is that the same set of prefixes can be used with any kind of unit—length, mass, time, energy, etc. Table 2-1 gives a list of the prefixes used with the International System, together with their abbreviations.

You can see that these prefixes give us the ability to handle very large or very small distances. The *centimeter* (1/100 meter) and the *millimeter* (1/1000 meter) are very common units of measurement. In dealing with small particles of

TABLE 2-1 Prefixes and Abbreviations of the International System

Prefix	Factor	Abbreviation
tera	10^{12}	T
giga	10^{9}	G
mega	10^{6}	M
kilo	10^{3}	k
deci	10^{-1}	d
centi	10^{-2}	c
milli	10^{-3}	m
micro	10^{-6}	μ
nano	10^{-9}	n
pico	10^{-12}	p

matter (atoms), the *nanometer* (a billionth of a meter) is a convenient unit of length. A unit that was often used in the past and is still found in many books is the *angstrom*. One angstrom is equal to 10^{-10} meter, or one-tenth of a nanometer. Angstroms are frequently used to measure wavelengths of light and diameters of atoms. (A single hydrogen atom—the smallest particle of hydrogen—is about one angstrom in diameter.)

In the past, the word "micron" was used to represent a millionth of a meter, and this term is still found in many books. However, to maintain the uniform prefix pattern that characterizes the International System, the word *micrometer* is now preferred. The fact that a mechanical device used for measuring precise lengths is also called a micrometer may be a source of confusion. The two words, however, are pronounced differently. In speaking of the mechanical device, we pronounce the word mi*cro*meter, with the emphasis on the second syllable. When we refer to a millionth of a meter, we wall it a *mi*crometer, with the emphasis on the first syllable. (Imagine pronouncing "microscope," and then replace the "scope" with "meter.")

The abbreviation for meter is simply m (without a period). Abbreviations for all the other units are made by joining symbols together; the abbreviation for kilometer is km, for millimeter is mm, for nanometer is nm, etc.

In measuring the area of a surface, we use square units. The square meter is the area of a square one meter on a side. Ten square meters is written as 10 m². Ten square centimeters would be written as 10 cm².

Example 2.9 How many centimeters are there in a kilometer?

Solution There are 10^2 cm in a meter and 10^3 m in a kilometer. Change kilometers into centimeters by using the conversion factors 10^2 cm/m and 10^3 m/km:

2.2 MEASUREMENTS IN THE METRIC SYSTEM

$$\text{no. of cm} = \frac{10^2 \text{ cm}}{1 \text{ m}} \times \frac{10^3 \text{ m}}{1 \text{ km}} \times 1 \text{ km}$$

$$= 10^5 \text{ cm}$$

Example 2.10 How many square centimeters are there in a square meter?

Solution There are 100 cm in a meter. One square meter is the area of a square 1 m on a side, so

$$\text{area} = \text{side} \times \text{side}$$
$$= 1 \text{ m} \times 1 \text{ m} = 100 \text{ cm} \times 100 \text{ cm} = 10^2 \text{ cm} \times 10^2 \text{ cm}$$
$$= 10{,}000 \text{ cm}^2 = 10^4 \text{ cm}^2$$

Units of Volume

Units of volume are found by cubing the units of length. A *cubic meter* is the volume occupied by a cube one meter on a side and is written as 1 m^3. This is a rather large volume, so the cubic centimeter (cm^3) is commonly used for chemical measurements. A cup of water (8 fl oz) has a volume of about 225 cm^3.

Another unit of volume commonly used in chemistry is the *liter*. One liter is equal to 1000 cm^3 and is a little bit larger than a quart. Since a cubic centimeter is one-thousandth of a liter, another name for the cubic centimeter is the *milliliter* (ml)—usually called "em-el" because the full name is such a mouthful. Chemists have traditionally favored the milliliter as a unit of volume, but consistency with the International System would suggest the use of the cubic centimeter.

Example 2.11 How many cubic centimeters are there in a cubic meter?

Solution A cubic meter is the volume of a cube 1 m on a side. There are 100 cm in 1 m, which means that there are 100 cm along each side of this cube. (That is, 1 m and 100 cm both represent the same length; see Figure 2-1.) If s represents the length of a cube's side, the volume V is given by

$$V = s \times s \times s = s^3$$
$$= (100 \text{ cm})^3$$
$$= (10^2 \text{ cm})^3$$
$$= 10^6 \text{ cm}^3$$

Remember that to raise 10^2 to the third power, we multiply the 2 by the 3. The result of this calculation tells us that a cubic meter contains a million (10^6) cm^3.

Example 2.12

How many liters are there in a cubic meter?

Solution

To convert cubic meters into liters, use the two conversion factors 10^3 cm^3/L and 10^6 cm^3/m^3. To arrange these two factors in such a way that the cubic centimeters cancel and the liters are in the numerator, use the reciprocal of the first factor, 1 L/10^3 cm^3. Then the following equation gives the number of liters in 1 m^3:

$$\text{no. of L} = \frac{1 \text{ L}}{10^3 \text{ cm}^3} \times \frac{10^6 \text{ cm}^3}{1 \text{ m}^3} \times 1 \text{ m}^3$$

$$= 10^3 \text{ L}$$

The result is 10^3 (1000) L per m^3.

Note: We read cm^3/m^3 as "cubic centimeters per cubic meter," and cm^3/L as "cubic centimeters per liter."

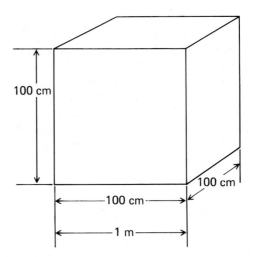

FIGURE 2-1 If a cube is 100 units on a side, then its volume is 100 × 100 × 100 = 10^6 cubic units.

Units of Mass

Sir Isaac Newton originally defined mass to be "the amount of matter in an object." But this definition is not adequate without a description of a method for *measuring* this amount of matter. Chemists usually measure the mass of an object by weighing it on a pan balance, which compares the object being weighed with a set of known weights. (See Figure 2-2.) For this reason, mass is often confused with weight, although they are entirely different things.

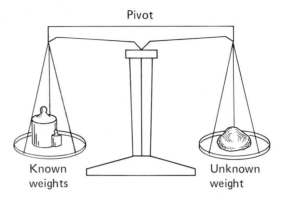

FIGURE 2-2 A pan balance consists of two pans suspended from an arm (or beam) that is free to rotate around a pivot. Sometimes the pivot is in the center of the beam, sometimes not. If the pivot is in the center of the beam, then when the unknown weight equals the known weights in the other pan, the pans are balanced and the beam will rest in a horizontal position.

The **weight** of an object is a measure of how strongly the earth's gravity pulls it down. You can *feel* how heavy the object is by holding it in your hand; you *measure* how heavy it is by putting it on a balance. If the object weighs the same as the known (or standard) weights, then the two pans of the balance in Figure 2-2 will hang suspended at the same level. If the object being weighed is heavier than the known weights, its pan will hang lower than the other pan. Yet the same object feels only one-sixth as heavy on the moon as on earth, because the gravity on the moon is one-sixth the strength of the gravity on earth. Furthermore, the same object in interplanetary space has hardly any weight at all, since it feels only the weak gravitational pull of the distant sun and planets.

Even so, the weightless object still has mass because it has a certain amount of "stuff" in it. The technical definition of **mass** has to do with the inertia of the object. **Inertia** is a physical property that measures how hard it is to start or stop the motion of an object. It is much harder to throw an iron ball than to throw a tennis ball of the same size. We say that the iron ball has more inertia, or mass, than the tennis ball. The importance of mass lies in the fact that it is a *constant* property of an object and does not change from one place to another the way weight does.

Mass is important in physics because there we are concerned with calculating the motion of objects such as rockets and planets. In chemistry we deal with quantities of substances reacting with each other. These quantities are measured by weighing the materials on a balance. Since the mass of an object is directly proportional to its weight, the weighing procedure gives us a measure of the mass. However, as we mentioned, this practice has led to a tendency to confuse mass and weight.

If we are being precise, we recognize that an object weighs a little bit less in Denver than in Washington because in Denver it is at a higher altitude where the pull of gravity is less. This difference in weight could be measured by a sensitive spring balance. However, if we do the weighing with a pan balance, we are comparing the weight of the *unknown object* with the weight of a *standard mass*, and this comparison gives us a direct measurement of the unknown mass, regardless of the location. The pull of gravity is the same on both the object being weighed and the standard mass, and so any effects due to differences in gravity in different parts of the world are canceled out.

The unit of mass in the metric system, as originally defined, was the *gram* (abbreviated g). The gram was intended to be the mass of 1 cm^3 of pure water measured at a temperature of 4°C. (This specification was necessary because the volume of 1 g of water varies slightly with temperature.) Later, with the adoption of the International System, the unit of mass was defined to be the *kilogram* (kg). Since the kilogram is 1000 g, its mass is approximately that of 1 L (1000 cm^3) of water, but it is defined more precisely as the mass of a certain platinum-iridium cylinder kept at the Bureau of Weights and Measures: the *standard kilogram*. A kilogram mass has a weight of 2.20 lb at sea level on the earth's surface. On the moon it weighs one-sixth as much (0.37 lb). In outer space it weighs nothing, but it is still a kilogram mass. A metric ton is 1000 kg, which has a weight of 2200 lb and so is approximately the same as an English long ton (2240 lb). In this book we will use the metric ton.

It is important to be aware that the kilogram is a unit of mass, not a unit of weight. (One kilogram weighs 9.8 newtons, a unit of weight not often used in chemistry.) Keep in mind that although people may talk of *weighing* a quantity of material and sometimes incorrectly speak of its weight as so many kilograms, they are really referring to the mass rather than the weight of the quantity.

Example 2.13 — What is the mass in kilograms of a car weighing 1.5 tons?

Solution — A ton is 1000 kg, so the mass of 1.5 tons can be found by figuring that

$$\text{no. of kg} = \frac{1000 \text{ kg}}{1 \text{ ton}} \times 1.5 \text{ tons} = 1500 \text{ kg}$$

Example 2.14 — We know that 1 cm^3 of water has a mass of 1 g. What is the mass of a water tank containing 50 L of water?

Solution — Since 1 L contains 1000 cm^3, 50 L contains

$$\frac{1000 \text{ cm}^3}{1 \text{ L}} \times 50 \text{ L} = 50,000 \text{ cm}^3$$

We then use the fact that each cubic centimeter has a mass of 1 g, so 50,000 cm^3 has a mass of

$$\frac{1 \text{ g}}{1 \text{ cm}^3} \times 50,000 \text{ cm}^3 = 50,000 \text{ g}$$

> Another way of looking at the problem is to start with the fact that 1 L of water has a mass of 1 kg. Therefore, 50 L has a mass of
>
> $$\frac{1 \text{ kg}}{1 \text{ L}} \times 50 \text{ L} = 50 \text{ kg} = 50,000 \text{ g}$$

Units of Time

The units of time used with the metric system are identical to the common time units we use every day. The fundamental unit of time is the second, which is based on the frequency of vibration of the atoms of the element cesium-133. There are 60 seconds in a minute and 60 minutes in an hour; thus there are 60 times 60, or 3600, seconds in an hour. There are 24 hours in a day and 365 days in a year, except in a leap year.

For work requiring the measurement of very small periods of time, we use the usual prefixes: a millisecond is a thousandth of a second, a microsecond is a millionth of a second, and a nanosecond is a billionth of a second. We generally do not use the large prefixes for talking about long periods of time; kiloseconds and megayears are not common terms.

2.3 English-Metric Conversions

The United States is the only major country in the world that is still using the English system for everyday measurements; the rest of the world uses the metric system. Most scientific work in the United States is done with the metric system, but many engineers continue to use the English system. For these reasons we are continually faced with the problem of converting from one set of units to another. Table 2-2 gives some of the commonly used conversion factors.

TABLE 2-2 Converting Units

Metric-to-English Conversions	English-to-Metric Conversions
1 meter = 3.28 feet	1 inch = 2.54 centimeters
1 meter = 39.37 inches	1 mile = 1.609 kilometers
1 kilometer = 3280 feet	1 ounce (avdp) = 28.35 grams
1 liter = 1.06 quarts	1 pint = 225 cubic centimeters
1 kilogram = 2.20 pounds	

The first item in Table 2-2 tells us that there are 3.28 feet per meter (abbreviated 3.28 ft/m). Suppose we want to find how many feet there are in 100 m. We reason as follows: If there are 3.28 ft in 1 m, then there must be 100 times that many feet in 100 m. Writing this statement in the form of an equation and using the unit factor method, we have

$$\text{no. of ft} = \frac{3.28 \text{ ft}}{1 \text{ m}} \times 100 \text{ m} = 328 \text{ ft}$$

The last item in the first part of Table 2-2 is somewhat peculiar, for it says that 1 k (weight) equals 2.20 lb. This seems to contradict what we said in the previous section about the kilogram being a unit of mass, not weight. However, people have gotten accustomed to measuring kilograms by weighing, so there is a tendency to use the kilogram as a unit of weight. But it must be kept in mind that what we really mean is that *a kilogram is a mass whose weight is equal to 2.20 pounds*.

Since the earth is not a perfect sphere, a kilogram mass has different weights at different locations on the earth's surface. It weighs 2.20462 lb at sea level at a latitude of 45 degrees. That same kilogram mass weighs 2.1988 lb at sea level at the equator, because the equator is farther away from the center of the earth than a point at 45 degrees latitude. At the mile-high city of Denver, the kilogram weighs 2.2035 lb. You can see that wherever you are, 2.20 lb is a reasonably good approximation to the correct weight.

Example 2.15 What is the mass, in kilograms, of a 150-lb person?

Solution We know that a 1-kg mass weighs 2.20 lb. Therefore, the conversion factor is 1 kg/2.20 lb. The equation is

$$\text{no. of kg} = \frac{1.00 \text{ kg}}{2.20 \text{ lb}} \times 150 \text{ lb} = 68.2 \text{ kg}$$

Here the lb units cancel, leaving the kg unit in the numerator. As always, making sure that the units cancel properly is the best way to determine whether the conversion factor has been used correctly.

Example 2.16 What is the weight in ounces of a kilogram of water?

Solution We see from Table 2-2 that there are 28.35 g in an ounce, so the number of ounces per gram is 1/28.35. Therefore, the number of ounces in a kilogram (1000 g) is

$$\text{no. of oz} = \frac{1.000 \text{ oz}}{28.35 \text{ g}} \times 1000 \text{ g} = 35.27 \text{ oz}$$

> One way of checking the result is to notice that since there are about 28 g in each ounce, the number of ounces must be *smaller* than the number of grams. Therefore you want to *divide* the number of grams by 28.35 g/oz.
>
> Notice that since a kilogram of water has a volume of 1 L (as stated in Section 2.2), this calculation shows that a liter of water weighs 35.27 oz. A liter, furthermore, has a volume of 1.057 qt (U.S. liquid). The weight of 1 qt is, therefore,
>
> $$\text{no. of ounces per quart} = \frac{35.27 \text{ oz/L}}{1.057 \text{ qt/L}} = 33.4 \text{ oz/qt}$$
>
> The ounces in this calculation are units of weight (avoirdupois). Since a quart contains, by definition, exactly 32 fl oz, we see that the ounce unit of weight is not the same as the ounce unit of fluid measure.
>
> The confusion is compounded by the fact that the British quart, the U.S. dry quart, and the U.S. liquid quart all have different volumes. No better reason for adopting the metric system need be given.

2.4 Density and Specific Gravity

Earlier in this chapter we discussed the concept of mass. Mass is a property of a particular object. A 10-cm^3 block of iron is 10 times more massive than a 1-cm^3 piece of iron, so when we speak of the mass of a piece of iron, we are dealing with a property of that one specific piece of iron. What do we do when we want to talk about the properties of iron in general?

No matter how big the piece of iron we are discussing is, there is one property that remains unchanged: the mass of each cubic centimeter of iron. It is a fundamental characteristic of matter that every sample of a pure material has certain unchanging, **intrinsic properties**. This characteristic is called *uniformity*. Every cubic centimeter of iron has the same mass: 7.87 g. Since the mass per cubic centimeter is a constant property of iron, it is given a specific name: the density of the iron.

In general, the **density** of any material is the mass per unit volume of that material. In chemistry it has been traditional to measure density in units of *grams per cubic centimeter*. However, in the International System it is consistent to measure density in units of *kilograms per cubic meter*. Often, especially when dealing with liquids, we measure density in units of *kilograms per liter*. (In the older system densities of liquids were measured in grams per milliliter.)

Whichever units we use, we can find the density of any material by making two measurements. First we weigh a sample of the material to find the mass, and then we find the sample's volume. The density is then given by

$$\text{density} = \frac{\text{mass}}{\text{volume}}$$

For example, we weigh 10.0 cm³ of iron and find its mass to be 78.7 g. The density is then found as follows:

$$\text{density} = \frac{78.7 \text{ g}}{10.0 \text{ cm}^3} = 7.87 \text{ g/cm}^3$$

The important feature of density is that it is a property of a particular kind of material, such as iron, aluminum, or water, and does not depend on the size, shape, or mass of the particular piece of material. Therefore, it is a number that can be tabulated in chemistry handbooks as a property distinguishing one kind of material from another. Frequently you can tell one metal from another just by measuring the density.

Because of the way the gram was originally defined, the mass of 1 cm³ of water is always 1 g. The density of water is therefore

$$\frac{1 \text{ g}}{1 \text{ cm}^3} = 1 \text{ g/cm}^3$$

(Strictly speaking, this statement is only approximately true, for the density of water or any other material depends to some extent on the temperature of the material.)

Furthermore, since there are 1000 cm³ in a liter, we can make the following calculation:

$$\frac{1 \cancel{\text{g}}}{1 \cancel{\text{cm}^3}} \times \frac{1000 \cancel{\text{cm}^3}}{1 \text{ L}} \times \frac{1 \text{ kg}}{1000 \cancel{\text{g}}} = 1 \text{ kg/L}$$

We see from this result that the mass of 1 L of water is just 1 kg, so the density of water is 1 kg/L. It is true in general that the density of any material in units of kilograms per liter is the same numerically as the density in units of grams per cubic centimeter.

Example 2.17   We found in Section 2.2 that there are 10⁶ cm³ in a cubic meter. What is the mass of a cubic meter of water?

Solution   We can answer this question by taking the definition of density,

$$\text{density} = \frac{\text{mass}}{\text{volume}}$$

and solving it for the mass. Multiplying both sides of the equation by volume, we find

$$\text{mass} = \text{density} \times \text{volume}$$

> Substituting the numerical values for density and volume, we have
>
> $$\text{mass} = \frac{1 \text{ g}}{\text{cm}^3} \times 10^6 \text{ cm}^3$$
>
> $$= 10^6 \text{ g}$$
>
> In other words, a cubic meter of water has a mass of a million grams—a number hard to comprehend. In the next example we find the same mass in kilograms.
>
> **Example 2.18** Find the mass of a cubic meter of water in kilograms.
>
> **Solution** We found in Section 2.2 that there are 1000 L in a cubic meter. We also know from the discussion in this section that the density of water is 1 kg/L. Using the same equation as in the previous example, we have
>
> $$\text{mass} = \text{density} \times \text{volume}$$
>
> $$= \frac{1 \text{ kg}}{\text{L}} \times 1000 \text{ L} = 1000 \text{ kg}$$
>
> Comparing the results of this example with those of the previous example, we see that 1000 kg is equivalent to 1,000,000 g, which is what we would expect, knowing that there are 1000 g in 1 kg.

From the results of Example 2.18, we see that the density of water in kilograms per cubic meter is

$$\text{density} = \frac{\text{mass}}{\text{volume}}$$

$$= \frac{1000 \text{ kg}}{1 \text{ m}^3}$$

$$= 1000 \text{ kg/m}^3$$

In other words, the density as measured in kilograms per cubic meter is a number 1000 times greater than the density as measured in grams per cubic centimeter. (This is true for the density of any material. Note that the actual density of the material does not depend on the units used; only the number used to express the density changes.)

One thing we can do with the concept of density is to calculate the mass of any object for which we know the density and volume. Problems involving water are easy. Since the density of water is 1 g/cm^3, we know immediately that 100 cm^3 of water weighs 100 g. Now consider a liquid such as ethyl alcohol, which has a density of 0.79 g/cm^3. The mass of 100 cm^3 of alcohol is found by the following method:

$$\text{mass} = \text{density} \times \text{volume}$$

$$= 0.79 \frac{\text{g}}{\text{cm}^3} \times 100 \text{ cm}^3$$

$$= 79 \text{ g}$$

We see that because the alcohol has a lower density than water, a given volume of alcohol weighs less than the same volume of water.

It is often convenient to compare the densities of various materials to the density of water, treating the density of water (1 g/cm³) as a standard. For this purpose, we define another quantity called the **specific gravity**, or the relative density, which is simply the density of the given material divided by the density of water.

$$\text{specific gravity} = \frac{\text{density(material)}}{\text{density(water)}}$$

Using alcohol as an example, we have

$$\text{specific gravity} = \frac{0.79 \text{ g/cm}^3}{1.0 \text{ g/cm}^3} = 0.79$$

Notice that the units g/cm³ cancel out completely and we are left with a nondimensional number (that is, a number without units) that represents the density of alcohol relative to the density of water. Also notice that the specific gravity of any material will always have the same numerical value as the density measured in grams per cubic centimeter when we use water as the standard comparison material. Table 2-3 lists the specific gravities of a number of common materials. (To be very precise, we must note that the specific gravities of all substances vary with temperature, so the temperature at which a measurement is made must be specified.)

TABLE 2-3 Specific Gravities of Some Common Materials

Material	Specific Gravity	Material	Specific Gravity
Gases		*Solids*	
hydrogen	9×10^{-5}	aluminum	2.70
nitrogen	1.25×10^{-3}	copper	8.96
oxygen	1.43×10^{-3}	gold	19.32
		ice	0.92
Liquids		iron	7.87
ethanol (alcohol)	0.79	lead	11.35
mercury	13.5	platinum	21.45
water	1.0	tungsten	19.3

2.4 DENSITY AND SPECIFIC GRAVITY

If we know the specific gravity of any material, we can always find its density by turning the above equation around so that

$$\text{density(material)} = \text{specific gravity} \times \text{density(water)}$$

Example 2.19 What is the volume of 500 g of aluminum?

Solution From Table 2-3 we find that the specific gravity of aluminum is 2.70. Therefore, the density of aluminum is

$$\text{density} = 2.70 \times 1 \text{ g/cm}^3 = 2.70 \text{ g/cm}^3$$

We also know that, by definition,

$$\text{density} = \frac{\text{mass}}{\text{volume}}$$

Multiply both sides of the equation by volume and by 1/density:

$$\text{density} \times \frac{1}{\text{density}} \times \text{volume} = \frac{\text{mass}}{\text{volume}} \times \text{volume} \times \frac{1}{\text{density}}$$

Cancel density on the left and volume on the right:

$$\text{volume} = \text{mass} \times \frac{1}{\text{density}}$$

Substitute the given quantities:

$$\text{volume} = 500 \text{ g} \times \frac{1}{2.70 \text{ g/cm}^3}$$

$$= \frac{500 \text{ g}}{2.70 \frac{\text{g}}{\text{cm}^3}}$$

Next, invert the fraction:

$$\text{volume} = \frac{500}{2.70} \times \text{g} \times \frac{\text{cm}^3}{\text{g}}$$

Performing the division and canceling gram units leaves

$$\text{volume} = 185 \text{ cm}^3$$

Table 2-3 gives a large amount of information about the range of specific gravities (and densities) among different materials. Notice that gases tend to have the lowest specific gravities. In general, liquids have specific gravities not too far from that of water, with the exception of mercury, an unusual substance. Mercury is a metal normally found as a liquid with an extremely high specific gravity. Metals such as gold, platinum, and tungsten are even more dense than lead. The densest metal is osmium, with a specific gravity of 22.6.

An important rule regarding specific gravities is as follows: Any material with a specific gravity greater than 1 will sink when immersed in water. On the other hand, any material that has a specific gravity less than 1 will float in water. This rule can be generalized to any liquid: *A material will float on any given liquid if the density of the material is less than the density of the liquid.* This rule even holds for one liquid floating on another liquid, as when oil floats on water. Using this rule, we can determine what substances will float on any liquid simply by comparing their specific gravities.

However, when we consider a solid object floating on a liquid, it is the *average* density of the solid that counts. This is why metal boats can float on water. Although the metal itself is more dense than water, the boat is mostly air space, so the average density (calculated from the entire volume of the boat) is less than the density of water. As is well known, a boat will tend to sink if it becomes filled with water.

2.5 Analysis of Graphs

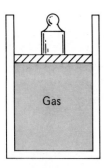

FIGURE 2-3 In this cylinder with a movable piston, the weight on the piston keeps the gas at a constant pressure.

A graph provides an easy way to visualize the relationship between two quantities. Consider the following example: A cylinder of gas has a piston on top that can move up and down. A weight on this piston keeps the gas at constant pressure while the volume varies (see Figure 2-3). A thermometer is used to measure the gas temperature, and the position of the piston is measured to find the volume of the gas. We now start increasing the gas temperature. As we do this, the piston rises because the gas expands as a result of the heating. By measuring the temperature and volume periodically, we can make a chart showing how the volume varies as the temperature changes. This chart might look something like Table 2-4. Here we have measured the gas volume at intervals of 100°F. The temperature is the quantity that we control; the volume is the quantity that varies in response to the changing temperature. We therefore call the temperature the *independent variable* and the volume the *dependent variable*. In mathematical terms, the volume is a *function* of the temperature.

TABLE 2-4 Volume Variance with Change in Temperature

Temperature (°F)	Volume (cm^3)
0	500
100	609
200	717
300	826

It is difficult to find much meaning in a chart of numbers. The only thing we can say after looking at these numbers is that the gas volume increases as the temperature increases. But *how* does the volume increase? Is there any kind of interesting relationship between the two variables?

The simplest way to answer these questions is to plot a graph, as shown in Figure 2-4. We start the graph by drawing vertical and horizontal axes. The horizontal axis (called the **abscissa**) always represents the independent variable—in this case the temperature—over which the experimenter has control. The vertical axis (called the **ordinate**) then represents the dependent variable—in this case the volume—which changes in response to our variation of the independent variable. We describe the graph as a plot of volume as a function of temperature, or volume versus temperature.

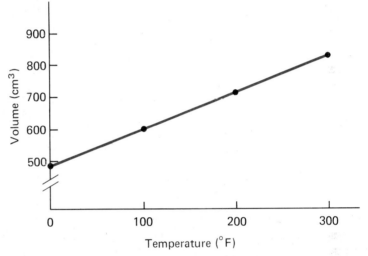

FIGURE 2-4 On this graph of the linear relationship between the volume of a gas at constant pressure and the temperature, the zero is suppressed.

The scales on each axis are usually drawn so that the intervals are uniform. That is, each mark on the scale represents an equal change in the quantity being represented. (There are times when it is useful to use a nonuniform scale, such as a logarithmic scale.) In Figure 2-4, the temperature is marked on the horizontal scale in intervals of 100 degrees: 0, 100, 200, 300, and so on. The vertical scale starts with 0 and jumps to 500, then to 600, 700, 800, etc. This would be a grave error if we had not put in a pair of wiggly lines to warn the reader that the zero is suppressed. This mark tells us that in the first interval we have skipped over 500 units on the scale. If we had started from zero and made all intervals the same size, all the interesting data would have been squeezed in at the top of the graph. We demonstrate this effect in Figure 2-5, in which the data from Figure 2-5 are again plotted on a scale with uniform intervals, but this time without suppression of the zero. In newspapers (especially on the financial page), suppression of the zero is a trick commonly used to magnify small fluctuations at the top of a curve. Watch for it.

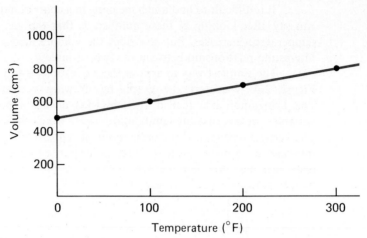

FIGURE 2-5 On this graph of the linear relationship between the volume of a gas at constant pressure and the temperature, the zero is not suppressed.

Notice that the scales in Figure 2-5 are clearly labeled. A label contains two pieces of information: the quantity being plotted and the units used for measurement (in parentheses).

Our graph shows a very simple relationship between the gas volume and the temperature, for a line drawn through the plotted points is simply a straight line. For this reason, the relationship between the volume and the temperature is called

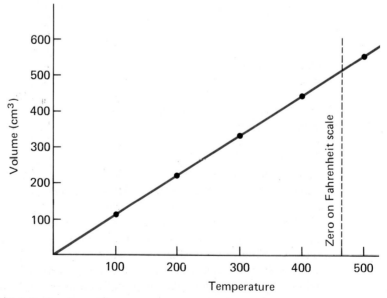

FIGURE 2-6 On this volume-temperature graph, the temperature scale starts at absolute zero (the Rankine scale).

a linear relationship. A **linear relationship** is one of the simplest relationships to be found between two quantities; it can be represented by the equation $V = mT + b$, where m and b are constants. (Notice the similarity to the linear equation $y = mx + b$, discussed in Section 1.5.)

The relationship becomes even simpler if we choose a temperature scale that makes the volume equal to zero when the temperature is zero. We can do this by defining a new temperature scale that starts at zero all the way over on the left where the straight line intersects the horizontal axis. (See Figure 2-6.) The old (Fahrenheit) zero occurs at about 460 on the new scale. (In Section 2.6 we will see that this new temperature scale has great significance in chemistry.) The volume-temperature relationship that results from the use of this new scale can be expressed by the simple equation

$$V = 1.086 \times T$$

where T is the temperature on the new scale. This relationship is still linear, but more significantly it is now a *direct proportion*. In a **direct proportion**, if one of the variables is doubled, the other variable also doubles.

Not all linear relationships are direct proportions. A direct proportion occurs only if both variables are zero at the same time. Not all relationships are linear either. The world would be simpler if they were, but the fact is that many other kinds of relationships occur in nature. Therefore, when you plot data on graphs, do not expect to produce straight lines every time.

2.6 Temperature Scales

Fahrenheit and Celsius Scales

We saw in the last section that when a container of gas is heated, the volume of gas increases with temperature in a linear manner if the pressure is held constant. Once we know what the volume-temperature relationship is, we can turn it around; by measuring the volume of a gas, we can deduce what its temperature is. This is the principle of the constant-pressure gas thermometer, used in the laboratory for very precise work.

The gas thermometer is inconvenient to use. Therefore, in most everyday applications we make use of the fact that liquids as well as gases expand when heated. The common mercury thermometer contains a bulb at the bottom filled with liquid mercury, as shown in Figure 2-7. The mercury can expand into a fine capillary tube above the bulb. The distance that the mercury moves up the tube varies in a linear way with temperature, so the tube can be marked with uniform divisions calibrated directly in degrees of temperature.

Temperature is a property that is never measured directly. It is always measured indirectly through its effects on a piece of material—in this case, the expansion or contraction of a mercury column as it is heated or cooled.

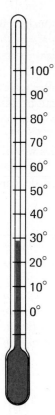

In order to calibrate a thermometer, it is necessary to know at least two temperature points in advance. We get these points by making use of a number of observable phenomena, each of which takes place at a particular, fixed temperature. Each phenomenon must be reproducible; that is, it must take place at the same temperature every time the measurement is made. At least two temperatures are needed to establish a temperature scale because two points are needed to determine the size of the division (the degree) and also to define where the zero of the scale is. (In a similar manner, to calibrate a meter stick you need the two points at the ends of the stick to establish the length of the meter. Then the stick can be divided into a hundred divisions to obtain a centimeter scale.)

One commonly used calibration point is the melting point of ice, for it is known that ice always melts at the same temperature (other conditions, such as pressure, being constant) and that a container filled with a mixture of ice and water will maintain that constant temperature as long as there is unmelted ice. (See Figure 2-8.) Another calibration point is the boiling point of water. Pure water at normal atmospheric pressure (760 mm of mercury) always boils at the same temperature. The temperature of a container of boiling water will not rise any higher than the boiling point as long as there is water remaining to be boiled away.

In the **Fahrenheit scale**, commonly used in the English system of measurement, the melting point of ice is defined to be 32°F and the boiling point of water is defined to be 212°F. This is a convenient scale for weather reports, since the temperature on a very cold day is about 0°F and on a very hot day is close to

FIGURE 2-7
This mercury thermometer consists of a bulb containing liquid mercury that is connected to a very fine tube into which the mercury can expand. The scale is calibrated in degrees.

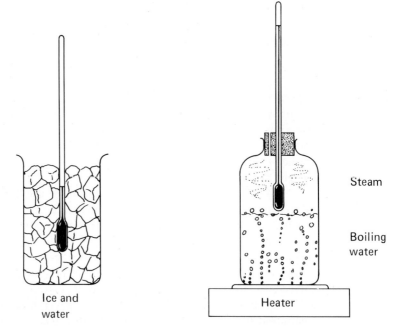

FIGURE 2-8 Measuring the freezing and boiling points of water

100°F. The ° stands for *degrees*, so 212°F is read as "212 degrees Fahrenheit." The term "degree" itself has no particular significance as far as size is concerned. It merely stands for one of the divisions on the temperature scale, and we see that between the melting point of ice and the boiling point of water there are 180 of these divisions.

The temperature scale used with the metric system is the **Celsius scale**, formerly known as the Centigrade scale. In the Celsius scale, the melting point of ice is defined to be 0°, and the boiling point of water is defined to be 100°. Most scientific temperature measurements use the Celsius scale, and in most foreign countries you will hear Celsius temperature readings being given in the weather reports.

Figure 2-9 shows a comparison of the Fahrenheit and Celsius scales. Between the melting point of ice and the boiling point of water there is a temperature range that includes 180 Fahrenheit degrees but only 100 Celsius degrees. In other words, 180 Fahrenheit degrees is equivalent to 100 Celsius degrees, so that the unit factor for converting from Celsius to Fahrenheit divisions is 180 Fahrenheit degrees per 100 Celsius degrees, or

$$\frac{180°F}{100°C} = \frac{9°F}{5°C}$$

This means that the size of one Celsius degree is greater than the size of one Fahrenheit degree by a factor of 180/100, or 9/5. That is, in the same scale distance there are 9 Fahrenheit degrees for every 5 Centigrade degrees.

For example, in the interval between 0° and 50° on the Celsius scale there

	K	C	R	F
Boiling water	373	100°	672°	212°
Melting ice	273	0°	492°	32°
Dry ice	195	−78°	351°	−109°
Liquid oxygen	90	−183°	162°	−297°
Absolute zero	0	−273°	0°	−460°

FIGURE 2-9 A comparison of the four temperature scales: Kelvin, Celsius, Rankine, and Fahrenheit

are 50 Celsius degrees, but on the Fahrenheit scale the number of Fahrenheit degrees is

$$\frac{9°F}{5°C} \times 50°C = 90°F$$

However, this does *not* mean that a temperature of 50°C is the same as a temperature of 90°F, because the zeros of the two scales do not coincide. Zero on the Celsius scale corresponds to 32° on the Fahrenheit scale, so we must add 32° to the 90° in order to get the actual Fahrenheit temperature corresponding to 50°C.

We can summarize the relationship for converting from 50°C to the temperature in degrees Fahrenheit by forming the equation

$$°F = \tfrac{9}{5} \times 50°C + 32 = 122°F$$

If we let °C be the Celsius temperature, this becomes a general formula. Then any Celsius temperature can be converted into the corresponding Fahrenheit temperature by the formula

$$°F = \tfrac{9}{5} \times °C + 32$$

The reverse conversion gives us the Celsius temperature if we have the Fahrenheit temperature. The formula for the reverse conversion is obtained by solving the above equation for degrees Celsius. We begin the solution by subtracting 32 from both sides of the equation:

$$°F - 32 = \tfrac{9}{5} \times °C + 32 - 32$$
$$= \tfrac{9}{5} \times °C$$

Now, to get the °C by itself, we multiply both sides of the equation by 5/9.

$$(°F - 32) \times \tfrac{5}{9} = \tfrac{9}{5} \times °C \times \tfrac{5}{9}$$

Then, after canceling terms and reversing the two sides of the equation, we have

$$°C = \tfrac{5}{9}(°F - 32)$$

Notice that in using this equation to find degrees Celsius, you first subtract 32 from the degrees Fahrenheit. Then you multiply by 5/9. (In general, any operations shown inside parentheses are performed before any other operations.)

Example 2.20   A mixture of ice and salt is at a temperature of −10°C. What is the corresponding Fahrenheit temperature?

Solution
$$°F = \tfrac{9}{5} \times °C + 32$$
$$= \tfrac{9}{5} \times (-10°C) + 32$$
$$= -18 + 32 = 14°F$$

Example 2.21   The temperature on a very hot day is 103°F. What is this temperature on the Celsius scale?

Solution

$$°C = \tfrac{5}{9}(°F - 32)$$
$$°C = \tfrac{5}{9}(103 - 32)$$
$$= \tfrac{5}{9} \times 71 = 39.4°C$$

Kelvin and Rankine Scales

Zero has no particular significance on either the Celsius scale or the Fahrenheit scale. The fact that 0°C is the melting point of ice, while of interest, is of no great importance. As you know, it is possible to find temperature readings below zero on either scale. Zero is not the bottom of the scale, but just an arbitrary point in the middle. However, it is possible to invent a temperature scale in which zero is actually the lowest temperature it is possible to reach. In Figure 2-6 we saw one way to find such a scale. This figure shows the volume of a container of gas (at constant pressure) as a function of temperature. If we extend the straight line far enough toward the left—to lower and lower temperatures—we reach a point where the volume as shown by the graph becomes zero. (If we actually reduce the temperature of a gas to such a low point, the gas first becomes a liquid and then freezes to a solid, so its volume does not actually vanish. Helium gas is an exception: it becomes a liquid but will not solidify.)

Ignoring the behavior of the real gas at low temperatures and extrapolating the straight line back to the point where it crosses the horizontal axis, we define that point to be zero on a new temperature scale. This point is called **absolute zero**, for it represents the absolutely lowest temperature that can be reached. It is a temperature that we can approach experimentally but cannot really reach. However, we can get to within a fraction of a degree of it.

Two temperature scales use absolute zero as their bottom point. One, the **Rankine scale**, uses the same size degree as in the Fahrenheit scale but shifts the zero over to the absolute zero point. This is the scale plotted in Figure 2-6. On the Rankine scale, approximately 460° corresponds to 0°F. That is, absolute zero is at −460°F. The Rankine scale is used by engineers (in particular, steam engineers) who use Fahrenheit degrees but must start the scale from absolute zero for their particular calculations.

The second scale that starts from absolute zero is the **Kelvin scale**, commonly used in chemistry. The Kelvin scale has the same size degree as the Celsius scale, but the zero point on the Kelvin scale is at approximately −273°C, putting 0°C at 273 on the Kelvin scale. Given a temperature on the Celsius scale, you merely add 273 to it in order to get the Kelvin temperature (sometimes called the absolute temperature):

$$K = °C + 273$$

In modern usage (within the International System of units), we do not use the word *degrees* with the Kelvin scale. Instead, we use *kelvins* as the unit of temperature (abbreviated K). Therefore, the temperature of melting ice on the Kelvin scale is written as 273 K, not 273°K.

The Kelvin scale is of great importance in science, especially in studying the behavior of materials as a function of temperature. Therefore, the ability to convert from Celsius degrees to kelvins is a necessity.

Example 2.22 Oxygen becomes liquid at a temperature of −183°C at normal atmospheric pressure. What is this temperature on the Kelvin scale?

Solution
$$K = °C + 273 = -183 + 273 = 90 \text{ K}$$

2.7 Accuracy and Precision

Error analysis is an important part of science. An **error** is the difference between a measured value and the correct or accepted value. Related to errors are the terms *accuracy* and *precision*, which are used interchangeably in everyday language. However, in relation to scientific measurements, accuracy and precision mean different things. The **accuracy** of a measurement refers to how closely the measurement approaches the correct value. On the other hand, the **precision** of a measurement describes how closely repeated measurements of the same quantity give the same result, correct or not. Let us examine the meaning of these definitions in more detail.

A meter stick is compared with a standard meter and is found to agree very closely with the true meter. We call a measurement made with that meter stick an accurate one because it is close to the correct value. However, all measurements have some error. If we know from experience that the length of the meter stick might be in error by as much as 0.01 cm, then a measurement of 100 cm made with that stick is written 100 ± 0.01 cm. We express the uncertainty, or possible error, in the measurement by saying that *the limits of error are plus or minus the possible error*—in this case, ± 0.01 cm.

An important and difficult part of experimental work is judging what the possible error might be in any given measurement. In the above example, we were able to judge the error by comparing the meter stick with a standard meter. However, very often we are required to measure quantities whose correct values are not known. Judging the possible error in such measurements becomes a fine art.

The first step in any accurate measurement is to calibrate all the instruments used by comparison with standards of known accuracy. (**Calibration** is a procedure used to set the scale of a measuring device.) Once the meter stick discussed above is calibrated, then any future measurement made with it should have a limit of error of ±0.01 cm, assuming that each measurement is made with equal care. However, one of the things learned from experience is that when measurements are repeated many times, each measurement is a little different from the others. That is, there is a variability in the measurements. If the user of a meter stick sees the markings from a different direction each time, a *parallax error* will occur. Because of this effect, there may be variation of as much as half a millimeter if the same measurement is repeated a number of times. Thus a measurement of 50 cm made in this way is written 50 ± 0.05 cm.

The uncertainty ± 0.05 cm represents the precision of the measurement. High precision means that if the same measurement is repeated many times, the results will be very close to one another. We see that in this example the accuracy of the measurement is greater than the precision, because the length of the meter stick is known to within 0.01 cm but the individual measurements may vary from the average by as much as 0.05 cm. On the other hand, if the meter stick had been calibrated incorrectly, so that its actual length was not what it was supposed to be, the precision of the measurement might be greater than the accuracy. Repeated measurements could be very close to one another, but they would all be wrong.

Two general types of errors can be made in any kind of measurement: random errors and systematic errors. **Random errors**, or "chance" errors, are due to unknown factors that vary in a random manner. During a weighing operation, stray air currents hitting the balance might introduce random errors. Random errors reduce the precision of the measurement. In the measurement of radioactivity, random errors can never be reduced below a certain amount because they are an intrinsic part of the phenomenon being measured. In fact, *all* measurements suffer from a certain amount of "noise" that puts a physical limitation on precision. There is no such thing as an *absolutely* precise measurement.

Systematic errors are those caused by the instrumentation or by mistakes in the method of measurement. Poor calibration, or standardization, of an instrument is one cause of systematic error. Without good calibration, the results will be incorrect even if care has been taken to eliminate random errors.

Very often the error in a measurement is expressed as a percentage. For example, suppose that a rod is cut to a length of 50.00 cm as measured by an accurate, or standard, meter stick. When measured by a less accurate device, its length is found to be 50.05 cm. The error is calculated as follows:

$$\text{error} = \text{measured length} - \text{standard length}$$
$$= 50.05 \text{ cm} - 50.00 \text{ cm} = 0.05 \text{ cm}$$

The percent error is then found:

$$\text{percent error} = \frac{\text{error}}{\text{standard quantity}} \times 100\% = \frac{0.05 \text{ cm}}{50 \text{ cm}} \times 100\% = 0.1\%$$

Example 2.23 A small sample of gold is weighed ten times, with the following results: 1.045 g, 1.005 g, 1.049 g, 1.042 g, 1.063 g, 1.055 g, 1.058 g, 1.017 g, 1.025 g, and 1.038 g.

(a) What is the average weight?
(b) What is the maximum deviation from the average?
(c) What is the percent deviation?
(d) If the correct weight is known to be 1.020 g, what is the absolute error?
(e) What is the percent systematic error?
(f) Which is greater, the random error or the systematic error?

Solution

(a)
$$\text{avg. weight} = \frac{(1.045 + 1.005 + 1.049 + 1.042 + 1.063 + 1.055 + 1.058 + 1.017 + 1.025 + 1.038)}{10}$$

$$= \frac{10.397}{10} = 1.0397 = 1.040 \text{ g}$$

We have rounded off the answer to 1.040 g because if each individual measurement has an uncertainty in the third decimal place, the average will perhaps also be uncertain in the third decimal place. The next section contains more details on rounding.

(b) The 1.005-g measurement has the greatest deviation, 0.035 g (1.040 − 1.005 = 0.035). This may be taken as a crude approximation to the random error.

(c) The percent deviation is

$$\frac{0.035}{1.040} \times 100\% = 3.4\%$$

(d) The absolute error is

$$1.040 - 1.020 = 0.020 \text{ g}$$

(e) The percent systematic error is

$$\frac{0.020}{1.020} \times 100\% = 2.0\%$$

(f) The random error of 3.4% is greater than the systematic error of 2.0%.

2.8 Significant Figures

Estimating Errors and Rounding Off

One of the most common arithmetical errors is illustrated by the following situation. You weigh the same object three times and get three different measurements: 2.45 g, 2.55 g, and 2.52 g. You want to find the average weight, and so you use your pocket calculator to add the three weights and then divide by 3:

$$\text{average} = \frac{2.45 + 2.55 + 2.52}{3}$$

$$= \frac{7.52}{3} = 2.506667 \text{ g}$$

The exact number of sixes obtained in the answer depends on which calculator is used. If the above division were done by an ideal computer, it would give an answer with an infinite number of sixes. However, a real computer or calculator "rounds up" the 6 in the last place to a 7 because the unseen number following it is greater than 5.

What is the fallacy in this solution? The trap into which we have fallen is the simple error of believing everything that our calculator tells us. We put into the calculation numbers that have great uncertainty in the value of the third digit (counting from left to right); the input numbers range from 2.45 to 2.55. As a result, there is some uncertainty in the third digit of the average value that we calculated. We don't really know whether the true weight is 2.49, 2.50, or 2.51, so it is meaningless to write down all seven digits in the answer. If three digits are used in the input data, only three digits are meaningful in the result. In scientific terminology, the meaningful digits are called **significant figures**.

In doing any kind of measurements or calculations, it is very important to know how to judge how many digits are significant in the numerical result. For example, suppose you make a rough measurement of the distance from the earth to the sun, good to the nearest million miles. That is, the probable error is $\pm 1,000,000$ mi. It makes no sense to write down a number such as 93,423,122 mi. It is necessary to round off the result to the nearest million miles; the result is $93,000,000 \pm 1,000,000$ mi. This number is good to two significant figures because the uncertainty is in the second digit; you know only that the correct distance lies between 92,000,000 and 94,000,000 mi.

The procedure for deciding how many digits are significant is simple: Determine which digit is uncertain because of errors in measurement, and round off all the digits after that one. ("Round off" means to replace the meaningless digits with zeros.) If you have obtained a number such as 432,460 as the result of a calculation and you know that this number is good only to three significant figures, round it off to 432,000. On the other hand, if the number is 432,560, then round it off to 433,000. The rule is that if the number following the last significant digit is equal to or greater than 5, then the rounding-off process increases the last significant figure by 1.

An ambiguous situation arises when you meet a number such as 45,000. This number might be the result of actually counting 45,000 items. In this case, the three zeros are significant figures, and the number has five significant figures altogether. On the other hand, this number might be the result of a measurement in which the error is ± 1000 units. In this case, there are only two significant figures, and the three trailing zeros merely tell where the decimal point is located. Thus, it is important to know the size of the probable error in any measurement.

The same procedures apply to small numbers. Consider a weight that is found to be 0.035 ± 0.001 g. The uncertainty is in the third decimal place. However, there are not three significant figures because the zero after the decimal point does not count as a significant figure. It merely tells where the decimal point is located. The above number is good to only two significant figures.

This result is more obvious if we write the number in scientific notation: 3.5×10^{-2} g. The 5 is the uncertain number, and so it is easy to see now that this number has only two significant figures.

Example 2.24

Round off the following numbers to four significant digits.
(a) 1,256,233 (b) 3,698,566

Solution

(a) 1,256,000 (b) 3,699,000

If the number to the right of the last significant digit is less than 5, just replace all the numbers to the right by zeros. If the number past the last significant digit is equal to or greater than 5, then raise the last significant digit by 1 and replace the numbers following it by zeros.

Example 2.24

Round off 3,699,582 to four significant figures.

Solution

3,700,000 (better expressed as 3.700×10^6)
The 9 in the fourth place is rounded up to 10, which in turn causes the third-place 9 to become 10, forcing the 6 to become 7.

Example 2.25

Round off 0.0025842 to three significant digits.

Solution

0.00258
The two zeros are not part of the three significant digits.

Addition and Subtraction

Very often we are required to estimate the number of significant figures in the results of a mathematical calculation. The rules for addition and subtraction differ from the rules for multiplication and division and are best explained by giving examples.

The matter is simple if the two numbers being added or subtracted have the same number of significant figures:

$$\begin{array}{r} 345 \pm 1 \\ + \ 678 \pm 1 \\ \hline 1023 \pm \ ? \end{array}$$

The uncertainty in the numbers being added is ± 1, and each of the two numbers has three significant figures. In the worst possible situation, the errors would add, making the uncertainty in the sum 2. Therefore, the sum may be written as 1023 ± 2, with four significant digits. In actuality the errors may sometimes be negative and sometimes positive. The subject is complex and is examined more thoroughly in books on statistics and error analysis.

Sometimes the following situation occurs:

$$\begin{array}{r} 345{,}000 \\ +2{,}670 \\ \hline 347{,}670 \end{array}$$

Does this answer make sense? That depends on how many significant figures there are in the numbers being added. If the last three digits of the top number are completely unknown (that is, if zeros are placed there because the correct numbers are not known), there is no point in adding anything to them. Thus, the final answer should be rounded off to 348,000.

In subtraction, another hazard occurs. Consider this example:

$$\begin{array}{r} 3456 \pm 1 \\ -3451 \pm 1 \\ \hline 5 \pm 2 \end{array}$$

Here we start with numbers good to four significant figures (0.01% error), and we end up with a number good to only one significant figure. The final answer has a 40% error (2 out of 5). This result indicates that when we subtract two numbers the percent error of the measurement is increased.

Example 2.27

Solution

Add the numbers 1.24 ± 0.01 and 0.6735 ± 0.0001.

$$\begin{array}{rl} 1.24 & \pm 0.01 \\ 0.6735 & \pm 0.0001 \\ \hline 1.91 & \pm 0.01 \end{array}$$

The first number is significant only to the second decimal place, so the third and fourth decimal places of the sum are not significant.

Multiplication and Division

Consider the product

$$111 \times 11{,}111 = 1{,}233{,}321$$

The first multiplier has three significant figures, and the second has five. Is the product actually good to seven significant figures? Let us see what happens if we change the 111 to 112:

$$112 \times 11{,}111 = 1{,}244{,}432$$

We see that a change (or uncertainty) in the third digit of one of the multipliers makes a change in the third digit of the product. This means that if one of the multipliers is good to three significant figures, the product is only good to three significant figures, even if the other multiplier has more significant figures.

The general rule for multiplication is this: *The product will be good to the number of significant figures in the factor that has the least number of significant figures.*

You can show by experiment that the same rule applies to division.

Example 2.28   Divide 73,489 by 2.3. Round off the answer to the correct number of significant figures.

Solution   The calculator gives 31,951.739 as an answer. However, the divisor, 2.3, has only two significant figures, so the quotient can have only two significant figures. Therefore, round off the answer to 32,000.

Important reminder: Do not be afraid to throw away extra digits given to you by your electronic calculator. They do not enhance your answer because they are worthless.

Glossary

abscissa The horizontal scale on a graph.

absolute zero Zero on the Kelvin scale; the lowest temperature that can be reached physically.

accuracy How close a measurement comes to the correct value.

calibration A procedure used to set the scale of a measuring device.

Celsius scale A temperature scale in which the freezing point of water is at 0°C and the boiling point of water is at 100°C; Centigrade scale.

conversion factor A number that converts quantities of one unit into quantities of another.

density Mass per unit volume.

direct proportion A relationship between two quantities such that when one quantity doubles, the other also doubles.

error The deviation of a measurement from its correct or accepted value.

Fahrenheit scale A temperature scale in which the freezing point of water is 32°F and the boiling point of water is at 212°F.

inertia A property of an object characterized by its resistance to change in its state of motion. The harder it is to start or stop the motion of an object, the more inertia it has. The mass of an object is measured by its inertia.

International System (SI) An international system of weights and measures based on the metric system. It is abbreviated SI because its French name is Système International.

intrinsic property A characteristic of a substance that does not depend on outside circumstances such as location, shape, or temperature.

Kelvin scale A temperature scale in which the freezing point of water is at 273.15 K and the boiling point of water is at 373.15 K. Zero on the Kelvin scale is known as absolute zero.

linear relationship A relationship between two quantities that is represented by a straight line on a graph.

mass The amount of matter in an object, as measured by its inertia.

metric system A system of measurements in which the kilogram, meter, second, and ampere are the fundamental units of mass, length, time, and electric current, respectively. A system of prefixes (centi-, micro-, kilo-, etc.) provides larger and smaller units, all differing by multiples of ten.

ordinate The vertical scale on a graph.

precision How closely a measurement can give the same result when it is repeated a number of times.

random errors Errors due to chance.

Rankine scale A temperature scale in which the freezing point of water is at 491.7°R and the boiling point of water is at 671.7°R. Zero on the Rankine scale is at absolute zero.

rate The speed with which something changes or is used.

specific gravity The density of a material compared with the density of water; relative density.

systematic errors Errors due to faults in observation, instrumentation, or calibration.

unit A precisely defined quantity in terms of which properties are measured (for example, the kilogram is a unit of mass).

unit factor method A method of solving problems in which a quantity in one set of units is multiplied by a unit factor so as to convert it into the corresponding quantity in another set of units. For example,

$$\text{no. of cm} = 2.54 \frac{\text{cm}}{\text{in}} \times \text{no. of in}$$

weight A measure of the pull of gravity on an object, proportional to the object's mass.

Problems and Questions

2.1 Problem-Solving Techniques

Write down all the steps needed to solve each problem. Then put in the numbers and find the answer.

2.1.P A person weighs 150 lb. How many ounces is that? (*Hint*: 1 lb = 16 oz.)

2.2.P A cubic foot of water weighs 62 lb. A waterbed has dimensions of 6 ft × 6 ft × 1 ft. How much does the bed weigh?

2.3.P A chemist can buy a gallon of alcohol at a cost of $7.50 or a pint at a cost of $2.00. Which would be the better buy per ounce?

2.4.P A label on a 1-lb can of tuna fish says that the contents provide 800 calories of energy value. A dieter wants to eat just 150 calories. How many ounces of the fish should he eat?

2.5.P A chemist wants to manufacture a drug that contains three ingredients, A, B, and C. The formula calls for 5 parts of A, 6 parts of B, and 16 parts of C. That is, 5 oz of A, 6 oz of B, and 16 oz of C would make 27 oz of the product. Each dose of the drug contains 0.5 oz of the mixture. How many ounces of each ingredient are needed to produce 1000 doses?

2.6.P You want to mix 1 part of sulfuric acid with 7 parts of water (by volume), and you want to make a gallon of the mixture. How much of each ingredient do you need?

2.7.P A certain type of concrete weighs 166 lb per ft^3. What is the weight of a cubic yard of this concrete?

2.8.P A person drives for 6 hr at 55 mi/hr. How far does she drive in that time?

2.9.P A writer has an obligation to produce a 540-page book in 15 months. How many pages must he write each day to complete the job? (Assume each month has 30 days.)

2.10.P A dry cereal manufacturing plant produces approximately 7×10^5 flakes each day. How many days would it take to produce 6×10^{23} flakes? How many years of production does this represent? (We will see in Chapter 9 why the number 6×10^{23} is important in chemistry.)

2.2 Measurements in the Metric System

2.11.Q What is the major advantage of the metric system?

2.12.Q What are the four fundamental units of measurement in the International System?

2.13.Q **a.** What was the original definition of the meter?
b. What is the present definition of the meter?

2.14.Q Name the metric system prefixes corresponding to these multiples of ten:
a. 1000 **b.** 1,000,000
c. 10^{-6} **d.** 10^{-3}

2.15.Q What is the basic unit of area in the metric system?

2.16.Q We know from the general meaning of the prefix that a megawatt is a million watts (even if we don't know what a watt is). How many units are represented by each of the following quantities?

a. megabyte **b.** kiloliter
c. centipede **d.** millipede
e. megaton **f.** microgram
g. gigadollar **h.** decimeter
i. nanosecond **j.** megabuck

2.17.P Think of an atom as a sphere with a diameter of 10^{-8} cm.
a. How many atoms would it take to make a row 1 cm long?
b. How many atoms would it take to make a row 1 m long?
c. How many atoms would it take to fill a block 1 cm square?

2.18.P How many square meters are there in a square kilometer?

2.19.P How many meter sticks would have to be laid end to end to reach from New York to San Francisco, a distance of about 5000 km?

2.20.P A single frame of one type of camera film has a length of 35 mm and a width of 25 mm.
a. What is the area of this frame in square millimeters and in square centimeters?
b. If a typical silver grain in this film has a diameter of about 10^{-4} cm, approximately how many grains would there be in a single picture?

2.21.P A hydrogen atom has a diameter of about 1 angstrom (10^{-10} m).

a. What is the diameter of this atom in centimeters?
b. What is the diameter of this atom in nanometers?
c. What is the diameter of this atom in micrometers?
d. Which is the most appropriate unit for measuring the size of this and similar atoms?

2.22.P Do the following conversions.
a. How many cubic meters are in 6 cm^3?
b. How many cubic millimeters are in 15 m^3?
c. How many cubic centimeters are in 18 L?
d. How many cubic centimeters are in 150 ml?
e. How many liters are in 127 cm^3?

2.23.P Imagine a spherical atom put into a cube whose side is equal to the diameter of the atom. Then picture many of these tiny cubes packed into a big cube (see Figure 2-10). If each of the small cubes is 10^{-8} cm on a side, how many of them can be packed into a large cube that is 1 cm on a side? (This is the way atoms in certain solids are arranged. As we shall see later, this arrangement is called a crystal lattice.)

2.24.P A large cylindrical oil tank is 20 m in diameter and 10 m high.
a. What is the volume of this tank in cubic meters? (Remember that the volume of a cylinder is given by $V = \pi r^2 h$, where r is the radius, h is the height of the cylinder, and $\pi = 3.1416$.)
b. How many liters will the tank hold?

2.25.P A certain oil furnace uses 50 cm^3 of fuel per minute. If it burns fuel at a constant rate, how many liters of fuel will it burn in a 365-day year? How many cubic meters is this?

2.26.P A plant manager needs a cylindrical tank capable of holding 10^6 liters of waste water for decontamination. What is the height of this tank if the diameter is 18 m? (Use the formula for the volume of a cylinder given in Problem 2.24.)

2.27.P How many grams of water are there in a beaker containing 250 cm^3?

2.28.P How many micrograms are there in a sample weighing 50 mg?

2.29.P a. When a piece of iron is placed on a metric scale on the surface of the earth the scale gives a reading of 60 kg. How many pounds does it weigh?
b. What is the mass of this piece of iron on the moon?
c. What is the weight in pounds of this piece of iron on the moon?

2.30.P a. If you mixed 600 cm^3 of water with 800 cm^3 of water, how many liters of water would you have?
b. How many grams of water are in each sample, and how many are in the total?
c. How many kilograms of water are in each sample, and how many are in the total?

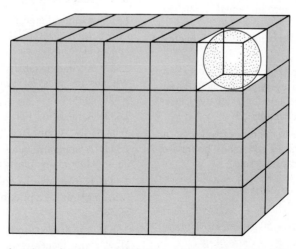

FIGURE 2-10 If we think of a spherical atom as just fitting into a little cubical box, then we can calculate the number of atoms fitting into a larger volume. In this case, the diameter of one atom equals the width of one box.

2.31.P
a. How many grams are there in a metric ton?
b. How many milligrams are there in a metric ton?
c. How many micrograms are there in a metric ton?

2.32.P Convert your own weight into
a. grams
b. kilograms
c. metric tons

2.33.P A coin is weighed on a balance and is found to have a mass of 5.526 g. Find its mass in
a. kilograms
b. milligrams
c. micrograms

2.34.P A bottle contains 50 g of aspirin. What is its mass in
a. milligrams
b. kilograms

2.35.P A bottle of water weighs 750 g. What is its volume in
a. cubic centimeters
b. liters

2.36.P A sample of metal weighs 7 lb. What is its mass in kilograms?

2.37.P How many seconds are there in a year?

2.38.P If 10 cm^3 of water are distilled each second, how many liters will be distilled in a year?

2.39.P If 10 million tons of iron are needed each year, how many tons must be mined every minute?

2.40.P
a. How many milliseconds equal 1,346,000 microseconds?
b. How many seconds equal 1,346,000 microseconds?

2.41.P A patient in a hospital is expected to receive 10 ml of intravenous fluid every hour. Each bottle contains 0.5 L. If the patient stays for 10 full days, how many bottles of fluid will be required?

2.42.P Every 24 hours, 3.8 million capsules of a certain medication are sold. In how many hours are the following numbers of capsules sold?
a. 7,500,000
b. 6 billion
c. 6.0×10^{23}

2.43.P An oil pipe can carry 50 gal/hr. How many hours are required to transmit 10,000 L of oil?

2.44.P A rocket travels at a speed of 10 km/sec. How many hours are required to travel 200,000 km?

2.45.P A certain reaction is completed in 7 milliseconds. Express this time in
a. seconds
b. hours
c. microseconds

2.46.P A strobe lamp flashes at a rate of 100 flashes per second.
a. How many flashes are there in an hour?
b. How many milliseconds are there between flashes?
c. How many microseconds are there between flashes?

2.3 English-Metric Conversions

2.47.P
a. How many feet are there in 30 m?
b. How many yards are there in 30 m?
c. How many centimeters are there in 3 ft? How many millimeters?
d. The distance between New York and Philadelphia is 90 mi. How far is this in kilometers?

2.48.P
a. How many cubic inches are there in a liter?
b. How many cubic feet are there in a liter?
c. How many liters are there in a cube that is 10 in on a side?

2.49.P What is the weight in pounds of 5 kg of salt?

2.50.P A bottle of photographic developer weighs 4.25 oz. What is its mass in grams?

2.51.P A clinical laboratory has 25 blood specimens, with a total mass of 184 g. What is the weight in ounces of 75 specimens?

2.52.P A vitamin tablet contains 13.5 mg of niacin. What is the mass of niacin in grams? What is its weight in ounces?

2.53.P A medicine bottle contains 375 ml of liquid. What is the volume first in liters and then in quarts?

2.54.P A standard wine bottle contains 750 ml of liquid. How many quarts does the bottle hold?

2.55.P Is it cheaper to buy 1 lb of sodium carbonate (washing soda) at a price of $1.89 or 680 g at a price of $2.38?

2.56.P A hydrogen atom has a diameter of about 1 Å. How many inches is this equivalent to? Why are angstroms more convenient than inches for measuring atomic distances?

2.4 Density and Specific Gravity

Refer to Table 2-3 for specific gravities.

2.57.P a. What is the mass of 1 L of mercury in kilograms?
b. What is the mass of 250 cm^3 of mercury in grams?

2.58.P What is the volume of 1 kg of nitrogen in cubic centimeters?

2.59.P What is the mass of 100 L of hydrogen first in grams and then in kilograms? What is the mass of the same volume of nitrogen? What conclusion can you draw about a hydrogen-filled balloon—will it sink or float in air? (Note that air is a mixture of nitrogen and oxygen.)

2.60.Q The specific gravity of ice is less than the specific gravity of water.
a. What does that information tell you about the floating properties of ice?
b. What would happen to the fish in a lake during a cold winter if the density of ice happened to be greater than the density of water?

2.61.P Compare the volume of 1 kg of aluminum with the volume of 1 kg of platinum. Which has the greater volume, and what is the ratio of the two volumes?

2.62.Q Will iron float on mercury? Give your reasoning, and show how you have used information about the specific gravities of the two substances.

2.63.Q Cork and wood float on water. What generalization can you make about these substances on the basis of this observation?

2.64.Q A manned balloon is seen traveling through the air. Later it is observed settling down to earth. Explain why at one time the balloon floats but later it sinks.

2.65.Q A ship made of steel floats in water, even though the specific gravity of steel is much greater than that of water. Explain this phenomenon.

2.66.P The acid in an automobile battery has a specific gravity that depends on whether or not the battery is charged. When the battery is charged, the specific gravity of the acid is 1.29; when discharged, it is 1.12.
a. What is the volume of 2 kg of battery acid in the charged state?
b. What is the volume of the same 2 kg of battery acid in the discharged state?
c. What happens to the volume of the acid as the battery becomes discharged?
d. If the battery is completely filled when charged, what is likely to happen to the acid if the battery is allowed to become discharged?

2.5 Analysis of Graphs

2.67.P There are two numbers that describe important properties of each kind of substance: the atomic number and the mass number. We will discuss the meaning of these numbers in a later chapter. For now, plot the relationship between the two sets of numbers given in the following table.

Atomic Number	Mass Number
11	23
13	27
15	31
17	35

a. Is the relationship linear?
b. Is it a direct proportion?

2.68.P Using Table 2-2 and the table of elements found on the inside cover of this book, graph the density of each metal versus its atomic number. Is the relationship linear?

2.69.P Plot a graph from the data below showing the relationship between atomic number and electron affinity. (The definitions of these terms will be presented in future chapters and need not concern you now.)

Atomic Number	Electron Affinity
9	81.0
17	84.8
35	79.1
53	72.1

a. Is the relationship linear?
b. Is it a direct proportion?

2.70.P Plot a graph from the data below showing the relationship between atomic number and atomic radius.

Atomic Number	Atomic Radius (in angstroms)
1	0.50
3	1.55
11	1.90
19	2.35
37	2.48
55	2.67

a. Is the relationship linear?
b. Is it a direct proportion?

2.71.P Using the data below, plot a graph showing the relationship between the two kinds of temperature scales.

Scale I	Scale II
0	273
20	293
50	323
100	373

a. Is the relationship linear?
b. Is it a direct proportion?

2.72.P The weather report says that the temperature is 95°F. What would the temperature be on the Celsius scale?

2.73.P While traveling in Europe, you hear a weather report that says the temperature is 20°C. What is the temperature on the Fahrenheit scale?

2.74.P The surface of the sun is at about 6000°C. What is its Fahrenheit temperature?

2.75.P Silver melts at a temperature of 1761°F. What is this temperature in Celsius units?

2.76.P There is one temperature at which both Fahrenheit and Celsius scales show the same number. What is this temperature? (*Note*: The use of some algebra is required to solve this problem.)

2.77.P Liquid oxygen, called Lox when used as a rocket fuel, has a boiling point of −183°C. What is this boiling point in degrees Fahrenheit?

2.78.P The average normal body temperature is 98.6°F. What is this temperature in degrees Celsius? (You would want to know this temperature if you bought a medical thermometer in Europe.)

2.79.P An auto tire warms up as it is driven over the highway. If its temperature increases by 25°C, how much does it increase on the Fahrenheit scale?

2.80.P The melting point of snow is decreased by 5°F when salt is added to it. How much is the melting point decreased on the Celsius scale?

2.81.P The temperature of the water in a bathtub is 100°F. What is this temperature in degrees Celsius?

2.82.P Liquid helium boils at about 4 K. What is the temperature on the Celsius scale?

2.83.P Dry ice (solid carbon dioxide) does not melt, but evaporates directly at a temperature of −109°F. What are the corresponding temperatures on the Celsius and Kelvin scales?

2.84.P The average normal body temperature is 98.6°F. What is that temperature in Kelvin units?

2.85.P a. A gas is collected at 395 K. What is that temperature on the Celsius scale?
b. The gas is cooled to 25°C. What is its new temperature on the Kelvin scale?

2.86.P We know that the boiling point of water is 100°C. We also know that blood is mostly water and that when a person has a fever the blood temperature may rise to 104° or even higher. Why doesn't blood boil under these conditions? (Be careful here!)

2.87.P A thermometer reads −200°F. What is the temperature on the Kelvin scale?

2.88.P A room thermostat is set for +32°C. What is the temperature on the Fahrenheit scale and on the Kelvin scale?

2.89.P A gas is collected at a temperature of 350 K. What is its temperature in degrees Celsius and in degrees Fahrenheit?

2.90.P A liquid at a temperature of 700 K is cooled 100 K to a temperature of 600 K. How much of a change was there on the Celsius and on the Fahrenheit scales?

2.91.P The melting point of silver is 962°C. What is this temperature in kelvins?

2.7 Accuracy and Precision

2.92.P Several samples of gas are collected by repeating a certain experiment. The volumes of the samples, which are at the same temperature and pressure, are

found to be 24.65 cm³, 24.69 cm³, 24.65 cm³, 24.70 cm³, 24.63 cm³, 24.68 cm³, and 24.72 cm³.
    a. What is the average volume?
    b. What is the maximum deviation from the average?
    c. What is the percent deviation?

2.93.P A sample of wire has its diameter measured several times, with the following results: 4.11 mm, 4.12 mm, 4.17 mm, 4.19 mm, 4.13 mm, and 4.18 mm.
    a. What is the average diameter?
    b. What is the absolute error (the maximum deviation)?
    c. What is the percent error?

2.94.P A sample of a chemical is weighed several times and the following results are recorded: 7.098 g, 7.105 g, 7.065 g, 7.003 g, 7.097 g, 7.099 g, and 7.100 g.
    a. What is the average weight?
    b. What is the maximum deviation from the average?
    c. What is the percent deviation from the average (using the maximum deviation as a basis)?
    d. If the correct weight is actually 7.098 g, what is the absolute error? That is, what is the difference between the average value and the true value?
    e. If the correct weight is actually 7.098 g, what is the percent systematic error? (Assume that the systematic error is the difference between the average value and the true value.)

2.95.P A sample of pure silver is weighed three times, with the following results: 3.334 g, 3.325 g, and 3.328 g. It is then weighed five more times, the results being 3.329 g, 3.335 g, 3.328 g, 3.330 g, and 3.337 g.
    a. Is the lowest percent deviation found in the three weighings, the five weighings, or all eight weighings combined?
    b. If the correct weight is 3.333 g, which of the three sets of data (three weighings, five weighings, or eight weighings) gives the smallest percent systematic error?

2.96.P A sample of sodium chloride is weighed several times, with the following results: 5.501 g, 5.505 g, 5.507 g, 5.506 g, 5.509 g, 5.500 g, 5.500 g, 5.500 g, 5.505 g, and 5.501 g. The correct value is determined to be 5.500 g. Which is the greater—the percent deviation from the average or the percent systematic error? (Use the maximum deviation.)

2.8 Significant Figures

2.97.P How many significant figures are in each of these numbers?
    a. 45 ± 1 **b.** 54.6 ± 0.2
    c. $45,300 \pm 100$ **d.** 0.00246 ± 0.00001

2.98.P Write out these products, rounding off to the correct number of significant figures.
    a. $35 \times 467 = ?$
    b. $3.67 \times 10^6 \times 4.876 = ?$
    c. $2.8 \times 1{,}834{,}200 = ?$

2.99.P Write out these quotients, rounding off to the correct number of significant figures.
    a. $308{,}200/4.5 = ?$
    b. $548.2/0.0023 = ?$
    c. $6.2 \times 10^9/1368 = ?$

2.100.P Round off the answers to these problems to the correct number of significant figures.
    a. $4.852 + 3.791 + 6.010 + 1.02 = ?$
    b. $23.9 + 19.854 + 0.02351 + 1.0001 = ?$
    c. $1056.5 - 457.25 = ?$

2.101.P Some of these answers have been properly rounded to the correct number of significant figures, and some have not. Correct those that need it.
    a. $\dfrac{(6.51 \times 10^3) \times (6.022 \times 10^{23})}{1.5 \times 10^4} = 2.614 \times 10^{23}$
    b. $(23.0 + 4.5) \times 6.02 = 165.6$
    c. $(3.0 \pm 0.1) + (4.10 \pm 0.01) + (5.67 \pm 0.05) = 12.8 \pm 0.1$

Self Test

1. The cost of a certain cleaning fluid is $3.89 per liter. How much would it cost to buy 250 L?

2. The price of gold is $12.35 per gram.
   a. How many kilograms of gold can you buy for $200,000?
   b. How many pounds does this gold weigh (at 454 g/lb)?

3. A manufacturing plant turns out 50 vitamin capsules each second.
   a. If the plant operates 8 hours per day, 5 days per week, how many capsules are manufactured in a year?
   b. It takes 3 people to operate the capsule-making machinery, each with a salary of $15 per hour. What is the labor cost for each bottle containing 100 capsules?

4. In 100 g of table salt (sodium chloride), there are 39.3 g of sodium and 60.7 g of chlorine. How much chlorine can be obtained from 20 kg of table salt?

5. a. How many centimeters are there in 20 m?
   b. How many feet are there in 20 m?

6. a. How many millimeters are there in 47 km?
   b. How many miles are there in 47 km?

7. a. How many microns are there in 4 mm?
   b. How many inches are there in 4 mm?

8. a. How many centimeters are there in 8.9 km?
   b. How many feet are there in 8.9 km?

9. Find the area of
   a. a square 5 mm on a side ($A = s^2$)
   b. a rectangle 6 cm by 1.4 m ($A = L \times W$)
   c. a circle with a radius of 27 cm ($A = \pi r^2$)
   d. a rectangle 2 km × 450 m

10. Find the volume of
    a. a cube 12 cm on a side ($A = s^3$)
    b. a sphere 10^{-6} cm in radius ($V = \frac{4}{3}\pi r^3$)

11. A warehouse must store 1500 boxes of chemicals, each 25 cm × 40 cm × 20 cm in dimension.
    a. What is the total volume of these boxes, first in cubic centimeters and then in cubic meters?
    b. Design a room to hold these boxes, allowing a meter of aisle space between adjacent rows of boxes and a floor-to-ceiling height of no more than 4 m.

12. A waterbed has dimensions of 2.0 m × 2.3 m × 0.30 m.
    a. What is its mass in kilograms?
    b. What is its weight in pounds?

13. A block of lead has a mass of 160 kg on the earth's surface.
    a. How much does the block weigh in pounds?
    b. How much does the block weigh on the surface of the moon?
    c. What is the mass of the block on the surface of the moon?
    d. What is the mass of the block in interplanetary space?
    e. If you had that block in a spaceship, where everything was "weightless," how would you perceive the mass of the lead block?

14. How many microseconds are there in 8 hr?

15. What is the mass of 125 cm³ of mercury?

16. A tank for the storage of oil has a diameter of 5 m and a height of 10 m. The oil has a specific gravity of 0.92.
    a. What is the volume of the tank in cubic meters and in liters?
    b. What is the mass of the oil in kilograms?

17. What would be the volume of 2 metric tons of iron?

18. The cost of an industrial chemical is $2.50 for 100 g, and $210 for 10 kg.
    a. Is the cost directly proportional to the quantity?
    b. Do we have enough information to know whether there is a linear relationship between the cost and the quantity?

19. What is the melting point of ice in degrees Celsius, in degrees Fahrenheit, and in Kelvins? (This is the same as the freezing point of water.)

20. Helium is normally a gas, but it becomes a liquid when cooled to a temperature of 4 K. What is this temperature in degrees Celsius and in degrees Fahrenheit?

21. A large number of measurements are made of the temperature of a reaction chamber. The measurements show an average temperature of 145°C, and the

fluctuations above and below this average value are about ±1.5°. When a more precise instrument is used to measure the temperature, it is found to be 148 ± 1.0°C.

   a. What is the percent random error of the first set of measurements?
   b. What is the percent systematic error?
   c. Which is better—the accuracy or the precision of the measurement?

22. a. Describe three causes of random errors in weighing small objects on a pan balance.
   b. What is meant by calibration?
   c. Is it possible to make an absolutely precise measurement? If your answer is no, give your reasons.

23. The volume of a gas sample is measured. Sample A has a volume of 15.3 cm^3, and sample B has a volume of 10.4 L. The probable error of each measurement is $\pm 0.5 \text{ cm}^3$. What is the percent error of each measurement?

24. Do the following calculations, and round off the answers to the correct number of significant digits:
   a. $3.2 \times 94.38 = ?$
   b. $6.423 \times 10^6 \times 1.05 \times 10^5 = ?$
   c. $11{,}200{,}000 + 679{,}000 = ?$
   d. $9.4301 - 9.420 = ?$
   e. $\dfrac{9{,}674{,}821}{4.2} = ?$

3
Matter and Energy

Objectives   After completing this chapter, you should be able to:
1. Understand some of the differences between ancient and modern ideas about matter.
2. Recognize the difference between a chemical and a physical property of matter.
3. Recognize the difference between elements and compounds.
4. Identify matter in liquid, solid, and gaseous states.
5. Recognize the difference between mixtures and compounds.
6. Recognize the difference between homogeneous and heterogeneous substances.
7. Recognize the common units for measuring energy.
8. Recognize some of the different forms of energy.
9. Apply the law of conservation of energy to practical problems.
10. Convert from units of mechanical energy to units of thermal energy.
11. Use the concept of specific heat in simple problems.
12. Understand the concepts of endothermic and exothermic reactions.
13. Solve simple problems involving heat of combustion.

3.1 The Nature of Matter

Ancient and Modern Ideas About Matter

Matter exists in a large number of forms, as a casual glance about you will verify. The gases of the atmosphere; the waters of the ocean and the rains; the minerals of the earth with their gleaming crystals, their gray granites, their colorful metallic ores—these are all varieties of matter.

A major characteristic of matter is the way it changes from one form to another. Mix a heavy red mineral dug from the ground with black sooty carbon and heat it to a high temperature. The result is a white-hot liquid that flows like water until it cools, when it becomes a dense gray metal: iron. Immerse this metallic iron in a corrosive liquid (sulfuric acid) and the iron gradually dissolves in a froth of bubbles, turning the liquid a dark green color. The bubbles, escaping into the air, contain a gas that is ignited by a flame. We ask: How does the heavy iron disappear into the liquid? Where does the gas come from?

The changes we have described are **chemical changes**—the changing of one substance into another. Chemistry involves not only learning about the many changes that can take place but also understanding why these changes happen.

The earliest settlers in the valleys of the Middle East knew about many of these changes. However, they were incapable of understanding why the changes occurred because they had no adequate notion of the basic structure of matter. No one realized that a piece of iron is a very dynamic object—that inside this rigid object are incredibly tiny things constantly moving about and the behavior of these tiny things explains why the iron looks the way it does and why it reacts with other substances the way it does (for example, dissolving in acids and not in water).

The Greeks, who had the most advanced civilization of the ancient era that preceded the Middle Ages, came very close to developing a proper theory of matter. Before the period in which the Greek philosophers flourished (starting about 600 B.C.), most of the events of nature (such as winds, rain, lightning, and the motion of the heavenly bodies) were thought to be the result of magical or supernatural causes. The Greek teachers and thinkers made a huge advance in ideas about the world with their devotion to the principle that the events of nature can be explained rationally with the help of natural laws.

Since the concept of experimentation had not yet been developed to any great extent, most of the speculations of the Greek philosophers bore little fruit. However, the ideas of Democritus (born about 460 B.C.) began to approach some of our modern ideas.

He asked: What would happen if you took a container of water and divided it into small drops and then took each drop and divided it into smaller drops? If you continued to divide each microscopic droplet into smaller drops, would there be a limit to the size of the drops, or could you continue to make them smaller and smaller until they were infinitely small?

Democritus's answer to the hypothetical question was no—you cannot make the drops indefinitely small. There comes a point when you reach the smallest possible particle of water and cannot divide it into anything smaller. Democritus called this smallest particle of water an *atom*. (Scientists now know that the smallest particle of water is a *molecule*, not an atom, but more of that later.) Democritus conceived the idea that all matter is mostly empty space, with tiny atomic particles moving about within the space.

Democritus pictured every kind of matter as being composed of different kinds of atoms. Atoms of water were smooth and shiny; atoms of fire were jagged, red, and hot; and so on. Chemical reactions were the results of the actions of atoms—moving about, being exchanged, joining in different combinations. The latter part of the theory comes very close to our present model of matter.

Unfortunately, most of Democritus's theories were speculation, based on very little hard evidence, and they were opposed vigorously by other philosophers, including the influential and renowned Aristotle. As a result, the ideas of Democritus did not play a large role in chemistry until the nineteenth century, when they were rediscovered by John Dalton.

At the present time, the concept of the atom is the most fundamental and central idea of modern science. The atomic theory makes it possible to understand why all forms of matter behave the way they do.

One important difference between our modern atomic theory and Democritus's speculation relates to the number of different kinds of atoms that exist. Democritus thought there was an unlimited number of kinds of atoms, one for each of the different varieties of matter.

However, modern theory provides for only a limited number of types of atoms. (At present over a thousand different atoms have been identified, but these represent only 108 substances.) How then do we explain the almost unlimited number of substances found in nature? In addition, how do we explain what happens when iron rusts or when a candle burns? How do we explain the creation of wood by a growing tree?

The explanation is simple: the millions of types of substances in nature result from the joining together of a few different atomic species into a great many combinations. These combinations, or clusters of atoms, are either molecules or crystals. A **molecule** is a cluster of a fixed number of atoms forming a self-contained

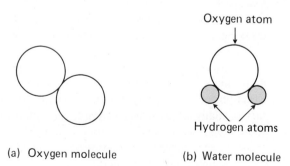

(a) Oxygen molecule   (b) Water molecule

FIGURE 3-1 (a) A molecule of oxygen contains two oxygen atoms, and so it is called a diatomic molecule. (b) A molecule of water contains two hydrogen atoms and one oxygen atom.

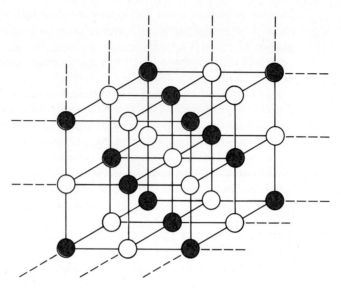

FIGURE 3-2   A cubic lattice of sodium atoms alternating with chlorine atoms forms a crystal of sodium chloride.

unit (see Figure 3-1). A **crystal** is an arrangement of an indefinite number of atoms or molecules in a regular pattern, as can be seen in Figure 3-2.

One of the fundamental tasks of chemistry is to study all of the ways in which atoms join together and to discover how the properties of matter arise from the way the individual atoms and molecules behave. However, before we can delve into atomic theory, we must first investigate the meaning of a physical or chemical property of matter. This is the subject of the next section.

Physical and Chemical Properties

A property of an object is a characteristic of that object which describes it, identifies it, and shows how it is either similar to or different from other objects. Roughly speaking, there are two main kinds of properties: **physical properties** and **chemical properties**.

Some physical properties describe a particular object: its size, geometric shape, mass, surface characteristics (rough, smooth, etc.), temperature, and so on. Other physical properties describe a particular material or substance that may be a part of many objects. For example, aluminum can be found in kitchenware, in house construction, and in automobile parts, but the properties of aluminum are the same everywhere. Such properties include hardness, density, color, melting point, boiling point, and whether the substance is normally a solid, a liquid or a gas.

Physical changes are changes in the physical properties of an object. Typical physical changes are heating, cooling, melting, boiling, breaking into smaller pieces, and changing shape. In any such physical change, the chemical substance itself does not change; the only thing that changes is the physical state of the substance.

The chemical properties of a substance describe its chemical behavior—that is, the chemical reactions that take place between this substance and other substances. For the time being, we define a chemical reaction (or change) as *a process in which one material changes into another or combines with other substances to form another material.* (When we study the different types of chemical reactions, we will learn a more precise definition.)

Some chemical properties are general. For example, materials such as helium are **inert** (they do not react with other substances), whereas other materials such as fluorine are highly **reactive** (they combine readily with other substances). The words "inert" and "reactive" describe substances in general.

Other properties are more specific. For example, the substance sodium combines with chlorine in a certain way to form a compound called sodium chloride. This is a specific property of both sodium and chlorine. And the substance potassium combines with chlorine to form potassium chloride. This is another specific chemical property.

If we had to remember the details of every such reaction, we would need an enormous amount of memory. But luckily many materials such as sodium and potassium (and also cesium) are very similar and behave in similar ways, so we can group them into a "family" of substances. Therefore, it is not necessary to memorize every single reaction individually. Once we know that sodium, potassium, and cesium are similar substances, it is easy to remember that the reactions of sodium are very much like the reactions of potassium and cesium. If we know the *general* chemical properties of groups of substances, we can make many predictions about the *individual* members of the group.

One of the functions of chemistry is to learn how to classify matter into groups and subgroups. We begin with a very broad division of all kinds of matter into elements and compounds. (A compound is made up of more than one kind of atom; an element is made up of only one kind of atom.) Then there is a division of compounds into organic and inorganic compounds. (Organic compounds have a molecular structure built around the carbon atom.) Inorganic compounds are roughly classified as acids, bases, or salts. In the following chapters, we will investigate these classifications in some detail.

Another kind of classification we make in chemistry is according to general conditions, or *states*. All the kinds of matter exist in just a few states, the most familiar of which are the solid state, the liquid state, and the gaseous state. A given kind of matter—water, for example—may exist either in the solid state (ice), in the liquid state, or in the gaseous state (steam), depending on its temperature and pressure (see Figure 3-3). Under normal room conditions, most substances exist in one particular state, and this normal condition is considered one of a substance's physical properties. In the next section, we will consider the characteristics of the various states of matter and learn what distinguishes one from another.

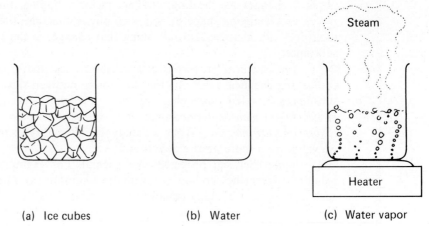

(a) Ice cubes   (b) Water   (c) Water vapor

FIGURE 3-3 Water can exist in three states: (a) solid, (b) liquid, and (c) gas.

3.2 States of Matter

The Solid State

A solid is an object that has a definite shape of its own and occupies a definite volume. A brick, a piece of wood, a crystal of salt—these are all solids. Generally speaking, a solid is an object that holds itself together and retains its own shape over a period of time.

Some solids have very regular internal structures composed of particles arranged in an orderly pattern. In common table salt, for example, the particles are arranged in a cubic pattern, shown in Figure 3-2. Such an arrangement is called a **crystal lattice,** and a material of this kind is called a **crystalline material.** Other common crystalline materials are ice, sugar, quartz, a great many minerals, and most metals. If you look at common table salt with a magnifying glass, you will see that each granule has a cubic shape, reflecting the structure of the crystal lattice (see Figure 3-4). (Table salt often comes finely ground, in which case you may not see full cubic crystals.) It is not so obvious that copper and iron have crystal structures, but a look with a powerful microscope would show that a piece of iron is composed of many small bits of crystal held tightly together. Advanced methods of analysis, such as x-ray diffraction, permit the scientist to determine the exact structure of any kind of crystal.

Other solids do not have a well-organized crystalline structure; the molecules in these materials are arranged in a more random fashion. Materials without definite structure are called **amorphous materials**. Common amorphous materials are glass, rubber, and many plastics.

FIGURE 3-4 Photograph of salt crystals (Courtesy Runk/Schoenberger, photo by Grant Heilman)

It is often difficult to assign a given material to one category or the other. Some plastics are mixtures of both crystalline and amorphous structures. Furthermore, a material such as sulfur can exist either in a crystalline state or in an amorphous state, depending on how it is prepared. (This characteristic makes for a simple and intriguing experiment. Sulfur, as it comes in the bottle, is in the form of a yellow powder. If it is heated in a dish, it will melt. Allowed to cool, the solidifying sulfur forms fine crystals. However, if the hot, molten sulfur is poured into cold water, it forms a rubbery mass: amorphous sulfur.) Making matters even more complicated is the fact that there are two separate crystalline states in which sulfur may exist, depending on such physical conditions as temperature.

For reasons such as these, when describing a given substance you must state the conditions (such as temperature and pressure) existing at the moment. You know that the appearance of water depends on temperature; at a temperature below its freezing point water will be a solid, and at a temperature above its freezing point it will be a liquid.

The Liquid State

A **liquid** is a substance that has a definite volume but not a definite shape; therefore, a liquid fills the bottom of any container into which it is put. In a liquid, the molecules are not bound together as tightly as they are in a solid, and so the liquid is free to flow from one place to another and to change its shape.

What determines whether a given material is liquid or solid? One important factor, as mentioned earlier, is the temperature. At extremely low temperatures, all materials are solid (with the exception of helium). If the temperature is raised to a high enough point, one of two things will happen: the material will *decompose* (break down into simpler substances), or it will *melt* (form a liquid).

A piece of wood, for example, does not melt when heated. If air is not present, it turns into charcoal; otherwise it burns. Substances such as ice or iron do not decompose when heated; instead, they change from the solid to the liquid state.

In a crystalline material, the change is fairly abrupt and happens at a particular temperature typical of the particular material. Measuring the **melting point** of an unknown material provides one clue to the identity of the substance.

Amorphous materials such as glass, on the other hand, do not have distinct melting points. Glass gradually softens as it is heated; the hotter it gets, the more easily it flows. Glass, because of its lack of a regular crystalline structure and because of its lack of a specific melting point, may be thought of as a very stiff liquid.

Melting is a physical change and does not alter the kind of atoms or molecules making up a substance. Melting is reversible; when a substance is cooled to a temperature below its melting point, it *solidifies* or *freezes*. For a given substance, freezing and melting take place at the same temperature.

The Gaseous State

A **gas** is a substance that will expand to fill any container that it occupies. The behavior of gases is one of the best arguments for the atomic model, for this model provides the simplest way to explain how a gas can be compressed in a cylinder with a movable piston or why the helium in a balloon expands to a greater volume when the balloon rises to a higher altitude. A gas does not have a fixed volume because the particles that make up the gas are not bound to one another, and so they can move closer together or farther apart. (See Figure 3-5.) Most of the space inside a gas is completely empty; its particles are much smaller than the distances between them.

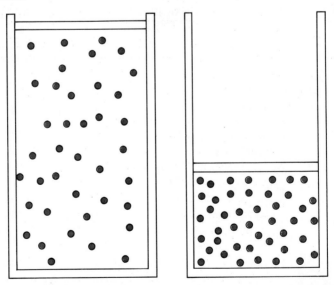

FIGURE 3-5 The particles that make up a gas are free to move farther apart or closer together. For this reason a gas can expand or be compressed over a great range of volumes.

The particles that make up a gas are molecules rather than atoms; in most cases, there are two or more atoms bound together in each molecule. These molecules move about with great speed, and, as we shall see in Chapter 12, it is the motion of the molecules that determines most of the properties of gases.

Any gas cooled to a low enough temperature will *condense* into a liquid. **Condensation** occurs at low temperatures because the molecules slow down in their motions, allowing the binding forces between them to hold them more firmly together. (The binding force is a force of attraction felt between two adjacent molecules.) Conversely, a liquid can *evaporate* into the gaseous state. **Evaporation** occurs when molecules escape from the surface of a liquid into the atmosphere. If the temperature is high enough, the liquid will boil, releasing bubbles of vapor from its interior (see Figure 3-6).

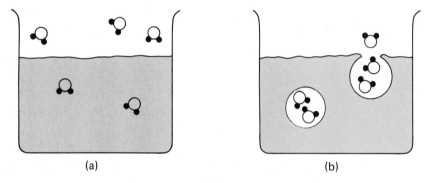

FIGURE 3-6 (a) Evaporation takes place when water molecules escape from the body of the liquid water. (b) Boiling takes place when the water is so hot that it turns into a gas within the body of the liquid, forming bubbles that rise to the surface.

Why are some things usually found in the solid state, some in the liquid state, and others in the gaseous state? It is all a question of molecular speed and the strength of the attractive force pulling molecules together. If the molecules are moving so fast that they are not able to stick together, the substance is in the gaseous state. If they are moving slowly enough for the forces between the molecules to have a chance to pull them together, the substance is a liquid. If the molecules are moving so slowly that the forces hold them rigidly together, the material is a solid. The important fact to remember is that *the higher the temperature, the faster the molecules move.* This is why solids melt as the temperature increases and vaporize at yet higher temperatures. The exact temperature at which a change takes place depends on the strength of the force between the molecules.

The Plasma State

At extremely high temperatures (thousands of degrees Celsius), all matter is vaporized. Such extremely hot matter becomes a kind of gas having properties so different from those of ordinary gases that it is considered to be in a fourth state of matter: the **plasma** state. The word "plasma" here is not to be confused with blood plasma but does refer to a kind of fluid. We will not be able to explain what a plasma is until we have delved further into atomic structure. But for the moment keep in mind that plasmas are quite common in the world. In fact, almost all the matter in the universe is in the plasma state: the hot gases of the sun and all the stars and the low-density gas filling the space between the stars are all plasma. On earth we see plasmas in electric arcs, in fluorescent lights, in neon signs—any place where electric current is sent through a gas-filled tube.

Liquid Crystals

The discovery of liquid crystals in recent years has led to development of a number of useful products. Liquid crystals represent a hybrid between the liquid and solid states. They consist of certain molecules with a long and narrow structure. Because of the shape of these molecules, the properties of these crystals *along* the molecules are different from their properties *across* the molecular width. In one direction the molecules are firmly fixed, as in a solid, but at right angles the molecules may be free to move, as in a liquid (see Figure 3-7).

Liquid crystals are used in displays for electronic watches and calculators. An electric field applied to a flat sheet of liquid crystal causes the molecules to rotate and in this manner changes their ability to reflect light.

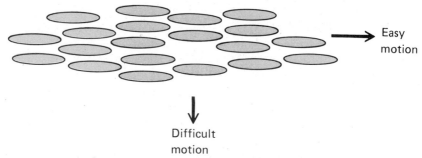

FIGURE 3-7 Liquid crystal molecules are long and slender, so they are free to move in their "long" direction but not so free to move at right angles to that direction.

3.3 Types of Substances

Pure and Impure Substances

A **pure substance** is made up entirely of one kind of atom or molecule, with no impurities present. In nature nothing is completely pure. However, scientists have learned to produce materials with an impurity content of less than one part per billion. (By one part per billion we mean that one out of every billion atoms is an impurity atom.) The matter of impurities is of great importance, for the operation of transistors and other solid-state devices depends on the presence of a few parts per million of certain impurity atoms within a highly purified crystal of silicon. (The silicon is "doped" with impurity atoms.) A tiny quantity of impurity in this situation makes a great difference in the way the material conducts electric current.

An **element** is a substance that contains only one kind of atom. At present, 108 different elements have been identified. (Not all of these elements exist normally in nature; several of them must be manufactured in the laboratory.) The names of some elements are already familiar to you: iron, aluminum, gold, oxygen. Each element has a particular set of properties that distinguishes it from every other element; no two elements are identical.

Atoms rarely exist separately; they are usually bound to other atoms to form molecules or crystals (except in outer space or at extremely high temperatures). The only atoms that exist separately under normal earthly conditions are those of the "noble" gases: helium, neon, argon, krypton, xenon, and radon. The noble gases consist of atoms that are very reluctant to join with each other or with other kinds of atoms. Some years ago these atoms were described as inert, and until recently it was believed that they did not form molecules at all. However, it is now known that some of them do combine with other atoms to form molecules.

The atoms of the common gases (hydrogen, oxygen, nitrogen, etc.) are not normally found alone. Usually these elements consist of diatomic molecules (molecules containing two identical atoms) rather than individual atoms (monatomic molecules). Oxygen also exists, under certain conditions, in the form of a molecule with three atoms—a triatomic molecule. Oxygen in this form is called ozone.

Compounds are substances formed when two or more different atoms join to form molecules or crystals. Because of the innumerable combinations of atoms possible, the number of compounds in nature is extremely large. In addition, chemists are expert at inventing compounds that do not exist normally in nature. The number of atoms in a molecule can range from two to hundreds of thousands, and the number of atoms in a crystal is indefinite. The existence of millions of compounds allows for the great variety of materials in the world and, indeed, provides for the existence of life itself.

A fascinating feature of molecules is that the properties of the compounds they form are quite different from the properties of the individual elements making up the compounds. For example, both hydrogen and oxygen are gases, but when combined they form water, a liquid. Carbon is a black solid, but it combines with hydrogen to form methane, a colorless gas. Add to the methane another gas—oxygen—and a liquid is obtained: methyl alcohol. Another example is the combination of copper (a reddish metal) with chlorine (a greenish gas), which yields cupric chloride, a yellowish-green crystalline solid. The possible combinations are endless.

Materials may be either homogeneous or heterogeneous. A material that has the same properties all through its volume is a **homogeneous material**. A beaker of pure water is a homogeneous object, as is a cup of tea with sugar dissolved in it.

On the other hand, a **heterogeneous material** contains two or more substances (or a single substance in different states) existing separately but side by side. A cup of tea with undissolved sugar at the bottom is heterogeneous. Concrete is easily seen to be a conglomerate of gravel, sand, and cement. A mixture of water and ice cubes contains nothing but water molecules, but since liquid and solid states exist side by side, the mixture is considered heterogeneous.

In a crystal of silicon doped with small amounts of impurities, the impurity atoms are an integral part of the crystal structure, so the crystal is considered a homogeneous structure. A crystal with impurities evenly distributed is called *uniform*; if there are more impurities at one end of the crystal than at the other, the crystal is said to be *nonuniform*.

Mixtures and Solutions

Many of the materials dealt with in everyday life consist of two or more substances mixed together. Examine the sand at the seashore, and you will find a heterogeneous mixture of grains with different colors and structures. Inside many medicine capsules is a mixture of tiny spheres of varying colors and compositions.

The quality that distinguishes a mixture from a compound is the fact that the different parts of the mixture can be separated by physical means. For example, you can sort the different grains of sand by their color. A mixture of sugar and iron filings can be separated with a magnet because there is no chemical bonding between the sugar molecules and the iron atoms; they sit there loosely, side by side. On the other hand, the elements within a compound cannot be separated by large-scale physical means—that is, by mechanical separation. The only way to alter the structure of a molecule is to loosen the bonds between the atoms in the molecule, and the study of chemistry is to a large extent the study of the reactions that can perform such alterations.

Another way of mixing different materials is to dissolve one in another, forming a solution (a homogeneous mixture). A solution contains two components: a *solvent* and a *solute*. When a teaspoon of sugar (the solute) is stirred into a glass of water (the solvent), the sugar crystals disappear, and the glass appears to contain nothing but water. The sugar is still there, however, as you can verify by tasting the solution. When sugar dissolves in water, the sugar molecules break their close connection within the sugar crystal and hook up (loosely) to the water molecules. The presence of the sugar in the solution changes many of the properties of the water; it changes the density, the boiling point, the way light passes through (the index of refraction), and so on. The solid sugar can be recovered from the solution by allowing the water to evaporate. (Boiling the solution will cause the sugar to decompose into a brown caramel.) The fact that the sugar can be recovered from the water by this simple means demonstrates that the sugar molecules were not bound strongly to the water molecules.

Water is often called the **universal solvent** because it dissolves more substances than does any other liquid. However, it is by no means the only solvent in common use. Chemical solvents are used for various purposes—for example, for cleaning clothes. Some substances (such as oil) that will not dissolve in water will dissolve in compounds such as trichloroethylene, found in many commercial cleaning fluids.

A substance that will dissolve in a given liquid is said to be soluble in that liquid, whereas a substance that does not dissolve readily is said to be insoluble. The degree of **solubility** of a substance is an important property. In the *Handbook of Chemistry and Physics* (an indispensable reference for any student of chemistry) are found tables listing the solubilities of the common compounds in water. For example, 35.7 g of sodium chloride will dissolve in 100 cm^3 of cold water, and 39.1 g will dissolve in 100 cm^3 of hot water. If you try to dissolve more than these amounts of sodium chloride in 100 cm^3 of water, the surplus salt will refuse to dissolve; the excess salt crystals will just sit at the bottom of the container. The solution is then said to be *saturated*.

Solutions can be made not only by dissolving solids in a liquid but also by dissolving one liquid in another. An alcohol solution is a solution of pure alcohol in water. In this case, we speak of *diluting* the alcohol with water; that is, the concentration of the alcohol is reduced. Gases can also be dissolved in liquids. A common example of a gas in water is soda water (or carbonated water), a solution of carbon dioxide in water.

In addition to solid-liquid, liquid-liquid, and gas-liquid solutions, there are solid-solid solutions, which are common especially among metals. Such solutions are called *alloys* and are generally made by melting two or more metals together, allowing the molten metals to mix thoroughly, and then letting the mixture cool. Pure metals are generally alloyed with other metals to improve their useful properties. Although copper itself is too soft to be very functional mechanically, its alloys are much stronger. Common brass is an alloy of copper and zinc, and bronze is an alloy of copper and tin. Similarly, pure iron is a fairly soft metal, but the addition of a small percent of carbon creates a much harder alloy—steel. Steel made from only iron and carbon is too brittle for many uses, so other metals such as chromium and molybdenum are added to improve the toughness.

The subject of solutions is a very large one and will be dealt with in more detail in Chapter 13. In the meantime, you will see that many of the reactions studied in chemistry take place in solutions, because when chemical substances are dissolved in water their molecules can interact with each other easily and parts of molecules can be interchanged.

3.4 Energy

Measuring Energy

Energy is a necessary requirement for all the motions and actions that take place in the universe. In order for anything to move, energy must be supplied to it. Indeed, for anything to happen, energy must be provided. Where does the energy come from? The energy to move our bodies comes from the metabolism of food, which is nothing more than the burning of fuel. The motion of a steam turbine or a gasoline engine results from the burning of fuel. Looking at all the moving things in the world, we can see that energy comes in most cases from the burning of fuel.

We can also see that energy is changed from one form into another as we use it. When a steam turbine spins the coils of a generator in a power plant, electrical energy pours out from the generator through transmission lines into homes and factories, where it operates electric motors, TV sets, light bulbs, and heaters. Energy thus manifests itself in many forms: mechanical motion, heat, electricity, light, radio waves, and so on.

But what is energy? It is hard to answer this question in an uncomplicated way, for energy is a very abstract concept and appears in a large number of disguises. Yet, in spite of its abstract nature, the concept of energy is one of the most useful ideas in science; with it we can solve many problems in a simple manner.

To understand what energy is, we begin with the concept of **mechanical work.** *Work is done whenever an applied force moves an object from one place to another.* A force is like a push or a pull, and it is a quantity that can be measured. If, for example, we lift a weight from the floor, we must pull up on it with enough force to overcome the downward pull of gravity. (The upward force must be greater than or equal to the weight of the object.)

Now, to lift the weight through any given distance requires that a certain amount of work be done. We may think of work as energy released from storage to make an object move from one place to another. The energy may come from the burning of food (fuel) in the body, or it may come from the burning of fuel in an engine. By definition, *the amount of energy needed equals the amount of work done in lifting the weight.* In turn, *the amount of work done equals the amount of applied force multiplied by the distance the object moves.*

In the English system, the basic unit of energy is the **foot-pound**, which equals the work done by a force of one pound moving through a distance of one foot. In the SI system, the unit of energy is the **joule**, equal to the work done by a newton of force (about 0.224 lb) moving through a distance of one meter.

In the study of chemistry, we are concerned with questions relating to the amount of energy released by certain chemical reactions or, in some cases, the amount of energy that must be provided in order to make a reaction take place. Therefore, our main interest will be directed at the way energy changes from one form to another, as described in the next section.

Forms of Energy

Energy comes in many guises. We divide these forms of energy into two major categories: kinetic energy and potential energy. **Kinetic energy** is the energy an object has because of its motion. The amount of kinetic energy possessed by a moving object depends on the mass of the object and its speed (velocity). The more mass it has and the faster it is going, the greater the amount of kinetic energy. (To be precise, the kinetic energy of a moving object is directly proportional to its mass and to the square of its velocity. For a detailed discussion of this relationship, the reader is referred to a physics text.)

The meaning of kinetic and potential energy can best be illustrated by going back to the situation used in the last section to illustrate the concept of work. There we saw that raising an object above the ground requires a certain amount of energy. What happens to this energy while the object is raised? It is useful to think of the energy as being stored temporarily in the earth's gravitational field. Even though it is invisible and intangible, the energy is returned when the object falls back to the earth's surface. We call the energy of the raised object **potential energy**—because it has the potential for being returned, like money in a savings account.

To visualize the transformation of energy from one form to another, picture the following story (see Figure 3-8). Climbing a ladder, you raise a load of bricks to the roof of a building. The energy for this process comes from the burning of fuel (food) in the body, which produces both heat and mechanical work. The heat maintains the body temperature, and the mechanical work represents the transfer of energy from the fuel to the bricks. Once the bricks are above the ground, the energy is stored as potential energy. However, if you drop the bricks from the roof, the potential energy is transformed back into kinetic energy as they fall to the ground. Now we go one step farther. What happens after the bricks hit the ground? Where does the energy go? If we investigate carefully, we will find that after coming to rest the bricks are warmer than they were before. (Anybody familiar with basketball knows that the ball is heated up considerably by repeated collisions with the floor.) The kinetic energy of a falling object is converted into heat when that object hits the ground.

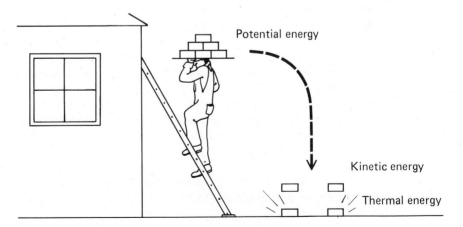

FIGURE 3-8 The work done in raising a load of bricks to the top of a building equals the potential energy of the bricks at the top. When the bricks fall to the ground, the potential energy is converted into kinetic energy. And when the bricks hit the ground, the kinetic energy is converted to thermal energy.

The transformation of kinetic energy into heat is an important one. Any time a moving object comes to rest because of friction or a collision, its kinetic energy is converted into heat. You can observe this effect every time you rub your hands together in order to warm them. The understanding of this phenomenon was one of the most important events in science during the first half of the nineteenth century. Prior to that time, it was believed that heat was caused by an invisible fluid called "caloric" which could pass from one thing to another. According to the caloric theory, when an object became heated it was because it had absorbed a certain amount of caloric fluid.

However, in 1798, Benjamin Thompson (Count Rumford), while overseeing the boring of a cannon, noticed that the metal being drilled continued to get hotter and hotter almost without end, until cooling with water was required in order to keep the metal from overheating. It seemed to him that the concept of "caloric" was insufficient to explain what was happening. He invented instead the theory that the mechanical energy from the drill was being converted into heat and that heat was simply another form of energy. This was one of the most important new ideas of its day, and it ushered in a whole new era in the understanding of nature. Yet another half-century had to pass before Rumford's explanation could be properly understood. Gradually it was realized that **heat** is simply an expression of the motion of the atoms and molecules within an object. If we put energy into an object, its molecules move about faster and faster. This increased motion is what we perceive as an increase in temperature.

Between 1840 and 1850, James Prescott Joule, an English physicist, performed numerous experiments to measure the amount of heat generated by various processes—by stirring water with a paddle wheel, by running electric current through a wire, by expanding and contracting gases, and so on. He found that a given amount of mechanical energy always produces a certain amount of heat. This evidence corroborated the theory that heat is just another form of energy. Furthermore, he discovered that when energy is changed from one form to another, none of the energy is ever lost and none is ever created. The total amount of energy in the system always remains the same. Joule's discovery became one of our most fundamental laws of nature: the **law of conservation of energy**. This law states that *in a closed system the total amount of energy always remains constant*; it never increases or decreases. In other words, energy can never be created or destroyed—it merely changes from one form to another. (A closed system is simply a system that does not allow energy to enter or escape through its boundaries.)

Experiments continue to verify the law of conservation of energy with greater and greater precision. The latest and most precise measurements relating to energy show that in certain reactions the amount of energy in the system remains unchanged to within one part out of 10^{15}. These measurements are among the most precise in the history of science and give us great confidence in the validity of the law of conservation of energy.

Conservation of energy is not purely of theoretical interest. It is a rule that allows us to answer certain important questions in a simple but powerful manner.

For example, it allows us to predict, with great confidence, that the perpetual motion machines described in the Introduction have not the slightest chance of success and so there is no point in wasting time on them. (Nevertheless, hoaxes still surface now and then—such as the claim that you can make an automobile fuel by dissolving a small pill in a tankful of water).

The law of conservation of energy is just one of a set of conservation laws—which tell us that there are certain physical properties of matter and energy that do not change in any kind of reaction. One of the functions of science is to determine from experience what these invariant quantities are. Knowing that energy is a conserved quantity allows us to look at any kind of proposed invention and ask ourselves whether the device (or chemical reaction) is supposed to put out more energy than it takes in. If so, we can say instantly, without further thought, that the invention will not work. It is an extremely powerful way of reasoning.

A few more questions remain to be answered. Remember we said that the energy needed to raise a weight above the ground comes from the burning of a fuel such as food. How does this energy arise? Is it created during the burning of fuel? Certainly not, for the law of conservation of energy says that energy cannot be created. Apparently the energy was in the fuel all the time and was simply liberated by the combustion. The energy stored in the fuel, called **chemical potential energy**, is converted into heat or other forms of energy during combustion. The chemical potential energy, in turn, arises from the binding forces between the atoms in the fuel compound; these binding forces are fundamentally electrical in nature. However, it is not necessary to worry about such atomic-level details in most practical situations. All we need to know when considering heat problems is the amount of chemical energy stored in a given amount of fuel.

Where did the energy in fuels originate? The fossil fuels (oil, gas, coal) all got their energy from the sun because such fuels are the remains of prehistoric plants and animals. The plants absorb radiant energy from the light of the sun and convert it into chemical energy. Animals, feeding on the plants, store some of the energy in their bodies. That stored energy remains in the earth for millions of years while the organic matter of which the bodies are composed changes to coal, oil, and gas. As we burn their fossilized remains, we convert the chemical energy into heat. Thus, almost all the energy at our disposal originated in the sun. (An exception is nuclear energy; the energy locked within the uranium atom got there by processes that took place even before the sun existed.)

Where does the sun get its energy? The sun gets its energy from reactions involving the central core of its atoms—*nuclear reactions*, which we cannot discuss at this point because we have not yet described the inner structure of the atom. These nuclear reactions differ in many ways from the chemical reactions that are the main interest of chemistry.

There are many forms of energy that we have not yet mentioned. Most ways of storing energy can be classified according to the type of potential energy involved. Winding a clock spring, charging a battery, and growing a tree are all ways of storing energy; the energy is stored in the form of mechanical potential energy, electrical potential energy, and chemical potential energy (or biomass), respectively. Even kinetic energy can be stored, as in a flywheel. Almost everything that happens in chemistry and physics involves a transfer of energy from one form to another.

Calculating the quantities of energy involved in chemical reactions is part of the science of thermodynamics. As you can imagine, it is of utmost importance to be able to calculate the maximum amount of useful energy obtainable by burning a given amount of fuel. Problems of this nature have dramatically increased in significance during the last quarter of the twentieth century because of the sharp increase in fuel costs starting in 1973.

The Mass-Energy Relationship

In this section, we are introduced to a concept that was new and startling when first proposed by Albert Einstein in 1905: the idea that energy has mass and that every bit of matter has a quantity of energy related to its mass. This idea is fundamental to science today.

Before this century, it was believed that the mass of an object was a constant, invariable physical property of that object. Indeed, this idea was embodied in a fundamental law of nature: the **law of conservation of mass**, according to which *the mass of an object or set of objects never changes, no matter what kinds of activities are engaged in by these objects.* However, the great German physicist Albert Einstein (1879–1955) showed that one of the consequences of his famous **principle of relativity** was that *the mass of an object depends on how fast it is moving.* The faster an object moves, the greater its mass becomes, until, as it approaches the speed of light (300,000 km/sec), its mass increases without limit. This prediction of Einstein has been verified by many experiments in which very tiny particles (atoms or parts of atoms) are accelerated to extremely high speeds. We find by direct measurement that the mass of each of these particles increases precisely as expected. Indeed, every experiment done in the physics of atomic particles verifies Einstein's principle in one way or another.

Einstein explained the mass increase of the moving particles by proposing one novel but extremely important idea: *the kinetic energy of the particle has mass of its own.* The mass of a moving particle thus has two parts: (1) the normal **rest-mass** that the particle always has when at rest and (2) the mass that belongs to the kinetic energy. The moving particle therefore has more mass than the particle at rest. We call this increased mass the **total mass** of the particle. For a particle traveling near the speed of light, the total mass may be thousands of times greater than the rest-mass.

Understand that everything we have been saying about atomic particles applies to all objects. However, if you were to travel in an airplane at a speed of 500 km/hr, the extra mass due to your kinetic energy would not be big enough to measure very easily. Atomic particles are so light to begin with that it is easy to make them move very fast, and so their mass increase is easily observed. You would have to travel with a speed of many thousands of kilometers per second before your increased mass would be readily noticeable.

One deduction from Einstein's theory is that no material object can reach the speed of light because an infinite amount of kinetic energy would be required to attain that speed. And, of course, if an object is unable to reach the speed of light, it certainly cannot go faster than the speed of light. This is a fact that is lamented by those who would like to travel to the distant stars and is often ignored by writers of science fiction.

Another consequence of the theory—one verified by many detailed and precise experiments—is the idea that matter and energy are two aspects of the same basic thing. We have already stated that the kinetic energy of a moving body has mass. We can extend this idea further, saying that *every* kind of energy has mass. Then we can turn the thought around and say that *for every quantity of mass, there is an equivalent quantity of energy*. The relationship between mass and energy is given in an equation that is one of the smallest and at the same time one of the most powerful equations in all of science:

$$E = mc^2 \qquad (3\text{-}1)$$

In this equation, E stands for the quantity of energy, m stands for the mass, and c stands for the speed of light (3×10^8 m/sec). What this equation means is that *if we have a quantity of any kind of mass, the amount of energy belonging to that mass is calculated by multiplying the mass by the square of the speed of light*. If the mass is measured in kilograms and the speed of light in meters per second, then the energy is expressed in joules.

None of the reactions we use for generating power convert more than a tiny fraction of the fuel's mass into energy. In chemical reactions, the change of mass is hardly measurable. In nuclear reactions such as those used in uranium fission power plants, less than a tenth of a percent of the fuel's mass is converted into energy; in conventional fossil fuel plants, the percentage is far less.

How does the relationship between mass and energy affect the law of conservation of energy? Does the conversion of matter into energy violate the law? Is this not a creation of energy? No, it is not, for we must remember that a certain amount of rest-mass energy belonged to this bit of matter in the first place. Therefore, whenever a bit of matter is replaced by a bit of energy, the energy does not come from nowhere; it merely changes from one form into another—from rest-mass energy into heat, light, or some other form of energy. The law of conservation of energy and the law of conservation of mass can be combined in the following broader law: *in a closed system the total amount of mass is constant and the total amount of energy is constant*.

The nature of the mass-energy relationship can be summed up as follows: All the energy in the system has an equivalent amount of mass, all the mass has energy, and both mass and energy are related by the Einstein equation given above. Therefore conservation of energy implies conservation of total mass.

3.5 Measuring Quantities of Heat

Heat is not to be confused with temperature. The concept of temperature is related to "how hot" an object is, whereas the concept of heat has to do with how much thermal energy (heat energy) went into the object to make it more or less hot. A kilogram of water at a temperature of 10°C has more thermal energy in it than does a gram of water at a temperature of 20°C, even though the gram of water is at a higher temperature. (Remember that you cannot assume that 20°C is twice as hot as 10°C! You must use the Kelvin scale to make that kind of comparison.)

In popular language such distinctions are often lost. In descriptions of the first space shuttle landing, we were told over and over again by newscasters (and even by space engineers) that "the shuttle was exposed to a heat of 2000 degrees." However, degrees are units of temperature, not of heat energy. In scientific writing, it is important to be careful about words. The temperature is analogous to the *intensity* of the thermal energy.

Quantities of heat are not measured directly. What we do is make use of the observation that when heat is added to a given mass of material, the rise in temperature is proportional to the quantity of heat added. The rise in temperature will also depend on the kind of material being used. For that reason, it is convenient to use pure water as a standard. A *unit quantity of heat* is then defined to be *the amount of heat needed to increase the temperature of a unit quantity of water by one degree.*

In the metric system, there are two closely related units of heat, the calorie and the kilogram calorie. The **calorie** (abbreviated **cal**) is the amount of heat needed to raise the temperature of one gram of water by one degree Celsius—from 3.5°C to 4.5°C, to be exact. This unit is also known as the small calorie, or gram calorie, to distinguish it from the large calorie, or Calorie, more properly known as the kilogram calorie. The **kilogram calorie** (abbreviated **kcal**) is the amount of heat needed to raise the temperature of one kilogram of water by one degree Celsius. Clearly, 1 kcal equals 1000 cal. The temperature change from 3.5°C to 4.5°C is specified because the amount of heat needed to make a one-degree change in temperature varies slightly as we go from one temperature to another. However, for many practical purposes it does not matter very much whether you make the measurement at 4°C or 20°C.

In the English system, the unit of heat is the **British thermal unit (Btu)**, which is the amount of heat needed to raise the temperature of one pound of water by one degree Fahrenheit.

Quantities of heat are measured in a device called a *calorimeter*, pronounced to rhyme with pe*ri*meter, with the emphasis on the "rim." (Since heat was formerly imagined to be a fluid called "caloric," many of the words dealing with heat are built around the root *calor*.) A calorimeter is a container with walls insulated so as to make it difficult for heat to go either in or out (see Figure 3-9). A thermometer measures changes in the temperature of the water or other material within the vessel as it is heated; in the calorimeter in Figure 3-9, the heating is done by

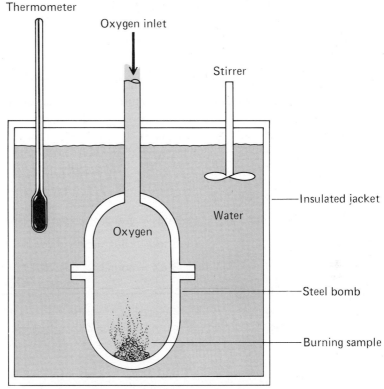

FIGURE 3-9 A calorimeter is an insulated container containing water and a thermometer to measure the change in temperature of the water. From such measurements, we can calculate the amount of thermal energy absorbed by the water.

burning some kind of fuel inside the chamber. In this way the amount of heat generated by a kilogram of fuel may be determined.

The definitions given above for the gram calorie, the kilogram calorie, and the Btu are based on the same principle: Put a unit mass of water in a calorimeter, add some heat to it, and measure the rise in temperature. For each degree of temperature rise, one unit of heat has been added. The units of mass and temperature used will determine whether the heat is measured in gram calories, in kilogram calories, or in Btu's. Traditionally, calorie units have generally been used in chemistry, whereas the Btu has been favored by power engineers in English-speaking countries.

The Calorie is certainly familiar to anybody interested in dietetics, since diet books carefully list the Calorie content of food samples. A slice of bread, when digested, will produce about 100 Calories of heat. This quantity is measured by burning the bread in a calorimeter, for it makes no difference whether the heat is produced by burning or by metabolism. It is important to know that the Calorie used by diet books is actually the kilogram calorie even though it is generally spelled with a lowercase c.

If you have more than one gram (or kilogram or pound) of water in the calorimeter, the amount of heat required for each degree change of temperature will be proportional to the mass of the water; the more water present, the more heat required. Therefore, we can write a general equation to relate change in temperature to amount of heat added. In this equation, Q is the quantity of heat, T_1 is the starting temperature, T_2 is the final temperature, and m is the mass of water being heated. The change in temperature is $T_2 - T_1$, so we write

$$Q = m(T_2 - T_1)$$

Abbreviating the change in temperature as ΔT (read as delta T), we have

$$Q = m\Delta T$$

In terms of units,

$$\text{calories} = \text{grams} \times \text{degrees Celsius}$$

$$\text{Calories} = \text{kilograms} \times \text{degrees Celsius}$$

$$\text{Btu's} = \text{pounds} \times \text{degrees Fahrenheit}$$

Example 3.1   A cubic foot of water weighs 62 lb. How many Btu's of heat does it take to warm this cubic foot of water from 70°F to 100°F?

Solution
$$\begin{aligned} Q &= m\Delta T \\ &= 62 \text{ lb} \times (100 - 70)°F \\ &= 62 \text{ lb} \times 30°F \\ &= 1900 \text{ Btu} \quad \text{(rounded off to two significant figures)} \end{aligned}$$

Example 3.2   How many calories equal 1 Btu?

Solution   In this problem we are asked to convert units from the English system to the metric system. We know that 1 Btu is the amount of heat that will increase the temperature of 1 lb of water by 1°F. We also know that a 1-kg mass weighs 2.205 lb, so 1 lb has a mass of $1/2.205 = 0.4535$ kg, and that $1°F = 5/9°C$. Therefore, we say that

$$\begin{aligned} 1 \text{ Btu} &= m\Delta T \\ &= 1 \text{ lb} \times 1°F \\ &= 0.4535 \text{ kg} \times \tfrac{5}{9}°C \\ &= 0.252 \text{ kcal} \\ &= 252 \text{ cal} \end{aligned}$$

This result shows that a Btu is roughly one-fourth of a kilogram calorie.

The above definitions of heat units originated long before it was realized that heat is a form of energy. At present the tendency is to replace the older units with the SI measurements. The basic unit of energy in the International System is the **joule, J**, which is defined in terms of mechanical work (force times distance). The joule is used for measuring heat as well as other forms of energy, but since many books still use the traditional units of heat measurement, we must become familiar with both the joule and the calorie, as well as with the relationship between them. The relationship between the calorie and the joule cannot be found by calculation, for the joule is defined by a mechanical experiment (it is the energy required to move a newton of force over a distance of one meter) whereas the Calorie is defined by a heat experiment. Therefore, the relationship between the Calorie and the joule is found by doing an experiment in which a known amount of work is converted into a measured amount of heat.

The English physicist James Prescott Joule performed a large number of such experiments during the 1840s. One of his measurements involved heating a container of water by friction with a rotating paddle wheel. In more modern experiments water is heated by passing electric current through a coil of wire immersed in the water. The best of these experiments have found that one kilogram calorie (kcal) is equal to 4184 joules. Acccordingly, one small calorie (cal) is equal to 4.184 joules. The number 4.184 J/cal is known as the mechanical equivalent of heat and is an important number in science. The fact that you get the same number no matter how you measure the amount of heat obtained from a given amount of energy is evidence that heat is one of the forms of energy.

To find out what heat really is, we must go to the atomic model of matter. The description that fits the data best is called the **kinetic theory of matter**. This theory states that all of the particles making up a piece of matter are in vigorous motion. **Thermal energy**, or heat, is nothing more than the *total* kinetic energy of all the particles in the object. **Temperature**, on the other hand, represents the *average* kinetic energy of the individual atoms or molecules. So an object is made hot by causing its atoms or molecules to move more and more rapidly. In a gas the molecules fly about in all directions, whereas in a solid they vibrate back and forth about their fixed positions within the solid. The hotter an object is, the more kinetic energy its molecules have.

What, then, is "coldness"? It is simply the absence of heat. An object is cold when its molecules are moving slowly; the faster the molecules move, the hotter the object appears. It is not correct to talk about moving "cold" from one place to another. An object is made cold by moving heat out of it. Thermal energy is what moves from one place to another.

3.6 Specific Heat

Basic Definitions

In the chemical industry, it is important to know how much thermal energy is required to heat a batch of material to the necessary working temperature. The concept of specific heat helps answer such questions.

We have already learned that it takes 1 kcal (4184 J) of thermal energy to raise the temperature of 1 kg of water by 1°C. Does it take the same amount of heat to change the temperature of other substances by one degree? The answer to this question is a straightforward *no*. Every material is different and requires a different amount of heat to raise its temperature by one degree. This characteristic of each substance is described by its **specific heat**, defined as the amount of heat needed to raise the temperature of unit mass of that substance by one degree. (By unit mass we mean one gram, one kilogram, or one pound, depending on the system of units we are using.)

In the meter-kilogram-second (MKS) system of measurement, the specific heat of a given material is the number of kilocalories needed to raise the temperature of 1 kg of that material by 1°C. In the centimeter-gram-second (cgs) system, the specific heat is the number of calories needed to raise the temperature of one gram of material by 1°C. Since 1 kcal equals 1000 cal and 1 kg equals 1000 g, the specific heat of a given substance has the same numerical value whether you measure it in the MKS or the cgs system. (For verification of this remark, see Example 3.3 below.)

In the more modern International System (SI), which uses joules as the unit of heat, specific heat is defined as the number of joules of thermal energy needed to raise the temperature of 1 kg of material by 1°C.

Finally, in the English system, the specific heat is the number of Btu's needed to raise the temperature of 1 lb of a substance by 1°F.

We can put these definitions into an equation by assigning symbols to the quantities involved: C is the specific heat, m is the mass of material being heated, and Q is the amount of heat needed to change the temperature from an initial temperature T_1 to a final temperature T_2. We abbreviate the *change* of temperature as ΔT (delta T), so $\Delta T = T_2 - T_1$. Then, by definition, we have

$$C = \frac{Q}{m\Delta T} \tag{3-3}$$

In words, the heat per unit mass (Q/m) divided by the change in temperature (ΔT) gives the specific heat. Suppose Q is the amount of heat used to raise the temperature of, say, 10 kg of a substance. Then $Q/10$ is the amount of heat that goes into each kilogram. If we divide this quantity by the total change of temperature, we get the number of kilocalories per kilogram per degree Celsius.

Example 3.3

We find in a calorimeter experiment that it requires 215 cal to heat 100 g of aluminum from 25°C to 35°C. What is the specific heat of the aluminum?

Solution

This problem is given in cgs units. The mass m is 100 g, the thermal energy Q is 215 cal, and the change of temperature is

$$\Delta T = 35°C - 25°C = 10°C$$

We put these numbers into the formula for specific heat:

$$C = \frac{Q}{m\Delta T}$$

$$= \frac{215 \text{ cal}}{100 \text{ g} \times 10°\text{C}}$$

$$= 0.215 \frac{\text{cal}}{\text{g-}°\text{C}}$$

If we want to find the specific heat in MKS units, we must convert the grams to kilograms and the calories to kilocalories. The conversion factors are 1000 g/kg and 1000 cal/kcal. We then say

$$\text{mass (kg)} = \text{mass (g)} \times \frac{1 \text{ kg}}{1000 \text{ g}}$$

$$= 100 \text{ g} \times \frac{1 \text{ kg}}{1000 \text{ g}}$$

$$= 0.1 \text{ kg}$$

In a similar manner, we find that 215 cal is equivalent to 0.215 kcal. (Once you get used to this procedure, you can convert from grams to kilograms or from calories to kilocalories simply by dividing by 1000.) We now go back to the definition of specific heat:

$$C = \frac{Q}{m\Delta T}$$

$$= \frac{0.215 \text{ kcal}}{0.1 \text{ kg} \times 10°\text{C}}$$

$$= 0.215 \frac{\text{kcal}}{\text{kg-}°\text{C}}$$

This result verifies our previous comment that the numerical value of C is the same whether we use kilograms or grams in our measuring system.

Example 3.4 What is the specific heat of aluminum in joules per kilogram per degree Celsius?

Solution We know that 1 kcal equals 4184 J. Therefore, making use of the results of Example 3.3, we have that

$$C = 0.215 \frac{\text{kcal}}{\text{kg-}°\text{C}} \times 4184 \frac{\text{J}}{\text{kcal}}$$

$$= 900 \frac{\text{J}}{\text{kg-}°\text{C}}$$

What is the specific heat of water? We know by definition that it takes 1 kcal to raise the temperature of 1 kg of water by 1°C. Therefore, the specific heat of water is simply 1 kcal/kg-°C, or 4184 J/kg-°C.

Notice that water has a specific heat nearly five times as great as that of aluminum; in other words, it takes five times more heat to raise a given mass of water from one temperature to another than it does to raise the same mass of aluminum through the same change of temperature. Water has an exceptionally high specific heat, which is why hot water cools down slowly compared to other materials. It also explains why water is a good heat transfer material and is often used in solar heating systems for storage of thermal energy.

Uses of Specific Heat

Sometimes it is necessary to find out how much heat is stored in a given amount of material when the temperature changes by a known amount. This question is easily answered by using our definition of specific heat. We solve Equation 3-3 for Q, to obtain

$$Q = Cm\Delta T \qquad (3\text{-}4)$$

In words, the amount of heat needed to change the temperature of a given body is proportional to the mass times the change of temperature. The specific heat C simply acts as a constant of proportionality.

If we divide both sides of Equation 3-4 by ΔT, we get

$$\frac{Q}{\Delta T} = Cm$$

This quantity is the amount of heat necessary to raise the temperature of a given body by one degree, and is called the **heat capacity** of the body. The specific heat is

TABLE 3-1 Specific Heats of Some Common Materials

	$\dfrac{\text{kcal}}{\text{kg-°C}}$	$\dfrac{\text{J}}{\text{kg-°C}}$
alcohol	0.586	2450
aluminum	0.215	900
carbon (diamond)	0.124	519
carbon (graphite)	0.170	711
copper	0.092	385
gold	0.0308	129
hydrogen	1.41	5900
iron	0.108	452
lead	0.038	159
oxygen	0.219	916
water	1.000	4184

the heat capacity of a single kilogram (or gram). For example, the specific heat of aluminum is 0.215 kcal/kg-°C. But the heat capacity of 100 kg of aluminum is 100 times the specific heat, or 215 kcal/°C.

Everything we have said about raising the temperature of an object by putting heat into it also applies to lowering its temperature by taking heat out of it. When we store heat in a tank of water, the water's temperature rises. When we want to use that stored heat, we remove it from the water, and in that process the water's temperature goes down. Exactly the same formulas apply to reducing temperatures as to increasing them, and the same values of specific heat apply. That is, if you must put 500 kcal into a pot of water to raise its temperature from 70°C to 80°C, then to reduce the temperature back to the original 70°C you must take 500 kcal out of the water.

Table 3-1 lists the specific heats of some common materials. This table will be used in solving problems in this chapter.

Example 3.5

We put 200 kcal into 25 kg of iron with a starting temperature of 20°C. What is the final temperature of the iron?

Solution

First solve Equation 3-4 for the change in temperature ΔT:

$$\Delta T = \frac{Q}{C \times m}$$

From Table 3-1, we see that the specific heat of iron is 0.108 kcal/kg-°C. Putting numbers into the above equation, we have

$$\Delta T = \frac{200 \text{ kcal}}{0.108 \frac{\text{kcal}}{\text{kg-°C}} \times 25 \text{ kg}}$$

$$= 74°C$$

The kcal and kg units cancel, and the °C comes up into the numerator.

The 74°C found here is not the final temperature—it is the *change* in temperature. The final temperature is the starting temperature plus the change in temperature:

$$T_2 = T_1 + \Delta T$$
$$= 20°C + 74°C = 94°C$$

Example 3.6

How many joules of thermal energy does it take to heat a tank containing 980 kg of alcohol from 20°C to 80°C?

Solution

$$\Delta T = 80°C - 20°C = 60°C$$

From Equation 3-4, we have

$$Q = Cm\Delta T$$

$$= 2450 \frac{J}{kg\text{-}°C} \times 980 \text{ kg} \times 60°C$$

$$= 1.44 \times 10^8 \text{ J}$$

3.7 Heat Changes in Chemical Reactions

Exothermic and Endothermic Reactions

Most chemical reactions involve a change in energy from one form to another. In most cases, the change is from chemical potential energy to thermal energy. There are two ways for reactions to go; a reaction may either generate heat or absorb heat. A reaction that generates, or gives off, heat is called an **exothermic reaction**; a reaction that absorbs heat from the surroundings is called an **endothermic reaction.** (In general, "exo" means outward and "endo" means inward.)

To understand why heat is given off or absorbed in chemical reactions, we must study what happens when two or more atoms join to form a molecule. Since we have not covered atoms or molecules in sufficient depth to go into these details, we will defer this topic until Chapter 15.

The most common kind of exothermic reaction is **combustion**, or burning. Examples of combustion include the combination of hydrogen with oxygen to form hydrogen oxide—which is just water—and the combination of carbon with oxygen to form carbon dioxide. So much energy results from these reactions that a great deal of heat and light is emitted; the combustion produces a flame.

We can explain this phenomenon in a rough way by considering the somewhat analogous situation of a weight falling to the ground. You may recall from our discussion in Section 3.4 that during the fall the weight's potential energy is converted into kinetic energy. When the weight hits the ground, its kinetic energy is released in the form of heat. Now imagine a pair of hydrogen atoms being attracted to an oxygen atom. These atoms rush together with great speed. When they collide, they either bounce apart or stick together. If they stick together, they form a water molecule, with two hydrogen atoms and one oxygen atom. When that happens, the kinetic energy of the colliding atoms is given to the surrounding molecules, and we observe the resulting emission of heat (see Figure 3-10).

Most fuels (such as coal, oil and gasoline) contain both carbon and hydrogen. Therefore, the burning of these fuels involves both the hydrogen-oxygen and the carbon-oxygen reactions mentioned above.

Endothermic reactions are not quite as common as exothermic reactions, but at least a few are familiar. The melting of ice might be considered an

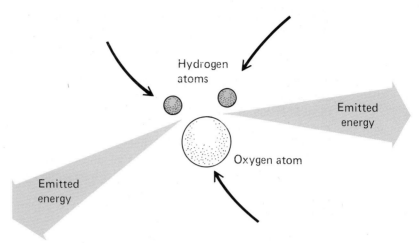

FIGURE 3-10   Hydrogen and oxygen atoms attracted to each other have extra energy to get rid of before they can stick to each other and form a water molecule. This extra energy heats up the surroundings, making all the molecules in the vicinity move faster.

endothermic reaction. To melt ice, you must add energy to the ice crystals in order to free the water molecules from their attachment to the crystal lattice so that they may move around. In fact, it requires 80 cal to melt each gram of ice. This quantity, 80 calories per gram (or 80 kilocalories per kilogram), is called the **heat of fusion** of ice. This heat must come from the surroundings, and so when ice melts it cools down everything around it.

Simply dissolving a chemical compound in water can cause either the production or the absorption of heat. The amount of heat given off or absorbed during this process is the **heat of solution** of the compound. The dissolving of sulfuric acid in water is highly exothermic; the reaction gives off a considerable amount of heat. Sodium chloride (table salt) and ammonium chloride, on the other hand, are endothermic and absorb heat; when they are dissolved in water, the solution becomes a little cooler. As a result, heat is absorbed from the surroundings. (Because of both the heat of solution and the heat absorbed by the melting ice, a mixture of salt, ice, and water has a temperature lower than that of pure ice and water. For this reason, such a mixture has long been used to cool the ingredients during the making of ice cream.)

Heat of Combustion

Combustion is a self-sustained exothermic reaction in which heat is given off so intensely that it keeps the reaction going. The common combustion reactions involve the combination of carbon and/or hydrogen with oxygen, although many other kinds of reactions are possible.

Generally the fuel that goes into a combustion reaction must be ignited—it must be heated to a temperature high enough for the reaction to begin. A piece of coal will do nothing until it is ignited, even though it is surrounded by oxygen. Even hydrogen, which is highly flammable, will mix quietly with oxygen until some spark sets it off.

This situation suggests the following model. In a mixture of hydrogen and oxygen, the molecules move around, occasionally bumping into one another. Surrounding each molecule is a kind of barrier—a *potential barrier*—that prevents them from actually reacting unless they are traveling fast enough to get past it (see Figure 3-11). Heating the gas to a high temperature makes the molecules move fast enough to get over the barrier. Then they start joining to form water molecules, giving off large amounts of energy as they do so. This energy heats the surrounding gas, and the reaction spreads through the remainder of the gas. If the reaction occurs extremely rapidly, we call it an explosion.

FIGURE 3-11 Two molecules coming together may repel each other, so they cannot actually come in contact unless they are traveling fast enough to overcome the repulsion. The situation is somewhat like that of a pair of balls rolling toward each other over a hill—they cannot reach each other and collide unless they are moving fast enough to get over the hump. In chemistry this hill or barrier is called the **activation energy** of the reaction.

An important quantity is the **heat of combustion** of a reaction. It is defined simply as the amount of heat given off when a unit mass of a given substance is burned. We measure heat of combustion in terms of the number of joules (or kilocalories) of thermal energy given off when a kilogram of the given material is burned. The symbol used for heat of combustion is ΔH (delta H). The delta refers to the *change* in the internal energy of the molecules due to the combustion. For example, the heat of combustion of hydrogen is given as

$$\Delta H = -1.21 \times 10^8 \text{ J/kg}$$

This is the number of joules of heat given off by the combustion of 1 kg of hydrogen in oxygen. The minus sign refers to the fact that the kilogram of hydrogen *loses* 1.21×10^8 J of energy when combining with oxygen to form water.

We might also call ΔH the **heat of formation** of the water molecule. However, we must be careful to describe the heat of formation as the number of kilocalories *per kilogram of water formed*, rather than per kilogram of hydrogen burned. (It is very important to be aware of what units are being used when you refer to any table of heat quantities.) Table 3-2 lists the heats of combustion of a few common materials.

A number of observations can be made about Table 3-2. First, hydrogen has by far the greatest heat of combustion of all the materials listed. In fact, it is near the head of the list of known combustibles. This is why hydrogen is commonly used as a rocket fuel. Octane (one of the constituents of gasoline) has much less energy per

TABLE 3-2 Heats of Combustion (energy per kilogram of material burned)

	kcal/kg	J/kg
carbon	7,800	3.26×10^7
ethyl alcohol (ethanol)	7,100	2.97×10^7
hydrogen	28,900	1.21×10^8
octane	11,400	4.77×10^7
TNT (trinitrotoluene)	3,610	1.51×10^7

kilogram than does hydrogen, and ethyl alcohol has even less. We see here one of the drawbacks of alcohol as a fuel—weight for weight it contains less energy than gasoline. TNT, usually considered a powerful explosive, makes a surprisingly weak showing. What makes TNT an explosive is not so much the amount of energy it releases but the speed of its combustion.

Example 3.7   How much energy is produced by the combustion of 5.6 kg of octane?

Solution   We write the equation for the production of thermal energy as follows:

$$\text{energy} = \text{heat of combustion} \times \text{mass of fuel}$$

or

$$E = \Delta H \times m$$

From Table 3-2, the heat of combustion of octane is 4.77×10^7 J/kg. Therefore, the thermal energy produced by 5.6 kg is

$$\text{energy} = 4.77 \times 10^7 \, \frac{\text{J}}{\text{kg}} \times 5.6 \text{ kg}$$

$$= 2.67 \times 10^8 \text{ J}$$

Example 3.8   How much hydrogen must be burned to obtain 5×10^{10} J of energy?

Solution   Solving the equation $E = \Delta H \times m$ for m, we have

$$m = \frac{E}{\Delta H}$$

$$= \frac{5 \times 10^{10} \text{ J}}{1.21 \times 10^8 \, \frac{\text{J}}{\text{kg}}}$$

$$= 4.13 \times 10^2 \text{ kg} = 413 \text{ kg}$$

Note that the units are arranged so that the joules cancel and the kg unit comes up to the numerator.

Glossary

amorphous materials Substances that lack an orderly internal structure.

British thermal unit (Btu) The amount of energy needed to raise the temperature of 1 lb of water by 1°F.

calorie (cal) The amount of thermal energy needed to raise the temperature of 1 g of water by 1°C (from 3.5°C to 4.5°C); gram calorie.

Calorie See **kilogram calorie**.

chemical change The changing of one substance into another, or the combining of two or more substances to make other substances; chemical reaction.

chemical potential energy Energy stored in the molecules of chemical compounds, resulting from the binding of the atoms within the molecules.

chemical properties Properties that describe how a substance behaves chemically.

combustion An exothermic reaction that releases enough energy to keep the reaction going by itself; a self-sustained reaction.

compound A substance composed of molecules that contain two or more different types of atoms.

condensation The process whereby gas turns into a liquid as it cools because the slowing down of the gas molecules allows the intermolecular attraction to have greater effect.

conservation of energy, law of The rule that in a closed system the total amount of energy is constant—that is, unchanging with time. (Since the mass is proportional to the energy, this means that the total mass is also constant.)

conservation of mass, law of The rule that in a closed system the total amount of mass is constant.

crystal A body with a symmetrical structure, whose atoms or molecules are arranged in an orderly geometrical pattern.

crystal lattice An orderly arrangement of atoms or molecules within a solid.

crystalline material A substance built of crystal lattices.

element A substance containing only one kind of atom; a material that cannot be put together from atoms of other substances.

endothermic reaction A reaction that absorbs heat from its surroundings.

evaporation A process whereby molecules of a substance (usually a liquid) separate from the surface of that material and form a vapor.

exothermic reaction A chemical reaction that releases thermal energy into the surroundings.

foot-pound The work done by a force of 1 lb acting through a distance of 1 ft.

gas A physical state of matter in which the molecules are relatively far apart and interact weakly (or not at all) with each other. A gas will expand to fill any container regardless of its size or shape.

heat See **thermal energy**.

heat capacity The amount of heat required to raise the temperature of an object by 1°C.

heat of combustion The amount of heat given off by the combustion (or burning) of a unit mass of a given substance.

heat of formation The amount of heat either given off or absorbed when a unit mass of a compound is formed out of its elements.

heat of fusion The amount of heat needed to melt a unit mass of a given substance.

heat of solution The heat given off or absorbed when one substance is dissolved in another substance.

heterogeneous material A substance having two or more kinds of materials mixed together.

homogeneous material A substance made up of only one kind of material; the opposite of heterogeneous.

inert substance A substance that does not react with other substances.

joule (J) A unit of work or energy in the metric system; the amount of work done by 1 N (newton) of force acting over a distance of 1 m.

kilogram calorie (kcal) The amount of thermal energy needed to raise the temperature of 1 kg of water by 1°C; 1000 cal (gram calories).

kinetic energy Energy an object has because of its motion.

kinetic theory of matter The idea that matter is made up of small particles (molecules) in constant motion, and that the higher the temperature, the more rapidly the molecules are moving.

liquid A physical state of matter characterized by the ability of a quantity of the substance to flow and change shape while keeping its volume constant.

liquid crystal A material whose molecules are relatively free to move in one direction but are not so free to move at right angles to that direction.

mass, total The mass of an object, consisting of its rest-mass plus the mass due to its kinetic and potential energy.

melting point The temperature at which a substance changes from a solid to a liquid when heated.

molecule A structural unit consisting of a definite number of atoms bound together.

physical properties Characteristics that describe an object or a substance without reference to chemical changes.

plasma A state of matter found in gases raised to extremely high temperatures.

potential energy Energy stored so that it can be recovered in the future.

pure substance A material made up of one kind of atom or molecule.

reactive substance A substance that reacts readily with other substances.

relativity, principle of A theory of matter and energy, space and time, two of whose most important consequences are (1) that matter and energy are equivalent and (2) that the mass of an object increases when it is in motion.

rest-mass The mass of an object when it is at rest.

solid A state of matter characterized by the ability of the substance to retain its size and shape because its atoms are bound closely together.

solubility The quantity of solute that will dissolve in a given amount of solvent.

solute A substance that is dissolved in another substance.

solvent A liquid in which another substance is dissolved.

specific heat The amount of thermal energy needed to raise the temperature of a unit mass of a substance by 1°.

temperature A property of an object proportional to the average kinetic energy of its molecules.

thermal energy The total kinetic energy of the random motion of the molecules within an object; heat.

work Transformation of energy by a force when it moves an object over a distance; also called mechanical work. The amount of work equals the force times the distance (when the direction of motion is parallel to the applied force).

Problems and Questions

3.1 The Nature of Matter

3.1.Q Define and explain the following:
  a. chemical change
  b. physical change

3.2.Q Give two examples of chemical changes with which you are familiar. Explain why they are considered to be chemical changes rather than physical changes.

3.3.Q Is the boiling of water a chemical change? If not, why not?

3.4.Q In what way did the Greeks indulge in supernatural thinking? How did this thinking differ from the more scientific thought of Democritus?

3.5.Q a. Describe how Democritus's speculations about water drops led to the concept of the atom.

  b. Use the concept of the atom to explain the behavior of water—how it flows, how it boils, etc.

3.6.Q a. How many different kinds of atoms does modern science recognize?

b. How do we explain the existence of millions of different kinds of substances?

3.7.Q List a number of physical properties of each of the following materials:
    a. a brick   **b.** a paper clip
    c. a glass of water

3.8.Q Which of the properties you listed in the answer to Question 3.7 are specific properties of the particular object and which are general properties of the material of which the object is made?

3.9.Q Which properties do the following substances have in common? What properties make them different?
    a. water   **b.** milk
    c. mercury   **d.** coffee
    e. chicken soup   **f.** gasoline

3.10.Q Classify each of the following as either a chemical change or a physical change, and give reasons for your choice.
    a. the melting of ice
    b. the change of dough into bread during baking
    c. the breaking of a glass bottle
    d. writing with a pen on a piece of paper
    e. the growth of an acorn into an oak tree
    f. the spreading of peanut butter on bread
    g. the digestion of food
    h. the slow change in the color of a bronze statue from gold to green

3.2 States of Matter

3.11.Q Define each of the following, and give an illustration.
    a. solid state   **b.** liquid state
    c. gaseous state   **d.** plasma state
    e. liquid crystal

3.12.Q Give the state of each of the following substances under normal conditions. If the substance is a solid, tell whether it is crystalline or amorphous.
    a. glass   **b.** butter
    c. salt   **d.** lubricating oil
    e. diamond   **f.** air
    g. outer space   **h.** the glowing vapor inside a neon sign

3.13.Q What is a molecule?

3.14.Q What happens to the motion of the molecules of a substance when it is heated?

3.15.Q Explain why a substance is a solid, a liquid, or a gas, depending on the motion of its molecules.

3.16.Q Substance A has molecules that bind together more strongly than the molecules of substance B. Which of these materials do you think would have the higher melting point? Explain your answer.

3.17.Q Explain why a gas does not have a fixed volume.

3.18.Q What kinds of molecules are found in liquid crystals?

3.19.Q Why do we buy milk by the quart rather than by the yard?

3.20.Q Several ice cubes are floating in a glass of soda. Describe the physical state of the glass and of each of the substances in it. Describe what changes are taking place in those substances.

3.3 Types of Substances

3.21.Q Define the following, and give illustrations of each.
    a. an element
    b. a compound
    c. a diatomic molecule
    d. a homogeneous material
    e. a heterogeneous material
    f. an inhomogeneous material

3.22.Q Classify each of the following as either homogeneous, heterogeneous, or inhomogeneous.
    a. the contents of an oxygen tank
    b. a chocolate bar with nuts
    c. an iron nail
    d. gelatin containing fruit salad
    e. a concrete block

3.23.Q The labels on some milk bottles claim that the milk inside is *homogenized*. What is the meaning of this statement? (Look in a dictionary, if necessary, for clues.)

3.24.Q How could you separate a mixture of sand and salt without picking out the separate grains?

3.25.Q What elements have atoms that do not combine with other atoms?

3.26.Q What elements have molecules that consist of two identical atoms?

3.27.Q Powdered iron and powdered graphite are stirred together in a crucible (a ceramic bowl). What is the physical state of the iron and graphite combination? What happens to this combination when the crucible is heated to such a high temperature that the iron melts?

3.28.Q Powdered coffee dissolves in hot water. Which is the solvent, and which is the solute?

3.29.Q What is the purpose of making alloys?

3.30.Q Name the following alloys:
    a. iron-carbon
    b. copper-zinc
    c. copper-tin

3.4 Energy

3.31.Q What is the relationship between energy and motion?

3.32.Q What is work?

3.33.Q Give three examples showing how work is done to set something in motion. Identify the motive force in each case.

3.34.Q It takes energy to set an object into motion. Once it is moving at a constant speed, is any more energy needed?

3.35.Q Define and give illustrations of each of the following:
    a. kinetic energy
    b. potential energy
    c. the law of conservation of energy

3.36.Q Which of the following are examples of potential energy, and which are examples of kinetic energy? Do any involve both?
    a. a wound-up clock spring
    b. a piece of burning wood
    c. an exploding bomb
    d. a flashlight battery
    e. a beaker of hot water
    f. a parachutist falling through the air

3.37.Q Where did most of the energy on the earth come from?

3.38.Q Explain what is meant by a perpetual motion machine. Why is such a machine impossible to build?

3.39.Q Consider the following device for operating a motor boat: A gasoline engine gets the boat started. Once the boat is moving, water comes into a funnel in the front end of the boat, causing a turbine to spin, and the shaft from this turbine is geared to the propeller in the back of the boat. The propeller pushes the boat through the water, and the gasoline engine is no longer needed.
    a. Explain what is wrong with this idea.
    b. Is this a perpetual motion machine?

3.40.Q Name three different fuels in common use. What is the function of a fuel?

3.41.Q What does food have in common with fuels?

3.42.Q What are the two main functions of food as far as energy is concerned?

3.43.Q A carbon battery will power a radio for 5 hr, whereas a nickel-cadmium battery will power a radio for 20 hr. Explain this difference in terms of energy concepts.

3.44.Q A microwave oven can boil water in a fraction of the time that it takes a conventional oven. Explain this effect in terms of the concepts of energy and molecular motion.

3.45.Q Describe the changes from one form of energy to another in each of the following examples.
    a. The driver of a car jams on the brakes, and the car quickly stops, its tires smoking.
    b. A coal furnace powers a steam locomotive that pulls a 100-car train.
    c. A camera takes a picture using the light from a photo-flash lamp powered by a small battery.

3.46.Q State some of the consequences of the principle of relativity.

3.47.Q What is meant by the equivalence of mass and energy?

3.48.Q Is the conversion of mass into energy in a nuclear power plant a violation of conservation of mass or of conservation of energy? Explain the reasons for your answer.

3.49.Q What is meant by rest-mass? How does it differ from the total mass of an object?

3.50.Q State the combined law of conservation of mass and energy. How does it differ from the earlier separate laws?

3.5 Measuring Quantities of Heat

3.51.Q Define the following and give illustrations of each.

a. the Calorie   b. the calorie
  c. the Btu   d. the joule
  e. heat   f. temperature
  g. hot and cold

3.52.Q Explain the difference between heat and temperature.

3.53.P How many joules are equivalent to 7500 kcal?

3.54.P A liter of water weighs 1000 g. How many of each of the following units are required to heat the liter of water from 0°C to 100°C?
  a. calories   b. kilogram calories
  c. Btu's   d. joules

3.55.P An 8-oz cup of tap water at 15°C is heated to the boiling point of water, 100°C. (*Note*: An 8-oz cup holds about 300 cm^3, or 300 g, of water.) How many of each of the following units are required?
  a. calories   b. kilogram calories
  c. Btu's   d. joules

3.56.P A pot containing 750 cm^3 of water at 0°C is heated to 90°C. The density of water is 1 g per cm^3. How many of each of the following units are required?
  a. calories   b. kilogram calories
  c. joules

3.57.P A cup of water (300 cm^3) is heated from 72°F to 212°F. How many of each of the following units are required?
  a. Btu's   b. kilogram calories
  c. joules

3.58.P A human body contains 100 lb of water. During an illness a certain person's body temperature goes from 98.6°F to 104°F. Calculate the number of each of the following units required to raise the temperature of the body water:
  a. Btu's   b. calories
  c. kilogram calories   d. joules

3.59.P The person in Problem 3.58 is placed in a tub of cold water, and the 104°F temperature is reduced to 99°F. If the cold water in the tub has its temperature raised by 2°F, what is the mass of water in the tub?

3.60.Q When a door is opened during winter months, we often hear: "Close the door; the cold is coming in." Is this a meaningful scientific statement?

3.6 Specific Heat

3.61.Q Define the following, and give illustrations of each.
  a. specific heat   b. heat capacity

3.62.P a. How many joules are equivalent to 250 kcal?
  b. How many joules are equivalent to 1980 cal?
  c. How many kilogram calories are equivalent to 3.87×10^9 J?

3.63.P A 100-g sample of gold is warmed from 18°C to 32°C. How much heat (in calories and in joules) is required to make this change?

3.64.P A 100-g sample of iron is warmed from 18°C to 32°C.
  a. How much heat is needed for this operation?
  b. Compare the amount of energy needed to heat the iron with the amount of energy needed to heat the gold in Problem 3.63. Give the answer as a ratio.

3.65.P If the amount of energy from Problem 3.64 was used to heat the 100 g of gold from Problem 3.63, what would be the final temperature of the gold?

3.66.P If 150 kcal raises the temperature of a 50-kg sample of material by 35°C, what is the specific heat of the material?

3.67.P A heater contains 2500 L of water at room temperature (20°C). How many joules of thermal energy are required to bring this water to its boiling point (100°C)?

3.68.P A 100-g sample of hydrogen and a 100-g sample of oxygen are both heated in such a way that each absorbs 5000 kcal of heat. If they start out at the same temperature, which one ends up at the higher temperature?

3.69.P A 400-g sample of lead is cooled from 75°C to 45°C in a container of water. What was the original temperature of the water? (*Hint*: The heat lost by the lead equals the heat gained by the water, and the lead and the water end up at the same temperature.)

3.70.P A solar heater warms water from 10°C to 80°C. How many joules of energy are absorbed from the sun by 1150 kg of water during this process?

3.7 Heat Changes in Chemical Reactions

3.71.Q Define the following terms, and give illustrations of each.
  a. exothermic reaction
  b. endothermic reaction
  c. combustion reaction

d. heat of fusion
　　e. ignition
　　f. heat of combustion
　　g. heat of formation

3.72.Q Which of the following are endothermic reactions, and which are exothermic reactions?
　　a. a log burning in a fireplace
　　b. salt dissolving in water
　　c. sulfuric acid mixing with water

3.73.P How much energy is produced by the burning of 4 kg of carbon? (Most of coal is carbon.)

3.74.P How much ethyl alcohol (ethanol) must be burned in order to obtain 6.8×10^{11} J of energy?

3.75.P
　　a. How much energy is obtained from burning 10 kg of octane?
　　b. How much ethanol must be burned to obtain the same amount of energy?
　　c. How much TNT must be burned to obtain the same amount of energy?

Self Test

1. Is each of the following processes a chemical change or a physical change? Explain the reasons for your answers.
　　a. the burning of a candle
　　b. the boiling of water in a kettle
　　c. the expansion of hot air in a balloon
　　d. the dissolving of salt in water
　　e. the production of oil in a refinery

2. Using the concept that a gas consists of molecules, describe the following events.
　　a. Fuel burns inside the cylinder of an automobile, producing hot gases.
　　b. The gas expands, pushing against the piston.

3. According to the atomic model of matter, what happens when two chemical compounds react with each other?

4. According to the atomic, or molecular, model of matter, what happens
　　a. when a liquid boils?
　　b. when a solid melts?

5. Describe what happens to the kinetic energy and to the potential energy of a ball when you throw it up into the air and it falls back to the earth.

6. The work of Albert Einstein revolutionized scientific thought early in the twentieth century. State two important results of his work, giving examples of a practical nature.

7. In common language, when we speak of "conserving energy," we mean saving energy by being efficient in its use. What do we mean in science when we speak of "conservation of energy"?

8. Describe in terms of the atomic model why the combustion of hydrogen generates heat.

9. A tank contains 2400 kg of water at a temperature of 90°C. This water is mixed with cold water whose temperature is 10°C. The final mixture has a temperature of 60°C.
　　a. How much thermal energy did the hot water lose while being cooled?
　　b. Assuming that all of this thermal energy went to warm up the cold water, how much cold water was there?

10.
　　a. How much energy is obtained from burning 5 kg of hydrogen?
　　b. How much octane must be burned to obtain the same amount of energy?

11. Explain why a mixture of hydrogen and oxygen does not burn at room temperature but will start burning if it is heated to a high enough temperature.

12. Two kilograms of octane is burned, and its thermal energy is used to heat a tank of water from 20°C to 75°C.
　　a. How much energy is obtained from the octane?
　　b. How much water is heated?

13. What actually happens when an object is cooled?

14. Do we measure quantities of heat directly? Explain how we go about measuring quantities of thermal energy.

4
Elements and Compounds

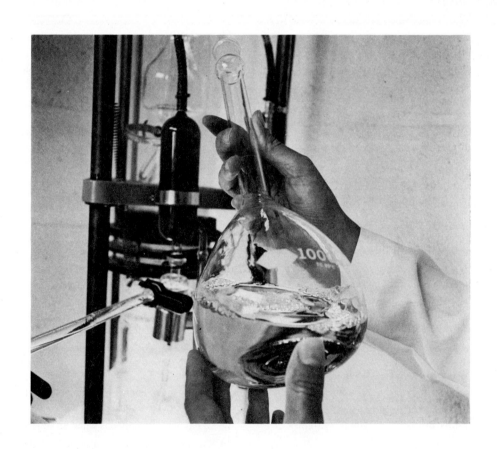

Objectives After completing this chapter, you should be able to:
1. Recognize the symbols for the more common elements.
2. Read simple chemical equations.
3. Recognize the Latin roots for names of elements.
4. Recognize the difference between an element and a compound.
5. Predict the consequences of the principle of conservation of mass.
6. Use the law of definite proportions to solve simple problems.
7. Calculate the percentage composition of simple compounds.
8. Use the law of multiple proportions to predict the possible formulas of simple compounds.

4.1 The Elements and Their Symbols

Chemists have developed a code for writing equations to represent chemical reactions. An example of such an equation is

$$C + O_2 \longrightarrow CO_2 \tag{4-1}$$

This equation tells us that when the element carbon burns, it combines with oxygen to form carbon dioxide. The letter C represents a carbon atom, and the letter O represents an oxygen atom. The combination O_2 stands for a molecule of oxygen (with two atoms); CO_2 represents a molecule of carbon dioxide (two atoms of oxygen joined to one carbon atom). Note how the equation gives information about the number of atoms taking part in the reaction: it says that during the combustion each atom of carbon combines with one molecule of oxygen to form one molecule of carbon dioxide. The left side of the equation shows the atoms or molecules entering into the reaction (the **reactants**), and the right side of the equation shows the results of the reaction (the **products**).

When we read this equation aloud, we say, "Carbon plus oxygen yields carbon dioxide." The arrow means "yields" rather than "equals." Equation 4-1 is not really an equation in the mathematical sense, but a statement describing an interaction between two substances. It tells us what ingredients go into the interaction and what products come out.

In general, each element in a chemical equation is represented by a symbol. For carbon, we use the letter C; for oxygen, the letter O. Since a molecule of oxygen has two atoms in it, the subscript 2 is used to show the number of atoms of oxygen that are present. To write the formula for carbon dioxide, we simply put the O after the C and use the subscript 2 to show that there are two atoms of oxygen attached to each atom of carbon. The order in which the symbols are written is important; you will never see O_2C for carbon dioxide. (Some of the rules for ordering the symbols will be discussed in Chapter 8.)

To write chemical equations successfully, you must first be familiar with the symbols for the elements. (However, some of the elements are much more common than others, so it is not necessary to memorize them all at once. Familiarity will come with use.) In many cases only the first letter of the element name is used. When there is a chance of confusion, another letter is added to the first. For example, C is enough to indicate carbon, but we must use Ca for calcium, Cl for chlorine, Cr for chromium, Cs for cesium, and Cd for cadmium. Note that the first letter is always a capital letter, and the second is always a lowercase letter. (A few of the most recently created elements have been assigned three letters, under a new system authorized by the International Union of Pure and Applied Chemistry.)

In some cases the symbol for the element is not obtained by abbreviating the element's English name but instead comes from the Latin name used during the Middle Ages, when Latin was the universal language of science. (The one exception is tungsten, whose symbol W comes from the Germanic *wolfram*.) Table 4-1 lists those elements whose symbols derive from their Latin or German names.

4.1 THE ELEMENTS AND THEIR SYMBOLS

TABLE 4-1 Derivation of Symbols of Some Common Elements

Symbol	Latin	English
Ag	argentum	silver
Au	aurum	gold
Cu	cuprum	copper
Fe	ferrum	iron
Hg	hydrargyrum	mercury
K	kalium	potassium
Na	natrium	sodium
Pb	plumbum	lead
Sb	stibium	antimony
Sn	stannum	tin

Symbol	German	English
W	wolfram	tungsten

Some of the Latin roots are also the source of familiar words. A plumber is so named because he uses solder made largely of lead, and the Latin word for lead is *plumbum*. The Latin word for iron, *ferrum*, has given us several words. Alloys that have iron in them are known as *ferrous* alloys. Ferroconcrete is concrete reinforced with iron rods. Ferromagnetism relates to the magnetic properties of iron. (The Ferris Wheel, however, is named for its inventor, G. W. G. Ferris, and not for the iron that makes up most of its structure.)

The fact that we use Latin names for certain elements indicates that these elements have been known for a long time. Indeed, many of the elements were known before the time of Rome. Gold (*aurum*, Au) was probably the first metal known, since it can be found in its pure state in the sand of some rivers. This is why people go panning for gold in riverbeds. Ornaments of gold and silver (*argentum*, Ag) dating from the latter part of the Stone Age (the Neolithic Age) have been found in Egypt. Copper (*cuprum*, Cu) and antimony (*stibium*, Sb) utensils have been found dating back to 3000 B.C., and mercury (*hydrargyrum*, Hg) was known to the Egyptians by 1500 B.C.

Prior to the seventeenth century there was no concept of "element" as we now know it. The Greek philosophers had the idea that matter is composed of fundamental or "primal" elements, but what they had in mind were such "elements" as earth, air, fire, and water. Later they added a fifth element—the heavens—which they called *quintessence* (fifth essence). The concept of a chemical element in the modern sense did not arise until much later.

Because nothing was known about the nature of the elements, it was reasonable for people to think that there ought to be some way to change common materials into precious gold. Changing a base element into gold was known as

transmutation, and during the Middle Ages there arose a widespread search for the "philosopher's stone," whose touch was supposed to effect this miraculous transmutation. This quest was a part of **alchemy**, which, along with astrology, had been a popular activity as far back as the era of the Babylonian civilization.

The alchemist was the forerunner of the chemist. We now think of alchemy as a pseudoscience, since it was based on a maximum of speculation and a minimum of observation—a poor foundation for a proper science. The study of alchemy did lead to some practical chemistry along the way, though. At the end of the Middle Ages, when all of science took on a more rational foundation, alchemy evolved into chemistry.

The concept of a "chemical element" took many centuries to take form, however. The English scientist Robert Boyle used the term in its modern sense for the first time in a book called *The Sceptical Chymist,* published in 1661. To Boyle, an element was a material that could not be decomposed into something simpler. Until the beginning of the twentieth century, when individual atoms began to be broken into smaller pieces, that definition was useful.

In 1774, an English chemist, Joseph Priestly, isolated pure oxygen for the first time, but he did not recognize it as an element. Shortly thereafter, in 1777, the great French chemist Antoine Lavoisier first formulated the correct theory of combustion: the idea that when substances burn they combine with the oxygen of the air. Lavoisier and his co-workers created a revolution in chemistry by using chemical equations for the first time and by inventing and using a rational system for naming the chemical compounds instead of calling them by the common names that had been used previously.

In his book *Elementary Treatise on Chemistry* (1789), Lavoisier included a list of 33 elements, among which were found such entities as light and heat. This may seem amusing to us, but the idea was not unnatural in a period when there was no real understanding of the difference between matter and energy and no knowledge of the fact that light and heat are simply two forms of energy.

One of Lavoisier's most important contributions to chemistry was his insistence that chemical experiments be done quantitatively—that the quantities of materials going into and coming out of a reaction be measured. It was by measuring that he was able to verify the principle of conservation of mass, the idea that the total mass of the materials going into a reaction is the same as the total mass of the materials coming out of the reaction. In other words, *matter cannot be created or destroyed.*

For example, when wood is burned in a closed container filled with pure oxygen, the mass of the container and its contents does not change during the reaction. The burning wood combines with oxygen to form carbon dioxide and water (because the wood contains both carbon and hydrogen). The mass of wood plus oxygen before the reaction equals the mass of carbon dioxide plus water plus ashes after the reaction.

In the next section, we will deal with some of the important consequences of Lavoisier's new quantitative approach to chemistry.

4.2 The Law of Definite Proportions

One of the important tasks of the chemist is to devise methods for determining the composition of the many materials found in the world. There are three kinds of questions involved: (1) What elements are present in the material? (2) How are the elements arranged? (What compounds are present and what is the crystal structure, if any?) (3) What are the quantities, or proportions, of the elements in these various substances?

Lavoisier, as we have mentioned, was one of the first to pay attention to the importance of measuring quantities; he was the founder of **quantitative analysis** (as opposed to **qualitative analysis**, in which we simply determine what elements are present in an unknown material). However, during Lavoisier's lifetime, methods of analysis were very imprecise, and the poor results obtained gave rise to many controversies and misunderstandings. One important controversy took place at the very end of the eighteenth century, when the science of chemistry was just taking its modern form and good facts were hard to find. The question was this: When a chemical compound is formed, what are the proportions, or relative amounts, of the elements within the compound? (Remember that this controversy took place before the idea of atoms and molecular structure had been clearly established.)

The cast of characters in the argument consisted of two French chemists: Claude Louis Berthollet and Joseph Louis Proust. Berthollet studied reactions such as the one that takes place when powdered copper is heated with sulfur in a test tube. The result of this reaction is the compound copper sulfide, a black, crystalline material. Berthollet made a large number of samples of copper sulfide, using various amounts of sulfur in the reaction vessel. When he analyzed the samples, Berthollet found that the amount of sulfur in a given amount of his copper sulfide varied over a wide range. As a result, he insisted that the percentage of sulfur in copper sulfide was a variable quantity. Proust, on the other hand, believed that the percentage of sulfur in a sample of *pure* copper sulfide would always be the same. He explained Berthollet's results by saying that there were perhaps two or three kinds of copper sulfide in existence, and that Berthollet was actually analyzing *mixtures* of those different kinds of copper sulfides in various proportions.

Proust turned out to be correct. As a result of numerous experiments, he formulated a law usually called the **law of definite proportions**, sometimes known as the law of constant composition. Proust's law states that *a pure compound is always composed of the same elements, which are always present in the same proportions by weight.* Consider the compound mercuric oxide, which played an important role in the discovery of oxygen. Mercuric oxide is a red powder that decomposes when heated, giving off oxygen and leaving pure mercury behind. If we weigh a sample of mercuric oxide, heat it, and then weigh the remaining mercury, the ratio of the mercury's mass to the original sample's mass is the proportion of mercury in the compound.

Consider specific numbers: 50.0 g of mercuric oxide is heated until decomposed, at which point 46.3 g of mercury is left in the container. The proportion of mercury in the mercuric oxide is found by taking the ratio of the two numbers, as follows:

$$\text{proportion of mercury} = \frac{\text{mass of mercury (part)}}{\text{mass of compound (whole)}}$$

$$= \frac{46.3 \text{ g}}{50.0 \text{ g}} = 0.926$$

The percentage of mercury is found by multiplying the proportion by 100:

$$\text{percentage of mercury} = 0.926 \times 100 = 92.6\%$$

The same percentage will be found if the measurement is repeated on other samples of pure mercuric oxide. The conclusion is that any sample of pure mercuric oxide will always have 92.6% mercury in it, as the law of definite proportions predicts.

Now go one step further. The percentage of mercury plus the percentage of oxygen in any sample must equal 100%. Thus, the percentage of oxygen in the above sample is found as follows:

$$\text{percentage of oxygen} = 100\% - 92.6\% = 7.4\%$$

Once the law of definite proportions was established, chemists had a clearer idea of what a pure chemical compound is. It is a substance that always has the same proportion (by mass) of its individual elements. Water, for example, always has 8 parts of oxygen to 1 part of hydrogen. (That is, every sample of water has 8 grams of oxygen to each gram of hydrogen.) Carbon dioxide always has $2\frac{2}{3}$ parts of oxygen to 1 part of carbon (32 g of O to 12 g of C).

The law of definite proportions is of great importance to chemical manufacturers, for it assures the manufacturers that when they make a ton of a particular chemical compound, the manufacturing process will always require the same amount of ingredients in the same proportion.

The fact that these proportions are rather simple numbers (for example, 8 parts of O to 1 part of H in water) gave rise to a most important piece of reasoning very early in the nineteenth century and ushered in the age of modern chemistry.

Example 4.1 The mineral quartz has the chemical name silicon dioxide (SiO_2) and contains 46.7% silicon by mass. (a) What is the proportion of oxygen in this compound? (b) How many kilograms of silicon are in 250 kg of quartz?

Solution (a) The percentage of oxygen is

$$100\% - 46.7\% = 53.3\% \text{ oxygen}$$

That is, 0.533 of the quartz mass consists of oxygen.

(b) By definition,

$$\text{proportion of silicon} = \frac{\text{mass of silicon}}{\text{mass of quartz}} = 0.467$$

Therefore, solving for the mass of silicon, we have

$$\text{mass of silicon} = \text{proportion of silicon} \times \text{mass of quartz}$$

$$= 0.467 \times 250 \text{ kg} = 117 \text{ kg}$$

Example 4.2 A sample of sodium chloride (NaCl) is analyzed and found to contain 69.0 g of sodium and 106.5 g of chlorine. How many kilograms of sodium can we obtain from 75.0 kg of sodium chloride?

Solution First, find the proportion of sodium in the compound:

$$\text{proportion of Na} = \frac{\text{mass of Na}}{\text{mass of NaCl}}$$

$$= \frac{69.0 \text{ g}}{69.0 \text{ g} + 106.5 \text{ g}} = 0.393$$

Then the amount of sodium in 75 kg of sodium chloride is

$$\text{mass of Na} = \text{proportion} \times \text{mass of NaCl}$$

$$= 0.393 \times 75.0 \text{ kg} = 29.5 \text{ kg}$$

4.3 Atomic Theory and the Law of Multiple Proportions

There were few theoretical ideas involved in the arguments concerning the composition of chemical substances described in the previous section. The arguments were empirical—that is, based on the results of measurements—and some of the measurements were not particularly accurate. Because they did not know which measurements were correct, it was hard for the chemists of the eighteenth century to agree on the composition of the substances with which they worked. (By composition, we mean the relative amounts of each of the elements making up a compound.) Part of the trouble was caused by the fact that scientists did not yet know that gases such as oxygen and nitrogen have diatomic molecules. Even the composition of water was unknown.

However, science is not just a matter of making measurements and finding patterns in the results of those measurements. An important part of science is the invention of theories whose underlying concepts give meaning to the observations.

Early in the nineteenth century, there came into existence a guiding principle that chemists could use to determine the composition of compounds. With this theory, scientists were no longer groping in the dark; they had a definite framework on which to fit their measurements.

The theory originated in the inventive mind of John Dalton, an English chemist who did much analytical work but whose greatest accomplishment was one of imagination. He visualized in great detail the microscopic world within matter, postulating that every substance is composed of atoms held together in little clusters called molecules.

As we have seen, Dalton was not the first to hypothesize the existence of atoms. Democritus had taught the atomic theory in the fourth century B.C., and in the seventeenth century the great Isaac Newton had entertained his own ideas about atoms as the basis of material structure. Dalton's theory differed from the previous speculations in one important way: it made specific predictions that could be compared with the results of experiments. Earlier theories had been so vague and general that they could not be related to any observable facts. *The ability to make specific predictions that can be tested against reality is the most important ingredient of a scientific theory.* The part of Dalton's theory that led to quantitative predictions was the idea that *each kind of atom has a particular weight and the weight of the molecule is simply the sum of the weights of the atoms in the molecule.* (Modern terminology speaks of the mass of the atom rather than its weight, but in this section we will use Dalton's language.) Dalton also believed that there were very few atoms of each kind in a molecule.

What can be predicted from Dalton's atomic theory? The law of definite proportions is one obvious consequence. If two elements are combined to make a compound, then the weights of the two elements in that compound will be in proportion to the weights of the atoms in the molecule. The concept of the weight of an individual atom is a central feature of Dalton's theory. Although Dalton was not able to weigh an individual atom in grams, he was able to *compare* the weights of different atoms. He would weigh the different elements making up a compound. Then, if he knew the number of each kind of atom in a molecule, he could compare the weights of the different atoms.

For example, Dalton weighed the amounts of oxygen and hydrogen reacting to form water. From these measurements he knew that 8 grams of oxygen combine with each gram of hydrogen. What could Dalton then say about the weight of a single oxygen atom compared with that of a hydrogen atom? In order to draw some kind of conclusion, Dalton made a bold assumption. He guessed that in each water molecule there is just one atom of hydrogen and one atom of oxygen, for it was his belief that the simplest assumption was the best. Since the weight of oxygen in the water was eight times the weight of hydrogen in the water, Dalton concluded that each oxygen atom weighed eight times as much as a hydrogen atom.

Dalton then developed a system of atomic weights based on such measurements. These weights were not the actual weights of the individual atoms (in grams) but the relative weights of the atoms: the weight of a given atom compared to that of a hydrogen atom (the lightest of all the atoms). Since hydrogen was assumed to have an atomic weight of 1, oxygen was assigned an atomic weight of 8, because an oxygen atom had been found to weigh eight times more than a hydrogen atom.

Unfortunately, Dalton fell into a trap with his assumption about the number of atoms in the water molecule. In this case the simplest assumption was not the best. As we now know, each water molecule contains two atoms of hydrogen and one of oxygen. This makes the atomic weight of oxygen equal to 2×8, or 16. The confusion over the number of hydrogen atoms in a water molecule persisted for many years after Dalton first proposed his theory.

In spite of these difficulties, Dalton's atomic theory was able to answer many questions. Consider again the case of copper sulfide, discussed in Section 4.2. Measurements show that there are two kinds of copper sulfide, each with a different proportion of sulfur. The atomic theory is helpful in determining the number of copper and sulfur atoms in each of these two compounds. The method we use demonstrates the interplay of theory and experiment.

First of all, we try to guess at the possible composition of these two molecules. We are allowed to make such a guess as long as we test the guess by comparison with experimental results. The first part of the guess is to assume that one of the two molecules is the simplest molecule possible: one atom of copper and one atom of sulfur. For the second part of the guess, there are several possibilities:

$$2 \text{ Cu to } 1 \text{ S}$$
$$1 \text{ Cu to } 2 \text{ S}$$
$$3 \text{ Cu to } 1 \text{ S}$$

and so on.

The only way to decide among the possibilities is to analyze samples of the two compounds. And so we ask: How many grams of copper are combined with a fixed weight of sulfur in each of the compounds? We decide to choose 10 g for the weight of sulfur, and list the results of the measurements as in Table 4-2.

We see that there is twice as much copper in Compound I as in Compound II, for the same amount of sulfur. In other words, the ratio of copper in the two compounds is 2 to 1. This is just the result we would expect if Compound I has 2 atoms of copper to each atom of sulfur and Compound II has 1 atom of copper to each sulfur atom. In this way we decide which of the several preliminary guesses is correct.

TABLE 4-2 *Analysis of Copper Sulfide Samples*

	Sulfur	Copper	Copper Ratio*
Compound I	10 g	40 g	2
Compound II	10 g	20 g	1

* The amount of copper in each compound divided by the amount of copper in Compound II. For example, the ratio of copper in Compound I to copper in Compound II is 2/1.

The above example demonstrates the law of multiple proportions, announced by John Dalton in 1804 as a consequence of his atomic theory. In general the **law of multiple proportions** says that *if two elements form more than one compound, the weights of the first element that combine with a fixed weight of the second element are in simple integer ratio to each other.*

We have seen that in the case of copper sulfide the ratio of the amounts of copper that combine with 10 g of sulfur in the two compounds is 2 to 1, which is a simple integer ratio. The law of multiple proportions does not tell us what the ratio will be; it merely says that the ratio will be simple (such as 2 to 1, or 3 to 2). This rule serves as a guide for interpreting measurements. If we get a ratio of 137 to 29 in an experiment, we must suspect that something is wrong with our measurements. We do not anticipate numbers such as 137 or 29 because the number of atoms in a molecule is usually small. (This reasoning does not hold for large, complex organic molecules.)

The law of multiple proportions does not answer the question of whether a molecule of Compound II should be written an CuS, $(CuS)_2$, or $(CuS)_3$. As a matter of fact, the concept of molecule does not really apply to certain types of crystalline compounds in which the entire crystal is essentially a single molecule consisting of billions of atoms connected together. In the case of CuS, however, the only thing that counts is the proportion of copper to sulfur atoms.

The law of multiple proportions has been dealt with at length in this section in order to emphasize that science is the product of human creativity and imagination. The law could not have been arrived at simply by examining the results of chemical analysis, for these measurements were not sufficiently accurate at the time of Dalton's work. The history of atomic theory is a reminder that the scientific method involves more than looking for patterns in long lists of data. It often involves daring jumps in logic and the creation of new hypotheses. Equally essential is the ability to test a new hypothesis by careful experimentation.

Example 4.3

There are two compounds of carbon and oxygen. In one compound, 13.3 g of oxygen is combined with 10 g of carbon. In the other, 26.6 g of oxygen is combined with 10 g of carbon.
  (a) How does this example illustrate the law of multiple proportions?
  (b) What are the probable numbers of atoms in each molecule?

Solution

(a) We see that the ratio of the weights of oxygen in the two compounds is $26.6/13.3 = 2/1$. This result agrees with the law of multiple proportions.

(b) The simplest assumption is that in the molecule with the lesser amount of oxygen there is 1 atom of oxygen to 1 atom of carbon, making the compound carbon monoxide (CO). Since the other compound has twice the weight of oxygen, it must have 2 atoms of oxygen to each atom of carbon, the ratio found in the compound carbon dioxide (CO_2).

Example 4.4 There are two kinds of iron chlorides (compounds of iron with chlorine). One sample is found to contain 63.5 g of chlorine to 50.0 g of iron, and another contains 381 g of chlorine to 200 g of iron. Show how these measurements agree with the law of multiple proportions.

Solution First, find the amount of chlorine associated with the *same* amount of iron in the two samples. The easiest way to do this is to find the number of grams of chlorine to each gram of iron. In the first sample, we have

$$\frac{63.5 \text{ g Cl}}{50.0 \text{ g Fe}} = 1.27 \text{ g Cl per g Fe}$$

In the second sample, we have

$$\frac{381 \text{ g Cl}}{200 \text{ g Fe}} = 1.91 \text{ g Cl per g Fe}$$

We now see that

$$\frac{1.91}{1.27} = \frac{3}{2}$$

The weight of Cl in one sample is 3/2 times the weight of Cl in the other sample, for the same amount of Fe. This result suggests that one compound has 2 atoms of Cl and the other has 3 atoms of Cl.

Glossary

alchemy A philosophy of science, popular during the Middle Ages and before, one of whose aims is to transmute base metals into gold.

definite proportions, law of A rule stating that a pure compound is always composed of the same elements, which are always present in the same proportions by weight; also called the law of constant composition.

multiple proportions, law of A rule stating that if two elements form more than one compound, the weights of the first element that combine with a fixed weight of the second element are in a simple whole-number ratio to each other.

products The substances resulting from a chemical reaction.

qualitative analysis Determining the composition of a material by identifying the constituents (the elements and compounds that make it up), without regard to the quantities of each constituent.

quantitative analysis Determining the composition of a material by identifying the constituents and their proportions by weight or volume.

reactants The starting materials of a chemical reaction.

transmutation Changing one element into another; specifically, the process sought by alchemists to transform common (or base) elements into gold. (Modern scientists know how to do this now, but their methods have no relationship to the speculations of the alchemists.)

Problems and Questions

4.1 The Elements and Their Symbols

4.1.Q
 a. What was the main goal of the alchemists?
 b. Were they successful in reaching this goal?
 c. What science arose from the efforts of the alchemists?

4.2.Q
 a. What was the Greek concept of an element?
 b. How did Robert Boyle define a chemical element?
 c. What was the important difference between the two concepts of elements?

4.3.Q Antoine Lavoisier noted the importance of quantitative experiments.
 a. What do we mean by quantitative experiments, as opposed to qualitative experiments?
 b. Why are quantitative experiments valuable?

4.4.Q What is the law of conservation of mass? Why is it important to modern chemists?

4.5.Q Which were the earliest elements known? From what language do we get their names?

4.6.Q Match the element name with the corresponding symbol:

 a. antimony   Al
    argon   As
    arsenic   Ar
    aluminum   Sb
 b. barium   Br
    beryllium   B
    bromine   Be
    bismuth   Bi
    boron   Ba

4.7.Q Give the chemical symbol for each of the following elements:
 a. aluminum   b. fluorine
 c. gold   d. hydrogen
 e. iodine   f. iron
 g. krypton   h. lead
 i. lithium   j. magnesium
 k. manganese   l. mercury

4.8.Q Write the name of the element corresponding to each of the following symbols. Watch the spelling.
 a. Ne   b. Pt
 c. Ni   d. K
 e. N   f. Si
 g. O   h. Na
 i. P   j. Sr

4.2 The Law of Definite Proportions

4.9.Q State the law of definite proportions.

4.10.Q How is the law of definite proportions useful?

4.11.Q What is the meaning of a "pure chemical compound" in light of the law of definite proportions?

4.12.P A sample of carbon dioxide contains 12 g of carbon and 32 g of oxygen.
 a. What is the percentage of carbon and the percentage of oxygen in the sample?
 b. What is the proportion of oxygen to carbon (by weight)?
 c. What is the percentage of carbon in 3 tons of CO_2?
 d. How many tons of carbon are there in 3 tons of CO_2?

4.13.Q What was Proust's explanation for the varying proportions of copper and sulfur in samples of copper sulfide made under differing conditions?

4.14.P Pure silicon is a valuable material obtained from sand, which consists mainly of silicon dioxide (SiO_2). (Quartz is the same compound but with a different crystalline structure.) The proportion of silicon in sand is 46.7% by weight.
 a. How many kilograms of silicon can be obtained from 2500 kg of sand?
 b. How many kilograms of sand are required to produce 500 kg of silicon?

4.15.P A simple sugar, $C_6H_{12}O_6$, consists of 40% carbon and 6.7% hydrogen (by weight).
 a. What is the percentage of oxygen in the sugar?
 b. How many grams of carbon are in 150 g of sugar?
 c. How many grams of oxygen are present if the sugar contains 75 g of carbon?

4.16.P The compound CH_4 (methane) contains 75% carbon by weight.
 a. What is the percentage of hydrogen?

b. A sample of methane contains 25 g of carbon. What is the weight of the sample?

c. A sample of methane contains 35 g of carbon. What is the weight of the hydrogen present?

4.17.P A sample of H_2SO_4 (sulfuric acid) weighs 98 g. The combined weight of sulfur and oxygen is 96 g.

    a. What is the percentage of hydrogen?

    b. The sulfur makes up 32.7% of the H_2SO_4. What is the percentage of oxygen?

4.18.P Two samples of compounds containing carbon and oxygen are analyzed. The analysis data are as follows:

	C	O
Sample 1	42.9%	57.1%
Sample 2	27.3%	72.7%

What can you say about these samples, taking the law of definite proportions into account?

4.3 Atomic Theory and the Law of Multiple Proportions

4.19.Q What was a key feature of John Dalton's atomic theory?

4.20.Q State the law of multiple proportions.

4.21.Q Describe how the law of multiple proportions is a consequence of the atomic model.

4.22.Q What is the best way to test a theory?

4.23.Q How can we test the atomic theory with the help of the law of multiple proportions?

4.24.P a. There are two samples of gaseous compounds; one contains 7 g of nitrogen and 4 g of oxygen, and the other contains 28 g of nitrogen and 48 g of oxygen. What are the possible formulas for these compounds?

    b. Show how these compounds illustrate the law of multiple proportions.

4.25.P Two liquid compounds contain hydrogen and oxygen. The first compound is found to contain 4 g of hydrogen and 32 g of oxygen. The second compound contains 0.5 g of hydrogen and 8 g of oxygen. Suggest a possible formula for each compound.

Self Test

1. **a.** What was the philosopher's stone, and what did it have to do with transmutations?
   b. What did the search for transmutation lead to?

2. Why are the symbols for many elements abbreviations of Latin names?

3. State the following laws and give an example of each.
   a. the law of conservation of mass
   b. the law of definite proportions
   c. the law of multiple proportions

4. If a sample of fuel is burned in a closed chamber, do you expect the weight of the contents of the chamber to change? Explain your answer.

5. Write the names of the following elements (spelling counts):
   a. He   **b.** Mn
   c. P   **d.** K
   e. Ag   **f.** Au
   g. Hg   **h.** Zn
   i. F   **j.** Cr
   k. W   **l.** U

6. Write the symbols for the following elements:
   a. hydrogen   **b.** sodium
   c. calcium   **d.** chlorine
   e. iron   **f.** cobalt
   g. radium   **h.** barium
   i. oxygen   **j.** nitrogen
   k. neon   **l.** carbon

7. Indicate the name and number of atoms of each element in the following compounds:
   a. CH_4 **b.** N_2O
   c. $CoCl_2$ **d.** $CaBr_2$
   e. NH_4Cl **f.** NaF
   g. FeI_2 **h.** H_2SO_4
   i. K_3PO_4 **j.** H_2O
   k. XeF_4

8. Three samples of a gas are analyzed, and the results show that two elements make up the gas. The results of the analysis are as follows:

	Element A	Element B
Analysis 1	3.1 g	15.5 g
Analysis 2	1.1 g	5.5 g
Analysis 3	0.12 g	0.60 g

What law is demonstrated by these data?

9. After the analyses of Question 8 (above), a second gas is examined, and it shows the same qualitative analysis as did the first gas. (That is, the same elements are present.) The quantitative results of the analysis, however, are as follows:

	Element A	Element B
Test 1	3.1 g	23.25 g
Test 2	1.1 g	8.25 g
Test 3	0.12 g	0.90 g

   a. What law does the second analysis demonstrate?
   b. What law is demonstrated by the data in Questions 8 and 9 combined?

10. Sulfur dioxide is a pungent-smelling gas, with the formula SO_2. Analysis of this gas shows that the amounts of sulfur and oxygen (by weight) are equal; that is, for each gram of sulfur there is 1 gram of oxygen. Sulfur trioxide is another compound of sulfur and oxygen, with the formula SO_3.
    a. How many grams of oxygen would you expect to find for each gram of sulfur in SO_3?
    b. What is the mass of a sulfur atom relative to that of an oxygen atom? That is, what is this fraction?

$$\frac{\text{mass of sulfur atom}}{\text{mass of oxygen atom}}$$

11. It has long been known that the compounds CO and CO_2 exist. Recently the existence of CO_3 has been reported. Using the information developed in this chapter, decide whether it is reasonable to expect CO_3 as a compound. Give your reasons.

12. What is the difference between qualitative analysis and quantitative analysis?

5
Atomic Structure

A cloud of condensation rising from a chemical research apparatus.

Objectives After completing this chapter, you should be able to:
1. Describe the forces acting between electrical charges.
2. Describe the meaning of an inverse-square force.
3. Describe the difference between electric and magnetic forces.
4. Describe how electrons were discovered.
5. Describe the nature of electrons.
6. Describe the three kinds of radiation emitted by radioactive substances.
7. Describe the relationship between radioactivity and our knowledge of atomic structure.
8. Describe the nuclear atom.
9. Do simple problems involving atomic dimensions.
10. Do simple problems involving atomic mass.
11. Explain the function of the neutron in determining atomic mass.
12. Explain the concept of isotopes.
13. Explain the formation of ions.
14. Describe the Bohr model of the atom.
15. Describe how the Bohr model is verified by optical spectra.
16. Describe the Schroedinger model of the atom.
17. Explain why the Schroedinger model is superior to the Bohr model.
18. Describe the structure of multielectron atoms in terms of shells and subshells.

5 ATOMIC STRUCTURE

5.1 Atomic Theory

Electric Charges

The motions and behaviors of material things are controlled by natural forces. A **force** is a push or a pull—an attraction or a repulsion. A force is a mysterious, invisible something that passes through the space separating two objects and causes them to affect each other. (Don't think that there isn't any space between your hand and the book in your hand. Between the atoms of your hand and the atoms of the book is a definite, though tiny, space, and across that space must travel the force used to pick up the book. All forces act through a distance.)

There are a very few fundamental kinds of forces, one of which is the gravitational force, by which you are held to the surface of the earth and by which the earth is attracted to the sun. The gravitational force is an attraction pulling together every pair of objects in the universe. It is such a weak force that it is hardly noticeable except in relation to massive objects, such as planets or stars. For this reason, the gravitational force plays no role in chemistry.

Another force is the electromagnetic force, which is responsible for holding solid matter together, for binding atoms into molecules, and for holding still smaller particles together within atoms. In the past, the electric force and the magnetic force were believed to be two separate forces, but since the work of James Clerk Maxwell in the 1870s, we know that electric and magnetic forces are simply two aspects of the electromagnetic force.

The existence of the electric force has been known for thousands of years, but its nature was a mystery until the end of the nineteenth century. The ancient Greeks knew that certain materials such as glass, ebonite (hard rubber), and amber would exhibit a strange attraction for bits of paper and fibers after being rubbed with silk or wool. Such objects were called *electrified*, or *electrically charged*.

After centuries of experimentation, a few simple rules were accumulated concerning the behavior of electrified objects. It was found there are two kinds of electrified objects. When a piece of glass is rubbed with silk and a piece of ebonite is rubbed with wool or fur, the glass and the ebonite attract each other. (See Figure 5-1.) On the other hand, two pieces of ebonite rubbed with fur repel each other, as do two pieces of glass rubbed with silk.

Scientists described (not "explained") these phenomena by saying that the glass possesses one kind of electric charge (called *positive*) and the ebonite has another kind of electric charge (called *negative*). Many substances can be electrically charged; a substance that behaves like glass is said to be positively charged, and a substance that behaves like ebonite is said to be negatively charged. Experiments show the following behavior:

1. Two positively charged objects always repel each other.

2. Two negatively charged objects always repel each other.

3. A negative charge always attracts a positive charge.

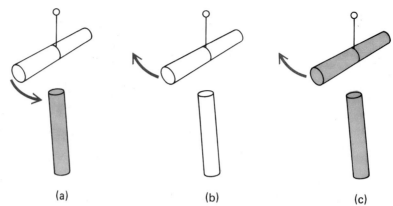

FIGURE 5-1 (a) A glass rod rubbed with silk and suspended from a string is attracted by an ebonite rod rubbed with fur or wool. (b) Two electrified glass rods repel each other. (c) Two electrified ebonite rods repel each other.

From these observations a general rule follows: *Like charges repel, whereas unlike charges attract each other.*

Notice an important difference between the electric force and the gravitational force. The gravitational force is *always* an attraction. There is no such thing as a gravitational repulsion. However, there is one way in which the electric and gravitational forces are similar. It has to do with the way the strength of the force varies with the distance between the two objects. When two charged objects are close together, the force is relatively strong; when they are far apart, the force is weak.

A quantitative relationship is found by measurement: if the distance between two objects is doubled, the force becomes one-fourth as strong, and if the distance is tripled, the force becomes one-ninth as strong. The gravitational force obeys the same formula, described by the equation

$$\text{force} \propto \frac{1}{(\text{distance})^2}$$

This equation says that the force is *inversely proportional* to the square of the distance between the two interacting objects. For this reason, we call the gravitational and the electric forces **inverse-square forces**. Although the force between two objects becomes weaker as they are separated, it never completely disappears, and so we call electric and gravitational forces "long-range" forces.

The magnetic force was noted prior to 600 B.C. by the Greeks, who recognized that certain minerals (lodestone) have the ability to attract pieces of iron. Later, people learned how to treat bars of steel to turn them into permanent magnets with a strong attraction for other magnets or for ordinary pieces of iron. It was learned that magnets have two poles—called north and south poles—and that these poles obey the same kind of rule as electric charges: *like poles repel each other, and unlike poles attract each other.*

The explanation for this behavior of magnetized iron is very complex and was not well understood until well into the twentieth century. We now know that the magnetism of an iron magnet is caused by the motion of electric charges inside the iron atoms. But before you can understand this concept, you must know what the inside of an atom is like, and that will be the subject of the rest of this chapter.

Electrons and Protons

In the last section, the concept of electric charge was introduced, and we saw how the identification of electric charges is based on their mutual attraction and repulsion. But what *is* an electric charge?

It was not until the last decade of the nineteenth century that a reasonable theory explaining those mysterious attractions and repulsions was created. In that decade several important discoveries laid the foundation for our modern knowledge of matter and its structure. These developments were a result of two technological advances. First, invention of a device called an induction coil made it easy to generate high-voltage electric currents. Second, improvements in vacuum pumps simplified the process of evacuating most of the air from a glass tube.

As a result, scientists began doing experiments to investigate the phenomena that occurred when electric currents were passed through tubes containing various gases at low pressures, as shown in Figure 5-2. In particular, Sir William Crookes and Heinrich Geissler studied the brightly colored light emitted by such gases and showed that the exact color and wavelength of the light depended on the nature of the gas. (Such tubes are still used in neon signs.)

The most interesting question was "What happens when you keep pumping the air out while electric current is passing through the tube?" As long as there is still air in the tube, a bright violet light is emitted. When most of the air is gone, the violet glow disappears. At this point a new phenomenon takes place: the glass of the tube begins to shine with a green **fluorescence**. Furthermore, a meter in the circuit shows a current still passing through the tube. How can an electric current pass through a vacuum?

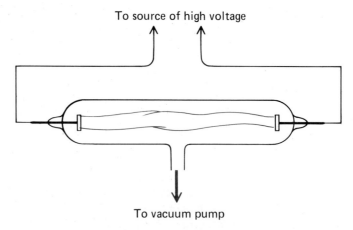

FIGURE 5-2 When high-voltage electricity is applied to the two electrodes, an electric arc streams through a glass tube that is being evacuated.

That mystery was solved by British scientist J. J. Thomson in 1897 while making the first discovery in the science of fundamental particles. Thomson built a tube like that shown in Figure 5-3. High-voltage electric current passes from one metal electrode to the other and forms a bright spot of light on the screen located at the far end of the tube. The spot of light is produced because the screen is coated with a chemical (zinc sulfide) that glows when struck by the cathode rays passing through the small hole in the positive electrode. (These rays are called cathode rays because they are emitted by the negatively charged electrode, called the *cathode*. The positive electrode, on the other hand, is called the *anode*.)

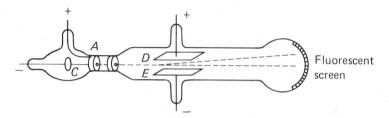

FIGURE 5-3 J. J. Thomson's cathode ray tube with deflection plates. A is the anode, C is the cathode, and D and E are the deflection plates. The path of the electrons is deflected upward if plate D is connected to the positive end of the power supply.

Thomson discovered that when the two plates parallel to the path of the cathode rays are connected to a battery, the cathode rays are attracted toward the positive plate. Thomson also found that the rays are deflected from a straight path if a magnet is brought near the tube. Both of these effects are explained by the idea that the cathode rays consist of tiny, electrically charged particles moving with high speed from the cathode to the anode.

These particles are extremely small and light and are called **electrons**. Since the electrons are attracted to the *positive* plate and since opposite charges attract each other, each electron must have a *negative* electric charge. We now look upon electrons as the *fundamental particles of negative electric charge*. That is, the charge on an electron is the smallest quantity of electric charge that can be found all by itself, and so it is considered one unit of electric charge. J. J. Thomson climaxed his proof that the electron is a particle by measuring the mass of each electron. He did this by measuring how far the electron beam was bent when it passed through electric and magnetic fields.

Strictly speaking, Thomson's measurement gave the ratio of the mass to the amount of charge. By making an educated guess about the amount of charge on an electron, Thomson was able to estimate the mass. It was not until 1911 that an American physicist, Robert A. Millikan, measured the electron charge directly, by means of a famous experiment involving small oil drops floating through a chamber under the influence of an electric field. Since each droplet bore a few extra electrons, observation of these drops moving in the electric field resulted in a measurement of the electron charge. The "Millikan oil drop experiment" nailed down the concept of the electron as a particle carrying the smallest possible quantity of electric charge.

Normal matter is electrically neutral; it has neither positive nor negative charge. Therefore, matter must contain positive charges to balance the negative charges of the electrons. The presence of such positive charges can be seen in the canal ray tube, invented by E. Goldstein in 1876 (see Figure 5-4). The canal ray (or positive ray) is produced by a cathode ray tube with a perforated cathode. If there is a small amount of hydrogen in the tube, a beam giving off red light can be seen streaming through the perforation. Since the red beam is accelerated by the negative cathode, it consists of positive charges.

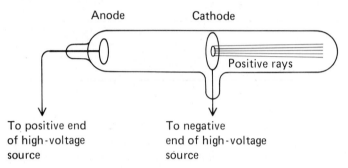

FIGURE 5-4 A positive ray, or canal ray, tube. The positive rays are protons attracted to the negative electrode (the cathode).

As a result of such observations, J. J. Thomson developed the theory that *the atoms of matter contain two kinds of particles: negatively charged electrons and positively charged* **protons**. (He did not actually call these positively charged particles protons; they were so named by Rutherford in 1914.) Thomson's logic was simple: When these particles are emitted from the electrodes and the gas in the tube, they are not created from nothing. They must already exist within the electrodes and the gas. The electric field merely pulls them out so they can be observed.

As we saw in Chapter 4, Thomson's was not the first atomic theory. In 1803, John Dalton thought of atoms as hard spheres with little hooks by which they cling to each other. The importance of Thomson's model was that it was the first theory to consider an *inner structure* of the atom. Also, it was the first to recognize that atomic structure and behavior are based on electric forces because the atom is built of electrically charged particles. Thomson's model was still a primitive one. He visualized an atom as a clump of positive electric charge in which the negative electrons are stuck like raisins in a pudding. Since the protons are much more massive than the electrons, most of the mass of an atom is in the protons. Thomson's raisin-pudding picture was only a temporary model. Soon it was replaced by the more realistic nuclear model developed by Rutherford in 1911, to be described in the next section. (In science a *model* is a mental image used to visualize things that cannot really be seen and to help explain the causes of observed events.)

Radioactivity and the Nuclear Model

French physicist Henri Becquerel made an important contribution to our understanding of matter in 1896, when he discovered that the element uranium emits a strange kind of radiation that is able to produce images on photographic film even after passing through layers of material opaque to ordinary light (see Figure 5-5). This emission of radiation by uranium was called **radioactivity**. Work done by Marie Curie, Ernest Rutherford, and others showed that this radiation consists of three components:

1. massive, positively charged particles, called alpha rays,
2. light, negatively charged particles, called beta rays, and
3. neutral radiation, called gamma rays.

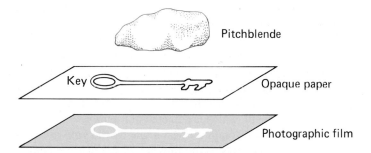

FIGURE 5-5 Gamma rays emitted by the pitchblende (uranium ore) pass through opaque paper to blacken a photographic film. The key stops some of the rays, however, so the film is not blackened as much in the shadow of the key.

The three kinds of rays can be separated by passing a beam of uranium radiation between the poles of a magnet; the negative particles are deflected one way, the positive particles are deflected in the opposite direction, and the neutral beam goes straight through, as shown in Figure 5-6. The negative particles (the beta rays) were found to be the same as electrons, and the positive particles (the alpha rays) were found to be nothing more than atoms of helium with their electrons removed. The gamma rays turned out to be identical in properties to the x-rays already discovered by Wilhelm Roentgen in 1895. The question of how such radiation could be created within a piece of uranium was one of the deepest mysteries surrounding the new phenomenon of radioactivity. This question was not answered for many years.

While investigating the problem, scientists learned how to use radiation for many purposes. Ernest Rutherford, who had identified the alpha particles as positively charged helium atoms, recognized that he could use them as probes to look inside individual atoms. By this means the inner structure of these atoms could be determined. This was a most novel and important idea.

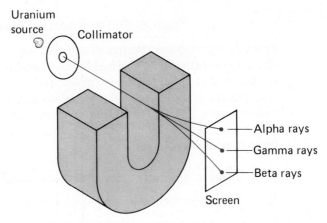

FIGURE 5-6 Rays emitted from uranium pass through a small hole (a collimator) to make a narrow beam. When they pass between the poles of a magnet, they separate into three beams corresponding to the positively charged alpha particles, the neutral gamma rays, and the negatively charged beta particles.

In an experiment devised by Rutherford in 1911, a narrow beam of alpha particles was shot through a thin gold foil and then observed when it emerged on the other side. The path of each particle could be followed by looking at the faint flash of light produced when it struck a zinc sulfide screen. The aim of the experiment, which is depicted in Figure 5-7, was to measure the number of particles scattered through a given angle by the gold atoms within the foil.

The raisin-pudding model of the atom common at that time predicted that the alpha particles would be scattered only through very small angles. To Rutherford's surprise, he found that many of the alpha particles were scattered through rather large angles. They behaved, in fact, as though they had collided with massive, solid spheres, like billiard balls colliding with one another. (See Figure 5-8.)

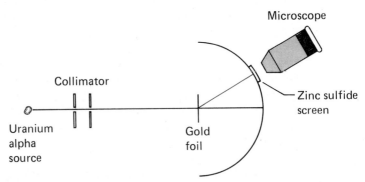

FIGURE 5-7 The Rutherford scattering experiment: Alpha particles from a uranium source are scattered by the atoms in a thin gold foil. They produce flashes of light on a zinc sulfide screen. These flashes are counted by someone observing through a microscope. The number of flashes per minute is recorded at various angles.

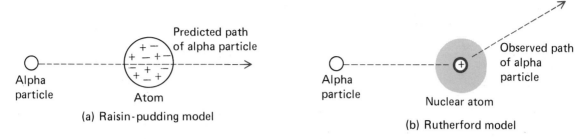

FIGURE 5-8 Interpretation of the Rutherford experiment: If the gold atom consisted of intermingled positive and negative charges, the alpha particles would only be deflected through small angles. Since some of the alpha particles are deflected through large angles, we conclude that the gold atom contains a small, dense nucleus.

Such a result could be explained only by a new interpretation of atomic structure. The experiment demonstrated that most of the mass of the atom is concentrated at the center. No longer could the atom be thought of as a bunch of electrons and protons intermingled. Instead, Rutherford developed a new model in which all of the protons (the massive particles) are clumped together in a tiny central core, the **nucleus**, with the electrons (the light particles) surrounding the nucleus like a cloud. (Note: *nucleus* is the singular form of the word; its plural is *nuclei*.)

Rutherford's scattering experiment was the first experiment in nuclear physics, and its interpretation was the core of our modern notion of atomic structure. However, one more ingredient was required before it could be understood how the atomic nucleus could even exist. The great mystery was this: If the nucleus contains nothing but protons, why doesn't the electric repulsion between all the positive charges blow the nucleus apart?

It was not until 1932 that the new ingredient was discovered by James Chadwick, a British physicist. This ingredient was a neutral particle with roughly the same mass as a proton but no electric charge. This neutral particle was named the **neutron**. With the help of the neutron, the atomic nucleus could be understood.

Our modern picture of the nucleus is as follows: Each nucleus consists of neutrons and protons clustered very closely together; these particles make up most of the mass of the atom. (The hydrogen nucleus has only a single proton.) The electrons, circling outside the nucleus, balance the electric charge of the protons within. Each electron carries exactly the same amount of charge as a proton (but of opposite sign), and there are always just as many protons as electrons in each neutral atom. The atom as a whole is held together by the electric attraction between the electrons and the protons. (See Figure 5-9 for an illustration of a nucleus.)

But what holds the nucleus together? A new kind of force must be postulated to answer this question—a force of attraction between neutrons and protons within a nucleus. Even protons attract each other when they get close

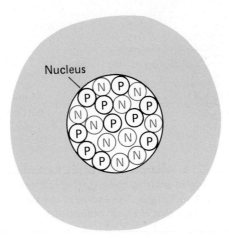

FIGURE 5-9 An atom consists of neutrons and protons packed into a small, dense nucleus, with the electrons surrounding the nucleus in one or more shells.

enough to touch, but that attraction is not strong enough to overcome the electrical repulsion between them. The neutrons are required in addition, as a kind of nuclear "glue" to stabilize the structure of the nucleus.

The **nuclear force** is an entirely new kind of force—a third fundamental force in addition to the electromagnetic and gravitational ones already mentioned. It is a "short-range" force, since it is only felt when the particles involved are very close together. It is always an attraction—a very powerful attraction—and is called the "strong nuclear force" to differentiate it from another kind of nuclear force called the "weak nuclear force."

In Section 5.3 we shall go into more detail about the way the neutrons and protons are arranged within the nucleus, and in Chapter 6 we will see how the electrons are arranged outside the nucleus to give each of the elements a different set of properties.

Atomic Dimensions and Masses

How small is an atom? Measurements of all kinds tell us that an atom is much smaller than anything visible through even the most powerful microscope. How then can we tell what an atom is like?

The answer is to use devices that send various kinds of signals into an atom. The incoming signals trigger responses that can be analyzed to deduce what is going on inside the atom. We have already seen how Rutherford bombarded atoms of gold with alpha particles to prove that most of the mass of the atom is located within its nucleus. The same experiment also gave Rutherford information about the size of the nucleus.

In modern experiments we use machines such as cyclotrons, synchrotrons, and other particle accelerators to shoot high-speed electrons, protons, or neutrons into the nuclei of the atoms being investigated. By detecting what comes out of the bombarded nuclei, we gain all sorts of information—not only about the size and mass of each nucleus but also about the arrangement of the particles inside, the amount of energy stored in the nucleus, and the nature of the force that binds the nucleus together.

In order to understand how small an atom is, we must understand the size of units used to measure atomic dimensions. We want to choose a unit that is conveniently close to the size of an atom. A hydrogen atom is about 10^{-10} m in diameter. This is one ten-billionth of a meter. Ten billion (10^{10}) hydrogen atoms can be lined up in a row 1 m long. Since a centimeter is $\frac{1}{100}$ of a meter, the number of hydrogen atoms required to make a line 1 cm long is

$$\frac{1 \text{ m}}{100 \text{ cm}} \times 10^{10} \frac{\text{H atoms}}{1 \text{ m}} = 10^{8} \frac{\text{H atoms}}{1 \text{ cm}}$$

That is, one hundred million hydrogen atoms can be lined up side by side along a 1-cm scale.

A unit close to the size of a hydrogen atom is called the **angstrom** (abbreviated Å), after Anders Jonas Ångström (1814–1874), a Swedish physicist. Table 5-1 summarizes some of the relationships between the angstrom and the more familiar metric units. Another unit commonly used is the nanometer (abbreviated nm), one billionth of a meter.

TABLE 5-1 Angstrom and Meter Conversions

1 angstrom	= 10^{-10} meter (m)
	= 10^{-8} centimeter (cm)
	= 10^{-1} nanometer (nm)
1 meter	= 10^{10} angstroms (Å)
	= 10^{9} nanometers
	= 10^{2} centimeters

Example 5.1 Show that the diameter of a hydrogen atom is 0.1 nm.

Solution We are given that the diameter of a hydrogen atom is about 10^{-10} m. Since $1 \text{ nm} = 10^{-9}$ m,

$$\text{diameter} = 10^{-10} \text{ m} \times \frac{1 \text{ nm}}{10^{-9} \text{ m}} = 10^{-1} \text{ nm} = 0.1 \text{ nm}$$

Since an angstrom is equal to 10^{-10} m, this result shows that 0.1 nm equals 1 Å, meaning that 1 nm = 10 Å.

It is not possible to give a precise value for the atomic diameter because the electrons surround the nucleus as a kind of cloud with fuzzy edges. However, we can give approximate dimensions. A hydrogen atom is about 1 Å in diameter, whereas the largest naturally occurring atom (uranium) is roughly 3 Å in diameter. Even though a uranium atom contains 92 electrons and the hydrogen atom has only 1, the uranium atom is not 92 times greater in volume than the hydrogen atom because the electrons are squeezed closer together in the uranium atom than in smaller atoms.

The nucleus of an atom is only a very small part of the entire atom. In the hydrogen atom, the diameter of the nucleus is about 3×10^{-15} m. This is only a very small fraction of the atomic diameter, as shown in the following example.

Example 5.2   In a hydrogen atom, what fraction of the atomic diameter does the nucleus occupy?

Solution   First, find the diameter of the nucleus in angstroms:

$$3 \times 10^{-15} \text{ m} \times \frac{10^{10} \text{ Å}}{1 \text{ m}} = 3 \times 10^{-5} \text{ Å}$$

Since the atomic diameter of hydrogen is 1 Å, the fraction of the atomic diameter occupied by the nucleus is calculated as follows:

$$\frac{\text{diameter of nucleus}}{\text{diameter of atom}} = \frac{3 \times 10^{-5} \text{ Å}}{1 \text{ Å}} = 3 \times 10^{-5}$$

To put this result into the form of a fraction, multiply and divide by 10^5:

$$3 \times 10^{-5} \times \frac{10^5}{10^5} = \frac{3}{10^5} = \frac{3}{100,000}$$

$$\cong \frac{1}{33,000} \quad \text{(Rounding off the answer)}$$

The diameter of the hydrogen nucleus is about 1/33,000 of the diameter of the atom itself.

Since the hydrogen nucleus consists of only a single proton, its diameter is the same as the diameter of the proton. Remember that this size is only a rough approximation, because the proton is not like a hard baseball. It is more like a puff of dense smoke, with indefinite edges. The same is true of the neutron, found in the nuclei of all atoms except hydrogen.

The size of an electron is even more indefinite than that of the proton or neutron. An electron cannot be thought of as a hard, solid particle like a grain of sand. In modern physics we think of it as a **wave packet**, having the properties of a wave localized into a bundle. (See the electron in Figure 5-10.) Electron waves are totally different from water waves, sound waves, and radio waves; they are called "matter waves."

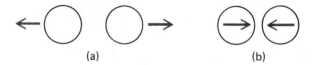

FIGURE 5-10
An electron behaves as though it is a packet or bundle of waves. The wavelength depends on the velocity of the electron but is usually smaller than the diameter of an atom. (See Figure 5-17 for a definition of wavelength.)

Although protons and neutrons are also described by wave packets, their sizes can be defined in a unique way. As two protons or neutrons are brought together, we can measure the distance at which they begin to feel the strong nuclear force pulling them together. At this point the two particles are essentially touching each other, and the distance between their centers is defined to be the diameter of the particle. (See Figure 5-11.)

Neutrons and protons have the same size and share many other properties. Modern physicists look upon a proton and a neutron as being essentially the same kind of particle except for the electric charge on the proton.

Neutrons, protons, and electrons are all fundamental particles—the particles of which atoms are built. The neutrons and protons make up the nucleus of the atom, and the electron is the particle that occupies the space outside the nucleus.

FIGURE 5-11 (a) When two protons are some distance apart, they repel each other by means of the electrostatic force. (b) When the same two protons are close together, they attract each other by means of the strong nuclear force. The distance (between centers) at which the strong nuclear force takes over is the same as the diameter of the proton.

The two most important properties of fundamental particles are their electric charge and their mass. As mentioned previously, the charge on the electron is the smallest quantity of electric charge found separately in nature. The charge on the proton is exactly equal in quantity to the charge on the electron, but is positive instead of negative. It is this equality that permits atoms, with equal numbers of electrons and protons, to be electrically neutral.

Although their charges are equal, the proton is much more massive than the electron. The masses of the three particles are listed in Table 5-2. Surveying the last column, we see that the proton is 1836 times more massive than the electron, and the neutron is just slightly more massive than the proton. (This difference in mass does not disprove the earlier statement that a proton is a neutron with an electric charge, but the theory behind the reconciliation lies in the realm of advanced physics.)

TABLE 5-2 Masses of Fundamental Particles

Particle	Mass	Ratio of Particle Mass to Electron Mass
electron	9.1095×10^{-31} kg	1
proton	1.6726×10^{-27} kg	1836
neutron	1.6750×10^{-27} kg	1839

Example 5.3 How many hydrogen atoms can be lined up along a 1-in distance?

Solution One hundred million (10^8) hydrogen atoms fit into a centimeter length. There are 2.54 cm to an inch, so the number of atoms per inch is

$$2.54 \frac{\text{cm}}{\text{in}} \times 10^8 \frac{\text{atoms}}{\text{cm}} = 2.54 \times 10^8 \frac{\text{atoms}}{\text{in}}$$

or 254,000,000 atoms per inch.

Example 5.4 How many hydrogen nuclei (protons) can fit into the volume of an entire hydrogen atom?

Solution The diameter of the atom is about 1.0 Å, so its radius is 0.50 Å. The nuclear diameter is 3.0×10^{-5} Å, for a radius of 1.5×10^{-5} Å. The volume of the spherical atom is found with the formula

$$V = \frac{4}{3}\pi r^3$$

Thus, the volume of one hydrogen atom is

$$V = \frac{4}{3}\pi(0.50 \text{ Å})^3 = 0.52 \text{ Å}^3$$

The volume of one proton is

$$V = \frac{4}{3}\pi(1.5 \times 10^{-5} \text{ Å})^3 = 1.4 \times 10^{-14} \text{ Å}^3$$

To find the number of protons that fit into the volume of a single atom, multiply the volume of an atom by the number of protons in a cubic angstrom. Cancellation of units gives protons per atom:

$$\frac{0.52 \text{ Å}^3}{1 \text{ atom}} \times \frac{1 \text{ proton}}{1.4 \times 10^{-14} \text{ Å}^3} = 3.6 \times 10^{13} \text{ protons/atom}$$

Conclusion: 37 million million nuclei can fit into the volume of the entire atom. The nucleus is just a small fraction of the whole atom, even though it contains almost all the mass.

Measuring Particle Masses

Measuring the masses of fundamental particles as well as of individual atoms and molecules is an important task of the chemist and physicist. A powerful tool used for this purpose is the **mass spectrometer**, which tracks the path of a charged particle moving through a magnetic field. A magnetic field is described by lines of force that go between the north and the south poles of a magnet, as shown in

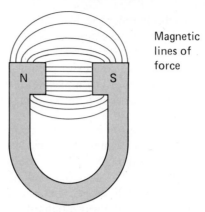

FIGURE 5-12 The lines of force from a horseshoe magnet go in curves from the north pole to the south pole. However, directly between the poles, the lines of force are relatively straight.

Figure 5-12. Just between those poles the lines of force are relatively straight, and the magnetic field has a constant strength.

When a charged particle travels through a magnetic field, the magnetic force acting on it causes its path to be deflected at right angles to the magnetic field lines, as shown in Figure 5-13. In a uniform magnetic field, a charged particle travels in a circular path, the radius of which is proportional to the mass of the particle. (This is because the more massive the particle, the harder it is for the magnetic field to bend the particle's motion away from a straight line.) Thus, by measuring the magnetic field strength, the charge on the particle, and the radius of curvature of the particle's path, we can calculate the mass of the electron, the proton, or another charged particle.

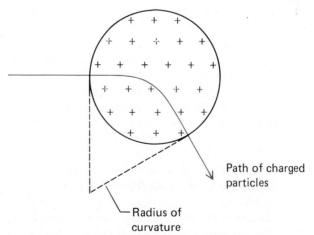

FIGURE 5-13 A charged particle traveling through a magnetic field moves in a circular path whose radius of curvature depends on the strength of the field and the mass of the particle. In this diagram the lines of force are going straight in and out of the page.

In this manner the mass spectrometer is used to measure the mass of entire atoms. In order to accomplish this, one or more electrons must first be removed from the atom's outer shell. The atom is left with a net positive charge; such a charged atom is called an *ion*. The ion can now be accelerated and sent through a mass spectrometer. Even molecules can be treated in this manner, so the mass spectrometer has become an important tool in the analysis of complex chemical compounds.

Measuring the mass of the neutron is a different problem, since it has no electric charge and cannot be ionized as an atom can. James Chadwick measured neutron mass originally by bombarding a block of paraffin with a beam of fast-moving neutrons. The neutrons, colliding with the hydrogen atoms in the paraffin (which consists of hydrogen and carbon), knocked these atoms right out of the paraffin molecules, the way a billiard ball knocks another ball out of a cluster. The ejected hydrogen atoms with their electrons left behind were at this point bare protons (hydrogen ions). By measuring the speed of these protons, Chadwick was able to calculate the mass of the neutrons. He found that the neutron mass was approximately equal to the proton mass. Since then, more precise measurements have shown the neutron to be slightly more massive than the proton.

5.2 Atomic Number and Atomic Mass

We have seen that most of the mass of an atom is in the nucleus, where the neutrons and protons reside. The electrons orbit about the nucleus, forming a kind of cloud of negative electric charge that exactly balances the positive charge on the nucleus. Since the electrons are on the outside, the arrangement of the electrons in orbit determines how the atom interacts with other atoms. In other words, *the electrons determine the chemical properties of the element, whereas the nucleus determines the mass of each atom.*

The number of electrons in an atom is the same as the number of protons in the nucleus, so the net charge on the atom is zero. The number of protons in the nucleus is called the **atomic number** of the element. Each element has a unique number of protons and, therefore, a unique number of electrons. Hydrogen has one proton and one electron, helium has two of each, lithium has three, and so on up to the "heaviest" element found in nature (uranium), which has 92. We define the atomic number to be the number of protons within the nucleus rather than the number of orbital electrons because under certain conditions an atom may change its number of electrons (becoming an ion, as we will see in the next section). The number of nuclear protons remains unchanged, however, and that determines the identity of the atom. An atom with an atomic number of 26 (26 protons) is always iron, no matter how many electrons it has.

Another number used to describe an element is the **atomic mass**, formerly called the atomic weight. As we saw in Chapter 4, measurement of atomic weights began with John Dalton. Dalton had no way of measuring the weights or masses of individual atoms, but he devised methods of measuring the *relative weights* of the atoms joined together in compounds—that is, the weight of one atom compared with that of another.

By repeatedly weighing the amount of hydrogen and the amount of oxygen that combined to make water, Dalton determined that they always combine in the same ratio: 8 g of oxygen always combines with 1 g of hydrogen (or 8 parts, by weight, of oxygen to 1 part of hydrogen). Assuming that water contains an equal number of atoms of hydrogen and oxygen, Dalton decided that the oxygen atom must be eight times heavier than the hydrogen atom.

In this way, the concept of relative atomic weight originated. Since the atomic weight of hydrogen was defined to be 1, Dalton assigned oxygen an atomic weight of 8. However, we now know that the water molecule has two atoms of hydrogen attached to one atom of oxygen. If each hydrogen atom has an atomic weight of 1, oxygen must have an atomic weight of 16, because the oxygen weighs eight times as much as the two hydrogens.

Atomic mass is now considered to be a more meaningful term than atomic weight, since mass is a more fundamental unit than weight and modern physical methods of measuring atomic mass are more accurate than older chemical methods of determining relative atomic weights. (However, the *Handbook of Chemistry and Physics* still uses atomic weight, and many scientists use the two terms interchangeably.) Also, the hydrogen atom is no longer used as the standard for measuring atomic masses. The modern standard is a particular type of carbon whose mass is defined to be exactly 12.0000 units. Carbon was chosen as a standard because it is stable and enters into many compounds, and so the relative masses of other elements can be measured conveniently. The standard carbon is called carbon-12. (Naturally occurring carbon has an atomic mass of 12.011 units.)

The **atomic mass unit** (abbreviated **amu**) is defined to be $\frac{1}{12}$ the mass of a carbon-12 atom. (It is called the *dalton* by some, in honor of John Dalton.)

When the atomic masses of the naturally occurring elements were measured in the traditional chemical manner (by analyzing their compounds and finding the proportions of each element in those compounds), a strange thing was observed: the atomic masses are generally not at all close to integral values (see Table 5-3). We might expect these numbers to be close to integral values equal (approximately) to the total number of neutrons and protons, since the neutron mass differs from the proton mass by only one part in a thousand and the electron mass is negligible.

For many years these fractional atomic weights were a mystery. The mystery was finally solved in 1912 with the discovery by J. J. Thomson of isotopes

TABLE 5-3 *Some Atomic Masses*

Element	Atomic Mass
hydrogen	1.008
oxygen	15.873
chlorine	35.17
iron	55.41

(different varieties of the same element). In the next section we will discuss what isotopes are and how they explain the nonintegral atomic weights of the elements.

Example 5.5

When carbon burns, it forms a colorless, odorless gas. Two oxides of carbon may result from this combustion: carbon monoxide and carbon dioxide. If the supply of oxygen is restricted, the toxic gas carbon monoxide tends to be formed. In this reaction we find that 1.00 g of carbon combines with 1.33 g of oxygen to form 2.33 g of carbon monoxide, containing one atom of oxygen to each atom of carbon. (*Note*: The prefix *mono* stands for "one" in chemical nomenclature.)

Natural carbon has an atomic mass of 12.011. What is the atomic mass of oxygen?

Solution

Since there is one atom of each element in the molecule, the atomic mass of oxygen is directly proportional to the mass of oxygen combining with a given mass of carbon. Let A stand for the atomic mass of oxygen. Then (to three significant figures),

$$A = 12.0 \text{ amu} \times \frac{1.33 \text{ g of oxygen}}{1.00 \text{ g of carbon}}$$

$$= 16.0 \text{ amu}$$

The atomic mass is a useful quantity because it allows the chemist to calculate the amounts of materials that will combine in any given reaction, as shown in the next example.

Example 5.6

How much oxygen is required to burn 5.00 kg of carbon to form carbon dioxide? (The prefix *di* stands for "two," indicating that carbon dioxide has two oxygen atoms for each carbon atom.)

Solution

The number of kilograms of each element is proportional to the atomic mass of that element. Therefore, we write

$$\text{mass of oxygen} = \text{mass of carbon} \times \frac{2 \text{ atoms} \times \text{at. mass of oxygen}}{1 \text{ atom} \times \text{at. mass of carbon}}$$

$$= 5.00 \text{ kg} \times \frac{2 \times 16.0 \text{ amu}}{1 \times 12.0 \text{ amu}}$$

$$= 13.3 \text{ kg}$$

(In Chapter 11 we will go into much more detail about such chemical calculations.)

5.3 Ions and Isotopes

Following his discovery of the electron, J. J. Thomson continued to study the properties of electrons and of the positive rays described in Section 5.1 and shown in Figure 5-4. Starting in 1906, he began to concentrate on the positive rays, devising ways to measure the masses of the positive particles by bending their paths in electric and magnetic fields. One of his investigations involved neon, one of the rare gases found in the atmosphere and now commonly used in advertising signs. In 1912 this investigation produced a most important result.

Figure 5-14 depicts Thomson's experiment with neon. A tube is divided into two parts. In the part on the left, electrons flow from the negative cathode to the positive anode through the neon gas (at low pressure) filling that part of the tube. The electrons, dashing with high velocity through the gas, act like bullets. Whenever one collides with a neon atom, it knocks an electron out of the atom. The freed electrons, accelerated toward the anode, pick up great speed and in turn knock more electrons out of other neon atoms encountered.

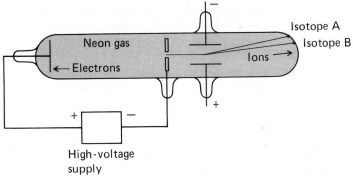

FIGURE 5-14 Separation of neon isotopes: Neon gas is ionized in the left section of the tube. Ions are pulled through the hole in the negative electrode (the cathode) and pass between the two charged plates. Ions of different masses are deflected through different angles and so are separated.

The result is an *electrical discharge* through the gas. (Other terms for the same phenomenon are *gaseous discharge* and *low-pressure electric arc*.) In the discharge is a mixture with three components: (1) electrons, (2) atoms that have lost electrons, and (3) neutral atoms that have not been affected. (See Figure 5-15.) The atoms that have lost electrons and are therefore positively charged are called **ions**; the process by which such ions are formed is called **ionization**. There are many ways for atoms to become ionized. In the gaseous discharge tube, the process is ionization by electron collision. (Such an ionized gas is called a *plasma*, one of the states of matter mentioned briefly in Chapter 3.)

The positively charged ions in the tube in Figure 5-14 are attracted to the negative cathode, and if the cathode is a solid plate, they simply collide with it. However, in the positive ray tube, there is a narrow slit in that cathode, through

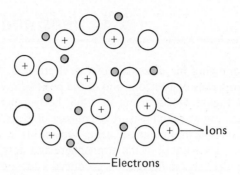

FIGURE 5-15 A plasma (ionized gas) consisting of positive ions, negative electrons, and neutral atoms. If the plasma is completely ionized, there will be no neutral atoms left.

which the positive ions pass. Once on the other side of the slit, they continue to drift through the remainder of the tube until at the far end they encounter a screen coated with zinc sulfide or some other compound that glows when struck by high-speed ions or electrons. (Instead of a fluorescent screen, a photographic film may be used. The film shows a black mark wherever the ions hit.)

In the Thomson experiment, the beam of ions travels between a pair of parallel plates connected to a battery. The positive ions are attracted toward the negative plate, and so their paths are deflected in that direction. The amount of deflection depends on the mass of the ion. The lighter ions are deflected through a larger angle, so they can be separated from the heavier ions. Thomson's instrument is, in fact, a variety of mass spectrometer. Thomson found in his experiment that a beam of neon ions was split by the mass spectrometer into two beams, and so a pair of spots showed up on the fluorescent screen. Evidently the beam of neon ions contained two kinds of neon with different masses. Careful measurement showed that, in a sample of neon, 90.9% of the atoms had an atomic mass of 20.0 amu, and most of the remainder (8.8%) had an atomic mass of 22.0 amu. Later examination showed that there were a few atoms (0.3% of the total) with an atomic mass of 21.0 amu. Conventional methods of measuring the atomic mass of neon (based on weighing a known volume of gas) gave an *average* atomic mass of 20.17 amu.

In this experiment it was demonstrated for the first time that atoms of the same element neon could exist with a number of different masses. It was subsequently found that all elements share this characteristic: they occur as mixtures of different forms having the same number of protons but different atomic masses. This discovery explained why the atomic masses of the elements, measured by chemical means, do not have integer values.

Even hydrogen, the simplest element, exists in nature in a second form with an atomic mass of 2.01 amu, twice that of the more common variety of hydrogen. The second form is called hydrogen-2, heavy hydrogen, or **deuterium**. Only 0.015% of all the hydrogen in nature is deuterium. (This percentage is called the *percent natural abundance* of deuterium.) The most abundant form of hydrogen, with an atomic mass of 1.008, is called hydrogen-1, or protium.

5.3 IONS AND ISOTOPES

The various forms of the elements, differing only in atomic mass, are called **isotopes** of the elements. Note that even the most common form of an element is considered an isotope; thus both hydrogen-1 and hydrogen-2 are isotopes of hydrogen. Every element is found to possess several isotopes. In fact, some elements have a great many isotopes. Calcium, for example, has 14 isotopes, of which 6 are stable, so they occur naturally. Some unstable isotopes are found in nature; most are manufactured in the laboratory. Unstable isotopes have a tendency to change to other, more stable isotopes over a period of time. A complete table of all known isotopes can be found in the *Handbook of Chemistry and Physics*. This table gives the percent abundance and other useful properties.

Why do isotopes exist and what makes one isotope different from another? The answer is simple. Remember that the chemical nature of an element is determined by its atomic number: the number of protons in the nucleus. Thus, in each element the number of protons is fixed: every atom of hydrogen has one proton, every atom of helium has two protons, every atom of oxygen has eight protons, and so on. However, the number of neutrons in the nucleus may differ, within limits, for the number of neutrons has no effect on the chemical properties of the atom.

The most common isotope of hydrogen has only a single proton in the nucleus. If a neutron is added to the single proton, an atom of deuterium (hydrogen-2) is formed. There is still a single electron outside the nucleus, for the number of electrons must equal the number of protons (in a neutral atom). That single electron ensures that the deuterium atom will behave chemically like hydrogen, although it has twice the mass of hydrogen-1. (See Figure 5-16.) What happens if a second neutron is added to the hydrogen nucleus? The result is an atom of hydrogen called **tritium**, with an atomic mass of 3.02 amu. Tritium is an unstable isotope and does not last indefinitely like the atoms of stable isotopes. (In any sample of tritium, half the atoms will change to one of the isotopes of helium during a 12-year period of time. This process is an example of radioactivity and will be discussed more thoroughly in Chapter 18.)

The total number of nucleons (neutrons plus protons) in the nucleus is the isotope's **mass number**:

$$\text{mass number} = \text{no. of neutrons} + \text{no. of protons}$$

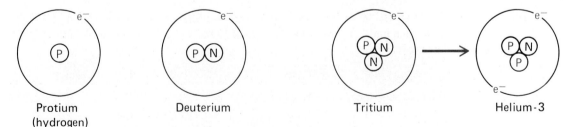

FIGURE 5-16 The structures of hydrogen, deuterium, and tritium. A tritium atom is shown changing into a helium-3 atom; this change is an example of radioactivity.

The mass number is an integer approximately equal to the isotope's atomic mass. (Atomic masses, even of isotopes, are generally not integers.) Individual isotopes are usually labeled with the mass number. Deuterium is hydrogen-2, tritium is hydrogen-3, the most commmon carbon isotope is carbon-12, and so on. The abbreviated symbol for deuterium is ^2H, for tritium it is ^3H, and for carbon-12 it is ^{12}C.

Study of a table of isotopes shows that the majority of the known isotopes are unstable. In this book we will be mainly interested in the stable isotopes found in nature. The lighter of these stable isotopes tend to have approximately equal numbers of neutrons and protons in the nucleus. However, as the number of protons in the nucleus (and hence the atomic number) increases, the electrostatic repulsion between the protons becomes very great and tends to blow the nucleus apart. This means that more neutrons are needed to keep the nucleus stabilized. (Remember that the neutrons contribute the "nuclear glue" holding the nucleus together.)

Examining the stable isotopes of mercury (atomic number 80), we find their mass numbers are 196, 198, 199, 200, 201, 202, and 204. The number of protons in the nucleus is 80, and the number of neutrons ranges from 116 to 124. Two conclusions result from this observation. First, the number of neutrons in the nucleus is considerably greater than the number of protons. Second, in order for an isotope to be stable, the number of neutrons must fall within a narrow range. If there are too few or too many neutrons, the nucleus will be unstable.

The matter of atomic masses is complicated by the fact that carbon, the standard of mass, has two natural isotopes; carbon-12 (98.89%) and carbon-13 (1.11%). By definition, the atomic mass of carbon-12 is exactly 12 amu. Natural carbon, being a mixture of carbon-12 and carbon-13, has an average atomic mass of 12.011 when measured on this scale. The atomic masses given in tables of the elements are averages over the several isotopes of each element, since that is the quantity of interest in chemical calculations.

Example 5.7   Prove that the average atomic mass of natural carbon is 12.011 amu. The composition of carbon is 98.89% carbon-12 (atomic mass 12.000 amu) and 1.11% carbon-13 (atomic mass 13.003 amu).

Solution   In every 10,000 atoms of carbon, there are 9889 atoms with a mass of 12.000 amu each and 111 atoms with a mass of 13.003 amu. The total mass of the 10,000 atoms is

$$(9889 \times 12.000 \text{ amu}) + (111 \times 13.003 \text{ amu}) = 120{,}110 \text{ amu}$$

The average mass of each atom is

$$\frac{120{,}110 \text{ amu}}{10{,}000} = 12.011 \text{ amu}$$

The various isotopes of a given element all behave the same way chemically. However, they have certain physical differences that make it possible to separate them in large quantities. For example, deuterium combines with oxygen to form heavy water. The density of heavy water is about 10% higher than that of ordinary (light) water. Its melting point is 3.8°C higher than that of ordinary water, and its boiling point is 1.4°C higher than that of ordinary water. These differences, though small, make it possible to separate heavy water from light water.

A very important difference among isotopes is the fact that the heavier atoms travel through small holes more slowly than do the light atoms. If a mixture of atoms of different masses is forced to go through a barrier with very tiny pores, the heavier ones will separate from the lighter ones. This process of gaseous diffusion is an important method in the separation of uranium-235 from the more abundant uranium-238. Uranium-235 is used as a fuel in nuclear power plants; uranium-238 is not quite as useful at the present time.

5.4 Electromagnetic Radiation and Photons

Electromagnetic Radiation

One of the fundamental questions in science is "What is light?" This is a question important to chemistry, since light is given off during many chemical reactions: for example, the light of a burning candle, the cold light of a firefly, the variety of brilliant colors formed by fireworks. What happens during such reactions that causes the emission of light?

Since the work of James Clerk Maxwell in the 1870s, it has been known that visible light consists of electromagnetic waves emitted by electrons vibrating to and fro. That is, light is made up of oscillations of an electromagnetic field. Maxwell predicted this result from purely theoretical calculations. He also predicted the properties of the waves that we now know as radio waves and showed that visible light and radio waves are basically the same thing.

It remained for Heinrich Hertz (in 1888) to create radio waves for the first time by the oscillation of electric currents in a wire loop. His experiment demonstrated that radio waves travel through space with the same speed as light waves: 300,000 km/sec (186,000 mi/sec). The fact that light waves and radio waves travel with the same speed is evidence that they are basically the same kind of wave.

In fact, the only way light waves differ from radio waves is in wavelength and frequency. (The wavelength is the distance between adjacent crests of the wave. See Figure 5-17.) The frequency, or the number of vibrations per second, is not an independent variable: the longer the wavelength, the lower the frequency. Short wavelengths have the most rapid oscillations. Visible light has a very short wavelength—between 3000 and 7000 Å ($3-7 \times 10^{-5}$ cm)—whereas radio waves have wavelengths from a millimeter up to many meters.

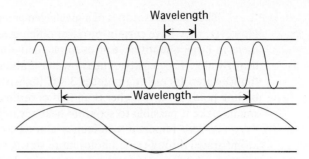

FIGURE 5-17 The wavelength of a wave is the distance between two adjacent crests. Here a short wavelength is contrasted with a long wavelength.

Furthermore, the only difference between lights of various colors is in the wavelength. Of the visible colors, violet light has the shortest wavelength, and red has the longest wavelength; all the other colors of the spectrum fall in between. There is another kind of light with wavelengths longer than those of red light (but shorter than those of radio waves). This *infrared light* is not visible to the unaided eye but can be detected by many kinds of instruments. At the other end of the spectrum, invisible *ultraviolet light* has wavelengths shorter than those of violet light. The x-rays and gamma rays mentioned earlier turn out to be electromagnetic waves with wavelengths even shorter than those of ultraviolet light. The range of all wavelengths is called *the electromagnetic spectrum* (see Figure 5-18), and all the members of the spectrum are classified under the general heading of *electromagnetic radiation*. An important way of transmitting energy through space is through electromagnetic radiation. All the energy arriving on earth from the sun travels in the form of electromagnetic radiation.

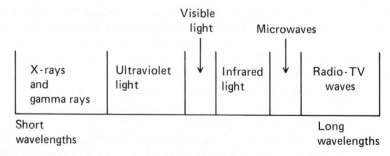

FIGURE 5-18 The range of all wavelengths of electromagnetic waves is called the electromagnetic spectrum.

Photons

In 1888 scientists were beginning to think they understood the nature of electromagnetic waves. However, just at that time they found something about these waves that they could not understand at all. The discovery was first made by Heinrich Hertz during his radio wave experiments. He found that when light (especially ultraviolet light) falls on a clean metal surface, electrons are knocked out of the metal. This phenomenon is now called the **photoelectric effect** and is the basis for many modern devices such as TV cameras and photocells.

A number of careful experiments showed that the energy of the electrons knocked out of the metal surface is directly proportional to the frequency of the light waves irradiating the surface. This result was completely unexpected and could not be explained by any of the classical theories about electromagnetic waves.

Since the old theory did not explain the observations, a new theory was needed. The new theory was provided by Albert Einstein, who, in 1905, devised an explanation for the photoelectric effect. (For this theory Einstein received the 1921 Nobel Prize in Physics.) It was a revolutionary new kind of theory, for it was based on the idea that a beam of light is not a continuous wave but rather a stream of **quanta**, or wave packets. These bundles of electromagnetic energy are now called **photons**.

The theory of the photoelectric effect is simple once the idea is accepted. The basic concept is that each photon carries with it a certain amount of energy. When a photon strikes a metal surface, all the photon's energy is transferred to one of the electrons in the metal. If that energy is enough to overcome the attraction of the metal for the electron, the electron is knocked out of the metal and flies off into the surrounding space. The notion that the energy of an electromagnetic wave is divided into little packets, the photons, revolutionized scientists' thinking. Just as matter is divided into particles, so is electromagnetic energy. The photon is now considered a fundamental particle on the same basis as the other particles.

An important question is "How much energy does a photon carry?" It turns out that the energy of the photon depends on the wavelength of the electromagnetic wave. The shorter the wavelength, the more energy is carried by the photon. This result has important consequences in chemistry. It explains why violet and ultraviolet light (short wavelengths) have more ability than red light to start chemical reactions. It explains why ultraviolet light causes tanning of the skin. It is also fundamental to understanding atomic structure by analysis of the light given off by atoms of the various elements.

An understanding of the theory of the photon is necessary to understand a great many kinds of chemical reactions. One of the chief ways in which energy is transferred from one atom to another is by the emission of electromagnetic radiation by one atom and its absorption by another atom. The creation of living plant tissue by the chlorophyll of a plant leaf is an example of a chemical reaction stimulated by the absorption of photons from the sun.

5.5 Electron Orbits

The Bohr Atom

Most of chemistry is determined by the configuration, or pattern, of the electrons in orbit around the nucleus. One of the fundamental problems of atomic theory is to decide how these electrons are arranged. The history of this subject is an example of scientific method at work, with an interplay of experiment and mathematical theory.

By the second decade of the twentieth century, certain facts about atoms were well known but had yet to be explained by a proper atomic theory. For example, it was known that when various gases are heated, or when they are ionized by passing electric currents through them, they emit light of different colors. (Today we encounter this fact every time we look at a "neon" advertising sign. A tube filled with neon gas produces a reddish-orange color, mercury vapor gives a greenish-blue light, xenon gives the white light emitted by photoflash tubes, and so on.)

During the nineteenth century, scientists made a systematic study of the light coming from electric arcs and discharge tubes. In this study spectroscopes (such as the one shown in Figure 5-19) were used to separate the light of different wavelengths and to measure these wavelengths. It was found that each element emits light with a pattern of particular wavelengths. An entire science of spectroscopy grew up, the purpose of which was to analyze materials by measuring the wavelengths in the **spectrum** of light emitted when those materials were vaporized and heated in an electric arc.

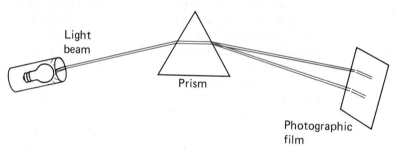

FIGURE 5-19   A spectroscope: Light passing through a narrow slit is separated by a glass prism into separate beams of different colors.

Hydrogen emits the simplest spectrum—a set of lines that are spaced more and more closely together as a certain limiting wavelength is approached (see Figure 5-20). The wavelengths of these lines had been measured and studied extensively for years, while scientists tried without success to explain why those particular wavelengths are emitted. Finally, in 1913, the Danish physicist Niels Bohr devised an atomic model which, though incomplete, did allow him to

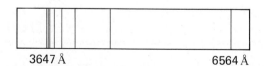

FIGURE 5-20   The spectrum of hydrogen consists of a series of lines (each corresponding to a different wavelength) that get closer and closer together as they go toward the shorter wavelengths.

calculate the correct wavelengths found in the hydrogen spectrum. Bohr chose to study hydrogen because it is the simplest atom.

The Bohr model made the following assumptions:

1. In the hydrogen atom, there is a single electron that travels in a circular orbit around the nucleus. (Later the theory was improved to include elliptical orbits.)

2. There are an infinite number of orbits, with different radii, that the electron may occupy. However, an electron is allowed to travel *only* in certain particular orbits located at certain distances from the atomic nucleus. (This is quite different from the situation of planets going around the sun. A planet can move in an orbit of any radius, depending on how fast it is going.) Each electron orbit is given a number, starting with the one closest to the nucleus. Thus the orbits in Figure 5-21 are labeled 1, 2, 3, ... in order of increasing sizes of the radius. Atomic scientists call the number labeling the orbit a **quantum number**, and it is represented by the symbol n.

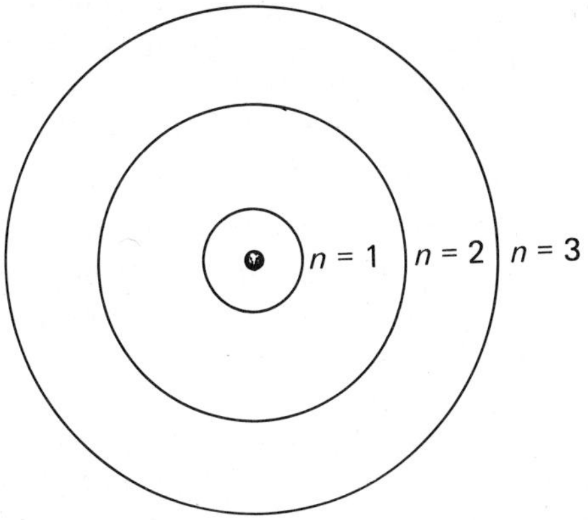

FIGURE 5-21 The orbits that an electron may have in a hydrogen atom. These orbits are labeled by the quantum numbers $n = 1, 2, 3, \ldots$.

3. The radius of a particular orbit and the velocity of the electron in that orbit are calculated by a mathematical procedure developed by Bohr. This procedure involves some assumptions that could not be explained initially but that were introduced because they enabled the correct wavelengths of the hydrogen spectrum to be calculated.

4. The first step in calculating the wavelengths of the light emitted by hydrogen is to calculate the energy of the electron in each of its allowed orbits. The electron energy in this context is the total energy—that is, the kinetic energy plus the potential energy. Kinetic energy is always a positive quantity, but potential energy in this situation is a negative quantity. The negative potential energy arises because of the *attraction* between the electron and the nucleus. We will not go through the details of Bohr's calculation but will simply quote the results, as follows.

The electron in each of its possible orbits has an amount of energy that depends on the quantum number n and can be calculated from a very simple formula:

$$E_n = -\frac{2.18 \times 10^{-18} \text{ J}}{n^2}$$

$$= -\frac{13.6}{n^2} \text{ eV}$$

In this formula E_n represents an allowed value of energy for the orbital electron, and n is the integer $(1, 2, 3, \ldots)$ that labels the orbit. For each value of n, there is a different value for the energy the electron is allowed to have. The **electron-volt (eV)** is a unit of energy commonly used in dealing with atomic energy levels. One electron-volt equals 1.60×10^{-19} J.

Figure 5-22 is a diagram showing the amounts of energy associated with the different orbits. This diagram is like a ladder, with each rung representing the energy of an electron in each of the possible orbits. The lowest rung has an energy of -13.6 eV, and the energy increases as the quantum number n (and the orbit radius) becomes greater.

Each rung in this ladder is called an **energy level**, and the diagram in Figure 5-22 is an *energy level diagram*.

5. The one electron of the hydrogen atom will normally be found in the orbit with the lowest possible amount of energy. It can go no lower. This bottom rung on the energy ladder is called the **ground state** of the electron. It is a stable state because any system always tends to fall into the state representing the least amount of energy.

The fact that there is a bottom rung on the energy ladder was a result unprecedented in science. All the older theories of electron behavior had predicted that an electron orbiting a nucleus would gradually lose energy and eventually fall into the nucleus. But electrons can fall no lower than the lowest allowable orbit—the ground state. If not for this fact, matter as we know it could not exist.

6. An electron can jump up into energy levels higher than the ground state if given extra energy. This is what happens when the atom is knocked about in an electrical discharge. The electron does not stay for long in its **excited state.** It immediately drops back to the ground state, and, in order to get rid of its extra energy, it emits a photon of light. Electrons tumbling down the energy ladder are the source of light emitted by the discharge.

7. The photon energy determines the wavelength of the emitted light. The greater the energy, the shorter the wavelength. Using this principle, Bohr was able to calculate the exact wavelengths of the light emitted from the hydrogen in a gaseous discharge tube.

Without going into the details of the calculation, we can see qualitatively what happens by referring to Figure 5-22. When an electron drops from level 2 to level 1, it loses a relatively large amount of energy, and as a result the emitted light is in the ultraviolet region. If it drops from level 3 to level 2, it loses less energy, and the emitted light is in the visible part of the spectrum.

All of the transitions from higher levels down to level 2 result in light with wavelengths in the visible part of the spectrum. Each transition gives rise to one of the lines in the spectrum, each of which corresponds to a different wavelength.

The wavelengths calculated by Bohr from the theory outlined above gave startling agreement with the experimentally measured wavelengths. For the first time in history, the wavelengths of the light emitted by an atom had been calculated from some kind of theory, crude as it was.

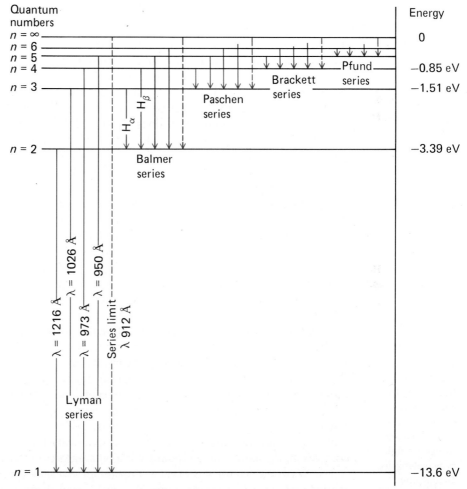

FIGURE 5-22 The energy levels of hydrogen: The levels get closer and closer together as the quantum number n becomes greater. Transitions between different sets of levels produce the various lines of the spectrum. The Lyman series is in the ultraviolet region, the Balmer series is in the visible light region, and the Paschen, Brackett, and Pfund series are in the infrared light region of the spectrum.

The Bohr theory was only a stepping stone. It was what is known as an *ad hoc* theory, a theory invented just to explain one set of facts. There was no underlying explanation for certain assumptions basic to the theory. It was simply a hypothesis that worked—it gave the right answer.

The Schroedinger Model

In 1923 the French physicist Louis de Broglie proposed a revolutionary theory of matter. It was based on the idea already proposed by Max Planck and Albert Einstein that light, which had been thought of as a wave phenomenon, was composed of particles (photons). De Broglie turned this idea around and said that perhaps an electron, which we think of as a particle, has some kind of wave associated with it. This wave is not an electromagnetic wave or any other kind of familiar wave but something new called a **matter wave**.

De Broglie's hypothesis turned out to be completely correct. Many experiments over the past half-century have demonstrated the presence of waves connected with electrons. The electron has a measurable wavelength and in every way acts like a wave. Yet the electron is also a particle. It is localized, has an electric charge, can bounce like a billiard ball off other particles, and altogether acts like a particle. Thus we call it a *wave packet*—a tiny bundle of waves.

The concept of matter waves was a purely theoretical idea to de Broglie, but the German physicist Erwin Schroedinger quickly realized how it could be put to use in making a new model of the atom. He imagined the single electron of a hydrogen atom to be a wave traveling in a circle about the nucleus (see Figure 5-23). In order for such a wave to exist in a circular orbit, there must be a whole number of wavelengths around the circumference of the circle—that is, the number of wavelengths must be 1, 2, 3, 4, and so on. Starting with these assumptions, Schroedinger constructed a complex equation representing the electron wave surrounding the nucleus. From this equation he was able to calculate what the allowed orbits and energy levels were for the hydrogen atom. Finally he arrived at an equation for the wavelengths of the emitted light (the hydrogen spectrum) that was *identical* to the equation obtained by Bohr. However, Schroedinger's equation was a great improvement over Bohr's theory. By thinking of the electron as a wave, Schroedinger arrived at a logical explanation for the properties of the hydrogen atom—an explanation based on deeper concepts and with fewer assumptions.

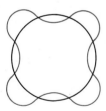

FIGURE 5-23
Circular waves: A model of the hydrogen atom, showing how standing waves (waves having an integral number of wavelengths around the circumference of the atom) can result in the observed energy levels.

The picture given above is a very oversimplified idea of the Schroedinger theory. To depict the hydrogen atom properly, we must think of a wave in three dimensions, surrounding the atom like a spherical cloud. The model involves complex mathematics, but it accounts for all the features of the hydrogen spectrum.

Wave mechanics is a branch of **quantum mechanics**, the modern study of matter and its behavior. Quantum theory is very abstract, but it is the most successful theory in the history of science because it has enabled us to understand the structure of matter in great detail. It has been extended to atoms with many electrons, to molecules, and to solid materials. Even the simplest molecule (the hydrogen molecule) cannot be understood without the use of quantum mechanics, because the classical theories have no way of explaining why two hydrogen atoms should hold together as a stable unit. Therefore quantum theory is basic to modern chemistry.

5.6 Multielectron Atoms and Shell Structure (Optional)

In a hydrogen atom with its single electron in the lowest (ground) state, the electron cloud surrounding the nucleus is simply a spherical blob. However, this cloud is not uniform. A plot of the density of this sphere as a function of radius (Figure 5-24) shows that the greatest density occurs about half an angstrom away from the nucleus. This distance is exactly the radius of the Bohr atom.

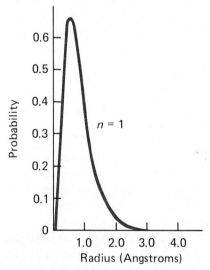

FIGURE 5-24 The probability function for the first level of hydrogen: The probability of finding the electron in a given region of space is greatest at a distance of 0.5 Å from the nucleus.

No microscope is able to show visually the details of atomic structure, since the atom is smaller than the minimum wavelength of visible light. Curves like those in Figure 5-24 are calculated mathematically from Schroedinger's wave theory. In addition there are experiments that verify the theoretical structure in considerable detail through indirect means, such as by scattering of electrons from the atom.

An important question is: what is the exact meaning of the electron cloud surrounding the atom? Is an electron actually a cloud of electric charge surrounding the nucleus? The answer is a firm NO! Whenever experiments are done to determine the electron's position at a given moment, the electron is found to exist at a particular point in the space around the atom. It is not really a cloud or a blob. However, each time the electron is detected, it is found to be in a different place.

The interpretation of the Schroedinger wave packet is one of the deepest problems in science—one that still has certain unanswered questions. The currently accepted interpretation of the cloudlike picture is that the density of the cloud in the diagram represents the *probability* of finding the electron at a given point in space. The curve of Figure 5-24 is called the **probability density** of the electron wave function. The electron is most likely to be found where the cloud is most dense, at the 0.5 Å distance. But it might also be found closer to the nucleus, or it might be found farther away. The height of the curve represents the chance that the electron will be detected at a given location.

Calculations about atomic electrons only give us information about probabilities and averages. The peak of the curve in Figure 5-24 is approximately the *average* position of the electron orbit; the electron itself may be found with equal likelihood *anywhere* around the circumference of the orbit.

An analogy may help clarify the concept of the electron probability density. Say there are 10 rows of seats in a classroom, and every day a particular student sits in a different seat, choosing his location at random. Suppose also that the student spends 100 days of the school year in that room. Five of those days he spends in the first row, 20 days in the second row, 30 days in the third row, 25 days in the fourth row, and so on. You do not know where the student is going to be on any particular day, but you can figure that the odds on his being in the third row instead of the first row are six to one. Putting it another way, the probability of finding him in the first row on any given day is 0.05 (5/100), whereas the probability of finding him in the third row is 0.30 (30/100). The curve in Figure 5-24 gives us much the same kind of information about the electron at various distances from the nucleus of the atom.

So far we have used only the energy of the electron as a variable to specify how one hydrogen atom differs from another. However, to describe an atom completely we must use several variables, especially when we are dealing with atoms containing more than one electron.

The energy is the most important variable, and, as we have seen, the energy for a given electron can take on any of a number of allowed values, depending on the quantum number n (1, 2, 3, 4, ...). The fact that the atomic electron can exist only with certain particular values of energy is a most important result. This idea represents a competely revolutionary break from classical physics. In the older physics, there was no reason for any rule prohibiting any particular amount or range of energy. In quantum physics it is the wave nature of the electron that forces it to take on only certain special values of energy when it is part of an atomic system.

In addition to the energy, three other variables are needed to give a full description of each electron in the atom. Two of these variables have to do with the motion of the electrons around the nucleus, and the last variable is related to the spin of the electron around its own axis. Taken as a whole, the atom is a complex system of orbiting and spinning motions—somewhat like the planets of the solar system spinning on their axes as they revolve around the sun, but in a more abstract manner.

There are four quantum numbers corresponding to the four variables. We have already seen that the energy of an electron depends on the quantum number n. The orbital motion depends on two quantum numbers called ℓ and m. The amount of spin depends on a quantum number called s. We will not go into detail about the meanings of these numbers, but in Chapter 6 we will show how these numbers determine the structure of multielectron atoms—atoms with more than one orbital electron. The quantum numbers act like a set of code numbers, regulating the number of electrons that can exist in each of the possible orbits around the nucleus. Knowledge of quantum numbers allows us to understand why each element reacts the way it does. In a real sense, the quantum numbers are the foundation of chemistry.

Chemists describe the structure of a multielectron atom using the concept of **shells** and subshells. The shells are related to the various orbits of the Bohr and Schroedinger models. In each shell the electron has an energy that depends mainly on the quantum number n and also to some extent on the number ℓ. For this reason each shell is labeled with the number n, and each subshell within a shell is labeled with the number ℓ. The fundamental rule governing the structure of atoms is that *no more than a fixed number of electrons can be located in a given subshell.*

Another fundamental principle determining atomic structure is that the electrons in an atom locate themselves within the various shells in such a way that their total energy will be a minimum. This rule follows from a general physical principle that any system will arrange itself in such a way that its energy is as low as it can get under the constraints that the surroundings put on the system.

Consequently, the single electron of the hydrogen atom normally is located in the lowest subshell, or energy level. This subshell is called the 1s subshell. In this notation the number 1 represents the quantum number n and is the label for the first shell. The letter s is the label for the subshell within the first shell.

It has become customary to label subshells with the letters s, p, d, and f. Each of these letters represents a different value of the quantum number ℓ, as shown in Table 5-4. In addition, Table 5-4 shows the maximum number of electrons that can exist within each subshell.

We can see how the subshell scheme works by considering the first few elements. We have already seen that in hydrogen the electron is in the 1s subshell. If another electron is added to the atom (and a proton and two neutrons are added to

TABLE 5-4 Subshell Values

Subshell	ℓ Value	Maximum Number of Electrons
s	0	2
p	1	6
d	2	10
f	3	14

the nucleus), an atom of helium is obtained. This second electron also goes into the $1s$ subshell. We see from Table 5-4 that no more than two electrons can reside in that shell. Therefore, with helium the $1s$ subshell is *completed*. (The reason two electrons are allowed in the $1s$ subshell has to do with the fact that one of the electrons spins in a clockwise direction while the other spins in a counterclockwise direction.)

If a third electron is added to form lithium, this new electron must go up into the $2s$ subshell (that is, the s subshell of shell number 2). We describe the *electron configuration* of the atom by writing out all the subshell symbols as follows: $1s^2 2s^1$. The superscripts tell us the number of electrons in each subshell. Thus, the expression $1s^2 2s^1$ says that there are two electrons in the $1s$ subshell and one electron in the $2s$ subshell. In Table 5-5 the electron configurations for the first 20 elements are listed. You can see that every time a subshell becomes filled, the new electrons go into a higher subshell.

TABLE 5-5   *Electron Configurations for Twenty Common Elements*

Element	Atomic Number	Electron Configuration
hydrogen	1	$1s^1$
helium	2	$1s^2$
lithium	3	$1s^2 2s^1$
beryllium	4	$1s^2 2s^2$
boron	5	$1s^2 2s^2 2p^1$
carbon	6	$1s^2 2s^2 2p^2$
nitrogen	7	$1s^2 2s^2 2p^3$
oxygen	8	$1s^2 2s^2 2p^4$
fluorine	9	$1s^2 2s^2 2p^5$
neon	10	$1s^2 2s^2 2p^6$
sodium	11	$1s^2 2s^2 2p^6 3s^1$
magnesium	12	$1s^2 2s^2 2p^6 3s^2$
aluminum	13	$1s^2 2s^2 2p^6 3s^2 3p^1$
silicon	14	$1s^2 2s^2 2p^6 3s^2 3p^2$
phosphorus	15	$1s^2 2s^2 2p^6 3s^2 3p^3$
sulfur	16	$1s^2 2s^2 2p^6 3s^2 3p^4$
chlorine	17	$1s^2 2s^2 2p^6 3s^2 3p^5$
argon	18	$1s^2 2s^2 2p^6 3s^2 3p^6$
potassium	19	$1s^2 2s^2 2p^6 3s^2 3p^6 4s^1$
calcium	20	$1s^2 2s^2 2p^6 3s^2 3p^6 4s^2$

In this discussion we have not tried to explain *why* each subshell becomes filled with a certain maximum number of electrons. In Chapter 6 we will go into more detail on this topic. The exact reasons for the limitation on the numbers of electrons allowed in each subshell belong in the highly mathematical realm of quantum mechanics.

An electron shell is not simply a circular orbit or a spherical ball. It may have a complex shape, depending on the quantum numbers labeling the shell. The shape of the shell represents the locations in space where the electrons are most

5.6 MULTIELECTRON ATOMS AND SHELL STRUCTURE

likely to be found and is related to the probability density curves that we spoke of earlier in this section.

Figure 5-25 shows some of the shapes belonging to a few of the simplest states of the hydrogen atom. Chemists call these forms **orbitals**. An orbital shows the electron probability density in three dimensions. In Chapter 7, when we deal with bonding of one atom to another, we will see that the shapes of the orbitals determine the way atoms interact with each other and also determine the form of the molecular and crystalline structures that may be formed.

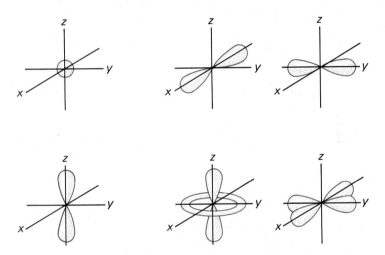

FIGURE 5-25 Orbital diagrams: The shapes depend on the l and m quantum numbers. When $l = 0$, the shape is a simple sphere.

Models in Science

In this chapter we have frequently mentioned models: the Bohr model of the atom, the Schroedinger model, and so on. In science models help us visualize abstract concepts, invisible objects, and complex systems. A model is not necessarily a solid, manufactured object like a model airplane. In science a model may be a set of mathematical equations, a diagram, a set of rules—any kind of picture or set of symbols that describes a thing or a system, such as the workings of an atom, a molecule, a living being, an economic system, a galaxy, or even the entire universe.

The Bohr and the Schroedinger models are not the only atomic models in use. Other models are even more abstract—for example, the Heisenberg model, which is no more than a set of equations that allows the scientist to predict such quantities as the wavelengths of the light emitted by an atom. The abstract models do not attempt to give you a visual picture, with shape or form. Instead, they describe what the atom does and aim to predict the results of the atom's actions. Indeed, only when a model is able to predict observable results can it be called a scientific model.

A good scientific model should have certain general features:

1. A model must be a *representation* of the thing or system being modeled. It is in some way a description of that thing or system. Often the model helps us visualize the thing being modeled, but in some cases the model is so abstract and mathematical that it cannot be translated into pictures or images. The model of the atom given by modern quantum theory is actually completely impossible to visualize, even though every textbook includes orbital diagrams to help the student imagine the unimaginable.

2. A good model must *predict* the behavior of the thing being modeled, and the prediction must be about some observable phenomenon. The reason the Schroedinger model of the atom is taken seriously is because it does give predictions of wavelengths that are in close agreement with the observed wavelengths of light emitted by atoms. (In certain sciences *retrodictions*, or predictions of the past, are permissible. For example, in geology we use our knowledge to "predict" what must have happened in the past.)

3. A good model should *explain* why things behave the way they do. In science, *explain* means to invoke the established laws of nature together with the assumptions of the specific model to demonstrate how the behavior of a given system follows from the workings of the inner parts of that system. For example, to explain why a metal conducts electricity we use our model of a metal as a substance containing a large number of free electrons. Knowing how electrons respond to the force of an electric field and knowing how they collide with the atoms in the metal when they move through it, we can deduce the behavior of the conductor, calculate its electrical resistance, and even understand why the resistance changes when the metal is heated.

4. Two models of the same system may differ in form but must give the same predictions if they are to be equally valid. As we have seen, the Bohr model and the Schroedinger model of the atom give the same predictions as far as the wavelengths of light emitted by the hydrogen atom are concerned. However, when we look into other predictions, we find differences appearing. The Bohr model does not correctly describe what happens when the atom is located in a magnetic field. It does not explain why two hydrogen atoms join together to make a molecule. It has no explanation for the way electrons remain in certain special orbits or for why there is a lowest stable energy level (the ground state). The Schroedinger model is very successful in explaining all these phenomena and therefore is preferable to the Bohr model, even though it is more complicated.

In later chapters of this book we will call upon our understanding of atomic models to explain a number of phenomena that take place in chemistry: Why are some substances soluble in water and others insoluble? Why is heat given off in some reactions and absorbed in others? The reason we devote so much attention to basic atomic structures is so that we can understand such phenomena.

Glossary

angstrom (Å) A unit of length defined as 10^{-10} m.

atomic mass The mass of an atom, measured in atomic mass units (amu) or kilograms.

atomic mass unit (amu) A unit of mass equal to 1/12 the mass of a carbon-12 atom.

atomic number The number of protons in the nucleus of an atom.

deuterium An isotope of hydrogen with an atomic mass of 2; it contains one proton and one neutron in its nucleus.

electron A negatively charged particle with a mass 1/1837 that of a hydrogen atom.

electron-volt (eV) A unit of energy equal to 1.60×10^{-19} J. It is the amount of energy gained by an electron when accelerated through a potential difference of 1 volt.

energy level The state of an electron in a particular orbit with a fixed amount of energy.

excited state Any energy level higher than the ground state (or lowest allowed state).

fluorescence The light emitted by materials as the electrons in their atoms fall from excited states toward the ground state.

force A push or pull causing an object to change its state of motion.

ground state The lowest energy level (the orbit closest to the nucleus) that may be occupied by an electron.

inverse-square force A force whose strength varies inversely as the square of the distance between the centers of the two interacting objects.

ion A charged atom or group of atoms; the charge results from the fact that the number of orbital electrons is different from the number of nuclear protons.

ionization The process of forming an ion from a neutral atom or molecule.

isotopes Different forms of an element, characterized by the same atomic number but with differing atomic masses.

mass number The total number of neutrons and protons in an atomic nucleus.

mass spectrometer A device used to measure the mass of ions by passing them through a magnetic (or electric) field. Ions of different masses are separated because the path traveled by the less massive ions has greater curvature than that of the more massive ions.

matter wave A wave associated with an atomic particle.

neutron An uncharged particle found in the nucleus of an atom, with approximately the same mass as a proton.

nuclear force A short-range force felt between neutrons and protons (as well as between protons and protons) that is strong enough to hold the nuclear particles together in spite of the mutual electric repulsion of the protons.

nucleus The central core of an atom, containing most of the atom's mass in the form of neutrons and protons.

orbital A region of space in an atom in which a given electron is most likely to be found. The shape of the orbital depends on the quantum numbers describing the particular electron.

photoelectric effect The ejection of electrons from a material surface by photons of light striking the surface.

photon A packet or bundle of light waves having mass as well as energy; a quantum of energy.

probability density A mathematical quantity representing the likelihood of detecting a particle within a given volume of space at a given location.

protons Positively charged particles found in the nucleus of an atom, each with a mass of 1.67×10^{-27} kg.

quanta (singular: **quantum**) See **photon**.

quantum mechanics The modern study of matter and its behavior; see also **wave mechanics**.

quantum number A number that determines the energy and other properties of an atom.

radioactivity A process by which an atomic nucleus changes its state (or changes into another nucleus) by emitting one or more particles.

shell The region of space surrounding an atomic

nucleus and containing electrons with a given value of quantum number n.

spectrum The range of wavelengths to be found in a beam of light, or in electromagnetic radiation in general.

tritium An isotope of hydrogen with an atomic mass of 3; it is unstable and contains one proton and two neutrons in its nucleus.

wave mechanics A theory based on the concept that atomic particles have matter waves associated with them, so the allowed forms and energies of atoms are determined by the possible patterns of standing waves.

wave packet A localized bundle of waves; may represent either a photon or an electron or other particle.

Problems and Questions

5.1 Atomic Theory

5.1.Q Name the natural forces that have been discussed in this section, and describe the action of each.

5.2.Q What are the two kinds of electric charges?

5.3.Q What is the law of attraction between electric charges? Why do some charges attract each other, while others repel each other?

5.4.Q How would you determine whether an object was electrically charged?

5.5.Q Describe one similarity and one difference between the gravitational force and the electric force.

5.6.Q What is meant by
   a. an inverse-square force?
   b. a long-range force?

5.7.Q What is the law of attraction and repulsion for magnetic poles?

5.8.Q Two cars traveling down the highway in the same direction are attracted to each other by the gravitational force, yet they do not collide with each other. Explain why.

5.9.Q A magnet does not attract a piece of paper. Yet the same magnet will hold a piece of paper to a refrigerator door even though gravity is pulling down on the paper. Explain what is happening.

5.10.Q A beaker contains a mixture of iron filings and sulfur. The iron filings can be separated from the sulfur by a magnet. Explain this effect in terms of fundamental forces.

5.11.Q Describe the electron and the proton.

5.12.Q Describe the action of a cathode ray tube.

5.13.Q What is the relationship between cathode rays and electrons?

5.14.Q What is a major difference between the atomic theory of John Dalton and the atomic theory of J. J. Thomson?

5.15.Q How do we know that electrons have negative electric charges?

5.16.Q How is it that atoms are electrically neutral?

5.17.Q Why is J. J. Thomson's theory called the "raisin-pudding" theory?

5.18.Q What is radioactivity?

5.19.Q Describe the three kinds of radiation that come from uranium.

5.20.Q Describe the Rutherford experiment.

5.21.Q Explain how the Rutherford experiment disproved the raisin-pudding model of the atom.

5.22.Q Describe the neutron. How is it similar to and how is it different from
   a. the electron?
   b. the proton?

5.23.Q Describe the strong nuclear force and explain why such a force is necessary to stabilize nuclear structure.

5.24.Q Where is most of the mass of an atom located?

5.25.Q Where are the electrons located in an atom, according to the Rutherford model?

5.26.Q Describe how scientists investigate the interior of an atomic nucleus.

5.27.Q Why is it not possible to give a precise number for the diameter of an atom?

5.28.Q How many protons can be lined up along the diameter of a hydrogen atom?

5.29.Q How do we know that the electric charge on a proton is exactly equal in magnitude to the electric charge on an electron?

5.30.Q a. What is a fundamental particle?
b. What are the two most important properties of fundamental particles?

5.31.Q a. What are the three fundamental particles that make up an atom?
b. What are the masses and electric charges of these particles?

5.32.Q What is an ion?

5.33.Q Describe a mass spectrometer.

5.34.Q Describe the Chadwick experiment for measuring the mass of a neutron.

5.35.P a. The head of a pin is a circle 1 mm in diameter. How many hydrogen atoms can fit into the area of the pin head? (Remember that the area of a circle is equal to πr^2. Assume that each hydrogen atom occupies a block 1 Å on a side.)
b. How many protons would fit onto the head of the same pin?

5.2 Atomic Number and Atomic Mass

5.36.Q Define the following terms:
a. relative atomic weight
b. atomic mass
c. atomic number
d. atomic mass unit

5.37.Q Refer to a table of atomic masses to answer the following:
a. What is the range of atomic masses found in the table of elements?
b. What is the range of atomic numbers?
c. Is the atomic mass of an element smaller than, equal to, or greater than its atomic number? Explain why.

5.38.Q Which is more important in determining the chemical properties of an element—the atomic number or the atomic mass? Explain your answer.

5.39.P If your instructor weighs 150 lb, and that weight is defined as 1 standard relative weight, what is your own relative weight?

5.40.P A molecule of the compound methane (CH_4) contains 4 hydrogen atoms and 1 carbon atom; 2.25 g of carbon reacts with 0.756 g of hydrogen to form this compound. What is the atomic mass of hydrogen?

5.41.P A molecule of the compound carbon tetrachloride (CCl_4) contains 1 carbon atom and 4 chlorine atoms; 25.0 g of carbon is combined with 293 g of chlorine to create this compound. Calculate the atomic mass of chlorine.

5.42.P Calculate the number of grams of iron in 500 g of iron oxide (FeO).

5.43.P How much sodium can be obtained from 2470 kg of sodium chloride (NaCl)?

5.3 Ions and Isotopes

5.44.Q Define or explain each of the following terms:
a. an ion
b. a gaseous discharge
c. a plasma
d. atomic number
e. mass number
f. atomic mass

5.45.Q a. What is an isotope?
b. What is the difference in the atomic structure of two isotopes of the same element?

5.46.Q a. Protium, deuterium, and tritium are isotopes of what element?
b. Give the atomic number and mass number of each of the above isotopes.
c. Describe the structure of the nucleus of each of the above isotopes.

5.47.Q a. Is the number of neutrons in a nucleus usually less than, greater than, or equal to the number of protons?
b. Give the reason for your answer to part a of this question.

5.48.Q What is the present standard of atomic mass?

5.49.Q Describe three ways in which the isotopes of a given element may differ from one another.

5.50.P A new element is found to consist of two isotopes. One, with an atomic mass of 107 amu, makes up 48.5% of all the atoms of this element. The other 51.5% of the atoms have an atomic mass of 109 amu. What is the average atomic mass of this element?

5.51.P Neon has three isotopes with the following atomic masses and natural abundances.

Atomic Mass	Percent Abundance
19.99	90.92%
20.99	0.257%
21.99	8.82%

What is the average atomic mass of this element?

5.52.P Magnesium has isotopes with the following atomic masses and natural abundances.

Atomic Mass	Percent Abundance
23.99	78.7%
24.99	10.1%
25.98	11.2%

   a. Describe the nucleus of each isotope.
   b. Calculate the average atomic mass of natural magnesium.
   c. How many electrons are there in each isotope?

5.4 Electromagnetic Radiation and Photons

5.53.Q What do radio waves and light waves have in common?

5.54.Q What is the electromagnetic spectrum?

5.55.Q List the various kinds of radiation in the electromagnetic spectrum in order, starting with the shortest wavelength.

5.56.Q Name three chemical processes that result in the emission of light.

5.57.Q Name two chemical processes requiring the absorption of light.

5.58.Q What is a photon?

5.59.Q Describe the photoelectric effect.

5.60.Q On what factor does the energy of a photon depend?

5.5 Electron Orbits

5.61.Q Describe the Bohr model of the atom.

5.62.Q Describe the process by which light is emitted by an atom in the Bohr model.

5.63.Q Define the following terms:
   a. wave packet   **b.** ground level
   c. excited state   **d.** energy level

5.64.Q Is the energy of an electron in the ground state greater than or less than the energies of electrons in the other states?

5.65.Q Describe the Schroedinger model of the atom. What is the basic assumption of this model?

5.66.Q Give two similarities between the Bohr and the Schroedinger models. Give two differences.

5.67.Q How does the wavelength of a photon depend on its energy?

5.68.Q
   a. When an electron makes a transition from an excited state to the ground state, in what part of the spectrum is the wavelength of the emitted photon?
   b. When an electron makes a transition from an excited state to the second level, in what part of the spectrum is the wavelength of the emitted photon? What is the name of the series of lines in the spectrum?
   c. Explain the difference between the photons in parts a and b of this question.

5.6 Multielectron Atoms and Shell Structure

5.69.Q Why do we not think of an electron as a cloud surrounding a nucleus?

5.70.Q What do we mean by probability density?

5.71.Q What is an orbital?

5.72.Q
   a. What are the four atomic quantum numbers?
   b. What physical property does each of the four quantum numbers determine?

5.73.Q What two rules determine the energy level of a new electron added to an atom?

5.74.Q Why is it not possible to put a third electron into the lowest electron shell?

5.75.Q What is meant by a closed, or completed, shell?

5.76.Q Which elements among the first 20 have complete, or filled, outer shells? What do these elements have in common?

5.77.Q List the electron configurations of the following elements:
   a. lithium
   b. sodium
   c. potassium
What do these elements have in common?

Self Test

1. What is a force?

2. Name three of the fundamental forces in nature.

3. What two laboratory devices led to the discovery of electrons and protons and to the development of atomic theory?

4. What is the fundamental particle of negative electric charge?

5. What is our evidence for thinking that most of the mass of an atom is located in its nucleus?

6. Describe the interaction between each of the following pairs:
   a. two protons
   b. two electrons
   c. an electron and a proton
   d. two neutrons

7. How does the force between two protons vary as the distance between them changes? What happens when they get very close together?

8. Explain why the element silver (Ag) has an atomic number of 47 but an atomic mass of 107.868. Why can't we find the atomic mass simply by adding the number of neutrons and the number of protons in the atom?

9. What percentage of the mass of a silver atom is made up of its electrons?

10. Zinc (atomic number 30) is found in nature with the following isotopes, natural abundances, and atomic masses:

	Percent Abundance	Atomic Mass
^{64}Zn	48.89%	63.93 amu
^{66}Zn	27.81%	65.93 amu
^{67}Zn	4.11%	66.93 amu
^{68}Zn	18.57%	67.92 amu
^{70}Zn	0.62%	69.93 amu

What is the average atomic mass of natural zinc?

11. What evidence leads us to think that light consists of photons?

12. Explain why the nucleus of an atom that contains many protons does not disintegrate from the mutual repulsion of the protons.

13. a. State the assumptions of the Bohr model of the atom.
    b. State the assumptions of the Schroedinger model of the atom.
    c. What are the important differences between these two models?

14. Why do the electrons surrounding the nucleus not come crashing into the nucleus because of the attraction between the electrons and the nucleus?

15. What can the Schroedinger model explain that the Bohr model cannot?

16. a. What is a quantum number?
    b. How many quantum numbers does it take to describe the state of an electron in an atom?
    c. Which quantum number determines the energy of the electron in a hydrogen atom?

17. a. How many electrons are allowed in the first atomic shell?
    b. In what element is the first shell filled with electrons?
    c. If we added another electron to the element described in part b of this question, into what shell would it go?
    d. What element is described in part c of this question?

18. What is an electron configuration?

19. a. What is meant by the symbol $1s^2 2s^2$?
    b. What element is represented by this configuration?

20. Explain why a model in science must be able to predict some observable occurrence.

21. Describe some of the observable phenomena predicted by the Schroedinger model of the atom.

6
The Periodic Table

Objectives After completing this chapter, you should be able to:

1. Describe some of the properties of the alkali metal and halogen families of elements.
2. Explain the meaning of the term "periodic."
3. Describe Lothar Meyer's contribution to the concept of periodic properties of elements.
4. Describe how Mendeleev constructed his periodic table.
5. Describe the predictions that made the periodic table a valuable tool.
6. Describe the modern periodic table.
7. Explain the meaning of groups and periods.
8. Explain the relationship between the atomic number and the number of protons in the atomic nucleus.
9. Explain the reason for the periodic table in terms of atomic structure.
10. Describe the various shells and subshells of a multielectron atom.
11. Describe the general properties of the elements in the various groups, and explain how these properties are related to the electron configurations of these elements.
12. Describe the general properties of metals, semimetals, and nonmetals.
13. Describe some of the more important trends in properties of the elements.

6.1 Historical Background

To the chemist the periodic table is a familiar and useful tool. But little more than a century ago, the world of chemistry scoffed at the idea that such an arrangement made any kind of sense. This skepticism was, in part, due to the fact that the chemists of the nineteenth century had no knowledge of atomic structure. They knew nothing about atomic electrons or how the number of electrons determines atomic properties. Chemists of that period knew that the atomic mass (or weight) of an element was a fundamental property, and they were vaguely aware that if you arranged all the elements in order of increasing atomic mass, you could find a few interesting relationships. However, atomic masses were difficult to measure accurately, and the relationships found did not seem very reliable or of great use.

What were some of these relationships? It was known, first of all, that certain elements are similar to one another, both physically and in chemical properties. These groups of elements were called "families."

One family of elements consists of lithium (Li), sodium (Na), potassium (K), rubidium (Rb), and cesium (Cs). This group is called the family of **alkali metals**. Each of the elements is a soft, silvery metal with a low melting point and a high degree of chemical activity. Sodium, for example, combines rapidly with the oxygen in the air (**oxidation**), and, if thrown into water, will react with the water quickly and explosively.

A different family is formed by the elements fluorine (F), chlorine (Cl), bromine (Br), and iodine (I). This group of elements is called the family of **halogens**. The elements are **nonmetals**. Fluorine and chlorine are highly poisonous, corrosive gases. Bromine is a dark red-brown liquid at room temperature, with a boiling point of 59°C. Iodine is a blue-black solid that vaporizes at room temperature into a violet gas. (When a solid turns directly into a gas, it is said to *sublime*. The process is called **sublimation**.)

Each of the halogen elements is highly active and will combine with any of the alkali metals to form a compound known as a *salt*. The most common of these salts is sodium chloride, ordinary table salt. Potassium fluoride, lithium bromide, and sodium iodide are also salts.

You can see that a remarkable simplification has taken place in the study of chemistry. Instead of speaking of 20 individual reactions (sodium with chlorine, cesium with bromine, etc.), we can discuss the reactions of "the alkali metals with the halogens," taking advantage of the great similarity among all of these reactions.

Another kind of pattern was noticed as early as 1839, when it was discovered that there were some interesting relationships among the atomic masses of the elements in a given family. For example, the atomic masses of Cl, Br, and I are 35.5, 79.9, and 126.9 amu, respectively. The average of the first and third numbers is 81.2 amu, which is not far from the atomic mass of the middle element. The same result is obtained for lithium, sodium, and potassium, as well as for calcium, strontium, and barium, part of the alkaline earth family. Such groups of three are called *triads*.

Unfortunately, this trick does not work with all similar elements. The metals iron, cobalt, and nickel are very similar in properties and seem to make a family, but their atomic masses are 55.8, 58.9, and 58.7 amu, respectively—almost identical as far as the inexact measurements of the early nineteenth century were concerned. So the rule of the average seems to work in some cases and not in other cases. It is hard to build a science on data of this nature.

Meyer and Mendeleev

It was not until 1869–1870 that two European chemists almost simultaneously showed how to arrange the known elements into a chart that would display the families of the elements in a natural way. These two men were Julius Lothar Meyer, of Germany, and Dmitri Mendeleev (pronounced Men-de-ley'eff), of czarist Russia. (Mendeleev's name is given a variety of spellings because the Russian alphabet cannot be transliterated exactly into English.) It was not simply a coincidence that the two of them struck upon the periodic table at the same time, for others had been discussing like ideas for many years. (John Newlands, for example, had proposed a similar scheme in 1864.) It was an idea whose time had come.

The contribution of Lothar Meyer and Dmitri Mendeleev was to put together convincing evidence to validate the theory that the properties of the elements are periodic in nature. In addition, Mendeleev's imaginative contribution was his arrangement of the elements in a way that most clearly brought out the similarities within the families.

What do we mean by the word *periodic*? A periodic property is a property that repeats itself at regular intervals as some variable changes. For example, the numbers exhibited by a digital clock are periodic in time (with a period of 12 hours), because at any instant the number on the face of the clock is the same as it was 12 hours ago. (A military clock would have a 24-hour period.) The atmospheric temperature is doubly periodic: it goes through a 24-hour cycle (warm during the day and cold at night) and also through an annual cycle (warm in summer and cold in winter).

Lothar Meyer demonstrated that repeating patterns are generated when a number of properties of the elements are plotted against their atomic masses (or weights). His most famous curve was a plot of atomic volume versus atomic weight, shown in Figure 6-1. The atomic volume is the volume of a specific number of atoms of a given element. Measuring an atomic volume is not difficult. The procedure is based on the following principle: Since one atom of oxygen weighs 16 times as much as an atom of hydrogen, a million atoms of oxygen weighs 16 times as much as a million atoms of hydrogen. In other words, *the mass of a given number of atoms is proportional to the atomic mass.*

One gram of hydrogen contains a specific number of atoms (or molecules). The exact number does not matter for this discussion, since we are not concerned with counting these particles. The only fact of importance at the moment is that 16 g of oxygen contains the same number of atoms as does 1 g of hydrogen.

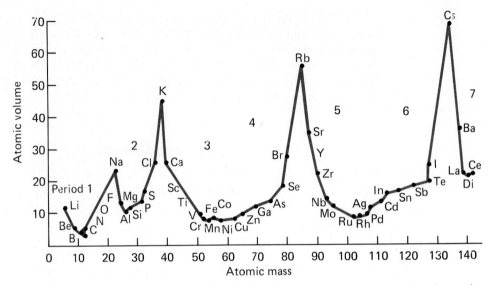

FIGURE 6-1 Lothar Meyer's plot of atomic volume versus atomic number shows the periodic peaks and valleys.

Because the atomic mass of iron is 55.8 amu, 55.8 g of iron contains the same number of atoms as does 1 g of hydrogen.

The number of grams equal to the atomic mass of an element is called the **gram-atomic mass** of that element. A gram-atomic mass of iron is 55.8 g of iron, whereas a gram-atomic mass of hydrogen is 1 g of hydrogen.

To find the atomic volume of any element, it is only necessary to determine the volume occupied by a gram-atomic mass of that element. As shown in the example below, this number is easily obtained from the density of the material.

Example 6.1

What is the atomic volume of carbon?

Solution

Starting with the density of carbon (2.0 g/cm^3), find the volume of 1 gram-atomic mass (12.0 g) of carbon.

By definition,
$$\text{density} = \frac{\text{mass}}{\text{volume}}$$

Solving for volume, we have
$$\text{volume} = \frac{\text{mass}}{\text{density}}$$

The volume of 12.0 g of carbon is
$$\text{volume} = \frac{12.0 \text{ g}}{2.0 \frac{\text{g}}{\text{cm}^3}} = 6.0 \text{ cm}^3$$

Note: Unit cancellation is as follows:

$$g \times \frac{1}{\dfrac{g}{cm^3}} = g \times \frac{cm^3}{g} = cm^3$$

The volume calculated in Example 6.1 contains a certain number of atoms of carbon. When we do the same kind of calculation for other elements, we obtain the volume of the *same* number of atoms of each of the other elements. This number is, therefore, proportional to the volume of a single atom of each element. The results of going through this procedure for all the solid elements are plotted on the graph in Figure 6-1. This curve clearly shows the rise and fall of the atomic volume. Obviously, the alkali metals have particularly large volumes compared with the other elements.

From a theoretical point of view, we would like to know what there is about the structure of the atom that causes this kind of behavior. The scientists of the nineteenth century were in no position to investigate this problem; they were engaged in amassing data that would be explained later on.

Dmitri Mendeleev was a chemist with an original and creative mind. He saw that it was difficult to find simple relationships among the atomic masses of the elements. Therefore, he invented a more useful scheme, arranging the elements in order of increasing atomic mass and giving each element a number showing its position in the list. This number was called the atomic number. At that time the atomic number had nothing to do with the number of protons or electrons in the atom, for these particles had not yet been discovered. The atomic number was simply a "sequence number."

Mendeleev then took the list of elements, arranged in order of atomic number, and broke it into a number of smaller lists, which he put side by side so that elements in the same family fell in the same row. Table 6-1 shows what Mendeleev's first periodic table looked like. From this table it was easy to see that if you start with lithium and increase the atomic number by 7, you come to sodium. If you go up another 7 places, you come to potassium. Things get more complicated further along. The simple rule of 7 does not work in getting from potassium to rubidium and to cesium. Nevertheless, the families do show up clearly—not only the alkali metals and the halogens but also the beryllium-cadmium family, the boron-aluminum family, the carbon-silicon family, and others.

One other feature of Mendeleev's table was extremely important. In order to make the families fall into place, he had to leave a few empty places, labeled with atomic weights 45, 68, and 70, plus a few unidentified slots among the heavier elements. A skeptic might have used these spaces as an argument to show how worthless the entire scheme was, but Mendeleev took a more positive point of view. He believed that the empty slots indicated elements that had not yet been discovered.

TABLE 6-1 Mendeleev's First Periodic Table

				Ti = 50	Zr = 90	? = 180
				V = 51	Nb = 94	Ta = 182
				Cr = 52	Mo = 96	W = 186
				Mn = 55	Rh = 104.4	Pt = 197.4
				Fe = 56	Ru = 104.4	Ir = 198
				Ni = Co = 59	Pd = 106.6	Os = 100
H = 1				Cu = 63.4	Ag = 108	Hg = 200
	Be = 9.4	Mg = 24		Zn = 65.2	Cd = 112	
	B = 11	Al = 27.4		? = 68	Ur = 116	Au = 197?
	C = 12	Si = 28		? = 70	Sn = 118	
	N = 14	P = 31		As = 75	Sb = 122	Bi = 210
	O = 16	S = 32		Se = 79.4	Te = 128?	
	F = 19	Cl = 35.5		Br = 80	I = 127	
Li = 7	Na = 23	K = 39		Rb = 85.4	Cs = 133	Ti = 204
		Ca = 40		Sr = 87.6	Ba = 137	Pb = 207
		? = 45		Ce = 92		
		?Er = 56		La = 94		
		?Yt = 60		Di = 95		
		?In = 75.6		Th = 118?		

Mendeleev then proceeded to predict the properties of these undiscovered elements by interpolating from the properties of elements already known in families where the empty places were located. He made predictions for three unknown elements, called "eka-aluminum," "eka-boron," and "eka-silicon." These elements were found in nature only a few years after Mendeleev announced his ideas. They are now known as gallium (discovered in 1875), scandium (discovered in 1876), and germanium (discovered in 1886).

The crucial test for any kind of theory is comparison of theoretical predictions with the results of actual measurements. When this comparison was done for germanium, the results, shown in Table 6-2, were very striking. Such close agreement is unexpected (and in fact the agreement was not as good for the other two elements), so there may have been an element of luck in Mendeleev's

TABLE 6-2 Mendeleev's Predictions

Characteristic	Property Predicted for "Eka-silicon"	Property Observed in Germanium
atomic weight	72.0	72.3
specific gravity	5.5	5.469
atomic volume	13.0	13.2
specific gravity of oxide	4.7	4.703
boiling point of chloride	100°C	86°C
specific gravity of chloride	1.9	1.887
boiling point of ethyl compound	160°C	160°C
specific gravity of ethyl compound	0.96	1.0

calculations. Nevertheless, the fact that a chemist could, for the first time, predict the presence of a new element and make some kind of judgment about its properties made Mendeleev's periodic table an instant popular success, and it continues to be a useful tool in the study of chemistry.

The Modern Periodic Table

The periodic table as it is used today differs in some details from Mendeleev's original version. It contains more elements, includes an entirely new family (the noble gases), and has been turned through 90 degrees so that the families of elements are in vertical instead of horizontal columns. A family is generally called a **group**, and in the table seen on the inside front cover, each family is labeled at the top of each column.

A number of systems of labeling the columns have been developed over the years by various groups of scientists. Some scientists prefer Roman numerals followed by an uppercase A or B (for example, IA, VIIB); other scientists prefer Arabic numerals followed by a lowercase a or b (for example, 2a, 5b). Still other scientists have suggested that the A and B designations of various columns be altered (for example, IIIB would become IIIA). In an attempt to bring order out of this chaos, the American Chemical Society has proposed an eighteen-column system, and it seems likely that this system will become the standard shortly. The periodic chart on the inside cover of this book shows two of the systems: 1a, 2a, ... , 8a; and 1, 2, ... , 18.

A row of the table is called a **period**, and we will see in a later section that the number of elements in each period is related to the way the electrons are arranged in their shells. The first column, for example, contains elements with only a single electron in the outer shell, and the last column contains elements with eight electrons in the outer shell.

The number on the left of the table is the *period number*. Period 1 contains only two elements; periods 2 and 3 each contain eight elements. The first two elements are split from the last six by a space because in the fourth, fifth, and sixth periods a group of ten elements (the **transition elements**) occupies that space. Furthermore, after element 57 a group of 14 elements fits into the space between lanthanum (57) and hafnium (72). These 14 were formerly called the **rare earth elements** and are now (since 1944) named the **lanthanides**. All the members of this group have very similar properties. Another group of 14 fits between actinium (89) and unnilquadium (104). Because these elements are radioactive, they are known as the **actinides**.

Until 1940 only the elements up to uranium (92) were known, and in fact they are the only ones found in nature. The elements beyond uranium (the *transuranium elements*) are not found in a natural state because they are radioactive, with lifetimes far shorter than the billions of years since the earth was formed. Uranium is also radioactive, but the average life of a uranium atom is 4.5 billion years, and therefore some of this element still remains in the crust of the earth.

Since 1940 a number of the transuranium elements have been created artificially in the laboratory. One of them, plutonium, is manufactured in fairly large quantities for nuclear reactors and weapons. The older chemistry textbooks insisted that elements were indestructible and that no element could be changed into another. Now, transmutation of one element into another is a routine matter. The dream of the alchemist—to turn lead into gold—has become reality. However, transmutation is generally an expensive business, and the aim of the alchemists to find a cheap and easy way of performing transmutations is, in the main, a fantasy.

In the remainder of this chapter we will look into the reasons for the existence of the periodic table and will survey group properties to see how they vary according to the periodic law.

6.2 The Periodic Law and Atomic Numbers

The periodic law states: *When the elements are arranged in order of their atomic numbers, they show periodic variations in their chemical and physical properties.* For example, starting with helium (atomic number 2) and advancing eight places to atomic number 10, we reach another gas—neon. Both helium and neon are extremely inert; they do not react with other elements and so are called **noble gases**. (They are found in small quantities in the atmosphere, and so they are sometimes called rare gases.) Starting with neon and advancing another eight places to atomic number 18, we encounter argon, another noble gas.

In the early days of the periodic table, the practice was to arrange the elements in order of increasing atomic mass. (The atomic number used at that time was simply the sequence number of the element when arranged in order of atomic mass, so atomic numbers and atomic masses followed the same ordering.) However, when the periodic table is arranged in that manner, a few elements seem to fall in the wrong places. For example, argon has a higher atomic mass than potassium, but if argon is to fall in the rare gas group, it must be placed before potassium. Similarly, iodine has a smaller atomic mass than tellurium, but if iodine is to take its proper position among the halogens, it must be placed after tellurium in the list.

Why do these exceptions to the periodic rule exist? This question was answered in 1914 by a brilliant young British physicist, Henry Moseley. Moseley was engaged in studying what happens when the various elements are bombarded by high-energy electrons. It was known that when solids are bombarded by high-speed electrons they give off x-rays with a broad spectrum of wavelengths. (Remember that x-rays are similar to ordinary light but have much shorter wavelengths.) Among the emitted x-rays, there is a continuous spectrum—a broad mixture of wavelengths. But in addition there are particular wavelengths, known as the *characteristic wavelengths*, which stand out as peaks in the spectrum. (See Figure 6-2.) The characteristic wavelengths vary from one element to another, and

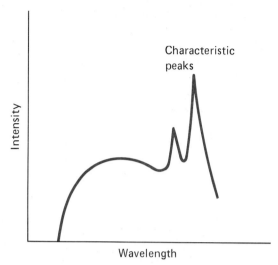

FIGURE 6-2 The x-ray spectrum emitted by one of the heavier elements when bombarded by energetic electrons: There is a continuous spectrum upon which is superimposed a pair of peaks corresponding to photons emitted when orbital electrons fall down to the two bottommost shells.

what Moseley did was to make precise measurements of these wavelengths for all the available elements.

Where do the characteristic x-rays come from? We can use the Bohr theory to develop the following model: Think of what happens when a high-energy electron hits an atom containing a number of orbital electrons. This high-speed electron is like a projectile. The first thing it does is to knock an electron out of one of the atomic orbits. Sometimes it knocks out the innermost electron. When that happens, one of the electrons in the next higher shell falls down into the innermost shell, emitting a quantum of energy as it falls. This quantum is a photon with much more energy than a photon of visible light, because it comes from an electron tumbling all the way down to the bottom energy level, whereas photons of visible light come from transitions among the upper levels. The high-energy photon emitted as a result of this electron bombardment of the atom is the characteristic x-ray photon.

Moseley started with this picture and used the equations of the Bohr theory to find a relationship between the wavelength of the x-ray photon and the number of protons in the nucleus. (The precise relationship he found was that the wavelength is inversely proportional to the square of the number of protons. That is, the more protons in the nucleus, the shorter the wavelength of the x-ray photon.) But the number of protons in the nucleus is what we now mean by atomic number. Moseley had, for the first time, found a way of measuring the atomic number of an element independently of its atomic mass.

Now, when Moseley lined up all the elements in order of the atomic numbers he had just measured with the x-ray method, he found that their positions were just right for the periodic table. Everything fell into the right place; argon did come before potassium in the atomic number scale, even though the two were reversed in the atomic mass scale. An important step had been taken in the understanding of atomic structure.

The story ended tragically. Moseley performed this very important research at the age of 27 and was considered to be the brightest young star in England's world of science. With the coming of World War I (called The World War at the time), Moseley enlisted in the British army, and in 1915 he was killed at the battle of Gallipoli.

6.3 Groups and Their Properties

Subshells and Groups

Chapter 5 described how the electrons in an atom are arranged in shells and subshells. In this section we describe how the arrangement of the atomic electrons into subshells results in the grouping of the elements into the periodic table.

The arrangement of the orbital electrons in any element is determined by starting with hydrogen and adding electrons one by one until each subshell is filled. Thus the order in which subshells get filled is important.

The order for filling subshells is given in Figure 6-3, which summarizes the relationship between atomic shell structure and the periodic table. Each block represents one of the elements in the periodic table, and the number in each block is the shell number. Each column represents a group of elements having similar arrangements of outer electrons. The symbol at the top of each column gives the electron configuration (the subshell number) for the *last* subshell to be filled as electrons are added to the atom. This is not necessarily the same as the *outermost* subshell in the atom.

Scanning across the top of the table, you can see that the first two elements (H and He) have electrons symbolized by the $1s^1$ and the $1s^2$ configurations. After two $2s$ elements (Li and Be) comes a group of six elements with $2p$ electrons. (This nomenclature means that in these six elements the last electron added falls into the $2p$ subshell.) There then follows a pair of $3s$ and another group of six $3p$ elements.

Each subshell may contain no more than a certain maximum number of electrons; when that number is reached, we say that the subshell is filled. As we shall see, the periodic properties of the elements are associated with the filling of subshells. The number of electrons required to fill a given subshell is related to the quantum numbers used to label that subshell. In the next (optional) subsection we will describe how to calculate the maximum number of electrons that can exist in a given subshell. For the present, we will concentrate on the order in which the subshells are filled, for that determines the form of the periodic table.

Having filled the $3p$ subshell, ending with the noble gas Ar, we might expect that the next electron added would fall into the $3d$ subshell. However, it turns out that adding an electron to the $4s$ subshell takes less energy than does putting it into the $3d$ subshell, even though the fourth shell is farther out than the third. (The

s^1																	s^2
1 H																	1 He

s^1	s^2											p^1	p^2	p^3	p^4	p^5	p^6
2 Li	2 Be				Transition Elements							2 B	2 C	2 N	2 O	2 F	2 Ne
3 Na	3 Mg	d^1	d^2	d^3	d^4	d^5	d^6	d^7	d^8	d^9	d^{10}	3 Al	3 Si	3 P	3 S	3 Cl	3 Ar
4 K	4 Ca	3 Sc	3 Ti	3 V	3 Cr	3 Mn	3 Fe	3 Co	3 Ni	3 Cu	3 Zn	4 Ga	4 Ge	4 As	4 Se	4 Br	4 Kr
5 Rb	5 Sr	4 Y	4 Zr	4 Nb	4 Mo	4 Tc	4 Ru	4 Rh	4 Pd	4 Ag	4 Cd	4 In	5 Sn	5 Sb	5 Te	5 I	5 Xe
6 Cs	6 Ba	5 La	5 Hf	5 Ta	5 W	5 Re	5 Os	5 Ir	5 Pt	5 Au	5 Hg	6 Tl	6 Pb	6 Bi	6 Po	6 At	6 Rn
7 Fr	7 Ra	6 Ac	6 Unq	6 Unp	6 Unh	6 Uns											

	f^1	f^2	f^3	f^4	f^5	f^6	f^7	f^8	f^9	f^{10}	f^{11}	f^{12}	f^{13}	f^{14}
Lanthanides	4 Ce	4 Pr	4 Nd	4 Pm	4 Sm	4 Eu	4 Gd	4 Tb	4 Dy	4 Ho	4 Er	4 Tm	4 Yb	4 Lu
Actinides	5 Th	5 Pa	5 U	5 Np	5 Pu	5 Am	5 Cm	5 Bk	5 Cf	5 Es	5 Fm	5 Md	5 No	5 Lr

FIGURE 6-3 Electron configurations and the periodic table: When a complex atom is built up by adding electrons to simpler atoms, the last electron added to a given atom occupies the subshell whose notation is given at the top of the column.

reason for this is complex and has to do with the amount of orbital motion.) There is one fundamental rule for deciding which shell the next electron will occupy: *Nature always does that which takes the least amount of energy.* For this reason the next two electrons fill the 4s subshell (K and Ca). Then it is time for the 3d subshell to be filled, one by one, until it contains ten electrons. The elements falling into that group are the transition elements, Sc to Zn.

It must be kept in mind that while electrons are being added to the 3d subshell, there are still two electrons farther out in the 4s subshell. Therefore, elements in the transition group usually have two electrons in the outer shell. It is possible, however, for electrons to be transferred from the 3d shell to the outer shell, so when such an atom is part of a molecule it may actually have three or more electrons in the outer shell, depending on the kind of compound that is formed.

The lanthanides are a group of elements that come after the $6s^2$ elements. The reason for the existence of this group is that after the 6s subshell is filled, the next electron added to the atom goes two levels down to the 4f subshell, where there is room for a total of 14 different electrons. Changing the number of electrons in that 4f subshell has very little effect on the chemical properties of the elements in that group, because the 4f subshell is two levels below the two outermost electrons in the 6s subshell. For this reason the lanthanides are all similar to one another.

The symbols at the top of each column in Figure 6-3 represent the particular subshell that gives the group of elements in that column its special characteristics. Of course, the complete electron representation of a given element must include all the other shells. For example, the electron configuration of the element in the fifth period of column 7a is

$$1s^2 2s^2 2p^6 3s^2 3p^6 4s^2 3d^{10} 4p^6 5s^2 4d^{10} 5p^5$$

This is the electron configuration of the element iodine. The configurations for all the other elements can be obtained in a similar way. Although the expression appears at first glance to be very complicated, if you take it one piece at a time, you will see that it simply tells how many electrons are in each subshell within this atom, as shown in Table 6-3.

TABLE 6-3 Number of Electrons in Each Subshell of Iodine

Subshell	No. of Electrons
1s	2
2s	2
2p	6
3s	2
3p	6
4s	2
3d	10
4p	6
5s	2
4d	10
5p	5

Note that the last column on the right in Figure 6-3 contains elements in which the outer shell has an $s^2 p^6$ configuration (except for He). These elements always have eight electrons in the outer shell. We will see in the next section that such a configuration is especially important and that the elements in this column have an unusual set of properties.

Atomic Structure and Quantum Numbers (Optional)

In this section we examine the reasons for the arrangement of electrons in their shells and subshells; a knowledge of this arrangement leads to an understanding of the periodic table. The key to this knowledge is the set of four quantum numbers n, ℓ, m, and s, used to label each atomic electron. (This subject was introduced in Chapter 5.)

Just as each person is recognized by a name, each atomic electron is identified by a four-number code. The quantum numbers are not arbitrary. Their values are restricted by rules given below:

1. The number n is allowed to be any positive whole number greater than zero ($n = 1, 2, 3, \ldots$). This number is the label for each electron shell.

2. Each shell contains a number of subshells, labeled by the quantum number ℓ. For a given value of n, the number ℓ can be any positive whole number from zero to $n - 1$. (That is, if $n = 4$, then ℓ may be 0, 1, 2, or 3.) In hydrogen the electron energy does not depend on the value of ℓ, but in the heavier elements it does, and we will see that this leads to some important consequences.

3. For a given value of quantum number ℓ, the number m can have any value from $-\ell$ to $+\ell$. For example, if $\ell = 2$, m may have any of the following values: $-2, -1, 0, +1$, or $+2$. (Notice that there are $2\ell + 1$ different values of m, and so in the above example, with $\ell = 2$, the number of different values of m is $2 \times 2 + 1 = 5$.)

4. Finally, there is the spin quantum number s, which can have only two possible values—either $+\frac{1}{2}$ or $-\frac{1}{2}$.

These rules are not just arbitrary or manmade rules. They emerge from the solution of the Schroedinger equation for atomic systems and are a fundamental and necessary part of quantum theory.

The four quantum numbers are used to determine the number of electrons that can be in each shell of a multielectron atom. The basic rule that acts as a guiding principle in the solution of this problem, known as the **Pauli Exclusion Principle**, states that *when an atom has a number of electrons, each of these electrons must have a different set of quantum numbers.* That is, every electron in an atom must be in a different state; hence it must have a four-number code different from that of every other electron. As a result, the maximum number of electrons in a shell is equal to the number of possible combinations of the four quantum numbers for a given value of n.

The use of this rule is illustrated by the chart in Figure 6-4, which shows how we may start with hydrogen and build more complex atoms by adding electrons one at a time. (It is understood that we add more protons and neutrons to the nucleus as we build heavier atoms.) This chart is similar to the one in Figure 6-3, but emphasizes the set of quantum numbers belonging to the last electron added. The arrows show the order in which the various subshells are filled.

In the first shell quantum number n equals 1. Since quantum number ℓ must be less than n, the value of ℓ must equal zero, and therefore there is only one possible combination of n and ℓ in this shell. However, there are two possible values of s: $+\frac{1}{2}$ and $-\frac{1}{2}$, corresponding to opposite electron spins. Therefore at most two electrons may exist in the first shell.

In the second shell n equals 2, so ℓ may have a value of either 0 or 1. The two electrons with $l = 0$ make up the $2s$ subshell. For those electrons with $\ell = 1$, there are three possible values of m ($-1, 0,$ and $+1$). For each of these there are two values of s, making a total of six electrons in the $2p$ subshell. Therefore there may be a maximum of eight electrons in the second shell.

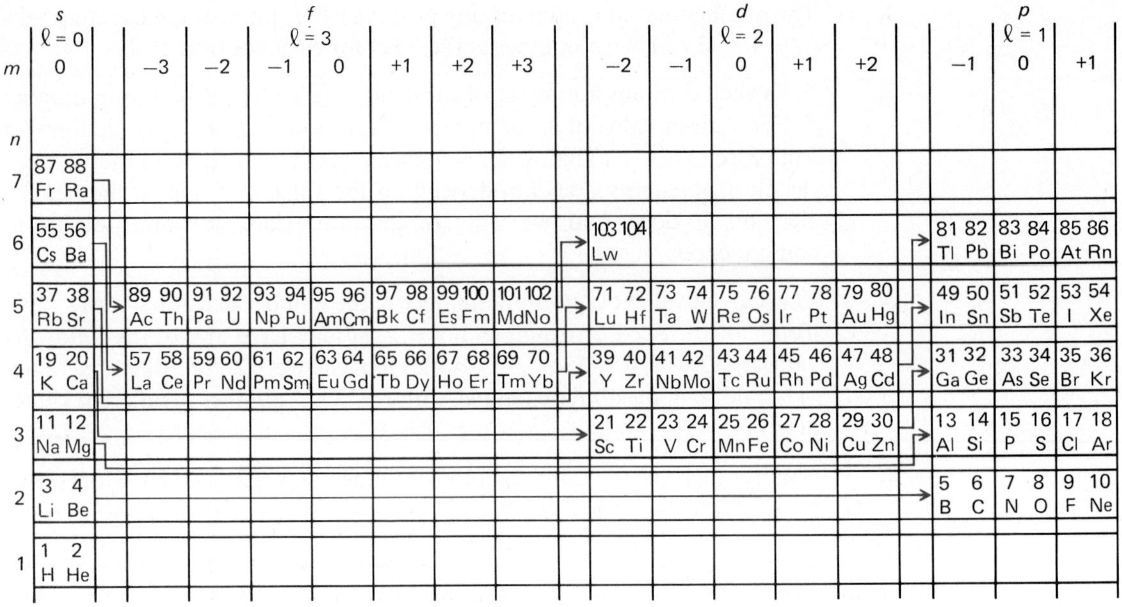

	s $\ell=0$		f $\ell=3$						d $\ell=2$				p $\ell=1$		
m	0	−3	−2	−1	0	+1	+2	+3	−2	−1	0	+1	+2	−1	0	+1
n																
7	87 88 Fr Ra															
6	55 56 Cs Ba								103 104 Lw					81 82 Tl Pb	83 84 Bi Po	85 86 At Rn
5	37 38 Rb Sr	89 90 Ac Th	91 92 Pa U	93 94 Np Pu	95 96 Am Cm	97 98 Bk Cf	99 100 Es Fm	101 102 Md No	71 72 Lu Hf	73 74 Ta W	75 76 Re Os	77 78 Ir Pt	79 80 Au Hg	49 50 In Sn	51 52 Sb Te	53 54 I Xe
4	19 20 K Ca	57 58 La Ce	59 60 Pr Nd	61 62 Pm Sm	63 64 Eu Gd	65 66 Tb Dy	67 68 Ho Er	69 70 Tm Yb	39 40 Y Zr	41 42 Nb Mo	43 44 Tc Ru	45 46 Rh Pd	47 48 Ag Cd	31 32 Ga Ge	33 34 As Se	35 36 Br Kr
3	11 12 Na Mg								21 22 Sc Ti	23 24 V Cr	25 26 Mn Fe	27 28 Co Ni	29 30 Cu Zn	13 14 Al Si	15 16 P S	17 18 Cl Ar
2	3 4 Li Be													5 6 B C	7 8 N O	9 10 F Ne
1	1 2 H He															

FIGURE 6-4   A chart showing how the periodic table results from the pattern of shells and subshells in each atom's electron distribution: The numbers to the left and at the top of the chart show the quantum numbers belonging to the last electron added to form each element.

In the third shell n equals 3, so ℓ may have a value of 0, 1, or 2. Using the same logic as in the above paragraphs, we find that the $3s$ subshell has 2 electrons, the $3p$ subshell has 6 electrons, and the $3d$ subshell may have up to 10 electrons, for a total of 18 possible electrons in the third shell. In the same manner we may calculate the number of possible electrons in all of the higher shells and subshells. The results are summarized in Table 6-4.

TABLE 6-4   *Number of Electrons in a Shell*

Shell No. (n)	Possible Values of ℓ	Possible Values of m	No. of Combinations	No. of Electrons
1	0	0	1	2
2	0	0	1	2
	1	−1, 0, +1	3	6
3	0	0	1	2
	1	−1, 0, +1	3	6
	2	−2, −1, 0, +1, +2	5	10
4	0	0	1	2
	1	−1, 0, +1	3	6
	2	−2, −1, 0, +1, +2	5	10
	3	−3, −2, −1, 0, +1, +2, +3	7	14

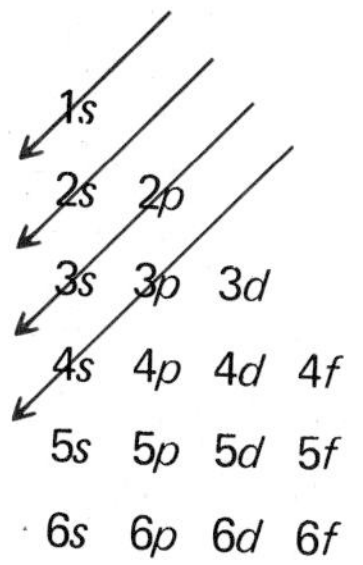

As described in the previous subsection, shells are not always filled in consecutive order. When the 3p subshell is filled, the next electron to be added does not go into the 3d subshell, as we might expect, but instead goes up into the 4s subshell. This is because an electron with $n = 4$ and $\ell = 0$ has less energy than an electron with $n = 3$ and $\ell = 2$. The high value of ℓ raises the energy of the state, making the state with $\ell = 0$ a preferred location for the new electron.

The principle that each new electron goes into the available subshell that requires the least amount of energy determines the order in which the subshells are filled. This order is summarized in Figure 6-5. In this diagram the horizontal rows contain the names of the various shells (for example, 3p means $n = 3$ and $\ell = 1$). Suppose, for example, you want to determine the electron configuration for calcium, which has 20 electrons. Draw diagonal arrows as shown in the diagram until you complete the count of electrons required. The arrows are drawn through $1s^2$, $2s^2$, $2p^6$, $3s^2$, $3p^6$, and $4s^2$, where the number in the exponent represents the number of electrons in each subshell.

FIGURE 6-5
By following the arrows in this table from top to bottom, we get the order in which the subshells are filled as we build up the table of the elements.

1s
2s 2p
3s 3p 3d
4s 4p 4d 4f
5s 5p 5d 5f
6s 6p 6d 6f

Columns of the Periodic Table

You are now ready to understand the meaning of the numerals at the heads of the columns in the periodic table. In the column labeled 1a are all the elements with one electron in the outer shell. The column labeled 2a contains the elements with two electrons in the outer shell. In general, the a columns are those in which the numeral gives the number of electrons in the outer shells of the atoms listed there. The b columns, on the other hand, contain the transition elements, and in these columns the number in the column head has no relation to the number of outer electrons.

The column numbers label the groups, or families, of elements. Before considering the properties of individual groups, note that all the elements can be divided into three broad classes: metals, semimetals, and nonmetals. The periodic table makes it simple to determine whether a given element is a metal, a semimetal, or a nonmetal. The chart on the inside cover shows the lines separating the three classes; the metals are on the left, the semimetals are in the right center, and the nonmetals are all the way over on the right. The most noticeable feature of this division is that most of the elements (including the lanthanides and the actinides) fall into the metal category.

Why does an element belong to a given class? This question will be answered in the following sections.

Metals

What is it that makes an element a metal? The atoms of a metallic element have one, two, or three outer electrons that are free to leave the atom with a little encouragement—that is, with the addition of a little extra energy. Thus, **metals** are defined as *elements whose atoms lose electrons easily and so form positive ions.*

Figure 6-3 helps us understand why there are so many more metals than nonmetals. The elements all the way on the left have either one or two outer electrons. As we move to the right along the rows of this table, each new electron is added to an inner shell, leaving the outer shell still with two electrons. It is not until we get all the way over to the right that we start adding the *p* subshell to the outer shell and so encounter the nonmetals.

All of the properties of metals are determined by the outer electrons. A metal exists as a crystalline solid, in which only a very small amount of energy is required to remove one or two outer electrons from each atom. The thermal energy of the solid (the energy of the atoms vibrating in the crystal lattice) is enough to separate at least one, and sometimes two, electrons from each atom in the crystal. These electrons float around inside the solid like the molecules of a gas. We speak of them, in fact, as an *electron gas*.

Thus, a metallic crystal (see Figure 6-6) consists of a tight structure of positively charged atoms (ions) immersed in the gas of free electrons. The crystal lattice is held together by the overall attraction between the electron gas and the positive ions. It is the electron gas that gives a metal its special characteristics:

1. *High electrical conductivity.* An electric current is a stream of electrons (or ions). Since there are many free electrons inside a metal, metal is a good conductor of electric current. When a battery is connected across the two ends of a piece of metal, the electrons inside the metal are attracted to the positive end of the battery and the current flows around the circuit, as shown in Figure 6-7.

2. *High heat conductivity.* When a metal is heated, its atoms vibrate about their positions in the crystal lattice. The higher the temperature, the more energetic the vibration. The electrons also move more rapidly, but whereas the ions are fixed to the crystal lattice, the electrons in the metal are free to move about. Therefore, if one end of a piece of metal is hotter than the other end, the rapidly moving "hot" electrons quickly flow toward the cooler end and deposit their energy there. In this way heat is conducted from one end of the metal to the other. A metal is a good conductor of heat for the same reason that it is a good conductor of electricity: because of the free electrons.

3. *High reflectivity.* Metals reflect light easily, so they tend to be bright and shiny. The reflection is due to the fact that light waves striking the surface of the metal cause the free electrons inside the surface to vibrate back and forth, reflecting the light. For that same reason, metals are *opaque* rather than transparent. Light does not easily pass through a medium where there are electrons free to vibrate back and forth.

4. Many metals have a high degree of **ductility** and **malleability**. That is, they can be drawn into fine wires, and they can be hammered or rolled into thin sheets. These properties come from a crystal structure that is bendable but not easily broken. A copper wire is easily bent into any form—a fact that makes copper especially easy to work with. The very high electrical conductivity of copper (second only to that of silver) also makes it a valuable material.

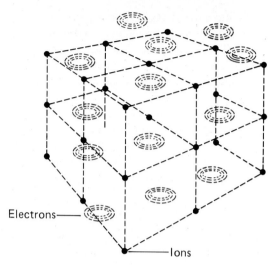

FIGURE 6-6 A metallic crystal consists of positive metal ions embedded in a sea of free electrons.

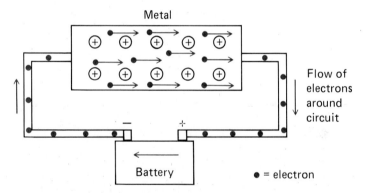

FIGURE 6-7 Conduction: Free electrons carry electric current through a metal bar when a battery is connected to the two ends.

We will now summarize the characteristics of the metal groups.

Group 1a is the group of elements with a single outer electron: hydrogen plus the alkali metals lithium, sodium, potassium, rubidium, cesium, and francium. Since elements in the same family should have similar properties, you might wonder why hydrogen is not an alkali metal. Hydrogen is an exception to the rule. It has a single electron that is bound tightly to the atom—more tightly than the outer electron of the alkali metals. Since an easily freed outer structure is required for a metal, hydrogen has no tendency to form a metallic, crystalline structure. However, it is believed on theoretical grounds that if hydrogen is subjected to extremely high pressures (as found at the center of planets Jupiter and Saturn), it turns into a metal. As in so many situations, the state of a given substance depends a great deal on the prevailing conditions of temperature and pressure.

The alkali metals are soft, silvery-white metals, of low density and high chemical activity. They are fairly common and have numerous uses in industry and medicine. Sodium chloride (table salt) is, of course, the most common compound of this group.

The term "alkali" comes from the Arab word *al-qaliy*, meaning the ashes of a certain plant. When mixed with water, these ashes form a bitter-tasting compound that forms a soap when cooked with fat or oil. Soap was made in this way for most of history. For centuries the alkalis were known only as compounds such as potash (potassium hydroxide) and lye (sodium hydroxide). It was not until 1807 that Sir Humphry Davy isolated the pure metals by the method of **electrolysis** which he developed.

Electrolysis is a common method of separating elements from their compounds by passing electric current either through a solution of the compound or through the molten compound itself. If two electrodes are immersed in molten sodium chloride and the electrodes are connected to a battery (see Figure 6-8), the positive sodium ions migrate toward the cathode (the negative electrode) while the negative chloride ions move toward the anode (the positive electrode). The sodium metal plates out on the cathode, while gaseous chlorine is released at the anode. Both elements can be collected for further use.

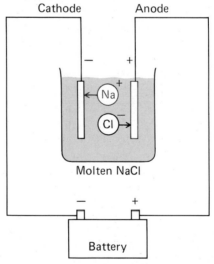

FIGURE 6-8 Electrolysis: Sodium ions and chloride ions carry electric current through molten sodium chloride, causing the sodium to be deposited at the cathode and the chlorine to be released at the anode.

Because of their high chemical activity, the alkali metals are never found in their pure state. Once extracted from their compounds, the alkali metals must be kept away from air and moisture if they are to be maintained in pure form. They are usually kept under a nonreacting organic liquid such as kerosene or oil.

Left in the open air, sodium quickly combines with oxygen to form sodium oxide (Na_2O). It also combines with the water vapor in the air to form sodium hydroxide (NaOH). Put a piece of sodium into water and a vigorous reaction ensues, in which a solution of sodium hydroxide is formed and hydrogen gas is liberated. The heat of the reaction is so great that the hydrogen will ignite, making it seem as though the sodium is burning on top of the water. (The density of sodium is 0.97 g/cm^3, so it floats on the water during the reaction. You see that the alkali metals go completely counter to our common notion of metals as hard, dense, and stable.)

Francium, at the bottom of column 1a, is one of the more recently discovered elements, first identified in 1939. Because of the short half-life of its isotopes, it is not found to a great extent in nature. It is estimated that in the entire crust of the earth there is probably less than an ounce of francium, created as a result of the radioactive decay of other elements.

Group 2a in the periodic table consists of the **alkaline earth metals**: beryllium, magnesium, calcium, strontium, barium, and radium. These elements all have two electrons in the outer shell. The name *alkaline earth* originated in the early days of chemistry, when certain oxide minerals were known as "earths." In particular, magnesium, calcium, strontium, and barium oxides combined with water to produce hydroxides that were bitter tasting and slippery. These compounds were considered alkaline in nature; hence the name alkaline earth.

The alkaline earth metals show a silvery surface when freshly cut; this surface quickly tarnishes and becomes coated with a white, powdery oxide. Beryllium and magnesium tarnish less quickly than the others and are also much harder than the soft calcium and barium.

These elements are all abundant in the earth's surface and are found in many minerals such as limestone (calcium carbonate), asbestos (calcium magnesium silicate), and gypsum (calcium sulfate). Calcium oxide and calcium hydroxide are important in the construction industry, since these compounds are the basis of cement, mortar, and concrete. They have the ability to be semifluid when mixed with water, but upon drying they crystallize into hard solids that hold together the aggregate of sand and gravel in concrete.

Radium was discovered by Madame Marie Curie in 1898 and because of its powerful radiation was the first of the radioactive isotopes used in cancer therapy.

The transition elements usually have one, two, or three electrons in the outer shell. Some of these elements are rare and have few important applications. However, in the top row are found the elements titanium (Ti), vanadium (V), chromium (Cr), manganese (Mn), iron (Fe), cobalt (Co), and nickel (Ni), which —together with molybdenum (Mo) from the second row—are strong metals with high melting points, commonly used in making iron alloys for many purposes. Pure iron by itself is fairly soft, but mix a little carbon into molten iron and you get, upon cooling, a harder material: steel. Plain carbon steel is rather brittle and prone to rust, but these qualities can be eliminated by adding some of the toughening metals listed above.

Niobium (Nb), formerly known as columbium, is a strong metal whose alloys are extensively used in the construction of space vehicles. In recent years niobium has become of great value because, when alloyed with zirconium (Zr), it displays a property known as **superconductivity**. When cooled to extremely low temperatures, this alloy shows absolutely no resistance to the flow of electric current; a current started in a superconducting ring will continue to flow indefinitely all by itself.

Among the transition elements is found a group of rare and valuable metals known as the noble metals, much used for jewelry as well as in important industrial applications: iridium (Ir), platinum (Pt), gold (Au), and silver (Ag). These metals tend to be relatively stable, do not oxidize easily, and are easily worked into many shapes because of their malleability and ductility.

Tungsten (W) is a very dense metal and is notable for its high melting point, which makes it a desirable material for the manufacture of incandescent bulb filaments. However, because of its brittleness special techniques are required for producing those fine filaments.

In the 3a column of the periodic table, we begin to encounter elements that have properties between those of metals and nonmetals. Aluminum (Al), gallium (Ga), indium (In), and thallium (Tl) are metals, but boron (B) belongs to the class of semimetals. All of these elements have three electrons in the outer shell.

Aluminum is the third most abundant element in the earth's crust and is relatively inexpensive. It is the main ingredient of many kitchen utensils and is found in many lightweight alloys used for construction purposes.

Semimetals

Semimetals are elements with properties between those of metals and nonmetals. Their atoms do not give up electrons as easily as do atoms of the metals. Although these elements conduct electricity to some extent, they are not good conductors. For this reason they are also called **semiconductors**.

As we scan across the periodic table, the first semimetal we encounter is boron (B), in the 3a column. Elemental boron is a crystalline solid with a very high melting point (2300°C). Fibers of pure boron have very great strength and in recent years have come into use in the manufacture of advanced aerospace structures. Boron is widely used in the form of borax (sodium tetraborate), which enters into many types of glass, enamels, and laundry products.

In the 4a column of the periodic table are found two very important elements: silicon (Si) and germanium (Ge). Silicon is the second most abundant element in the outer part of the earth's crust (oxygen is the first). It is found in numerous minerals, and the oxide—silica (SiO_2)—is the most common ingredient of sand. Sand is one of the raw materials in the manufacture of glass, cement, and concrete and is therefore important in the construction industries.

Since the 1950s the use of highly purified silicon as a semiconductor has revolutionized the electronics industry, for it has made possible the manufacture of transistors and other devices used in *microelectronics*—the complex electronic circuits fabricated on a single small chip of crystalline silicon. These silicon chips are the basis of the modern computer industry (hence the name "Silicon Valley" used to describe the region near San Francisco where many computer companies have built their headquarters).

The electrical properties of silicon are the result of the structure of the silicon crystal. Silicon has four electrons in its outer shell, and, by sharing one electron with each of the four neighboring silicon atoms, it forms a strong crystalline lattice. The reason silicon is not a good electrical conductor is that its electrons are bound to the atoms of the lattice, so they do not form an electron gas floating within the crystal, as electrons do in a true metal. It takes about 1 eV of energy to loosen an outer electron from each silicon atom. At room temperature only a few electrons have enough energy to spring free (perhaps one for every billion atoms). Because of these few free electrons, silicon is not an insulator but a semiconductor. (What makes silicon useful is the way its electrical behavior changes when tiny amounts of certain impurity atoms are introduced to increase the number of free electrons in the crystal. But this is a separate topic more suitable for a book on electronics or solid-state physics.)

The other important semimetal in column 4a is germanium. Germanium is a gray-white metallic substance with a melting point (937°C) considerably lower than that of silicon. Germanium also has four electrons in the outer shell, and its crystalline structure is similar to that of silicon. It is a semiconductor and has electrical properties much like those of silicon. It was the first of the semiconductors to be used in electronics, but since silicon withstands high temperatures better than germanium, silicon is now preferred as the base for microelectronic circuits.

Another useful semimetal is arsenic (As), which in the pure state is a grayish, somewhat metallic substance. Its compounds are important in agriculture, particularly as insecticides, since they are very poisonous in nature. This element is not to be confused with the common rat poison called arsenic, which is actually arsenic trioxide.

In the same family is antimony (Sb), which has many uses in alloys, paints, and semiconductors. In the pure state, it is a bluish-white, brittle, metallic-looking substance with a flaky texture. It does not oxidize in air the way sodium does, but it will burn when heated, giving off a bright light and copious fumes of antimony oxide (Sb_2O_3). An important use of antimony is in the manufacture of storage batteries; the lead plates used in these batteries would be too soft for practical purposes if they were not alloyed with about 10% antimony. Antimony is also a common ingredient in pewter and in the antifriction alloys used in making bearings.

Nonmetals

Nonmetals are on the right side of the periodic table. These elements come in a variety of forms and have a diverse set of chemical properties. All of the nonmetals belong to families in which there are between four and eight electrons in the outer shell. Generally speaking, these elements do not look like metals when pure, and they are **insulators**, or poor conductors of electricity. Over half of them are gases at room temperature. Even though there are relatively few nonmetals, they help form most of the chemical compounds in existence.

The chemistry of carbon has some similarities to the chemistry of silicon, since they both have four electrons in the outer shell. However, because carbon has a smaller atom, it is more flexible in the types of compounds that it forms. The number of known carbon compounds is in the millions—greater than the number of compounds of all the other elements combined.

The study of carbon compounds makes up an entire branch of chemistry: *organic chemistry*. The name comes from the fact that many of the carbon compounds occur in living organisms. In fact, all living things are composed mainly of carbon compounds. However, modern chemists are able to manufacture thousands of carbon compounds that have never seen a living organism, so in current usage the term **organic compound** simply refers to a compound based on the carbon atom, usually with hydrogen, often with oxygen, and sometimes with nitrogen and other elements. In Chapter 17 we will explore in some detail the chemistry of carbon compounds.

Although nitrogen (N) and phosphorus (P) both inhabit the 5a column of the periodic table, their pure forms differ greatly, proving that members of the same group are not always similar in appearance—especially among the nonmetals. Nitrogen is a colorless, odorless gas, making up about 78% of the atmosphere. Its compounds are of great importance in industry and in agriculture. In addition, nitrogen is an essential ingredient of protein molecules. Nitric acid (HNO_3), one of the most important nitrogen compounds, is a pungent, fuming liquid that dissolves almost all metals (except the noble metals Au, Pt, Rh, and Ir). When cooled to a temperature of $-196°C$, pure nitrogen becomes a colorless liquid used in large quantities as a refrigerant.

Phosphorus exists in four distinct atomic arrangements, known as **allotropic forms**. There are red, black, and two forms of white phosphorus. It is very active chemically; the white form, which is highly toxic, will ignite and burn by itself when exposed to air. Phosphorus is used in the manufacture of fertilizer, pesticides, fireworks, matches, and smoke bombs. It is also an essential ingredient in living cells, particularly nerve tissue and bones.

The 6a column shows, once again, a contrast between two elements in the same group. Oxygen (O) is a colorless, odorless gas making up 21% of the atmosphere, whereas sulfur (S) normally appears as a yellow crystalline or powdery solid that burns to form noxious sulfur dioxide (SO_2). Oxygen is most important, of course, as the element that supports burning, or combustion. We breathe oxygen in order to metabolize (or burn) carbohydrates to produce energy for living.

Sulfur has many important industrial uses. Sulfuric acid is used in automobile storage batteries and also in the manufacture of paper and the bleaching of dried fruits. Sulfur is an ingredient in gunpowder and also appears in a number of organic materials. The odor of rotten eggs is due to hydrogen sulfide (H_2S).

There are several forms of selenium (Se), ranging from amorphous red to crystalline gray. Its compounds resemble those of sulfur. It was one of the first materials used to make photoelectric cells, since it has the ability to convert light into electric current. (Photons of visible light release electrons from the outer shells of selenium atoms.) Properly treated, silicon also serves the same purpose. In recent years selenium has found great use in xerography, for the photocopying of printed material.

The 7a column of the periodic table contains the important halogen family. As has already been mentioned, these elements range from the highly reactive gas fluorine (F) to the less active solid iodine (I), with the radioactive element astatine (At) at the bottom. Fluorine is so active that almost everything reacts with it; finely divided metals, glass, ceramic, and even water will burn in fluorine gas. Hydrofluoric acid (HF) is commonly used to etch glass. It is one of the few materials that cannot be kept in a glass bottle.

Chlorine (Cl) is similar to fluorine but is not quite as active chemically. It is a pungent, poisonous, greenish gas, infamous for its use in chemical warfare during World War I. Hydrochloric acid (HCl) will dissolve many metals, and a mixture of hydrochloric and nitric acid (known as *aqua regia*, the "royal water") will even dissolve platinum and gold, metals that resist most other acids.

Finally, in the last column of the periodic table, we come to the noble gases, so called because they are so inert that until fairly recently it was believed that none of them entered into combination with any other element. However, during the 1960s it was found that krypton and xenon do form compounds with fluorine and oxygen.

The inert behavior of the noble gases is a result of the fact that the outer shells of these elements are completely filled with electrons, so addition of another electron requires starting a new shell. An outer shell of two electrons for helium and eight electrons for the other elements is a particularly stable configuration. Such an outer shell is reluctant to give up electrons and has no empty spaces that can accept more electrons. For this reason elements with filled outer shells form compounds with great difficulty.

Helium and neon form no compounds at all, but toward the bottom of the column the noble gases become increasingly active. This trend results from the increase in atomic volume as we go from the top to the bottom of the periodic table. The next section will survey a few of the trends that we can identify within the table, moving from top to bottom within a column and also from left to right across a period.

6.4 Trends in Properties

Atomic Radius

We have already seen that significant changes in the properties of the elements can be observed as we scan from the left to the right side of the periodic table. Most important is the change from metal to nonmetal. Furthermore, within a single family—the alkali metals—the chemical activity of the elements increases as we go down the column from lithium to cesium. (By "chemical activity" we mean the ease with which a given element reacts with other elements. Generally, the more active the element is, the more energy is given off during the reaction.)

The noble gases show the same sort of behavior. Helium forms no known compounds; it has so little activity that it is completely inert as far as chemistry is concerned. Xenon, on the other hand, can form a number of compounds under special conditions. Strangely, the opposite behavior is true of the halogens. Fluorine, the lightest of the halogens, is the most strongly reacting of the nonmetals, whereas iodine is much less active.

We should be able to explain such behavior on the basis of fundamental atomic properties. In particular, the radius of the atom should be an important factor in determining how the atom behaves. An electron close to the nucleus is attracted to the nucleus more strongly than is a distant electron. Therefore, the outer electron of a large atom is more easily removed than is the outer electron of a small atom.

Figure 6-9 displays a plot of atomic radius versus atomic number. The periodicity and the trends are striking. Every time we go from left to right across one period (say from lithium to neon), the radius becomes smaller. This is because

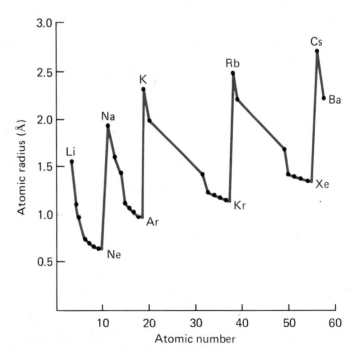

FIGURE 6-9 A plot of atomic radius versus atomic number shows the periodic properties of the atomic radius.

an increase in the atomic number means an increase in the number of positive charges in the nucleus, and the increased charge pulls all the electrons in a given shell closer to the nucleus, as shown in Figure 6-10(a). On the other hand, when an electron is added into a new outer shell, as in going from neon to sodium, the radius of the atom abruptly increases [see Figure 6-10(b)]. The alkali metals are seen to have the largest atomic radii of all. The peaks and troughs of the curve clearly show the periodic behavior of the atomic radius: there is an increase whenever an electron goes into a new shell, and there is a gradual decrease as more electrons are added within the same shell.

Comparing the radii of the elements in one family (the alkali metals at the peaks or the halogens in the troughs), we note a gradual increase in radius as the atomic number increases. The increase occurs simply because with more electrons there are more electron shells in each atom.

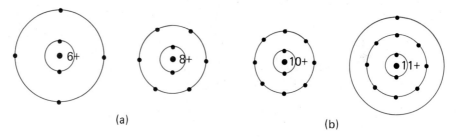

FIGURE 6-10 (a) As the number of electrons in a shell increases (together with the number of protons in the nucleus), the radius of the shell decreases. (b) When a new shell is started, the radius of the new shell is greater than that of the older shell, so the addition of one electron can cause a sudden increase in the diameter of the atom.

Ionization Energy

The ionization energy of an atom is the amount of energy needed to remove an electron from one of the shells of the atom. The ionization energy of the outermost electron is usually of greatest interest; Figure 6-11 shows a plot of this energy as a function of atomic number. The behavior displayed is just the opposite of that of the atomic radius. The alkali metals, which have the largest radii, have the smallest ionization energies. The noble gases, which have small radii, have the greatest ionization energies.

The variation of ionization energies can be understood on the basis of our knowledge of electrical forces. An electron close to the atomic nucleus is bound more strongly than is an electron farther away. Therefore, it takes less energy to remove from the atom an electron that starts from a greater distance. The result is that the electrons in shells with large radii have relatively small ionization energies. This is an important consequence, for it allows us to explain chemical activity on the basis of ionization energy. *An element that has a small ionization energy—an element whose outer electron is easily removed—is a chemically active element.*

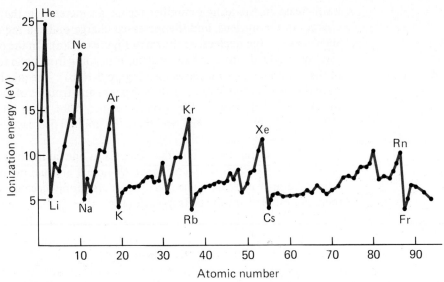

FIGURE 6-11   A plot of ionization energy versus atomic number shows the periodic properties of ionization energy.

This behavior can be observed by scanning across the bottom of the ionization energy curve. From lithium to cesium, as the atom becomes larger, the ionization energy becomes smaller; thus cesium is more active than lithium. The noble gases, which have very large ionization energies, are not active at all.

A similar argument explains why the halogens behave in an opposite manner. From fluorine to bromine, the radius of the outer shell increases. The halogens are *acceptors* of electrons because the outer shell has an empty space. In bromine the outer shell is a relatively weak attractor of electrons because it is far from the nucleus. On the other hand, the outer shell in fluorine is very close to the nucleus, and so there is a very strong tendency to attract any electron that might be in the neighborhood. As a result, among the halogens, the smaller the radius of the outer shell, the greater the chemical activity.

The relationships discussed in this section are of great importance to the chemist. First of all, the trends throughout the periodic table help us organize great masses of information that would otherwise be scattered and incoherent. A knowledge of relationships reduces the number of disorganized facts that must be memorized. Second, relating the observations to the theory of the atom does two things: it helps support the theory, and it puts the observations into a logical framework that makes them easier to understand. The fact that the activity of the halogens decreases as the atomic number increases would be an intriguing, but isolated, bit of information if we did not have a theory to explain why this happens.

Glossary

actinides A group of radioactive elements in the periodic chart with atomic numbers 90–103. Some of these elements do not occur naturally and are created in the laboratory.

alkali metals The elements appearing in column 1a of the periodic chart.

alkaline earth metals The elements appearing in column 2a of the periodic chart.

allotropic forms Different crystalline or molecular forms of the same element.

ductility The ability of a material to be pulled into a wire.

electrolysis A process of decomposing chemical compounds by sending an electric current through either the molten compound or a solution of the compound in order to separate the positive ions from the negative ions.

fission reaction A nuclear reaction in which the nucleus of any atom splits into two approximately equal parts.

gram-atomic mass The number of grams equal to the atomic mass of an element.

group The elements belonging to one of the columns of the periodic chart; family.

halogens Elements belonging to column 7a of the periodic chart.

insulators Materials that conduct electric current poorly.

lanthanides A group of elements in the periodic chart with atomic numbers 58–71.

malleability The ability of a material to be pressed or hammered into a thin sheet.

metals Elements whose atoms lose electrons easily, forming positive ions. As a result of the tendency to lose electrons, they are good electrical conductors.

noble gases The group of elements occupying column 8a in the periodic chart, containing eight electrons in the outer atomic shells; inert gases.

nonmetals Elements that tend to accept electrons, forming negative ions.

organic compound A compound having carbon as a base, usually combined with hydrogen (and often with oxygen, nitrogen, and other elements).

oxidation A reaction in which an element or compound combines with oxygen; more generally, a reaction in which an atom raises its oxidation state by giving up electrons to another atom.

Pauli Exclusion Principle A principle, or rule, enunciated by Wolfgang Pauli in 1925, stating that two electrons cannot be in the same state; that is, all electrons in a given atom must have different sets of quantum numbers.

period A horizontal row in the periodic chart.

rare earth elements See **lanthanides**.

semiconductors Materials that do not conduct electric currents as easily as do metals because their atoms give up electrons with some difficulty.

semimetals Elements with properties between those of metals and nonmetals; formerly called metalloids.

sublimation A process in which a solid changes directly into a vapor without going through the liquid state.

superconductivity A property of certain materials that allows them to conduct electric current without resistance at very low temperatures.

transition elements Elements in the periodic chart with incomplete inner shells; columns are designated by "b."

Problems and Questions

6.1 Historical Background

6.1.Q Define the following terms:
   a. periodicity
   b. atomic volume

6.2.Q Give two examples of periodic quantities, and state the period.

6.3.Q Describe Meyer's periodic law of the elements.

6.4.Q What key factor led to the immediate success of Mendeleev's periodic table?

6.5.Q Name three families, or groups, of elements.

6.6.Q To what family does each of the following elements belong?
   a. potassium
   b. iodine
   c. neon

6.7.P Calculate the atomic volume of iron, and compare it with the atomic volume of carbon. (The density of Fe is 7.85 g/cm^3.)

6.8.P There are 6×10^{23} atoms in a gram-atomic mass of any substance. Using this number, calculate the volume of a single atom of carbon and a single atom of iron.

6.9.Q Describe the differences between the following:
   a. a group and a period
   b. Mendeleev's original periodic table and the modern periodic table

6.10.Q Why does uranium exist on the earth, even though it is radioactive?

6.11.Q What characteristic do all the transuranium elements have in common?

6.2 The Periodic Law and Atomic Numbers

6.12.Q Give two definitions of atomic number.

6.13.Q Which of the two definitions of atomic number is in current use?

6.14.Q Explain why the other definition of atomic number is unsatisfactory.

6.15.Q What happens when you bombard matter with high-energy electrons?

6.16.Q What did Moseley measure with his x-ray experiment?

6.17.Q Which nuclear property did Moseley calculate from the x-ray measurements?

6.18.Q Why are x-rays (rather than visible light) emitted when iron is bombarded with high-energy electrons?

6.19.Q Describe two methods of determining atomic numbers of elements. Which is the more accurate method?

6.20.Q What mystery concerning the periodic table did Moseley's experiment solve?

6.3 Groups and Their Properties

6.21.Q What do the Roman numerals in the a columns of the periodic chart mean?

6.22.Q Define the following terms:
   a. metal
   b. semimetal
   c. nonmetal
   d. organic compound
   e. allotropic forms

6.23.Q An unknown element has one electron in its outer shell. Is it a metal or a nonmetal?

6.24.Q Name the element that has each of the following numbers of electrons (total), and tell whether it is a metal, a semimetal, or a nonmetal.
   a. 56   b. 16
   c. 32   d. 44

6.25.Q Label each of the following as either a metal, a semimetal, or a nonmetal.
   a. Na   b. Fe
   c. Bi   d. F
   e. U   f. Pb
   g. Si   h. C

6.26.Q What tests or observations could you devise to differentiate between a metallic and a nonmetallic element?

6.27.Q Describe the most important properties of metals.

6.28.Q What is the basic cause underlying all the properties that distinguish metals from nonmetals?

6.29.Q What are some of the more important uses of the following elements and their compounds?
   a. oxygen
   b. silicon
   c. selenium
   d. nitrogen
   e. calcium

6.30.Q Why is silicon called a semiconductor? What are semiconductors used for?

6.31.Q What process is used for the manufacture of sodium?

6.32.Q Where did the name "alkali" originate?

6.33.Q What happens when sodium is put into water?

6.34.Q Name the most important uses of calcium.

6.35.Q Name at least six of the transition elements. What are they used for?

6.36.Q What is the most valuable property of tungsten?

6.37.Q What three elements are most abundant on the earth's surface?

6.38.Q What element has the most compounds?

6.39.Q Name two elements that form no compounds. What group do they belong to?

6.40.Q Do elements in the same column of the periodic table always have the same appearance? Give examples to illustrate your answer.

6.41.Q Write the electron configuration schemes for these elements:
   a. fluorine   b. calcium
   c. xenon   d. uranium

(Optional)

6.42.Q Give the number of electrons in each subshell for the following elements:
   a. H   b. Na
   c. Cl   d. Zn
   e. Si

6.43.Q Write out the quantum numbers for each shell and subshell of each of the elements listed in Question 6.42.Q.

6.44.Q Why do we start occupying the fifth shell before the fourth shell is complete?

6.45.Q a. How many electrons can be in a subshell with $\ell = 4$?
   b. Write out all the possible values of m for $\ell = 4$.
   c. Show that the answer to part b of this question can be calculated from the expression
   $$\text{no. of values of } m = 2\ell + 1$$

6.46.Q a. How many electrons are in the outer shell of an He atom?
   b. How many electrons are in the outer shell of the other noble gas atoms?

6.47.Q How many electrons are always in the outer shell of an alkali metal?

6.48.Q What is the rule that determines the maximum number of electrons that can exist in each shell?

6.49.Q What are the two rules that determine the order in which the shells are filled as we add more electrons to the atomic structure?

6.4 Trends in Properties

6.50.Q Describe the trends in the following properties as you go down the periodic chart:
   a. atomic radius
   b. ionization energy

6.51.Q Describe the trend in the properties listed in Question 6.50.Q as you go from left to right across the periods of the periodic chart.

6.52.Q Which element in each of the following groups has the greatest value of the property listed?
   a. radius: Na, Li, K, Rb
   b. radius: F, Cl, Br, I
   c. radius: Na, Mg, Al, P
   d. ionization energy: Na, Mg, Al, P
   e. ionization energy: Na, Li, K, Rb

6.53.Q **a.** How is the ionization energy related to the radius of the atom?
b. How is the activity of an atom related to its ionization energy?
c. How is the activity of an atom related to its radius?

6.54.Q Why is cesium more active than lithium?

6.55.Q Why is fluorine more active than bromine?

Self Test

1. Describe the appearance of each of the halogens.

2. Describe two properties of the elements that are periodic in nature.

3. a. How would you measure out a gram-atomic mass of carbon?
   b. How much chlorine contains the same number of atoms as does the sample of carbon in part a of this question?

4. Mendeleev's periodic chart had three empty spaces.
   a. Why did those spaces occur?
   b. What elements were found to fill those spaces?

5. When the elements are arranged in order of increasing atomic mass, their order is not exactly the same as the order in which they are listed in the periodic table.
   a. What number determines the order of the elements in the periodic table?
   b. What does this number represent within the atom?
   c. Who discovered a direct method of measuring this number?
   d. Describe the method.

6. How many electrons are there in the outer shell of each of these elements?
   a. hydrogen   b. helium
   c. lithium   d. neon
   e. cadmium   f. chlorine
   g. krypton   h. potassium
   i. iodine   j. carbon

7. What is the basic difference between a metal, a semimetal, and a nonmetal?

8. What rule determines the maximum number of electrons allowed in a given atomic shell?

9. If we start with a given atom and add another electron, what rule determines the shell into which the new electron falls?

10. What is meant by each of the following symbols?
    a. $1s^2 2s^1$
    b. $1s^2 2s^2 2p^4$
    c. $1s^2 2s^2 2p^6$

11. What elements are represented by the electron configurations shown in Problem 10?

12. Why are there more metals than nonmetals?

12. Name two areas of industry in which silicon is important.

14. Silicon, germanium, and carbon all have four electrons in the outer shell. Silicon and germanium are semiconductors, but pure carbon (in the form of diamond) is not.
    a. Explain what makes silicon and germanium semiconductors.
    b. What would prevent carbon from being a semiconductor?

15. Explain why cesium is an active element and helium is inert.

7
Bonding

Objectives After completing this chapter, you should be able to:
1. Explain the binding of atoms in terms of electron transfer.
2. Explain the attraction of positive and negative ions for each other.
3. Explain the concept of electronegativity.
4. Distinguish between ionic and covalent bonds.
5. Describe the physical differences between ionic and covalent compounds.
6. Explain the difference between molecular solids and covalent solids.
7. Explain the difference between a covalent bond and a polar-covalent bond.
8. Describe what kinds of compounds have polar-covalent bonds.
9. Explain the relationship between a polar-covalent bond and an ionic bond.
10. Explain the concept of oxidation.
11. Explain the concept of reduction.
12. Find the oxidation numbers of the elements in simple compounds.
13. Draw diagrams of molecules using the Lewis electron-dot model.
14. Draw diagrams of simple molecules using the tetrahedron model.

7.1 The Nature of the Chemical Bond

Early Concepts

We have seen that the force that holds an orbital electron in place within its atomic shell is the **electromagnetic force**—the attraction between the negative electrons outside the nucleus and the positive protons inside the nucleus. The forces that hold atoms together in clusters—either in molecules or in crystals—are also electromagnetic in nature. They make themselves felt in a variety of ways and are classified under the general heading of *chemical bonds*.

The concept of the chemical bond originated early in the nineteenth century, before electrons, protons, and electromagnetic forces were known to exist. The older theories treated a chemical bond simply as a kind of hook holding two atoms together and did not go into further detail.

Along with the concept of the chemical bond arose the notion of **valence**. The valence of an atom is a number representing how many "hooks" the atom has available to connect with another atom. For example, in the compound hydrogen chloride [see Figure 7-1(a)], there is one atom of hydrogen for each atom of chlorine. If we assume that the hydrogen atom, as the simplest, has a valence of one, we then know that the chlorine atom also has a valence of one. In water, on the other hand, there are two atoms of hydrogen to each atom of oxygen. This means that each oxygen atom has two "hooks," and thus oxygen has a valence of two [see Figure 7-1(b)].

In modern practice the term "valence" has been replaced by the more descriptive term *oxidation number* (or *oxidation state*), which will be discussed in Section 7.3. However, the word "valence" still remains part of the nomenclature. We call hydrogen and sodium *monovalent* elements, whereas oxygen is a *bivalent* (or *divalent*) element. In addition, *covalent* compounds are compounds held together by a particular type of bond, to be discussed in the next section.

Since an atom is a neutral particle, with no net electric charge, you might wonder why two atoms should attract each other at all. To understand the bonding of atoms into molecules and crystals, we must dig deeply into the detailed events that take place when two or more atoms approach each other. One of the great triumphs of the atomic theory is its ability to provide a detailed explanation of the nature of chemical bonding. All chemical bonds result from the sharing of orbital electrons by neighboring atoms. (It is as though two people were bound to each other by having to juggle two baseballs back and forth.) In the following sections, we will investigate why atoms have a tendency to share electrons with each other, and you will see how this sharing results in the binding together of atoms.

Electron Affinity and Electronegativity

In Chapter 6 we saw that the sodium atom has two electrons in the first shell, eight electrons in the second shell, and one electron in the third shell. That single

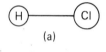

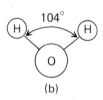

FIGURE 7-1 (a) There is one bond between the Na atom and the Cl atom in NaCl. (b) Two bonds are attached to the O atom in H_2O, so it holds two H atoms.

outermost electron is easily separated from the rest of the atom by the addition of a little energy; this fact makes sodium a metal.

Now consider the element chlorine. The chlorine atom has two electrons in the first shell, eight in the second shell, and seven in the third shell. Remember that a shell with eight electrons is very stable and is considered to be "filled." The outer shell of chlorine has one less than eight electrons. For this reason it behaves as though it had a "hole" in it [see Figure 7-2(a)].

The unfilled outer shell is not satisfied until it finds another electron to complete its quota of eight electrons. To do so, it tries to pull in any loose electron that is in the neighborhood. It will even attract an electron that already belongs to another atom. We say that the unfilled outer shell has an *affinity* for nearby electrons.

The result of the above situation is this: when a sodium atom is close to a chlorine atom, the chlorine atom steals an electron from the sodium atom. (During

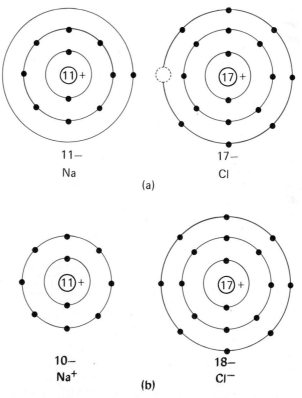

FIGURE 7-2 (a) An atom of Na has 11 protons in the nucleus and 11 orbital electrons. An atom of Cl has 17 protons and 17 orbital electrons, with one vacant place in the outer shell. (b) In NaCl the outer electron of the Na atom goes to fill the empty space in the outer shell of the Cl atom, forming a positive Na ion (with 11 protons and 10 electrons) and a negative Cl ion (with 17 protons and 18 electrons). (The drawing is only a symbolic diagram; in an actual atom the electrons are not evenly spaced around circular orbits.)

this theft the number of protons in the nucleus is unchanged.) The sodium atom then has one electron less than normal, and the chlorine atom has one electron more than normal. Since the number of protons in the sodium atom is one more than the number of electrons, the sodium atom has an excess of positive electric charge compared to negative charge. It is, therefore, a *positive ion* rather than an atom [see Figure 7-2(b)]. (Recall from Chapter 5 that, in general, an **ion** is a particle—either atom or molecule—with one or more unbalanced electric charges.)

The chlorine atom, having captured an electron from the sodium atom, has one extra negative electric charge and so is a *negative ion* called a chloride ion. (The suffix "ide" will be discussed in a later chapter.) Since they have opposite electric charges, the sodium and chloride ions attract each other strongly. This attraction is referred to as **bonding**.

In the compound sodium chloride, the chemical bond has two causes: (1) the affinity of the chlorine atom for the sodium atom's outer electron, causing the production of positive sodium and negative chloride ions, and (2) the electrical attraction between the oppositely charged ions.

The amount of affinity an atom has for an electron, called the **electron affinity**, is measured by the energy change in the reaction. A certain amount of energy is required to remove the electron from the sodium atom; this energy comes from the chlorine atom's capture of the electron. *The amount of energy given up by the chlorine atom to become a chloride ion is its electron affinity.* The more energy given off, the more vigorous the reaction and the more stable the final compound.

The electron affinities for the halogens are listed in Table 7-1. Note that elements at the top of the table, such as fluorine and chlorine, have greater affinities than elements at the bottom. This means that elements at the top are more active and release more energy in their reactions than do elements at the bottom. These observations are in agreement with the trends discussed in connection with the periodic table (Chapter 6).

A concept closely related to electron affinity is that of electronegativity. **Electronegativity** is a measure of the attraction of one atom for the electrons of

TABLE 7-1 Electron Affinities

Element	kJ/mole	eV/atom
F	333	3.45
Cl	349	3.61
Br	325	3.36
I	295	3.06

This table shows the amount of energy released when each of the listed elements undergoes a reaction in which the outer shell of the halogen atom becomes filled with an electron. The middle column gives the energy in units of kilojoules per gram molecular weight of the element. The last column gives the energy in electron-volts per atom.

another atom to which it is bonded. The greater the electronegativity, the greater the attraction. In the case of sodium chloride, discussed above, the chlorine atom is much more electronegative than the sodium atom, so it pulls the sodium atom's outer electron all the way over to itself. In this way it acquires a negative charge; hence the term "electronegativity."

The electronegativities of a few of the elements are shown in Table 7-2. The numbers in this table express electron attractions on a scale invented by the well-known American chemist Linus Pauling (the only scientist to win the Nobel Prize both for chemistry and for peace). The higher the number, the more strongly the atom attracts a neighboring electron into its outer shell.

TABLE 7-2 Electronegativities

H	2.1		
Li	1.0	F	4.0
Na	0.9	Cl	3.0
K	0.8	Br	2.8
Rb	0.8	I	2.5

We see from this table that chlorine has a much higher electronegativity number than does sodium. Therefore the chlorine atom is able to pull the sodium's outer electron completely away from the sodium atom.

On the other hand, if we compare chlorine with hydrogen, we find that the chlorine electronegativity number (3.0) is just slightly larger than the hydrogen number (2.1). For this reason, when chlorine combines with hydrogen to form the compound hydrogen chloride, the chlorine atom is not able to steal the hydrogen atom's electron completely away from it. The electron moves in an orbit that surrounds both atoms, and we say that this electron is *shared* by the two atoms. It spends more time near the chlorine atom than near the hydrogen atom, but it is not completely bound to the chlorine atom.

The concept of electronegativity helps to explain why there are different types of chemical compounds with different kinds of bonds. An *ionic* compound (such as sodium chloride) is formed when the electronegativity of one atom in a compound is much greater than that of the other. One or more electrons are transferred from one atom to the other, and the two elements of the compound are ionized. When each atom contributes one or more electrons and the electronegativity numbers are not very different, the electrons are shared by the two atoms and we call the compound a *covalent* compound. In the next section we will consider ionic and covalent compounds in some detail.

7.2 Ionic and Covalent Bonding

Ionic Bonds

In the last section we showed how the compound sodium chloride is formed when a chlorine atom pulls an electron from the outer shell of a sodium atom. The chlorine atom becomes a chloride ion with a single negative charge and the sodium atom becomes a sodium ion with a single positive charge. The strong attraction of the positive ion for the negative ion is a type of chemical bond known as an **ionic bond**. For this reason sodium chloride is called an ionic compound.

An important feature of this bond is that it is not an "exclusive" bond. A given sodium ion feels an attraction for all the chloride ions in its vicinity, because the electrostatic force is a long-range force. Similarly, each chloride ion is attracted to all the sodium ions in its neighborhood. The net result is that the sodium and chloride ions form a cubic crystal **lattice** in which the two types of ions alternate in a three-dimensional network (see Figure 7-3).

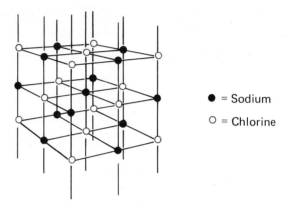

FIGURE 7-3 In a crystal of NaCl each Na^+ ion is attached to 6 Cl^- ions, and each Cl^- ion is attached to 6 Na^+ ions. This is an illustration of a bond that is not saturated—that is, more than one Cl^- ion is attracted to one Na^+ ion.

The sodium and chloride ions within the crystal do not make up what we commonly think of as a molecule. A molecule is a cluster of atoms in a unit capable of existing by itself. The atoms within a molecule have their bonds completely satisfied, so there is no great tendency for them to reach out and join other atoms. For example, when two atoms of hydrogen join to make a hydrogen molecule, their bonds are said to be "saturated." There is no tendency for a third atom to join them in the molecule. In sodium chloride, on the other hand, each sodium ion is joined to six chloride ions, and each chloride ion is joined to six sodium atoms. The entire crystal, containing billions of atoms, is the unit structure of sodium chloride.

Even though sodium chloride is made of electrically charged ions, it is not a good electrical conductor, because its ions are fixed in place within the crystal lattice. In order for a substance to conduct electric current, it must contain electrically charged particles that are free to move about. However, if sodium chloride is heated until it melts, its ions are free to move within the molten compound, and it can conduct an electric current.

Thus, one way to determine whether a compound is ionic is to test the electrical conductivity of the molten compound. This is done by dipping a pair of electrodes into the melt and measuring the amount of electric current that flows through the compound when the electrodes are connected to the terminals of a battery (see Figure 7-4). If the compound is ionic, a large electric current will flow.

This current consists of positive ions flowing through the molten material toward the negative electrode (the **cathode**) and negative ions flowing toward the positive electrode (the **anode**). Because the positive ions are attracted toward the cathode, they are often called **cations**. Similarly, the negative ions are called **anions**, because they are attracted to the anode. (*Cathode* is pronounced kath'ohd; *cation* is pronounced kat'eye-on; *anion* is pronounced an'eye-on; and *anode* is pronounced an'ohd.)

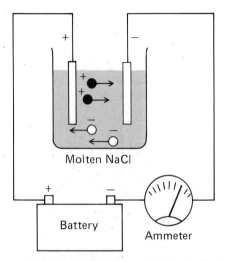

FIGURE 7-4 When a battery is connected to a pair of electrodes dipped into molten NaCl, the Na ions travel toward the negative electrode (the cathode), while the Cl ions travel toward the positive electrode (the anode).

The process described above is known as electrolysis. As you saw in Chapter 6, the electrolysis of NaCl is the chief method for the manufacture of metallic sodium, as well as gaseous chlorine.

Another characteristic of ionic compounds is that they have relatively high melting points, because the binding force between the ions is very strong. This property of ionic compounds is sometimes used in deciding whether a given compound is ionic or nonionic.

It is important to understand the difference between the bonds in an ionic compound and the bonds in a metallic crystal. You have seen that in an ionic compound, such as sodium chloride, positive and negative ions have alternate locations within the crystal lattice. Their mutual attraction gives the structure its strength. In the case of a metal such as copper, only copper ions fill the lattice. These ions must repel each other, because they all have positive charges. What holds them together?

The answer lies in a fact mentioned in Chapter 6: a metallic crystal consists of positive ions immersed in a "gas" of free electrons. The electrons are shared by all the atoms in the lattice. These electrons are not thought of as belonging to any particular atom; they belong to the crystal as a whole. It is the electrical force between all the electrons and all the ions that locks the crystal structure into place. (A complete explanation of metallic crystals requires the use of quantum theory involving the wave nature of the electron.)

Calculating the Charge on an Ion

The total (or net) charge on any particle is the number of positive charges minus the number of negative charges. The number of electrons is equal to the number of protons in a neutral atom. Its net charge is, therefore, zero.

When an electron is removed from an atom, the atom is left with more nuclear protons than orbital electrons. For example, removing one electron from a sodium atom results in an ion with eleven positive charges and ten negative charges. The net charge is

$$(11+) + (10-) = 1+$$

The sodium ion is seen to have a single positive charge.

In everyday practice it is not necessary to go to the trouble of adding electrons and protons. Use the following simple rule: *Each electron removed from an atom gives it one net positive charge. Each electron added to an atom gives the atom one net negative charge.*

Example 7.1	What is the net charge on a chloride ion?
Solution	The chlorine atom has an atomic number of 17; therefore the nucleus contains seventeen protons. The chloride ion has one extra electron, for a total of eighteen electrons. $$\text{net charge} = (17+) + (18-) = 1-$$ In short, adding one electron to the atom results in an ion with a net charge of $1-$.

Covalent Bonds

As stated earlier, a bond formed when two atoms share a pair of electrons more or less equally is called a **covalent bond**, and compounds held together by such bonds are called covalent compounds. Covalent compounds do not have the same properties as ionic compounds. Pure water, a covalent compound, is not a good conductor of electricity (which proves that there are few loose electric charges floating around in it), and it has a low melting point (0°C). The same can be said of other covalent compounds such as hydrogen (H_2) and carbon tetrachloride (CCl_4).

These properties contrast with those of ionic compounds, which have high melting points and are good conductors of electric current (when in the liquid state).

In many common gases such as hydrogen, oxygen, and nitrogen, the atoms are tied together by covalent bonds into **diatomic molecules** ($di = 2$; thus a diatomic molecule is a molecule containing two atoms). Once two hydrogen atoms join together as a pair, they show little tendency to combine with other hydrogen atoms to form larger clusters (in contrast to NaCl, in which millions of atoms join to form an ionic crystal). This saturation property accounts for the fact that hydrogen, at room temperature, exists as a gas rather than as a crystalline solid.

The properties of covalent substances are explained by the nature of the covalent bond. The hydrogen molecule provides the simplest example of covalent bonding. In hydrogen the covalent bond results from the sharing of outer electrons by the two atoms. A true explanation of this bond cannot be developed in terms of the Bohr atomic model, in which the electron is treated as a hard particle. It is necessary to use the Schroedinger model, in which the two electrons are pictured as a combined cloud surrounding both protons. (See Figure 7-5.) These two orbital electrons are shared by the two atoms—it is as though a rubber band were pulling the two atoms together.

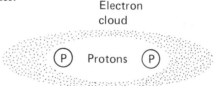

FIGURE 7-5 Quantum theory tells us that the electrons in a hydrogen molecule are shared by both atoms, so they form a "shell" around both nuclei. It is the average force between the electrons and the protons that holds the molecule together.

In this configuration, each hydrogen atom behaves as though it has two electrons in its first shell. Recall that according to the shell model discussed in Chapters 5 and 6 there can be no more than two electrons in the first shell. With those two electrons, the shell is completely filled. A filled shell is particularly stable; it does not like to give up an electron, and it is not eager to accept another electron. For this reason the hydrogen molecule is complete; there are no sites free to accept or share electrons from other atoms or molecules.

Covalent bonds produce molecules that stand alone. These molecules are the smallest units of the compounds formed by them. However, it is not true to say that there is *absolutely* no attraction between them. If two molecules get very close together, they feel a weak mutual attraction due to the so-called **van der Waals force**. This force is basically electrical—it is the result of averaging all the attractions and repulsions between the electrons and protons in the two molecules, and it only makes itself felt when the molecules are essentially touching each other.

Because of the weakness of the van der Waals force, covalent compounds tend to be gases at high temperatures. But when a covalent compound is chilled to a sufficiently low temperature, the molecules slow down enough that the van der

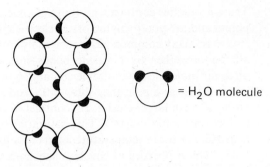

FIGURE 7-6 In a hexagonal ice crystal, the unit of structure at each lattice point is a molecule of water.

Waals force can pull them together to form a liquid or a solid. This is the reason for condensation and freezing. The solid is formed of crystals in which the molecules are the units of the crystalline lattice (see Figure 7-6). Ice, solid hydrogen, solid CO_2 (dry ice), and frozen alcohol are examples of such *molecular solids*. Usually, molecular solids have low melting points and are relatively soft materials because of the weakness of the van der Waals force.

FIGURE 7-7
The molecule of methane has one carbon atom with four bonds, each one attached to a hydrogen atom.

Many molecular solids are compounds of carbon and hydrogen. Carbon has two electrons in the first shell and four in the second, or outer, shell. A carbon atom can accept four additional electrons in its outer shell to form a stable shell of eight electrons. When carbon combines with hydrogen, each hydrogen atom shares its electron with the carbon atom. Because the carbon atom has four empty spaces in its outer shell, it combines with four hydrogen atoms to form methane (CH_4), as shown in Figure 7-7. Once the methane molecule has been formed, there are no more electrons to share. Therefore methane is a molecular compound that is gaseous at room temperature and has a low melting point.

Certain other substances joined by covalent bonds have quite different characteristics: they have very high melting points and are extremely hard solids.

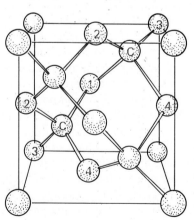

FIGURE 7-8 In the diamond crystal, each of the four carbon bonds is attached to another carbon atom.

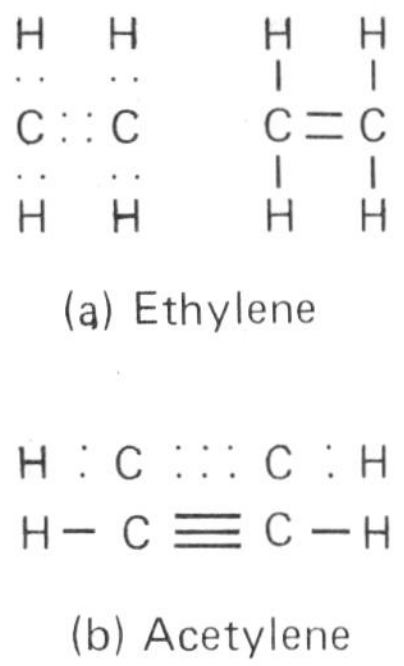

```
H  H      H   H
··  ··     |   |
C ::  C    C = C
··  ··     |   |
H  H      H   H
```
(a) Ethylene

```
H : C ::: C : H
H — C ≡ C — H
```
(b) Acetylene

FIGURE 7-9
(a) In ethylene, the carbon atoms are connected by double bonds (sharing of two pairs of electrons). (b) In acetylene, the carbon atoms are connected by triple bonds (sharing of three pairs of electrons).

7.3 POLAR COVALENT BONDS

Examples of such covalent solids are carbon (in the form of diamond, one of the hardest materials known), silicon, and silicon carbide—which, under the trade name of Carborundum, is sold as an abrasive for grinding metals.

The properties of diamond are explained by the four electrons in the outer shell of the carbon atom. (See Figure 7-8.) Each of these electrons can be shared with an electron from another atom. With four outer electrons to share, each carbon atom is joined to four other carbon atoms, producing a covalent structure of great strength. The entire crystal is essentially one large molecule. (Two kinds of crystalline structures can be formed by carbon. Diamond is an extremely hard, transparent crystal, whereas graphite has flat, platelike crystals that can slide over each other, so graphite appears to be a very soft material.)

Silicon also has four electrons in its outer shell. Therefore its crystalline structure is identical to that of diamond. In the silicon carbide crystal, silicon atoms alternate with carbon atoms. The result is a substance with a hardness between that of diamond and silicon.

So far we have discussed covalent compounds in which a single electron pair is shared by two neighboring atoms through a **single bond**. There are also compounds in which the sharing of two electron pairs by adjoining atoms results in a **double bond**. Such situations occur commonly in organic compounds. The simplest organic compound with a double bond is ethylene (C_2H_4), diagrammed in Figure 7-9(a). A **triple bond** is formed when three electron pairs are shared by neighboring atoms, as in ethyne, commonly called acetylene (C_2H_2), shown in Figure 7-9(b).

A double bond is stronger than a single bond, and the distance between the bonded atoms is shorter. Table 7-3 shows interatomic distances and bond strengths for the single, double, and triple carbon bond. The **bond strength** is represented by the energy necessary to separate the two atoms from each other. The stronger the bond, the more energy required. We see that the closer the atoms are to each other, the more energy is required to break up the molecule.

TABLE 7-3 Bond Distances and Strengths for Carbon Atoms

Bond Type	Distance (angstroms)	Bond Energy (electron-volts)
Single	1.54	3.6
Double	1.33	6.5
Triple	1.21	8.5

7.3 Polar Covalent Bonds

In the last section we described two types of chemical bonds: ionic bonds and covalent bonds. These categories are but two extreme cases of the more general polar covalent bond. Water (H_2O) and hydrogen chloride (HCl) are polar covalent compounds, and hydrogen (H_2) is a purely covalent molecule.

We can understand the meaning of the polar covalent bond by examining the difference between the H_2 and the HCl molecule. The H_2 molecule has a symmetrical shape because the two hydrogen atoms are identical and have equal attractions for the two shared electrons. (That is, the two hydrogen atoms are equally electronegative.) Therefore, as we saw in Figure 7-5, the electron orbital is an elliptical cloud surrounding both atoms equally. As a consequence there are no unbalanced electric charges to be found anywhere in the molecule.

In the HCl molecule, on the other hand, the two atoms are not identical. The electron orbital diagram takes the form shown in Figure 7-10. Theory and experiment both show that the two shared electrons tend to cluster closer to the chlorine atom than to the hydrogen atom, the reason being that the chlorine atom attracts the electrons more strongly than the hydrogen atom does. (That is, the chlorine is more electronegative than the hydrogen atom.) Therefore the chlorine end of the molecule has a small negative charge, and the hydrogen end is slightly positive. We say that the molecule is **polarized**.

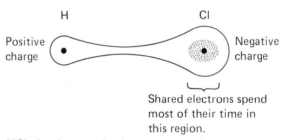

FIGURE 7-10   In HCl, the electron cloud representing the outer shell surrounds both the H and the Cl atoms. The Cl atom has a stronger attraction for the electron than does the H atom; as a result the electron spends most of its time near the Cl atom.

Any object with a positive charge at one end and an equal negative charge at the other end is called an **electric dipole**. The dipole molecule is neutral as a whole—it has an equal amount of positive and negative charge. It is a dipole simply because the average position of the electrons is shifted from the center toward the more electronegative atom, giving that end of the molecule a net negative charge.

The bond that results whenever there is a covalent bond between atoms with different electronegativities is a **polar covalent bond**. It is a covalent bond in which the electrons are, on the average, closer to one of the atoms than to the other. A molecule with an electric dipole is called a *polar molecule*. A polar covalent bond is stronger than a pure covalent bond because of the additional electrical attraction between the positive and negative ends of the molecule. Furthermore, polar molecules attract each other more strongly than do nonpolar molecules, because the positive end of one molecule is able to attract the negative end of another molecule. This kind of attraction is called a **dipole-dipole interaction** (see Figure 7-11).

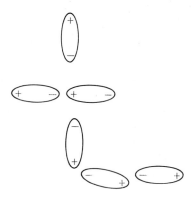

FIGURE 7-11 In a dipole-dipole interaction, the positive end of one dipole attracts the negative end of a neighboring dipole. By this means, long chains of dipole molecules can form.

The strength of the intermolecular attraction determines such properties as the melting point and the boiling point of the compound. We find, for example, that the boiling point of nonpolar liquid hydrogen ($-253°C$) is much lower than the boiling point of polar HCl ($-85°C$). The stronger the intermolecular force, the higher the boiling point will be, other things being equal. This is because the stronger the force, the more energy it takes to separate one molecule from another.

A dipole is described by two quantities: the distance between the two atoms and the amount of electric charge located at the position of each atom. (Even though the electrons are constantly moving about, we can speak of the *average* amount of charge at each end of the dipole.) The amount of electric charge will vary from one kind of molecule to another. For HCl the effective electric charge on each end of the dipole is about 17% of the unit electron charge, whereas for hydrogen fluoride (HF) the effective electric charge is 60% of an electron charge. This difference in part accounts for the fact that the boiling point of HF (19.5°C) is higher than the boiling point of HCl ($-85°C$).

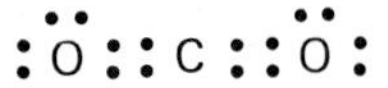

:Ö::C::Ö:

FIGURE 7-12 In CO_2, the center of the molecule has a positive charge, and the two ends are negative. The electric fields cancel each other out, and the molecule behaves as though unpolarized.

An ionic bond may be considered an extreme case of a polar covalent bond, one in which the valence electron spends *all* of its time in the vicinity of one of the atoms, so instead of being shared it is completely captured by the atom with the greatest electronegativity.

The shape of a molecule is important in determining its polarization. The carbon dioxide (CO_2) molecule, for example, is *linear:* its three atoms are strung out in a straight line, with the two oxygen atoms on either side of the carbon atom. (See Figure 7-12.) The two ends of the molecule have negative charges, and the center is positive; the polarizations of the two bonds cancel each other, since they point in opposite directions. Therefore CO_2 as a whole is not polarized.

In the case of water, on the other hand, there is an angle of 104° between the two O—H bonds. (See Figure 7-1.) As a result the water molecule is highly polarized, with the oxygen end of the atom bearing a negative charge. In Chapter 13 a number of important consequences arising from this fact are examined.

7.4 Oxidation Numbers

As we saw in Chapter 6, the number of electrons occupying the outer atomic shell strongly influences the properties of the element and determines the types of chemical reactions into which it enters. This is because the formation of a compound takes place through the sharing of the outer electrons between two or more atoms.

When a hydrogen atom combines with oxygen to form water, it forms a polar bond with the oxygen atom. Since the shared electrons are on the average closer to the oxygen atom than to the hydrogen atom, we say that the hydrogen atom *donates* its electron to an oxygen atom. (The electron is not completely transferred, as in an ionic compound, but is transferred for the greater part of the time.) Because the oxygen atom has two empty spaces in its outer shell, it is able to accept two electrons. As a result, the oxygen atom must combine with two hydrogen atoms to form a stable water molecule:

$$2H_2 + O_2 \longrightarrow 2H_2O$$

The number of electrons transferred, or donated, from one atom to another is given by the *oxidation number*. Before we look at a more specific definition of oxidation number, let us consider the meaning of oxidation.

The original, historic definition of oxidation was simply "the combination of an element with oxygen." All combustion reactions are oxidation reactions, including the combination of hydrogen with oxygen discussed above. In this reaction hydrogen is oxidized to form the compound hydrogen oxide (water). The element oxygen is called the **oxidizing agent** because it causes the hydrogen to be oxidized.

In modern terminology oxidation has a more general meaning that can best be illustrated in terms of the events that take place during the oxidation of hydrogen. As described above, each hydrogen atom gives most of its orbital electron to the more electronegative oxygen atom. The oxygen atom, having gained two electrons, now possesses extra negative charge. (That is, it has more electrons than protons.) Because of this unbalanced negative charge, we say that the oxygen atom has an oxidation number (or oxidation state) of 2−. (This is not the same as the electric charge on the oxygen atom; the atom is not ionized.)

When the oxygen atom is in its normal state (in an oxygen molecule, O_2), it is electrically neutral, with as many electrons as protons. In that neutral state, it has an oxidation number of 0. When it forms water its oxidation number is *lowered* from 0 to 2−; consequently, we say that the oxygen is *reduced*.

The hydrogen atom, on the other hand, is called the **reducing agent**, since it facilitates the reduction of the oxygen. Each hydrogen atom in the water molecule has an oxidation number of 1+, because after it loses an electron it has more positive charge than negative charge. The hydrogen atom, which goes from an oxidation number of 0 to an oxidation number of 1+, is said to be oxidized. So you see that *oxidation is the opposite of reduction*.

We now generalize the concept of oxidation and reduction to include all kinds of reactions, even those that do not involve oxygen. *Whenever an element, compound, or ion accepts electrons, it is an oxidizing agent, and whenever an element, compound, or ion gives up electrons, it is a reducing agent.* The number of electrons transferred to or from the atom, compound, or ion is the **oxidation number**. The species whose oxidation number is raised because it gives up electrons is said to be oxidized. The species whose oxidation number is lowered because it gains electrons is said to be reduced.

It is clear from this definition that the reducing agent is the material that becomes oxidized, while the oxidizing agent becomes reduced. For example, in the reaction

$$2\,Na + Cl_2 \longrightarrow 2\,NaCl$$

a sodium atom gives up a single electron, which is accepted by a chlorine atom. The roles of the various components of this reaction are summarized in the following table:

Sodium	Chlorine
is oxidized	is reduced
reducing agent	oxidizing agent
charge: 1+	charge: 1−

In this reaction the sodium becomes oxidized, even though there is no oxygen involved in the reaction at all.

The oxidation numbers may be included in the reaction equation as superscripts:

$$2\,Na^0 + Cl^0_2 \longrightarrow 2\,Na^{1+}Cl^{1-}$$

Note that the oxidation number of the Na (1+) is the same as its ionic charge; the same is true of the chlorine atom. The reason the oxidation number is the same as the ionic charge in this case is that NaCl is an ionic compound, so the transferred electron is completely separated from the sodium atom, which becomes an ion. The oxidation number of an element in a compound is not necessarily the same as its ionic charge, as we shall see in Chapter 8.

In Chapter 8 you will see that it is necessary to know the oxidation number of each element in a compound in order to write the formula for that compound correctly. You will learn the rules for determining the oxidation number of any element. That information will enable you to decide how many atoms of a given element can combine with one atom of another element. Then Chapter 16 will introduce you to a special system for writing equations involving oxidation and reduction reactions.

Note: Mnemonic devices are sometimes helpful in remembering which reaction is connected with a gain or loss of electrons:

LEO = Loss of Electrons means Oxidation

GER = Gain of Electrons means Reduction

OLE! = Oxidation is Loss of Electrons

7.5 The Lewis Electron-Dot Model

In 1916 at the University of California, G. N. Lewis proposed a model of covalent bonding based on the Bohr atom. The **Lewis electron-dot model** is still useful for visualizing the structure of many molecules, even though it does not include the features of Schroedinger's wave model of the atom and so is incomplete.

The electron-dot model, used to draw pictures of molecular structure, is simple to apply. The atomic nucleus is represented by the symbol for that atom: H for hydrogen, O for oxygen, and so on. Each electron in the *outer* shell is represented by a dot outside the nucleus. (The inner shells are not important in discussions of molecular binding.) These dots are drawn as though each atom were inside a box:

$$\boxed{\text{Na}}\cdot \qquad \cdot\boxed{\ddot{\text{Cl}}\!:}$$

Electron dots are placed on the sides of the invisible box, never more than two to a side. Following this procedure, we can represent hydrogen, sodium, carbon, and chlorine by the following symbols:

$$\text{H}\cdot \qquad \text{Na}\cdot \qquad \cdot\dot{\underset{\cdot}{\text{C}}}\cdot \qquad \cdot\ddot{\text{Cl}}\!:$$

When a pair of electrons is shared by two atoms, the two electrons are placed between the symbols for those two atoms. Accordingly, the hydrogen molecule is drawn as follows:

$$\text{H}:\text{H}$$

As has been emphasized, the outer shell of an atom is especially stable when it contains eight electrons (except for the first shell, which is stable with two electrons). This principle leads to the **octet rule**, which states that *the atoms in covalent molecules tend to share electrons in such a way that there are eight electrons in the outer shell of each of the atoms (except for those in the first row of the periodic table, which are filled with two electrons)*. Following this rule, we diagram the methane (CH_4) molecule and the chlorine (Cl_2) molecule as follows:

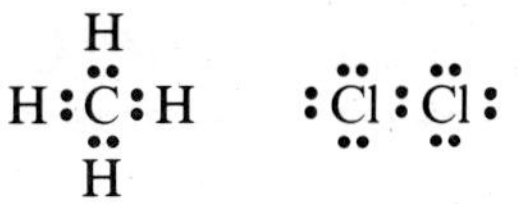

$$\begin{array}{cc} \text{H} \\ \text{H:}\overset{..}{\text{C}}\text{:H} & \text{:}\overset{..}{\underset{..}{\text{Cl}}}\text{:}\overset{..}{\underset{..}{\text{Cl}}}\text{:} \\ \text{H} \end{array}$$

We recall that with the exception of helium, the noble gas atoms (argon, neon, krypton, etc.) each have eight electrons in their outer shells. It is clear that each of the atoms in a covalent molecule tries to make its outer shell look like that of a noble gas atom.

Since oxygen has six electrons in its outer shell, it takes two additional electrons to fill that shell. For that reason oxygen is able to form double bonds, in which two pairs of electrons are shared. The diagrams for carbon dioxide (CO_2) and formaldehyde (CH_2O) are shown below:

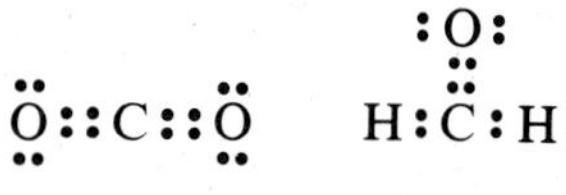

$$\overset{..}{\underset{..}{\text{O}}}\text{::C::}\overset{..}{\underset{..}{\text{O}}} \qquad \begin{array}{c} \text{:O:} \\ \text{::} \\ \text{H:C:H} \end{array}$$

Because carbon has four outer electrons, it is able to form compounds with many combinations of bonds: four single bonds, two double bonds, a double bond and two single bonds, or a triple bond and a single bond. Nitrogen, with five outer electrons, is able to form triple bonds by accepting three electrons into its outer shell. These structures are shown below in the nitrogen molecule (N_2) and in the poisonous gas hydrogen cyanide (HCN):

$$\text{:N:::N:} \qquad \text{H:C:::N:}$$

The bond resulting from the sharing of a pair of electrons can be represented by a straight line, as well as by pairs of dots. Accordingly, the above molecules can also be drawn in the following way:

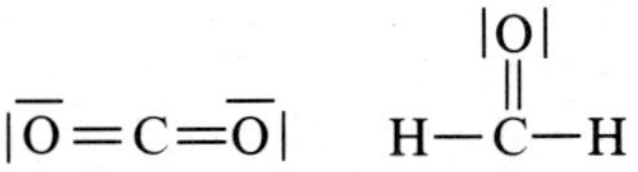

$$|\overline{\text{O}}=\text{C}=\overline{\text{O}}| \qquad \begin{array}{c} |\underline{\text{O}}| \\ \parallel \\ \text{H}-\text{C}-\text{H} \end{array}$$

$$|\text{N}\equiv\text{N}| \qquad \text{H}-\text{C}\equiv\text{N}|$$

The practice of drawing bonds with straight lines is more convenient than the dot method, and so is used more often in diagramming molecular structure. This is especially true in organic chemistry—the chemistry of carbon compounds—in which molecular diagrams are necessary to show clearly the structure of the molecule.

7.6 The Geometry of Molecules

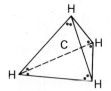

FIGURE 7-13
The Lewis electron-dot model shows a molecule only in one dimension; this tetrahedral model of methane shows the positions of the atoms in the molecule more realistically.

The Lewis electron-dot model is not realistic, since all the diagrams are drawn in a flat plane and most real molecules are not flat. A more realistic diagram must be drawn in three dimensions. Let us consider how to visualize the form of the methane molecule, whose planar diagram is

$$\mathrm{H : \overset{..}{\underset{..}{C}} : H}$$
(with H above and below)

Each of the paired electrons is a unit with two negative electric charges. Each pair repels the other three pairs. As a result, the four pairs of electrons get as far away from one another as they can, while tied to the molecule by the attraction of the nuclear protons.

The geometric figure that allows the electron pairs the greatest separation from one another is the **tetrahedron**—a figure composed of four equal triangular sides. (A triangular pyramid is an example of a tetrahedron.) Figure 7-13 shows a tetrahedral model of a methane molecule, with the carbon atom at the center of the figure and each of the hydrogen atoms in one of the four corners. The hydrogen atoms are equidistant from one another, and the four C—H bonds are of equal lengths. In this geometry all of the internal H—C—H angles are 109.5°.

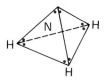

FIGURE 7-14
A tetrahedral model of the ammonia (NH_3) molecule

Figure 7-14 displays an ammonia molecule (NH_3) in a tetrahedral model. The nitrogen atom is at the center, and hydrogen atoms occupy three of the corners. The fourth corner holds the electron pair not shared with any hydrogen atom. The repulsion between this lone pair and the others forces the H—N—H angles to be approximately 107° instead of 109.5°.

In the water (H_2O) molecule, seen in Figure 7-15, there are two pairs of electrons not connected with a hydrogen atom. The oxygen atom is at the center of the tetrahedron, the hydrogen atoms are at two corners, and the other two corners are occupied by unshared electron pairs. These unshared pairs cause the H—O—H bond angle to be reduced to 104.5°, so the tetrahedron is not quite symmetrical.

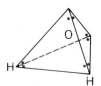

FIGURE 7-15
A tetrahedral model of the water (H_2O) molecule

It must be understood that the tetrahedral model is only a much simplified way to show what a molecule is like—to show the positions of the atoms in relation to one another. The straight lines are only part of the diagram; they do not exist in the actual molecule. And the electrons are not actually fixed at the points where dots are seen in the diagram, but rather are constantly in motion; the dots only show where the electrons are most likely to be found at any given moment.

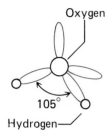

FIGURE 7-16
An orbital model of the water molecule

A slightly more realistic model of the water model is shown in Figure 7-16. This diagram represents the orbitals—the regions of space where the probability of finding the electrons is greatest.

The tetrahedron, it must be understood, is not the only form that can be used for molecular models. It is useful in the examples we have chosen, where symmetry requires that the four angles between the bonds be equal or almost equal.

Glossary

anion A negative ion; an ion attracted to the anode of an electrolysis apparatus.

anode The positive electrode in an electrolysis apparatus.

bonding The attraction of atoms for each other causing them to form a molecule or polyatomic ion.

bond strength The amount of energy required to separate two atoms in a molecule or polyatomic ion.

cathode The negative electrode in an electrolysis apparatus.

cation A positive ion; an ion attracted to a cathode.

covalent bond A bond formed by the sharing of electrons between two atoms in a molecule or polyatomic ion.

diatomic molecules Molecules consisting of two atoms.

dipole-dipole interaction The attraction of the positive end of a polar molecule to the negative end of an adjacent molecule (and vice versa).

double bond The bond formed when two adjacent atoms share four electrons (two pairs of electrons).

electric dipole A molecule in which the positive charge center is not concentric with the negative charge center, so there is a positive charge at one end and a negative charge at the other end; any object with a positive charge at one end and a negative charge at the other.

electromagnetic force A fundamental physical interaction between particles, responsible for holding electrons and protons together within an atom.

electron affinity The attraction one atom has for the electrons of another atom.

electronegativity A comparative measure of the amount of attraction an atom has for electrons from another atom.

ion An atom or group of atoms with an unbalanced electric charge.

ionic bond A bond between atoms in which one atom completely captures an electron from the other atom so the two atoms are ionized and are held together by an electrostatic attraction.

lattice A three-dimensional network of atoms in a crystalline solid.

Lewis electron-dot model A molecular model in which the outer electrons of each atom are shown as dots surrounding the atom's symbol and electron pairs shared are drawn between the symbols of two atoms.

octet rule A generalization that predicts that the outer shell of an atom will be most stable when it contains eight electrons.

oxidation number The number of electrons transferred to or from an atom when a molecule is formed; hence, the number of electron charges on an atomic or polyatomic ion.

oxidizing agent An element or group that accepts electrons from another element or group during the course of a reaction.

polar covalent bond A covalent bond in which one atom is more electronegative than the other, so the shared electrons remain closer to the first atom than to the second, causing the molecule to be polarized.

polarized Has positive electric charge at one end and negative charge at the other; in chemistry, applied to molecules with polar covalent or ionic bonds.

reducing agent An element or group that donates electrons to another element or group during the course of a reaction.

single bond A bond in which one pair of electrons is shared between two atoms.

tetrahedron A four-sided solid figure in which each side is a triangle.

triple bond A bond in which three pairs of electrons are shared between two atoms.

valence The number of hydrogen atoms (or equivalent atoms) that can be bound to one atom of a given element. The concept of valence has been superseded by that of oxidation number.

van der Waals force A weak force of attraction between two neutral molecules, resulting from the average electromagnetic force among all the electrons and protons.

Problems and Questions

7.1 The Nature of the Chemical Bond

7.1.Q How was the chemical bond visualized before the discovery of the electron?

7.2.Q What is a bivalent element?

7.3.Q Water contains two hydrogen atoms for each oxygen atom. What was the classical explanation of this fact?

7.4.Q Name a trivalent and a tetravalent element.

7.5.Q What is the meaning of electron affinity, and how is it measured?

7.6.Q What is electronegativity?

7.7.Q What kind of compound is formed when the electronegativity number of one element in the compound is much higher than that of the other element?

7.8.Q If two elements have electronegativity numbers that are close in value, what kind of compound will they form? Describe the location of the electrons in such a compound.

7.2 Ionic and Covalent Bonding

7.9.P How many electrons, protons, and neutrons are found in each of the following ions?
    a. Br^- **b.** H^+
    c. H^- **d.** Na^+
    e. Ca^{2+} **f.** Fe^{3+}

7.10.P What is the charge on an ion that contains twenty-eight electrons and thirty protons? What is the name of this ion?

7.11.Q Define the following:
    a. electron affinity (or affinity for electrons)
    b. ionic bond
    c. covalent bond
    d. cation
    e. anion
    f. van der Waals force
    g. double bond

7.12.Q
    a. What kind of bond holds a hydrogen molecule together?
    b. Explain why the bond in a hydrogen molecule is saturated.

7.13.Q Is NaCl normally found in a molecule with one atom of Na and one atom of Cl? In what form is it found?

7.14.Q Describe the difference between the bond that holds Na^+ to Cl^- in NaCl and the bond that holds two hydrogen atoms together in H_2.

7.15.Q Describe the difference between the ionic bond in a crystal of NaCl and the metallic bond in a crystal of copper.

7.16.Q Explain the difference between a liquid and a solid, using the concept of binding force.

7.17.Q Explain why an ionic solid has a higher melting point that a covalent solid.

7.18.Q Describe an experiment that will determine whether a compound is an ionic compound.

7.3 Polar Covalent Bonds

7.19.Q Define the following:
    a. polarization
    b. polar covalent bond
    c. dipole molecule

7.20.Q What conditions are necessary for a molecule to be polarized? (*Hint*: What is the relationship between electronegativity and polarization?)

7.21.Q Describe a dipole-dipole interaction. What is the effect of this interaction on the molecule it forms?

7.22.Q Explain how water molecules can stick together to form the drops you see on the side of a glass.

7.23.Q Explain how water bugs can walk on water.

7.24.Q How does the boiling point of a polar covalent compound depend on the amount of polarization?

7.25.Q Why is the boiling point of water, H_2O, much higher than that of H_2S? (*Hint*: What kind of bonding must exist in H_2O to produce a higher boiling point?)

7.26.Q Explain why CO_2 is not a polarized molecule, even though it is held together by polar bonds.

7.27.Q Would you expect liquid HCl to be a good conductor of electricity? (*Hint*: Does the HCl molecule have a net electric charge?)

7.28.Q What effect does the angle between the two OH bonds in the water molecule have on the polarization of the molecule?

7.4 Oxidation Numbers

7.29.Q **a.** What is the original definition of oxidation?
b. What is the modern definition of oxidation?
c. How is the new definition of oxidation more general than the old?

7.30.Q **a.** What is an oxidizing agent?
b. What is the difference between oxidation and an oxidizing agent?

7.31.Q What is the modern definition of reduction?

7.32.Q What is the relationship between an oxidizing agent and a reducing agent?

7.33.Q **a.** If an atom in a compound has three excess electrons, what is its oxidation number?
b. If an atom in a compound has two excess protons, what is its oxidation number?

7.34.Q In the reaction $2K + Br_2 \rightarrow 2KBr$, which element is the oxidizing agent?

7.35.Q In the reaction $Fe_2O_3 + 3H_2 \rightarrow 2Fe + 3H_2O$,
a. what is the oxidation number of the Fe on the left side of the equation?
b. what is the oxidation number of the Fe on the right side of the equation?
c. has the Fe been oxidized or reduced?

7.36.Q In the reaction $Zn + 2HCl \rightarrow ZnCl_2 + H_2$,
a. which element is the oxidizing agent?
b. which element is the reducing agent?
c. what is the oxidation number of each of the following?
(1) Zn (2) H in H_2
(3) Zn in $ZnCl_2$ (4) H in HCl

7.37.Q Is it possible to have oxidation without reduction in a chemical reaction? Explain your answer.

7.5 The Lewis Electron-Dot Model

7.38.Q How many electrons are required to make the first atomic shell most stable?

7.39.Q What is the octet rule?

7.40.Q How many electrons are shared by each of the elements in the following compounds?
a. water (H_2O)
b. methane (CH_4)
c. ammonia (NH_3)
d. carbon dioxide (CO_2)

7.41.P Insert the dots into the following molecular diagrams:

```
       H H                      H  H
a. H C  C  O H        b.    C  C
       H H                      H  H

       H  H                     H  O
c. H C  O  C  H       d. H C  C  O H
       H  H                     H
```

7.42.P Redraw the same diagrams in Problem 7.41.P, using lines instead of dots to represent the bonds.

7.43.P Draw the Lewis electron-dot diagram for each of the following:
a. CCl_4 **b.** CS_2
c. H_2O **d.** NH_3
e. N_2 **f.** CH_3CH_3

7.44.P Redraw the diagrams from Problem 7.43.P, using lines instead of dots to represent the bonds.

7.45.Q In a real covalent molecule, are the shared electrons actually found between the neighboring atoms? What is a more accurate way of picturing the electrons in motion?

7.46.Q Why is oxygen able to form double bonds?

7.47.Q What are the various combinations of bonds that carbon can form?

7.6 The Geometry of Molecules

7.48.Q Why does the molecule of methane fit into a tetrahedral shape?

7.49.Q What kind of bond is there between the carbon and hydrogen atoms in methane?

7.50.Q Draw both the electron-dot diagram and a three-dimensional diagram for each of the following molecules:
    a. NH_3 **b.** H_2O
    c. CH_4 **d.** N_2

7.51.Q What is one drawback of the tetrahedral diagram?

7.52.P Draw an orbital diagram for the following:
    a. CH_4
    b. NH_3

7.53.Q Referring back to Section 7.2 (polar covalent molecules), discuss the way in which the tetrahedral shape of the water molecule leads to a polarized molecule.

7.54.Q Compare the geometrical structures of the following pairs:
    a. CH_4 and CCl_4
    b. H_2O and H_2S

7.55.P a. Propose geometrical structures for the following ions:
    (1) NH_4^+
    (2) H_3O^+
b. Use Lewis electron-dot diagrams to show why these ions have positive electric charges.

7.56.P a. Propose geometrical structures for the following ions:
    (1) OH^-
    (2) NH_2^-
b. Use Lewis electron-dot diagrams to show why these ions have negative electric charges.

7.57.P Draw a structural diagram for formaldehyde (CH_2O). (*Hint*: Oxygen is at the center of its own tetrahedron, as is carbon.)

Self Test

1. Define or explain each of the following terms:
   a. cation
   b. anion
   c. covalent bond
   d. ionic bond
   e. polarization
   f. van der Waals force
   g. octet rule
   h. oxidation number

2. Describe the relationships between covalent, polar-covalent, and ionic bonds.

3. What causes a covalent bond to become polarized?

4. Why is the hydrogen molecule unpolarized, whereas the HCl molecule is polarized?

5. Suggest a testing procedure to determine whether a molecule has covalent or ionic bonding.

6. Why does hydrogen generally have an oxidation number of 1^+ in its compounds with nonmetals?

7. Write the oxidation number for each of the atoms in the following compounds:
   a. NaI   b. Na_2O_2
   c. MgH_2 d. $FeBr_3$
   e. CO_2

8. Draw the Lewis electron-dot formula for each of the following compounds:
   a. H_2S b. HF
   c. NH_3 d. NCl_3
   e. H_2O f. CH_4

9. What is the fundamental reason for the octet rule?

10. Draw a three-dimensional structure for each of the compounds in Question 8.

11. Why is the tetrahedron a likely shape for many molecules?

8
Nomenclature

Objectives After completing this chapter, you should be able to:
1. State the rules for naming compounds.
2. Name simple compounds, given their formulas.
3. Write the formulas for simple compounds, given their names.
4. State the rules for assigning oxidation numbers to elements and groups.
5. Describe the classification of compounds into acids, bases, and salts.
6. State the rules of nomenclature for each class of compounds.

8.1 Names and Formulas

Up to this point, we have dealt with fundamentals: atomic structure and molecular bonding. We are now ready to pursue *macroscopic chemistry* (chemistry on a large scale): naming compounds, writing formulas and equations, understanding chemical reactions, and calculating quantities of substances involved in reactions.

Compounds often answer to two or more names. Those substances that have been in use for a very long time may have common names, or **trivial names**, that have been handed down through history. In addition, all chemical compounds have **technical names** that follow rules set forth at various times in history. The most recent attempt to systematize chemical nomenclature is incorporated in the rules laid down by the International Union of Pure and Applied Chemistry (**IUPAC**) in 1957.

The IUPAC cites three major goals in the formulation of rules for naming compounds:

1. The rules must include all cases. That is, by using the rules, you should be able to name all possible compounds.

2. There should be no uncertainty or ambiguity. That is, each name should apply to one and only one compound. There should be no possibility of giving two compounds the same name.

3. The name should describe the formula. That is, the name should let you know what elements make up the compound and what proportions of these elements are in the compound.

The fact that some chemical compounds go by a common name as well as a technical name may confuse the unwary. For example, common table salt is a chemical by the name of sodium chloride (NaCl), baking soda is sodium hydrogen carbonate ($NaHCO_3$), sugar is sucrose ($C_{12}H_{22}O_{11}$), and the main ingredient in vinegar is acetic acid (CH_3CO_2H). There is widespread distrust of chemical food additives, but actually all food contains chemicals, including many traditional additives, as the above list shows.

The health-food enthusiast often views "natural" chemicals as good and "artificial" chemicals as bad. But even with natural chemicals, reasonable care is needed. Adding too much of a natural chemical like salt or sugar to food is not good for all people. Too much salt is bad for those with high blood pressure, and sugar is not recommended for diabetics. Many people think honey is a safe substitute for sugar; actually the biochemical behavior of the two is the same. The only criterion that counts is the effect of the chemical on the body, and knowledge obtained by careful scientific research is more useful than folklore. A chemical compound put together by artificial means behaves exactly the same way as the identical compound obtained from plants and animals growing in nature.

Table 8-1 shows some traditional common names, together with their preferred scientific names. (The common names tend to be the older ones; the IUPAC names are relatively recent.) Many of the common names give no idea of what elements are in the compound or how they are joined together. The technical names, on the other hand, are designed to tell us what elements are in the compound and also something about the structure of the compound. We will give more details about these names in the following sections.

A **chemical equation** is a statement that tells in shorthand what elements and/or compounds go into a reaction and what elements and/or compounds come

TABLE 8-1 Common and Technical Names of Compounds

Common Name	Technical Name	Formula
ammonia	trihydrogen nitride	NH_3
water	dihydrogen oxide (hydrogen hydroxide)	H_2O
sal ammoniac	ammonium chloride	NH_4Cl
calcite, limestone	calcium carbonate	$CaCO_3$
chlorinated lime (bleaching powder)	calcium hypochlorite	$Ca(ClO)_2$
fluorite	calcium fluoride	CaF_2
lime	calcium oxide	CaO
plaster of Paris	calcium sulfate	$CaSO_4 \cdot \frac{1}{2}H_2O$
diamond, graphite	carbon	C
heavy water	deuterium oxide	D_2O
laughing gas	dinitrogen oxide (nitrous oxide)	N_2O
muriatic acid	hydrochloric acid	HCl
Prussian blue	ferric hexacyanoferrate(II) (ferric ferrocyanide)	$Fe_4[Fe(CN)_6]_3$
Epsom salt	magnesium sulfate	$MgSO_4 \cdot 7H_2O$
quicksilver	mercury	Hg
Calomel	mercury(I) chloride	Hg_2Cl_2
potash	potassium hydroxide	KOH
saltpeter	potassium nitrate	KNO_3
washing soda	sodium carbonate	$Na_2CO_3 \cdot 10H_2O$
baking soda	sodium bicarbonate	$NaHCO_3$
salt	sodium chloride	$NaCl$
lye, caustic soda	sodium hydroxide	$NaOH$
hypo	sodium thiosulfate (sodium hyposulfite)	$Na_2S_2O_3 \cdot 5H_2O$

Note: Many salts contain molecules of water as part of their crystal structure. This water is known as *water of hydration*, or *water of crystallization*. In cases where the water of hydration is important, the above table shows the number of water molecules attached to each salt molecule. A dot is used between the formulas for the salt and the water molecules to indicate that the water part of the crystal is just water of hydration. Very often, drying the crystal will cause the water to be released.

out of the reaction. It would be awkward to use the full chemical names of the compounds in the equations, so we make use of the formulas that represent those compounds.

A **formula** is a group of letters that represents the elements in a compound. If the formula contains numbers that show how many atoms of each element are present in the molecule, it is called a **molecular formula**. We have already encountered such formulas for water (H_2O), sodium chloride (NaCl), and methane (CH_4).

Another kind of formula is the **structural formula**, which shows how the elements are arranged in a molecule. You have already encountered in Chapter 7 the structural formulas for compounds such as methane:

$$\begin{array}{c} H \\ | \\ H-C-H \\ | \\ H \end{array}$$

Structural formulas are valuable in organic chemistry, because there are many cases in which two organic compounds have the same molecular formula but different structural formulas.

In this chapter we will concentrate on molecular formulas of inorganic compounds; organic compounds will be treated in Chapter 17. We will see how the inorganic compounds are commonly divided into three major groups: acids, bases, and salts. The compounds of each group have their own structures, rules of nomenclature, and types of reactions.

8.2 Writing Formulas

Before you can learn to write formulas properly, two preliminary matters must be taken up. You must learn (1) the names of certain groups of elements that appear in many compounds and (2) the rules for assigning oxidation numbers to the elements in compounds, for these are the numbers that determine how many atoms of each element are found in a given compound.

Polyatomic Ions

In many molecules certain groups of atoms are bound together so closely that they form a subunit within the molecule. In the past such groups were called **radicals**. (The term "radical" now has a more specialized meaning.) In sulfuric acid (H_2SO_4), the group SO_4 acts as a building-block of the molecule. The same group appears in the following compounds:

sodium sulfate	Na_2SO_4
aluminum sulfate	$Al_2(SO_4)_3$
platinum sulfate	$Pt(SO_4)_2$

These examples show that when a group (in this case, the sulfate group) appears more than once in the molecule, it is surrounded by parentheses and followed by a subscript that tells how many times it appears.

The above compounds are all ionic compounds. Sodium sulfate consists of two Na^+ and one SO_4^{2-} ion. The SO_4^{2-} ion is a **polyatomic ion**—an ion formed from a closely knit group of atoms. Since it carries two extra electrons and therefore has two negative charges, the symbol $2-$ appears in the exponent. The sulfate ion in sodium sulfate, being an anion, behaves just like the chloride ion in sodium chloride, and so it is treated as a single unit.

Another common ion is the hydroxide ion (OH^-), found in solutions of compounds such as

Name	Molecular Formula	Ions in Solution
sodium hydroxide	NaOH	Na^+ OH^-
calcium hydroxide	$Ca(OH)_2$	Ca^{2+} OH^-
thorium hydroxide	$Th(OH)_4$	Th^{4+} OH^-

In the formula for an ion, the superscript shows the amount of unbalanced electric charge, either positive or negative. The $+$ in Na^+ indicates that Na has lost an electron; the $2-$ in SO_4^{2-} shows that the SO_4 ion has two extra electrons. We could write the superscript on the Na as $1+$, but the convention is simply to write a $+$ or a $-$ when a single charge is involved. A calcium ion has two positive charges and so is Ca^{2+}. (It is understood that when we speak of the charge of an ion, we mean the number of *extra* charges, either positive or negative.)

The structure of polyatomic ions is a complex subject and cannot be treated in detail here. However, Figure 8-1 gives a rough idea of the sulfate ion arrangement. The ion is in the form of a tetrahedron, with the sulfur atom in the center. The sulfur and oxygen atoms are held together by covalent bonds—by the sharing of electrons. The sulfur atom has six electrons to share and four oxygen atoms with which to share them. Each oxygen atom has two acceptor sites, making a total of eight acceptor sites. As a result, the SO_4 ion tends to capture a pair of electrons from a hydrogen or metal atom. In this way, the sulfate ion ends up with two extra electrons and a pair of ionic bonds connecting the sulfate ion to a pair of positive ions.

A very common polyatomic ion with a positive charge is the ammonium ion NH_4^+. The compound ammonia (NH_3) is a colorless, pungent gas with a boiling point of $-33°C$. When dissolved in water, it forms the compound ammonium hydroxide (NH_4OH), which consists of two polyatomic ions. (The chemical called "ammonia" that is sold for cleaning purposes is actually a dilute solution of ammonium hydroxide.)

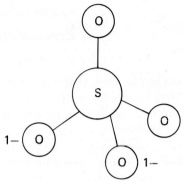

FIGURE 8.1 The tetrahedral structure of the sulfate ion

TABLE 8-2 Table of Selected Polyatomic Ions

Formula and Charge	Name
$C_2H_3O_2^-$	acetate ion
NH_4^+	ammonium ion
AsO_4^{3-}	arsenate ion
AsO_3^{3-}	arsenite ion
HCO_3^-	bicarbonate ion (hydrogen carbonate ion)
CO_3^{2-}	carbonate ion
CN^-	cyanide ion
ClO_4^-	perchlorate ion
ClO_3^-	chlorate ion
ClO_2^-	chlorite ion
ClO^-	hypochlorite ion
H_3O^+	hydronium ion
OH^-	hydroxide ion
NO_3^-	nitrate ion
NO_2^-	nitrite ion
MnO_4^-	permanganate ion
PO_4^{3-}	phosphate ion
SO_4^{2-}	sulfate ion
SO_3^{2-}	sulfite ion
O_2^{2-}	peroxide ion
CrO_4^{2-}	chromate ion
$Cr_2O_7^{2-}$	dichromate ion
SiO_3^{2-}	silicate ion

Note: The numbers in the exponent represent the quantity and sign of the electric charge on the ion. Where there is only a single charge, it is represented by a plus sign or a minus sign (+ or −) without a number.

Numerous groups become polyatomic ions in ionic compounds. Some of these ions are listed in Table 8-2, together with their charge numbers. You should be completely familiar with the names, formulas, and charges of these ions, since they will be encountered in a large number of common compounds. In later sections we will discuss the IUPAC rules for naming these ions, and you will learn how their oxidation numbers are determined.

Rules for Oxidation Numbers

To write the formulas for compounds, it is necessary to know how many atoms of each type are found in the molecule. The number of each type of atom is not the result of chance or accident. The number of hydrogen and oxygen atoms in a water molecule, for example, is determined by the fact that hydrogen has an oxidation number of $+1$ and oxygen has an oxidation number of -2. (Remember, the oxidation number represents the number of electrons transferred to or from an atom when it forms a compound. The number $+1$ means the hydrogen has lost an electron, whereas the number -2 means the oxygen has gained two electrons.) Knowledge of the oxidation number of each element in a compound helps us write the formula for that compound.

In this text the oxidation number is written above and to the right of the symbol for the element. The $+$ or $-$ is written to the *left* of the oxidation number. This convention is in contrast to the rule for writing the ionic charge number, which calls for the $+$ or $-$ sign to be to the *right* of the number.

It is easy to confuse the meanings of *oxidation number* and *charge number*. The **charge number** gives the number of charges on an ion, whereas the oxidation number may apply to compounds formed with polar covalent bonds, in which there are no ions at all. For example, gaseous HCl is a covalent compound, so there are no ions in this compound. However, since the Cl is more electronegative than the H, the shared electrons are closer to the Cl than to the H. For this reason we say the H gives its electron to the Cl; accordingly, H has an oxidation number of $+1$ and the symbol H^{+1}. Similarly, Cl has an oxidation number of -1 and is written Cl^{-1}.

Several rules simplify the learning of oxidation numbers and the writing of formulas. In a compound the oxidation number of each element is determined (in part) by the number of electrons in the outer shell, as Rules 2–8 demonstrate.

1. An element in its natural state (not combined with another element) always has an oxidation number of 0.
Examples: Na^0, $H_2^{\,0}$, $Cl_2^{\,0}$

2. The alkali metals in column 1a of the periodic table almost always have an oxidation number of $+1$, because each of these elements has a single outer electron that is easily given up.

Example: $NaCl = Na^{+1}Cl^{-1}$

(Here the formula displays the oxidation numbers.)

3. Elements in column 2a of the periodic table (the alkali earths) almost always have an oxidation number of $+2$, because each of them has a pair of electrons in the outer shell; these electrons are easily given up by the atom.

Example: $MgO = Mg^{+2}O^{-2}$

(Exceptions to Rules 2 and 3 are so rare that you will probably never encounter them.)

4. For elements in column 7a of the periodic table (the halogens), the oxidation number depends on the type of compound.

    a. A **binary compound** is one containing two different elements. In binary compounds the oxidation number of the halogen atom is usually -1, because in such a compound one electron has been transferred to the halogen atom in order to fill its outer shell. Exceptions to this rule will not concern us.

        Example: $CaCl_2 = Ca^{+2}(Cl^{-1})_2$

    b. A **ternary compound** contains three different elements. There is no simple rule for oxidation numbers in ternary compounds. For example, in sodium chlorate ($NaClO_3$) the chlorine atom has an oxidation number of $+5$, in sodium chlorite ($NaClO_2$) the chlorine atom has an oxidation number of $+3$, and in sodium hypochlorite ($NaClO$) the chlorine atom has an oxidation number of $+1$.

5. Oxygen usually has an oxidation number of -2, with the following exceptions:

    a. In compounds called **peroxides**, the oxidation number of the oxygen atom is -1.

        Example: The common antiseptic hydrogen peroxide is $H_2O_2 = (H^{+1})_2(O^{-1})_2$.

    b. In **superoxides** the oxidation number is -0.5.

        Example: Potassium superoxide is $KO_2 = K^{+1}(O^{-0.5})_2$.

        Note: It is not necessary for an oxidation number to be a whole number.

6. Hydrogen always has an oxidation number of $+1$ except in hydrides, where the number is -1. Hydrides are formed when hydrogen combines with a metal whose atom gives up an electron more easily than does the hydrogen atom (usually a column 1a or 2a metal).

Example: Potassium hydride has the formula $KH = K^{+1}H^{-1}$.

7. The sum of the oxidation numbers in a molecule always equals 0.

Example: In magnesium bromide, the Mg is a column 2a element, with an oxidation number of +2, and the Br is a column 7a element, with an oxidation number of −1. Hence the formula is

$$MgBr_2 = Mg^{+2}(Br^{-1})_2$$

We verify that this is the correct formula by adding the oxidation numbers, taking into account the fact that there are two bromide ions, so the −1 must be multiplied by 2:

$$(+2) + 2(-1) = 0$$

This rule comes from the fact that the number of electrons in the molecule is conserved, so the number of electrons given up by the Mg atom must equal the number of electrons accepted by the Br atoms. Since the Mg atom has two electrons to give up, there must be two Br atoms, because each Br can handle only one electron.

8. The sum of the oxidation numbers in a polyatomic ion is equal to the charge on the ion.

Example: The formula for the sulfate ion is

$$SO_4^{2-} = [S^{+6}(O^{-2})_4]^{2-}$$

In this ion the sulfur atom has an oxidation number of +6 (meaning that it has given up six electrons), and each oxygen atom has an oxidation number of −2, so the sum of all the oxidation numbers is

$$(+6) + 4(-2) = -2$$

This number (−2) is equal to the ionic charge and is the number found in the exponent of the SO_4 group.

The above example demonstrates the difference between oxidation number and ionic charge. The sulfate ion as a whole carries a 2− charge. Since the sulfur atom is joined to the oxygen atoms with covalent bonds, it does not carry six electron charges. But because it shares six electrons with the oxygen atoms, it has an oxidation number of +6.

Another kind of problem can be solved with Rule 8: Find the oxidation number of one of the atoms inside a polyatomic ion if the charge on the ion is given. Consider the phosphate ion PO_4^{3-}. We know that oxygen always has an oxidation number of −2 in this kind of situation, and the formula says that the charge on the ion is 3−. Let x stand for the oxidation number of the phosphorus atom. The ionic formula can be written

$$PO_4^{3-} = [P^x(O^{-2})_4]^{3-}$$

Using Rule 8, write the sum of the oxidation numbers:

$$x + 4(-2) = -3$$
$$x = -3 - 4(-2) = -3 + 8$$
$$= +5$$

This result means that the phosphorus atom has an oxidation number of $+5$ in the phosphate ion.

Writing the Formula

You are now ready to learn the rules for writing formulas. The first rule specifies the order in which the symbols for the elements are written, for their arrangement is not at all random.

A compound can generally be divided into two parts: one with a *positive* oxidation number and the other with a *negative* oxidation number. As can be seen from the examples given in the last section, chemical formulas are always written with the positive part first, the negative part second. If one of the components of the molecule is a polyatomic group or ion, then within that group the elements are in order, with positive first and negative second. For example,

$$KClO_4 = K^{1+}[Cl^{+7}(O^{-2})_4]^{1-}$$

The next rule deals with the important question of how many atoms of each species are present in the molecule. This question is answered using the rule for adding oxidation numbers: *The sum of the oxidation numbers must be zero in a neutral (uncharged) molecule.*

In the case of NaCl, the matter is simple. Na has an oxidation number of $+1$, and Cl has an oxidation number of -1. The sum of the oxidation numbers is

$$(+1) + (-1) = 0$$

Therefore, one atom of sodium is needed for each chlorine atom. This result verifies that the correct formula for sodium chloride is NaCl.

In the compound magnesium bromide, magnesium has an oxidation number of $+2$, and each bromine ion has an oxidation number of -1. How can we get these numbers to add up to zero? This problem is solved by letting x stand for the number of bromine atoms. Then, setting the sum of the oxidation numbers equal to 0, we have

$$1(+2) + x(-1) = 0$$
$$2 - x = 0$$
$$x = 2$$

Therefore, the formula for magnesium bromide is $MgBr_2$. (Recall that this result follows from the principle of conservation of electrons.)

A compound like potassium phosphate is handled in a similar manner. Potassium (K) has an oxidation number of $+1$, and the phosphate ion (PO_4^{3-}) has a charge of -3. (*Note*: Oxidation numbers apply to atoms, whereas charges are used for polyatomic ions. For purposes of determining formulas, the two numbers are treated as equivalent.) If each phosphate ion accepts three electrons and each potassium atom donates one electron, there must be three potassium atoms to each phosphate ion. The formula for potassium phosphate is thus written K_3PO_4.

The formula for calcium phosphate presents a more complicated situation. Here Ca has an oxidation number of $+2$, and the phosphate ion has a charge of -3. Let us try using the above method to find how many calcium ions go with each phosphate ion. Let the number of Ca ions in the molecule be represented by x. If one phosphate ion is assumed to be in the molecule, summing the charges gives

$$x(+2) + 1(-3) = 0$$

The result of this equation is $x = 3/2$. Of course, there cannot be a fractional number of calcium ions in the molecule. The number x must be a whole number. Since the equation produced a fraction, we must have done something wrong. The error was in assuming that the compound contains only *one* phosphate ion. We can correct this error by multiplying the assumed number of phosphate ions by the denominator of x. We therefore assume that there are two PO_4^{3-} ions in each molecule and solve this equation:

$$x(+2) + 2(-3) = 0$$
$$2x = 6$$
$$x = 3$$

The result indicates that for every two PO_4^{3-} ions there are three Ca^{2+} ions in the molecule. The formula for calcium phosphate is, accordingly, $Ca_3(PO_4)_2$. Notice that the PO_4 group is inside parentheses. The subscript 4 indicates the number of oxygen atoms in the phosphate group. The subscript 2 outside the parentheses tells how many phosphate groups are in the molecule. (Therefore, there are eight oxygen atoms and two phosphorus atoms in each molecule.)

We will now show a simple short-cut method for obtaining the same results with less labor. Start by writing the formula of the compound, with the oxidation or charge numbers in the exponents. At first, leave blank the subscript following each atom or ion. Then transpose the exponent of each ion to the subscript of the other ion.

Try this method with the compound sodium sulfate. First write the formula with the charges in the exponents:

$$Na^{+1}SO_4^{-2}$$

Now transpose the numbers in the exponents to the subscripts (ignoring the $+$ or $-$ sign):

$$Na^{+1}SO_4^{-2}$$

$$Na^{+1}_2(SO_4^{-2})_1$$

Now eliminate the 1 in the subscript and all the characters in the exponents, to get the final formula for sodium sulfate, Na_2SO_4. (The parentheses are not needed unless there is more than one sulfate ion in the molecule.)

Let us apply this method to a compound containing two polyatomic ions:

ammonium chromate. The ammonium ion is NH_4^{+1}, and the chromate group is CrO_4^{-2}. Write the trial formula with the charges in the exponents; then transpose the exponents:

$$NH_4^{+1} CrO_4^{-2}$$

$$(NH_4^{+1})_2 (CrO_4^{-2})_1$$

Eliminate unnecessary superscripts to get the final formula, $(NH_4)_2CrO_4$.

This method of *crossing the exponents* will always work for compounds consisting of pairs of ions with known charges or oxidation numbers.

Conversely, given the charge or oxidation number of one component, you can always find the charge or oxidation number of the other component if the formula is known. Consider the compound Fe_2O_3. We know that O usually has an oxidation number of -2. Let X equal the oxidation number of the Fe atom, and set the sum of the oxidation numbers in the molecule equal to 0:

$$2(X) + 3(-2) = 0$$

$$2X - 6 = 0$$

$$2X = 6$$

$$X = 3$$

Each of the Fe atoms is found to have an oxidation number of $+3$.

Example 8.1

Write the formula for aluminum sulfate, in which Al has an oxidation number of $+3$ and the sulfate ion (SO_4^{2-}) has a charge of -2.

Solution

Transposing exponents, we have

$$Al^{+3} SO_4^{2-}$$

$$Al^{+3}_2 (SO_4^{2-})_3$$

Finally, eliminating exponents, we have $Al_2(SO_4)_3$.

8.3 Naming Compounds

Chemical nomenclature is based on a system of roots, prefixes, and suffixes. The **root** is the part of the name that identifies a particular element. The root of oxygen is *oxy*; the root of chlorine is *chlor*. Many elements have names that end in the suffix *-ine*, as in *chlorine*. If that suffix is changed to *-ide*, as in *chloride*, the name means "of chlorine," or "pertaining to chlorine." Accordingly, an ion of chlorine is called a *chloride ion*, whereas a binary compound containing a chloride ion and one metallic ion is a *chloride* (e.g., calcium chloride).

Learning to name compounds is simplified by the fact that compounds are classified into three major groups, for each group has its own rules of nomenclature. In inorganic chemistry the major groups are the **acids**, the **bases**, and the **salts**. You have already encountered some of these and are now ready to make your knowledge more systematic.

Acids

An acid is a compound in which a positively charged hydrogen ion is combined with a negatively charged nonmetallic ion. Common acids are

hydrochloric acid	HCl
hydrofluoric acid	HF
sulfuric acid	H_2SO_4
nitric acid	HNO_3

The first two are simple binary compounds, consisting of two different ion species—a hydrogen ion and a halogen ion. The second two examples are more complex. In those compounds the hydrogen ions are joined to polyatomic ions—$(SO_4)^{2-}$ and $(NO_3)^{1-}$. These oxygen-containing acids are called **oxo-acids**. They are ternary compounds because they contain three different elements.

As described in Chapter 7, hydrogen chloride in its pure liquid or gaseous form is a polar covalent compound. However, when hydrogen chloride dissolves in water, the hydrogen ion and the chloride ion separate, forming a solution of positive and negative ions suspended in water. The ionization is a *result* of the reaction with the water. (The mechanism for this ionization will be discussed in Chapter 14.) In writing the formula, we sometimes differentiate between pure HCl and the water solution of HCl by writing the subscript *aq* (abbreviation for *aqueous*) after the formula. Accordingly, the formula for a water solution of HCl is written HCl_{aq}.

The formation of the hydrogen ion is what makes an acid. Pure hydrogen chloride is not ionized, does not conduct an electric current, and is therefore not an acid. Similarly, pure hydrogen sulfide (H_2S), a foul-smelling, poisonous gas, does not conduct an electric current and so is not considered an acid. But when dissolved in water, it ionizes and becomes hydrosulfuric acid (H_2S_{aq}).

The rules for naming acids make it possible to deduce the structure of a molecule from its name, and vice versa. The rules include the following:

1. *Binary acids.* Hydrogen chloride and hydrogen sulfide are binary compounds. Binary compounds have two-part names: the first part is *hydrogen*, and the second is simply the root name (chlor- or sulf-) of the nonmetallic element with the -*ide* suffix added. In very old books, you might see "chloride of hydrogen," but the modern nomenclature is simply "hydrogen chloride."

As discussed above, the pure compounds hydrogen chloride and hydrogen sulfide are not acids. But when dissolved in water so that they ionize, they become hydrochloric acid and hydrosulfuric acid. The name *hydrogen* condenses to the root *hydro* and combines with the root *chlor* or *sulfur*, and the suffix *-ic* turns the word into an adjective modifying the noun *acid*. (You might wonder why hydrosulfuric acid is used instead of "hydrosulfic acid"; there is apparently some flexibility in the choice of names. Very often when a name has been in use for a long time, the accepted usage wins out over the formal rules.)

2. *Ternary acids.* A common ternary acid is H_2SO_4. The ion SO_4^{2-} is the sulfate ion, so the pure compound might be called "hydrogen sulfate." In naming the ion, we combine the root *sulf* with the suffix *ate*, meaning a group in which sulfur is combined with oxygen. Similarly, pure HNO_3 might be called "hydrogen nitrate."

However, these compounds are always found in water solution, so they become *sulfuric acid* and *nitric acid*. The rule in this case is to take the root of the ion name and add *-ic* to it. The prefix *hydro-* is not used, since that prefix applies only to binary acids.

A common organic acid is $HC_2H_3O_2$. The group $C_2H_3O_2^{1-}$ is the acetate group; the pure compound is *hydrogen acetate*. In water it becomes *acetic acid*.

Often different groups contain the same root element, but with different numbers of oxygen atoms. For example, the nitrate group is NO_3^{1-}, and nitric acid is HNO_3. However, there is also a nitrogen-containing group with one less oxygen atom: NO_2^{1-}. This group is called the *nitrite* group. The hydrogen compound, HNO_2, is called *hydrogen nitrite*, and the corresponding acid is *nitrous acid*.

The rule is that the most common group for a given root takes the ending *-ate*, and the group with one less oxygen atom takes the suffix *-ite*. The acids of the *-ate* group use the suffix *-ic*, and the acids of the *-ite* group use the suffix *-ous*.

For example, chloric acid is $HClO_3$, corresponding to the compound *hydrogen chlorate. Hydrogen chlorite*, on the other hand, has the formula $HClO_2$ and forms *chlorous acid* in water.

There is yet another ternary chlorine acid, containing only one oxygen atom. It is *hypochlorous acid* (HClO), corresponding to *hydrogen hypochlorite*. Here the prefix *hypo-* means *fewer*, or *less*. (Biomedical scientists use the same prefix in such words as "hypoglycemia" and "hypoallergenic.")

At the other extreme is an acid with one more oxygen atom than is found in chloric acid. This acid is called *perchloric acid* ($HClO_4$). The corresponding pure compound is *hydrogen perchlorate*. The prefix *per-* is a short form of *hyper-*, meaning "extra" or "more" (as in *hyperactive*, or *hyperventilation*).

You see that there can be a confusing array of acids, all with the same root but different numbers of oxygen atoms. The system of prefixes and suffixes not only

8.3 NAMING COMPOUNDS

TABLE 8-3 *Names of Some Common Acids and Salts*

Acids		Salts	
Formula	Name	Formula	Name
HCl	hydrochloric acid	NaCl	sodium chloride
H_2SO_4	sulfuric acid	Na_2SO_4	sodium sulfate
HNO_3	nitric acid	$NaNO_3$	sodium nitrate
HNO_2	nitrous acid	$NaNO_2$	sodium nitrite
HClO	hypochlorous acid	NaClO	sodium hypochlorite
$HClO_2$	chlorous acid	$NaClO_2$	sodium chlorite
$HClO_3$	chloric acid	$NaClO_3$	sodium chlorate
$HClO_4$	perchloric acid	$NaClO_4$	sodium perchlorate

keeps order in the system but allows you to determine what the formula is simply by inspecting the name.

Table 8-3 summarizes the rules for naming acids.

Bases

Bases make up a class of compounds formerly known as alkalis. They include such classic compounds as sodium hydroxide (NaOH; common name: lye or caustic soda) and potassium hydroxide (KOH; common name: potash), originally made by cooking wood ash with water. The characteristic feature of a base is the formation of the hydroxide ion OH^-. When the OH^- ion is combined with a metallic ion, the resulting compound is a base named a **hydroxide.**

To understand the nature of this group, recall that the oxygen atom has two empty spaces (or sites) in its outer shell, and so wants to accept two electrons. When a hydroxide group is formed, one of these spaces is filled with the electron from the hydrogen atom; the other site is still open to receive another electron. (See Figure 8-2.) That other electron can come from another hydrogen atom. In that event, a molecule of water is formed (HOH, or hydrogen hydroxide). On the other hand, the open site may be filled by an electron from a metal ion. When that happens, the result is a metal (M) combined with one or more hydroxide groups: $M(OH)_x$. This compound is a *metal hydroxide*. When hydroxides dissolve in water, the solution contains positive metal ions M^{+x} and negative hydroxide ions $(OH)^-$.

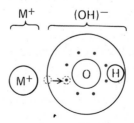

FIGURE 8-2 The electron structure of a metal hydroxide

It is the hydroxide ion that gives the base its distinctive character, just as the hydrogen ion is the chief feature of an acid.

The number of hydroxide ions in the molecule equals the oxidation number of the metal, since the charge of the hydroxide ion is -1. Lithium hydroxide is LiOH, since lithium, as an alkali metal, has an oxidation number of $+1$. (The parentheses are omitted if there is a single OH group.) Calcium hydroxide (slaked lime) is $Ca(OH)_2$, since the oxidation number of calcium is $+2$. To name these compounds, simply append the term *hydroxide* to the metal name.

There are groups that contain only nonmetals but behave like metals in a compound. For example, the ammonium ion (NH_4^{1+}) contains nothing but hydrogen and nitrogen, but because the oxidation number of the group is $+1$, the ammonium group takes the place of a metal in a compound such as ammonium hydroxide (NH_4OH). A solution of ammonium hydroxide acts as a base.

Many metals have more than one oxidation number, depending on the compound. For example, the oxidation number of iron is sometimes $+2$ and sometimes $+3$. Two systems exist for the purpose of distinguishing by name the oxidation number $+2$ compounds from the $+3$ compounds. The older, more traditional system uses the suffix *-ic* for the higher oxidation number and *-ous* for the lower oxidation number. These suffixes are added to the root of the metal name. Since iron was formerly called *ferrum*, the Latin root *ferr-* is used to name iron compounds. For example, a compound in which iron has an oxidation number $+3$ is *ferric hydroxide*: $Fe(OH)_3$. The corresponding compound with oxidation number $+2$ iron is *ferrous hydroxide*: $Fe(OH)_2$. There is no consistency in the use of Latin roots, however. Even though the symbol for mercury is Hg, its compounds employ *mercuric* (Hg^{+2}) and *mercurous* (Hg^+) as first names. Of course, if there is no Latin name, the English root is used.

In the newer IUPAC system of nomenclature, the oxidation number, expressed in Roman numerals, is inserted into the compound name immediately after the English name for the metal. Thus, $Fe(OH)_2$ is *iron(II) hydroxide*, and $Fe(OH)_3$ is *iron(III) hydroxide*. The IUPAC system is preferable for formal writing, but chemists often say "ferric hydroxide" rather than "iron(III) hydroxide."

Beginning students frequently have great difficulty with the multiple names for compounds. With experience and practice, you will gradually become familiar with the names.

Salts

We have seen that an acid is a compound in which a nonmetallic element or group is combined with one or more positive hydrogen ions. Hydrochloric acid (HCl) is an example. We have also seen that a base is a compound in which a metallic element or group is combined with a negative OH^{-1} ion. Sodium hydroxide (NaOH) is one of the most common examples.

A salt is a compound in which a metallic element or group (other than hydrogen) is combined with a nonmetallic element or group (other than the OH group). A salt is formed from the acid HCl and the base NaOH when the metallic ion (Na^+) from the base is linked to the nonmetallic ion (Cl^-) from the acid. The resulting compound is sodium chloride (NaCl), the common salt that flavors the oceans of the earth and the food on your table.

Note how the formation of NaCl comes about when NaOH and HCl react with each other. (See Figure 8-3.) The reaction takes place in an aqueous solution. Since NaOH ionizes when it dissolves, a solution of NaOH consists of Na^+ and OH^- ions suspended in the water. Similarly, a solution of HCl consists of H^+ and Cl^- ions intermingled with water molecules. When one of these two solutions is poured into the other, the result is a mixture of Na^+, H^+, Cl^-, and OH^- ions knocking about in the water.

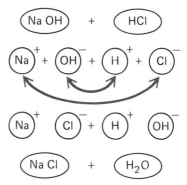

FIGURE 8-3   When solutions of NaOH and HCl are mixed, the H^+ and OH^- ions join to form water, leaving Na^+ and Cl^- ions in the solution. If the solution is evaporated, solid NaCl is left behind.

The H^+ and the OH^- ions tend to join together to form water molecules: H_2O, or HOH. This leaves a solution of Na^+ and Cl^- ions in water. If the water is evaporated away, the crystalline NaCl is left behind. The equation for this reaction is

$$NaOH + HCl \longrightarrow NaCl + H_2O$$

In other words, the sodium hydroxide and the hydrochloric acid exchange ions to form sodium chloride plus hydrogen hydroxide, which is simply water. This kind of reaction is a general one: *a base combines with an acid to form a salt plus water.*

$$base + acid \longrightarrow salt + H_2O$$

As you can see, it is possible to form an enormous number of salts. From 10 bases and 10 acids, 10 × 10, or 100, different salts can be produced, because each of the bases can combine with every one of the acids.

The rules for naming salts are straightforward. First the salts are divided into two major types.

1.   *Binary salts.* Binary salts are compounds containing a metallic ion (or a metal-like group such as the ammonium ion) and a nonmetallic ion. The names of these compounds simply consist of the name of the metal, followed by the nonmetal root with *-ide* as the ending. For example:

Name	Formula	Root
potassium sulfide	K_2S	sulf-
barium chloride	$BaCl_2$	chlor-
ammonium iodide	NH_4I	iod-
iron(III) bromide	$FeBr_3$	brom-
copper(II) fluoride	CuF_2	fluor-

Note: The Roman numerals are only used when multiple oxidation states for the metal are possible. This situation generally occurs with transition elements.

2. *Ternary salts.* Salts derived from ternary (oxo) acids use the name of the metallic ion together with the name of the acid group. Several examples are

silver sulfate	Ag_2SO_4
silver sulfite	Ag_2SO_3
silver perchlorate	$AgClO_4$
silver chlorate	$AgClO_3$
silver chlorite	$AgClO_2$
silver hypochlorite	$AgClO$
iron(III) nitrate	$Fe(NO_3)_3$
iron(II) nitrate	$Fe(NO_3)_2$

In cases where the negative ion has an oxidation number of 2 and the metal has an oxidation number of 1, two forms of the salt are possible. Consider these three common compounds:

carbonic acid	H_2CO_3	
sodium carbonate	Na_2CO_3	(washing soda)
sodium hydrogen carbonate (sodium bicarbonate)	$NaHCO_3$	(baking soda)

Notice that in sodium carbonate both hydrogen atoms have been replaced by sodium atoms, whereas in sodium hydrogen carbonate only one of the hydrogens has been replaced by a sodium. In the older nomenclature the prefix *bi-* was used to name compounds containing a hydrogen ion and a metallic ion within the same salt. In the newer nomenclature the use of *hydrogen* as part of the name is preferred. *Sodium hydrogen carbonate* means simply a carbonate with both sodium and hydrogen as the positive ions. The sodium ion is listed first, since it is more positive than the hydrogen.

When the negative group has an oxidation number of -3, three different salts are possible. Phosphoric acid is H_3PO_4. The acids and the three potassium salts containing the phosphate group are

trihydrogen phosphate (phosphoric acid)	H_3PO_4
potassium phosphate	K_3PO_4
potassium monohydrogen phosphate	K_2HPO_4
potassium dihydrogen phosphate	KH_2PO_4

In the first salt, all three hydrogen atoms are replaced with potassium atoms. In the second salt, one of the hydrogens is left in the molecule; in the third, only one of the hydrogen atoms is replaced with a potassium atom. The prefix *mono-* indicates that there is a single hydrogen ion in the molecule, the prefix *di-* indicates that there are two hydrogen ions, and the prefix *tri-* indicates that all three hydrogen ions are present.

Another possibility is a *base salt*, a salt that contains one or more hydroxyl ions (OH^-) in addition to the negative ion. A base salt can occur with a metal having an oxidation number of $+2$ or more. For example, cupric iodate is $Cu(IO_3)_2$; the copper ion has an oxidation number of $+2$ and the iodate ion is -1. In cupric basic iodate, one of the iodate ions is replaced by a hydroxide ion, and the result is $Cu(OH)IO_3$.

Also common are salts with more than one metal, such as sodium potassium sulfate ($NaKSO_4$). In addition, there are many polyatomic ions with metals in them. A well-known case is the permanganate ion (MnO_4^-). The salts formed from these groups have two metals in them; one example is potassium permanganate ($KMnO_4$), a compound with a very strong purple color, used for many chemical and medical purposes.

Covalent Compounds

A number of binary covalent compounds use a system of nomenclature similar to that of the binary salts. In many of these compounds, the various elements involved can exist in more than one oxidation state. Therefore a system of prefixes is used to indicate how many atoms of each element are present in the compound.

The following examples illustrate the use of these prefixes:

carbon monoxide	CO
carbon dioxide	CO_2
carbon tetrachloride	CCl_4
phosphorus trichloride	PCl_3
phosphorus pentachloride	PCl_5
sulfur dioxide	SO_2
sulfur trioxide	SO_3
nitrogen dioxide	NO_2
dinitrogen tetroxide	N_2O_4

Prefixes representing the numbers from one to ten are listed in Table 8-4. These prefixes derive from the corresponding Greek numbers; some of them are familiar. For example, a dilemma is a choice between two options. A trio is a group of three people or a musical number written for three instruments. A quadrangle is an area with four sides. The Pentagon is the name of a building with five sides. A hexagon has six sides, and an octagon has eight sides.

TABLE 8-4 Numerical Prefixes

1. mono-	6. hexa-
2. di-	7. hepta-
3. tri-	8. octa-
4. tetra-	9. ennea-*
5. penta-	10. deca-

*Ennea- is the prefix preferred by the IUPAC, but nona- is the more common prefix.

Glossary

acids Ionic compounds yielding hydrogen ions in solution, e.g.,

$$HCl \longrightarrow H^+ + Cl^-$$

bases Ionic compounds yielding hydroxide ions in solution, e.g.

$$NaOH \longrightarrow Na^+ + OH^-$$

binary compound A compound consisting of two elements, e.g., $CaBr_2$.

charge number The number of unbalanced electric charges on an ion, measured in electrons.

chemical equation A symbolic statement describing the reaction of elements or compounds to form chemical products, e.g.,

$$NaOH + HCl \longrightarrow NaCl + H_2O$$

formula A symbolic statement of the elements found in a molecule and their proportions, e.g., $K_2Cr_2O_7$.

hydroxide A negatively charged ion with the formula OH^- whose presence characterizes basic solutions.

IUPAC International Union of Pure and Applied Chemistry, which established a set of rules to systematize chemical nomenclature.

molecular formula The formula of a compound showing the kind and number of each species of atom present, such as $KClO_3$.

oxo-acids Acids containing oxygen in their formulas, e.g., HNO_3.

peroxides Metal (or hydrogen) and oxygen compounds in which the oxygen has an oxidation number of -1, e.g., H_2O_2.

polyatomic ions Ions made up of two or more atoms, e.g., CO_3^{2-}.

radicals See **polyatomic ions**.

root The core part of the name of an element or group, e.g., "chlor—."

salts Ionic compounds composed of a positive ion other than hydrogen and a negative ion other than the hydroxide ion, e.g., K^+Br^-.

structural formula A diagram indicating the kind, amount, and arrangement of the atoms in a compound, e.g.,

$$\begin{array}{c} H \\ | \\ H-C-O-H \\ | \\ H \end{array}$$

superoxide A metal-oxygen compound in which the oxygen has an oxidation number of -0.5, e.g., NaO_2.

technical name The name of an element or compound that follows IUPAC or other formal rules of nomenclature, e.g., iron(II) nitrate.

ternary compound A compound consisting of three kinds of atoms, e.g., $KClO_3$.

trivial name A common name of a compound, often of long historical usage, e.g., muriatic acid.

Problems and Questions

8.1 Names and Formulas

8.1.Q What are the three major purposes behind the rules for naming compounds according to the IUPAC system of nomenclature?

8.2.Q In what situation might a compound go by two different names?

8.3.Q Is it possible for two compounds to have the same name?

8.4.Q What two features should the name of a compound describe?

8.5.Q What are the technical names of the following compounds?
  a. table salt
  b. sugar
  c. baking soda
  d. washing soda
  e. calcite
  f. lime
  g. plaster of Paris

8.6.Q What are the technical names of the following elements?
  a. diamond
  b. quicksilver
  c. graphite

8.7.Q What are the common names of the following compounds?
  a. dihydrogen oxide
  b. trihydrogen nitride
  c. potassium hydroxide
  d. calcium hydroxide
  e. sodium hydroxide

8.8.Q a. What is a chemical equation?
  b. What is a chemical formula?

8.9.Q What are the two essential parts of a chemical formula?

8.10.Q What is the principal difference between a molecular formula and a structural formula?

8.11.Q In what branch of chemistry are structural formulas most often used?

8.2 Writing Formulas

8.12.Q List three compounds in which the group OH exists as a unit.

8.13.Q List three compounds in which the group SO_4 exists as a unit.

8.14.Q What is a polyatomic ion?

8.15.Q What is the name of the SO_4^{2-} ion?

8.16.Q Count the number of protons and electrons in the SO_4^{2-} ion. Explain why the ion has a charge. Where do the extra electrons come from?

8.17.Q What is the charge on each of the following polyatomic ions?
  a. $C_2H_3O_2$ b. NH_4
  c. SO_4 d. MnO_4
  e. PO_4 f. CO_3

8.18.Q Name the following polyatomic ions:
  a. HCO_3^- b. ClO^-
  c. OH^- d. NO_3^-
  e. SO_3^{2-}

8.19.Q Write the formula and give the charge number of each of the following polyatomic ions:
  a. perchlorate ion
  b. chlorate ion
  c. hydronium ion
  d. nitrate ion
  e. peroxide ion
  f. chromate ion

8.20.Q Name the polyatomic ions found in the following compounds, and indicate the charge on each:
  a. Na_3PO_4 b. $Ca(OH)_2$
  c. H_2SO_4 d. NH_4NO_3
  e. $KMnO_4$ f. $NaHCO_3$

8.21.Q What is the oxidation number of the molecule O_2 or NH_3?

8.22.Q What is the oxidation number of metallic Fe or Al?

8.23.Q What is the usual oxidation number of a column 1a element? Of a column 2a element?

8.24.Q What is the usual oxidation number of each of the following elements (in a binary compound)? (Use a periodic chart for reference.)
  a. Na   b. Cs
  c. Ba   d. Sr
  e. Ca   f. Cl
  g. Br   h. O
  i. H   j. C

8.25.Q In the compounds CrO_2 and Cr_2O_3, oxygen has an oxidation number of -2. What is the oxidation number of Cr in each of those compounds?

8.26.Q Give the oxidation number of each atom in the following list.
- a. NH_3
- b. O_2
- c. KCl
- d. $CaBr_2$
- e. H_2O
- f. $KClO_3$
- g. H_2O_2
- h. H_2SO_4
- i. SO_2
- j. Fe_2O_3

8.27.Q Give the oxidation number of each atom in the following list of ions. (The ionic charge is given as a superscript.)
- a. NO_3^{1-}
- b. CO_3^{2-}
- c. SO_4^{2-}
- d. NH_4^{1+}
- e. H_3O^{1+}
- f. ClO_3^{1-}
- g. BrO_4^{1-}
- h. PO_4^{3-}
- i. HCO_3^{1-}
- j. OH^{1-}

8.28.Q In a chemical formula, which appears first—the positive ion or the negative ion?

8.29.Q What must be the sum of the oxidation numbers in a molecule?

8.30.Q The formula for the compound iron(III) ferrocyanide is $Fe_4[Fe(CN)_6]_3$. (The group CN is the *cyanide* group.) What is the function or meaning of each of the subscripts in this formula?

8.31.Q Write the formula for the molecule created from each of the following combinations:
- a. Na^{+1} and CrO_4^{-2}
- b. Mg^{+2} and I^{-1}
- c. NH_4^{+1} and CO_3^{-2}
- d. K^{+1} and ClO_4^{-1}
- e. Rb^{+1} and SO_4^{-2}
- f. S^{+4} and O^{-2}
- g. S^{+6} and O^{-2}
- h. H^{+1} and S^{+6} and O^{-2}
- i. H^{+1} and PO_4^{-3}
- j. Ca^{+2} and $Cr_2O_7^{-2}$

8.32.Q Check each of the following molecular formulas to see whether it is written correctly. If it is not, rewrite it in correct format.
- a. PO_4K_3
- b. HSO_4
- c. NH_4SO_3
- d. KNO_2
- e. $NaClO_3$

8.33.Q Place the correct subscript after the closing parenthesis:
- a. $Na(C_2H_3O_2)$
- b. $Mg(CO_3)$
- c. $K_2(Cr_2O_7)$
- d. $Na_2(SiO_3)$
- e. $Ca(OH)$

8.3 Naming Compounds

8.34.Q
a. What is a binary compound?
b. Give the name and formula of a binary compound containing chlorine.
c. Give the name and formula of a binary compound containing iodine.

8.35.Q What is an acid?

8.36.Q What is the difference between a binary acid and a ternary acid?

8.37.Q The prefix *hydro-* is used in naming what kind of acid?

8.38.Q
a. If the name of a compound ends in *-ate*, what ending does the corresponding acid have?
b. If the name of a compound ends in *-ite*, what ending does the corresponding acid have?

8.39.Q Using the named compounds in parentheses as a guide, determine the name of each compound in the list below.
- a. H_2SO_4 (Na_2SO_4 is sodium sulfate)
- b. KCl (HCl is hydrochloric acid)
- c. Na_3PO_3 (H_3PO_3 is phosphorous acid)
- d. HNO_3 ($NaNO_3$ is sodium nitrate)
- e. HNO_2 ($NaNO_2$ is sodium nitrite)
- f. $HClO_4$ ($KClO_4$ is potassium perchlorate)

8.40.Q What is the difference between an acid and a base?

8.41.Q How do you name a base?

8.42.Q What is the oxidation number of the OH radical?

8.43.Q What is the difference between a base and an alkali?

8.44.Q In some compounds of iron the Fe ion has an oxidation number of $+3$, and in other compounds the Fe ion has an oxidation number of $+2$. In naming these compounds, how do you distinguish between the two types of compounds? (Give two methods.)

8.45.Q Give the common and IUPAC names of the following compounds:

a. NaOH b. Mg(OH)$_2$
c. NH$_4$OH d. KOH
e. Ca(OH)$_2$ f. Fe(OH)$_3$
g. Fe(OH)$_2$

8.46.Q Using Table 4-1, write the "ic" and "ous" names as well as the IUPAC names for the following compounds:
   a. SnCl$_2$ and SnCl$_4$
   b. HgO and Hg$_2$O
   c. CoF$_2$ and CoF$_3$
   d. CuCN and Cu(CN)$_2$
   e. Pb(OH)$_2$ and Pb(OH)$_4$

8.47.Q a. What is a salt?
   b. How do salts differ from acids and bases?

8.48.Q Describe how an acid and a base react to form a salt.

8.49.Q How do you name a salt that has both a metal and a hydrogen ion, such as NaHCO$_3$?

8.50.Q How do you name a salt that has both a hydroxide ion and a negative acid ion, such as Cu(OH)IO$_3$?

8.51.Q Name the following compounds:
   a. Na$_2$CO$_3$
   b. NaHCO$_3$
   c. H$_2$CO$_3$

8.52.Q Name the following compounds using IUPAC rules:
   a. CuHAsO$_3$
   b. Fe(OH)(C$_2$H$_3$O$_2$)$_2$
   c. PbHAsO$_4$
   d. Pb(OH)NO$_3$
   e. LiH$_2$PO$_4$

8.53.Q Give the names of the following compounds using the IUPAC system:
   a. K$_2$SO$_4$
   b. NaH$_2$PO$_4$
   c. CoCl$_2$
   d. KMnO$_4$
   e. FeBr$_2$
   f. Na$_2$HPO$_4$
   g. KHSO$_3$

8.54.Q Write the formula for each of the following compounds:
   a. silver periodate
   b. iron(II) nitrate
   c. sodium hydrogen sulfate
   d. potassium chlorate
   e. sodium acetate

8.55.Q Name each of the following compounds using IUPAC rules:
   a. KNO$_2$ and KNO$_3$
   b. MgSO$_3$ and MgSO$_4$
   c. KClO and HClO
   d. NaHSO$_4$ and H$_2$SO$_4$

8.56.Q For each of the following acids, name the acid and the corresponding sodium salt. Also give the formula of the sodium salt.
   a. H$_2$SO$_3$
   b. H$_3$PO$_4$
   c. HI
   d. HClO$_3$
   e. HClO$_2$

8.57.Q Why is it necessary to use prefixes such as "mono" and "di" when naming covalent compounds?

8.58.Q Name the first ten numerical prefixes.

8.59.Q Name the following compounds:
   a. CO b. CO$_2$
   c. N$_2$O d. NO$_2$
   e. SO$_2$ f. SO$_3$
   g. CCl$_4$ h. PF$_5$

8.60.Q Write the formula for each of the following compounds:
   a. dinitrogen oxide
   b. dinitrogen trioxide
   c. chlorine monofluoride
   d. chlorine trifluoride
   e. phosphorus dichloride trifluoride
   f. tetraphosphorus triselenide

Self Test

1. Write the formula for each of the following compounds:
   a. carbon dioxide
   b. potassium hydrogen sulfate
   c. potassium dichromate
   d. ammonium chloride
   e. iron(III) sulfite
   f. aluminum sulfate
   g. lithium nitrite
   h. magnesium bromate
   i. nitric acid
   j. lead(III) hydroxide

2. Name the following compounds using IUPAC names:
   a. H_2SO_4
   b. $Ca(OH)_2$
   c. $Mg_3(PO_4)_2$
   d. NaI
   e. $KClO_2$
   f. HF_{aq}
   g. NH_3
   h. $CaCO_3$
   i. SrO
   j. $HC_2H_3O_2$

9
Avogadro and the Mole

Amedeo Avogadro (1776–1856)

Objectives After completing this chapter, you should be able to:
1. Explain Gay-Lussac's law of combining volumes.
2. Describe the role of Amedeo Avogadro in the development of ideas concerning chemical reactions.
3. Explain the significance of Avogadro's hypothesis.
4. Describe at least one method of measuring the number of molecules in a quantity of material.
5. Explain the meaning of Avogadro's number.
6. Define a mole (as used in chemistry).
7. Calculate the number of moles in a given amount of any substance.
8. Calculate the gram-atomic mass and the molar mass of any element or compound.
9. Calculate the mass of a single atom or molecule of a given element or compound.

9.1 Historical Review

In this chapter we continue to explore the history of our ideas concerning atoms and molecules.

Gay-Lussac and the Law of Combining Volumes

Chapter 4 described how John Dalton's atomic theory, published in 1808, clarified many problems in chemical theory. Many of Dalton's ideas were incorrect, however, and years of effort and argumentation were needed to sort them out. Among the incorrect parts of Dalton's theories were the notions that (1) the fundamental particle of a gas such as hydrogen is a single atom (rather than a diatomic molecule), (2) there is little empty space between the particles of a gas, and (3) a molecule of water consists of one atom of hydrogen and one atom of oxygen. These incorrect ideas are mentioned because efforts to correct them played an important part in the history of chemistry, and a recounting of the logic required to correct them makes a useful case study in the scientific method.

In an attempt to verify the law of multiple proportions, John Dalton devoted much effort to the measurement of quantities of materials reacting with each other. His attention was focused on the masses of the substances entering into chemical reactions, and his entire theory was based on the idea that the *combining masses* were proportional to the masses of the individual atoms. His logic started with the observed fact that 8 g of oxygen combines with 1 g of hydrogen to form 9 g of water. Once he assumed that one atom of oxygen combines with one atom of hydrogen to make a molecule of water, it followed that the mass of one atom of oxygen must be eight times greater than the mass of one atom of hydrogen.

However, Dalton had no evidence to back up his assumption that one atom of oxygen combines with one atom of hydrogen to form the molecule HO, and, indeed, this was not the only assumption (or hypothesis) capable of explaining the same facts. He might have guessed, for example, that the oxygen and hydrogen atoms have equal mass, but that there are eight oxygen atoms to every hydrogen atom. That would make the formula for water HO_8. It is not as neat a hypothesis, but how could he know it was not just as valid?

Dalton argued that the simplest hypothesis was the best. Yet his simple hypothesis turned out to be just as wrong as the more complicated one. The trouble was that *there was no evidence to verify any hypothesis*. What was needed was an independent set of experimental facts that would tell how many atoms of each element were actually in a given compound.

Independent evidence concerning the composition of molecules came from the work of the French chemist Joseph Louis Gay-Lussac. In 1802 (six years before Dalton announced his theory), Gay-Lussac began researching the properties of gases. One important piece of work for which he is remembered is the discovery that all gases expand by 1/273 of their original volume for every degree increase in temperature (at constant pressure). This fact is the basis for the Kelvin (absolute)

TABLE 9-1 Combining Volumes of Some Gases

2 volumes of hydrogen + 1 volume of oxygen	⟶	2 volumes of water (H_2O)
1 volume of nitrogen + 3 volumes of hydrogen	⟶	2 volumes of ammonia (NH_3)
2 volumes of nitrogen + 1 volume of oxygen	⟶	2 volumes of nitrous oxide (N_2O)
1 volume of nitrogen + 1 volume of oxygen	⟶	2 volumes of nitric oxide (NO)
1 volume of nitrogen + 2 volumes of oxygen	⟶	2 volumes of nitrogen dioxide (NO_2)

temperature scale. Gay-Lussac's interest in gases led him into a line of research involving measurement of the *volumes* of gases entering into chemical reactions.

Gay-Lussac's results were striking. He found, for example, that 2 L of hydrogen always combines with 1 L of oxygen to form 2 L of water vapor (gaseous water). In several other reactions, he found that the *volumes* of the combining gases are always in simple numerical ratios. Some of these combining volumes are shown in Table 9-1. The results of this work were summarized in Gay-Lussac's **law of combining volumes**, which states that *under comparable conditions of temperature and pressure the ratios of the volumes of gas entering into chemical reactions are simple whole-number ratios.* In the case of hydrogen combining with oxygen, the volume ratio is 2:1, clearly a ratio of simple whole numbers.

The law of combining volumes is the statement of an experimental result. But what is the meaning of this statement? Why does 1 L of nitrogen combine with 3 L of hydrogen to form ammonia gas? Why are there only 2 L of ammonia vapor after the reaction, instead of the 4 L of gas we started with? There was nothing in Dalton's atomic theory to explain these phenomena (which is, perhaps, why Dalton refused to acknowledge the validity of Gay-Lussac's observations).

A clearer theoretical idea about the nature of molecules was needed. The solution was not to throw out all of Dalton's theory, but merely to remove those parts which came entirely from his imagination with no experimental verification.

We see in this story one of the fundamental problems of scientific research. Imaginative theories are necessary in order to explain the complex and mysterious universe about us. Yet theories not verified by precise experiments are merely speculations. Theories have value only when they are able to explain observed facts. In the next section you will see how Dalton's imperfect theory was modified to fit all the observed facts.

Avogadro's Hypothesis

The clarification of molecular theory came from a professor of physics at the University of Turin in Italy. While considering the implications of Gay-Lussac's law of combining volumes, Amedeo Avogadro realized that everything about these experiments could be understood if he made two modifications of Dalton's atomic theory.

The first modification required a major change in viewpoint. Dalton had always believed that the atom was the *smallest* particle of any element found free in nature. This was why he insisted that, in the reaction of hydrogen with oxygen, one atom of hydrogen reacted with one atom of oxygen to form one molecule of water. Avogadro asked, with great clarity, *Why is this a necessary hypothesis?* He then proposed an alternative hypothesis: The smallest free particle of a gas is a *molecule* instead of an atom, and this molecule can have more than one atom in it. Thus Avogadro was the first person to distinguish between the atoms and the molecules of a gas.

The second part of Avogadro's hypothesis was related to the number of molecules in a given volume of gas. He realized that scientists had no idea how small a molecule really is. Suppose, he reasoned, a molecule is much smaller than the space *between* molecules (at least in a gas), just as stars are much smaller than the spaces between them. If that is true, then the actual size of a molecule has no relation to the number of molecules in a given volume, as can be seen in Figure 9-1.

Given that premise, the simplest assumption that can be made about the number of molecules in a given volume is this: *Equal volumes of gases at the same temperature and pressure contain the same number of molecules.* This statement is now called **Avogadro's hypothesis**. It implies that each molecule in a gas is an extremely tiny particle dashing about and colliding with other molecules, and that the distance between molecules is much greater than the diameter of the molecule itself. It also implies that there is no interaction between molecules—that each molecule ignores all of the others except during an occasional collision. If that is so, the total amount of space occupied by all of the molecules has nothing to do with the nature of each individual molecule. That is, in samples of hydrogen and oxygen at the same temperature and pressure, there is no reason for the hydrogen molecules to be farther apart than the oxygen molecules.

When Avogadro first published his idea in 1811, it was just a guess. There was hardly any experimental basis for it, because there was still no way to measure the actual number of molecules in a gas. For this reason—and because of John Dalton's continuing rejection of the hypothesis—there was no great enthusiasm for Avogadro's hypothesis among the chemists of the world.

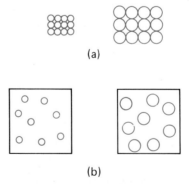

FIGURE 9-1 (a) If molecules are packed closely together, the size of two packages of equal numbers of molecules depend on the size of each molecule. (b) If the molecules are moving around freely, with large distances between them, the size of the packages of molecules does not depend on the size of each molecule. It only depends on the number of molecules and the distance between them.

9.1 HISTORICAL REVIEW

It was not until 1860 that general recognition came to Avogadro—four years after his death. In that year the First International Chemical Congress was held in Karlsruhe, Germany. The purpose of the meeting was to undo some of the confusion that existed with regard to atomic weights and chemical formulas. The correct formula for water and the correct atomic weight for oxygen were still not known. At that meeting the Italian chemist Stanislao Cannizzaro gave a lecture showing how Avogadro's hypothesis, together with the law of combining volumes, could make order out of chaos, thereby allowing chemists to make sense out of chemical formulas and equations.

To understand Cannizzaro's logic, look at the hydrogen-oxygen reaction:

2 volumes of hydrogen + 1 volume of oxygen $\longrightarrow$ 2 volumes of water vapor

(The term 1 *volume* means 1 L, 1 cm^3, or any other arbitrary unit of volume.)

What happens if we write a similar relationship describing the number of molecules reacting, using the hypothesis that equal volumes of gases contain equal numbers of molecules? Let the letter N represent the number of molecules in 1 volume of gas. Then $2N$ is the number of molecules in 2 volumes of gas. From the above equation, it follows that

$2N$ molecules of hydrogen + N molecules of oxygen $\longrightarrow$
$\qquad\qquad\qquad\qquad\qquad\qquad\qquad\qquad$ $2N$ molecules of water vapor

Therefore it must be true that

2 molecules of hydrogen + 1 molecule of oxygen $\longrightarrow$
$\qquad\qquad\qquad\qquad\qquad\qquad\qquad\qquad$ 2 molecules of water vapor

This means that in the water molecule there must be twice as many atoms of hydrogen as there are atoms of oxygen.

Thus the simplest formula for water is H_2O. Furthermore, since two molecules of water come out of the reaction, two atoms of oxygen must go into the reaction. Therefore, the single oxygen molecule contains two atoms. Likewise, the two water molecules contain four hydrogen atoms, and so each of the two hydrogen molecules must contain two hydrogen atoms. The equation for the reaction then becomes

$$2H_2 + O_2 \longrightarrow 2H_2O$$

You can see that by logically applying Avogadro's hypothesis and the fact that a molecule may contain two atoms of the same element, we are able to clarify many mysteries and discover the correct composition of water. After 1860 the science of chemistry was on a much firmer footing than it had been previously.

Example 9.1   How many liters of nitrogen (N_2) and how many liters of oxygen (O_2) are required to produce 50 L of nitrogen dioxide (NO_2)?

Solution   Table 9-1 tells us that 1 L of N_2 combines with 2 L of O_2 to form 2 L of NO_2.

$$N_2 + 2O_2 \longrightarrow 2NO_2$$

The conversion factor between liters of NO_2 and liters of N_2 is

$$\frac{1 \text{ L } N_2}{2 \text{ L } NO_2}$$

Therefore the number of liters of N_2 needed to make 50 L of NO_2 is calculated as follows:

$$\text{no. of liters of } N_2 = \frac{1 \text{ L } N_2}{2 \text{ L } NO_2} \times 50 \text{ L } NO_2$$

$$= 25 \text{ L } N_2$$

Note: This problem can be done without an equation; since the volume of N_2 is half the volume of NO_2, the answer is 50/2. However, knowing how to write the equation is useful for more complicated situations.

9.2 Avogadro's Number

Measuring Avogadro's Number

In 1860, when Cannizzaro was expounding on Avogadro's theory, scientists had no knowledge concerning the size of a molecule or the distance between molecules in a gas. In 1865, Joseph Loschmidt made the first estimate of these numbers. His method was based on the fact that two gases in contact with each other require some time to spread into each other. This process, known as **diffusion**, can be seen with the naked eye when a colored gas such as iodine vapor is released in a container of air. The purple iodine vapor slowly diffuses outward from its source as the iodine molecules bounce back and forth among the air molecules. By measuring the speed of this diffusion and then applying some advanced mathematics, we can calculate how many collisions per second the iodine molecules make as they travel through the air; from this information the number of air molecules per cubic centimeter can be determined.

Loschmidt found that there are approximately 2.7×10^{19} molecules per cubic centimeter in a gas at a temperature of 0°C and a pressure of one atmosphere. (These conditions of temperature and pressure are called STP—Standard Temperature and Pressure.) In recognition of this monumental accomplishment, the number of molecules per cubic centimeter of a gas is now called **Loschmidt's number**.

With Loschmidt's measurement, it began to appear that the molecule was a real thing, with a definite mass and a definite size—something that could be counted, though the numbers involved were too large to visualize. Even with this evidence, however, some scientists were reluctant to believe in the existence of molecules, and the battle of molecular theory continued for two more decades. Habits of thought are hard to change.

Since the time of Loschmidt, more straightforward and accurate methods of measuring numbers of molecules have been developed. One method involves the electrolysis of water. In electrolysis an electric current passed through a container of water (with a little acid added to aid the conduction of electric current) causes oxygen to be released at one electrode and hydrogen to be given off at the other electrode. (See Figure 9-2.) The mechanism for this reaction involves the presence in the acid solution of numerous H^+ ions. (The reasons for the formation of ions in solutions will be given in Chapter 14.) When the electric current is turned on, each H^+ ion near the cathode (the negative electrode) picks up an electron from the cathode and becomes a neutral hydrogen atom; each neutral hydrogen atom then quickly joins another hydrogen atom to form a hydrogen molecule:

$$2H^+ + 2e^- \longrightarrow 2H \longrightarrow H_2$$

As a result, gaseous hydrogen is released at the cathode.

The purpose of the experiment is to measure the number of molecules per gram of hydrogen. This is done by measuring the amount of electric current passing

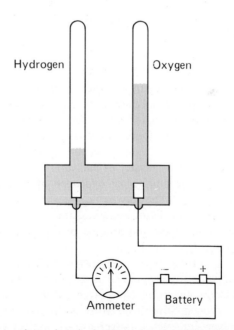

FIGURE 9-2 Water is electrolyzed—broken down into hydrogen and oxygen—by passing an electric current through it. (A little acid is added to the water so that it will conduct an electric current.) The volume of hydrogen released is twice as great as the volume of oxygen released, because two molecules of hydrogen are formed for each molecule of oxygen.

through the water during the electrolysis and then weighing the hydrogen gas released. Knowing the amount of charge on each electron and knowing the total charge carried by the electric current, we can calculate how many electrons are required to electrolyze 1 g of hydrogen. Since one electron is required to free each atom of hydrogen from the water, we then know the number of atoms in the gram of hydrogen.

Such a measurement can be made with a high degree of accuracy. As described in Section 5.1, the amount of electric charge on a single electron was first measured by Robert A. Millikan in 1911. Once that number is known, the number of electrons passing through an electric circuit can be determined simply by measuring the electric current with an ammeter and timing the current with a clock. (The number of electrons is proportional to the amount of electric current multiplied by the amount of time the current flows.)

It is found that just 6.02×10^{23} electrons are needed to release 1 g of hydrogen from 9 g of water. Since each electron releases one hydrogen atom, these experiments demonstrate that 1 g of hydrogen contains 6.02×10^{23} atoms of hydrogen. One gram of hydrogen is of special interest, because the atomic mass of hydrogen is 1.008 amu and so 1 g of hydrogen is, approximately, the number of grams equal to hydrogen's atomic mass. As we saw in Section 6.1, this quantity is given a special name: the gram-atomic mass of hydrogen. In general, the gram-atomic mass of a substance is the number of grams equal to the atomic mass of the substance. Accordingly, a gram-atomic mass of helium is 4.003 g of helium, a gram-atomic mass of copper is 63.5 g of copper, and so on.

The gram-atomic mass is an important unit of measurement for the following reason: If an atom of helium weighs (approximately) four times as much as an atom of hydrogen, then 4 g of helium must have the same number of atoms as 1 g of hydrogen. Likewise, 63.5 g of copper must have the same number of atoms as 1 g of hydrogen, for each atom of copper weighs 63.5 times as much as an atom of hydrogen. The following general rule emerges: *A gram-atomic mass of any element always contains the same number of atoms.* For this reason the gram-atomic mass is a significant quantity to measure.

The number of hydrogen atoms in 1 gram-atomic mass of hydrogen is 6.02×10^{23}. The above rule states that there is the same number of helium atoms in 4.003 g of helium, and the same number of copper atoms in 63.5 g of copper. Because of the importance of the number 6.02×10^{23}, it was given the name Avogadro's number.

By definition, **Avogadro's number** is the number of atoms in a gram-atomic mass of any element. It is not surprising, then, that it takes the same amount of electric charge to electrolyze 107.9 g of silver as it does to electrolyze 1.008 g of hydrogen. Silver is electrolyzed by passing electric current through a solution of a silver salt, such as silver nitrate, and plating out metallic silver on the negative electrode. In fact, Avogadro's number can be more accurately measured through the electrolysis of silver than through the electrolysis of hydrogen, since the silver deposited on the cathode can be weighed more precisely than can the hydrogen gas released in the electrolysis of water.

The Meaning of Avogadro's Number

Avogadro's number is simply the number 6.02×10^{23}. Just as we talk about a dozen eggs or a gross of paper clips, we can talk about an Avogadro's number of atoms. In chemistry the term *mole* has become synonymous with Avogadro's number. (In biology a mole might be a small animal, but in chemistry it is something quite different.)

But first let us get acquainted with the number. Just how big a number is 6.02×10^{23}? To answer this question, imagine 6.02×10^{23} of salt, one-tenth of a millimeter on a side, in a pile. What would be the volume of that pile?

Let s equal the length of each side of one of the small cubes. Then,

$$s = 0.1 \text{ mm} = 10^{-4} \text{ m}$$

So the volume occupied by each cube is

$$V_c = s^3$$
$$= (10^{-4} \text{ m})^3 = 10^{-12} \text{ m}^3$$

This means that a million million (10^{12}) grains would be needed to fill a cubic meter of space.

The volume of 6.02×10^{23} grains is found by multiplying the volume of a single grain by the number of grains:

$$\text{total volume of cubes} = 6.02 \times 10^{23} \text{ cubes} \times \text{volume of 1 cube}$$
$$= 6.02 \times 10^{23} \text{ cubes} \times 10^{-12} \frac{\text{m}^3}{\text{cube}}$$
$$= 6.02 \times 10^{11} \text{ m}^3$$

Our pile of grains must have a volume of 602 billion m³. Now suppose the pile were spread over the entire city of Los Angeles, which covers an area of 464 mi², or 1.2×10^9 m². How high would this pile be?

Given a definite area and the volume of the pile of grains, we can calculate its height, because the volume equals the area multiplied by the height (see Figure 9-3):

$$\text{Volume} = \text{area} \times \text{height}$$
$$V = A \times H$$
$$H = \frac{V}{A}$$
$$H = \frac{6.02 \times 10^{11} \text{ m}^3}{1.2 \times 10^9 \text{ m}^2} = 5.0 \times 10^2 \text{ m}$$

We have found that the height of the pile is 500 m, half a kilometer, the approximate height of a 160-story building. An Avogadro's number of salt grains makes a pile whose size is within our powers to visualize. However, an Avogadro's

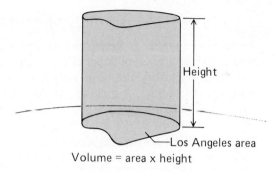

FIGURE 9-3   The volume of the space above Los Angeles that would be occupied by an Avogadro's number of salt grains is equal to the area of the city multiplied by the height of the space.

number of pinheads, each ten times greater in diameter than a salt grain, would make a mountain whose top would reach above the earth's atmosphere. And yet, 1 g of hydrogen—the amount of hydrogen containing that number of atoms—is not a very large quantity of hydrogen. It occupies a volume of only 11.2 L, which is a cube 22.4 cm (less than 9 in) on a side—roughly 3 gal. It is really difficult to visualize the magnitude of Avogadro's number; at best we can only make fantastic analogies to demonstrate its unimaginable dimensions.

The Mole

The atomic mass of hydrogen is the mass of a hydrogen atom, measured on a scale in which the mass of one carbon-12 atom is defined to be exactly 12.000 amu, as defined in Chapter 4. The *molecular mass* of hydrogen (formerly called the molecular weight) is defined to be the mass of one molecule of hydrogen on the same scale. Since there are two hydrogen atoms in a molecule, the molecular mass of hydrogen is twice the atomic mass, or 2.016 amu.

The **gram-molecular mass** of a substance, or **mole** (abbreviated **mol**), is defined as the number of grams equal to the molecular mass of that substance. Thus, a mole of hydrogen is 2.016 g of hydrogen—twice the gram-atomic mass. A mole of oxygen is 32.0 g of oxygen. A mole of natural carbon is 12.02 g. (The Latin word *moles* means a heavy bulk or mass; the word *molecule* is the diminutive of mole, and so means a very small mass. Now, with the modern use of the word, *mole* once more means a mass.)

Notice that in the case of carbon (and many other solid elements) the molecular mass is the same as the atomic mass, so for carbon the gram-molecular mass is the same as the gram-atomic mass. The gram-atomic mass of carbon refers to naturally occurring carbon—a mixture of isotopes rather than the single isotope carbon-12.

We have seen that a gram-atomic mass of hydrogen contains 6.02×10^{23} atoms. A gram-molecular mass (mole) of hydrogen must contain the same number of hydrogen molecules, or *pairs* of atoms. Thus, a mole of hydrogen contains 6.02×10^{23} molecules of hydrogen—an Avogadro's number of hydrogen molecules. The same is true of oxygen: a mole of oxygen molecules (32.0 g) contains an Avogadro's number of oxygen molecules. In general, a mole of any element or compound contains the same number of molecules—6.02×10^{23} molecules.

For this reason, modern chemists have redefined the mole to be that quantity of any substance containing 6.022045×10^{23} molecules (the current, more accurate value). The symbol N_A is generally employed to represent this number. The definition can be further generalized to include all kinds of particles: A mole of any kind of particle is N_A such particles. According to this new definition, a mole of atoms is N_A atoms. A mole of electrons is N_A electrons. A mole of salt grains, encountered in the previous section, is N_A grains.

The **molar mass** of a substance is now defined to be the mass of a mole of that substance—that is, the mass of an Avogadro's number of molecules (or atoms)—in units of *grams per mole*. For example, the molar mass of Na is 22.99 g/mol.

The reasons the mole is a convenient unit of measurement will become clear as you learn to do chemical calculations. The following examples illustrate how the mole concept applies to the interpretation of equations.

The equation

$$H \longrightarrow H^+ + e^-$$

tells us that when a hydrogen atom is ionized, it yields a hydrogen ion plus an electron. However, the equation also applies to any number of hydrogen atoms. Therefore the equation can be interpreted as follows: 1 mol of hydrogen atoms, when ionized, yields 1 mol of hydrogen ions plus 1 mol of electrons.

Now consider the reaction in which hydrogen combines with chlorine to produce hydrogen chloride:

$$H_2 + Cl_2 \longrightarrow 2HCl$$

Since there are two atoms in each of the molecules on the left side of the equation, the result of the reaction is two molecules of HCl. In terms of molecules, the equation says that one molecule of hydrogen combines with one molecule of chlorine to produce two molecules of HCl. But the equation can also be read to say that 1 mol of hydrogen molecules combines with 1 mol of chlorine molecules to yield 2 mol of HCl molecules. In general, a chemical equation gives information about the number of moles of the various compounds entering into the reaction. This is most valuable information, since it enables us to calculate the quantities of materials entering into the reaction, as we shall see in the following sections.

9.3 Molar Mass and the Mole

Calculation of Molecular Masses

The atomic mass of an element is the mass of one atom of that element, measured on a scale in which an atom of carbon-12 has a mass of exactly 12.000 amu. The **molecular mass** of a compound is the mass of the entire molecule, measured on the same scale. The molecular mass is simply the sum of the masses of all the atoms in that molecule. Therefore, the molecular mass of a compound is calculated by *adding up the atomic masses of all the elements, taking into account how many atoms of each element are in the compound.*

Example 9.2

Calculate the molecular mass of H_2O.

Solution

Make a table listing the atomic masses of the elements in the compound. Multiply each atomic mass by the number of atoms of that element in the molecule, to get the total atomic mass of that element.

Element	At. Mass		No. of Atoms		Total At. Mass
H	1.008	×	2	=	2.016 amu
O	15.999	×	1	=	15.999 amu
	Molecular mass of H_2O			=	18.015 amu

The molecular mass is the sum of all the numbers in the last column.

Example 9.3

What is the molecular mass of H_2SO_4?

Solution

Element	At. Mass		No. of Atoms		Total At. Mass
H	1.008	×	2	=	2.016 amu
S	32.064	×	1	=	32.064 amu
O	15.999	×	4	=	63.996 amu
	Molecular mass of H_2SO_4			=	98.076 amu

Example 9.4

What is the molecular mass of $Fe_2(SO_4)_3$ (ferric sulfate)?

Solution

Here there is a choice of two methods. One possibility is to use the straightforward method of the previous examples; the other is to calculate the molecular mass of the sulfate group first, and then use that value in the table as one of the ingredients. We will adopt the second approach. The molecular mass of the sulfate group is

$$32.064 + (4 \times 15.999) = 96.060$$

Then the table is as follows:

Element	At. Mass	No. of Units		Total At. Mass
Fe	55.847 ×	2	=	111.694 amu
SO_4	96.060 ×	3	=	288.180 amu
	Molecular mass of $Fe_2(SO_4)_3$		=	399.874 amu

Each of the above solutions involves counting atoms and multiplying the number of each atom by its particular atomic mass, then summing up to find the molecular mass. The same procedure applies to the most complicated molecules. There are large organic molecules in living organisms with molecular masses of many hundreds of thousands. Unraveling the structure of such complex molecules is one of the important and uncompleted tasks of chemistry at the present time.

Solving Problems Using Moles

The following statements summarize the essential points covered in this chapter:

1. A *gram-molecular mass* of any substance is the number of grams of that substance equal to its molecular mass (on a relative scale in which the mass of carbon-12 is exactly 12.0000 amu).

2. The number of molecules in a gram-molecular mass of any substance (N_A) is the same for all substances and is called *Avogadro's number*. Measurements of many kinds show that this number has the value 6.022045×10^{23}.

3. A *mole* of any substance contains an Avogadro's number of molecules (or atoms, if it is a monatomic substance). The mass of this quantity of substance is the *molar mass*.

4. A mole of any kind of particle is 6.022045×10^{23} of such particles.

Statements 2, 3, and 4 are simply equivalent definitions of the mole. Now let us look at some of the problems we can solve with these definitions.

Statements 1 and 2 together tell us that the mass (in grams) of N_A molecules of any substance is numerically equal to the substance's molecular mass. It must be understood that in most cases the natural material consists of a mixture of isotopes, so the molecular mass will be the average for all the isotopes. For example, the molecular mass of natural carbon is 12.011 amu. Therefore, N_A atoms of carbon have a mass of 12.011 g. If m_C represents the average mass (in grams) of a single carbon atom, the mass of N_A atoms is given by the equation

$$N_A \times m_C = 12.011 \text{ g} \tag{9-1}$$

The same can be said of any other element. Letting m_{Fe} represent the mass of a single iron atom, we can write the similar equation

$$N_A \times m_{Fe} = 55.9 \text{ g} \qquad (9\text{-}2)$$

Since 55.9 g is the average atomic mass of natural iron, m_{Fe} in this equation represents the average mass of the various iron isotopes.

We can solve Equation (9-1) for m_C to find the mass of a single carbon atom:

$$m_C = \frac{12.011 \text{ g}}{N_A} = \frac{12.011 \text{ g/mol}}{6.02 \times 10^{23} \text{ atoms/mol}}$$

$$= 1.99 \times 10^{-23} \text{ g/atom (average mass)}$$

You see how easy it is to find the mass of a single atom now that the value of Avogadro's number is known. The same procedure, using Equation (9-2), gives the mass of an iron atom.

When large amounts of materials are involved, the kilogram is often a more convenient unit than the gram. In that case we use the **kilomole (kmol)** (also called the *kilogram mole* or the *kilogram molecular mass*), which is the number of kilograms of a substance equal to its molecular mass (12 kg of carbon, 18 kg of water, etc.). It is important to keep in mind the fact that *the number of molecules in a kilomole is 1000 times greater than the number of molecules in a mole.*

Use of the concept of the mole simplifies numerous problems in chemistry. The examples below show how to do simple problems involving the mole concept; the next chapter will present further details concerning more complicated problems.

Example 9.5   Calculate the mass of 20.00 mol of HCl.

Solution   Add the atomic masses of H and Cl to find the molecular mass of HCl:

$$1.008 \text{ amu} + 35.453 \text{ amu} = 36.461 \text{ amu}$$

The molar mass of HCl is 36.461 g/mol. In general, the mass of M moles equals M times the molar mass. Therefore the mass of 20.00 mol is

$$20.00 \text{ mol} \times \frac{36.461 \text{ g}}{1 \text{ mol}} = 729.2 \text{ g}$$

Example 9.6   How many molecules are there in 20.00 mol of H_2SO_4?

Solution   The number of molecules in 20.00 mol of any compound is

$$20.00 \text{ mol} \times 6.02 \times 10^{23} \frac{\text{molecules}}{\text{mol}} = 120 \times 10^{23} \text{ molecules}$$

$$= 1.20 \times 10^{25} \text{ molecules}$$

It is not necessary to know anything about the molecular mass of the compound because 1 mol of *any* compound has the same number of molecules.

9.3 MOLAR MASS AND THE MOLE

Example 9.7 How many atoms of oxygen are there in 20.00 mol of H_2SO_4?

Solution Example 9.6 showed that there are 1.20×10^{25} molecules in 20.00 mol of H_2SO_4. Since each H_2SO_4 molecule contains four oxygen atoms, the total number of oxygen atoms is

$$1.20 \times 10^{25} \text{ molecules} \times \frac{4 \text{ atoms}}{1 \text{ molecule}} = 4.80 \times 10^{25} \text{ atoms}$$

Example 9.8 When sodium hydroxide (NaOH) reacts with hydrochloric acid (HCl), 1 mol of NaOH combines with 1 mol of HCl:

$$NaOH + HCl \longrightarrow NaCl + H_2O$$

(a) How many moles of NaOH combine with 20.00 mol of HCl?
(b) How many grams of NaOH combine with 20.00 mol of HCl?
(c) How many grams of HCl combine with 150.0 g of NaOH?

Solution (a) The number of moles of NaOH is

$$20.00 \text{ mol HCl} \times \frac{1 \text{ mol NaOH}}{1 \text{ mol HCl}} = 20.00 \text{ mol NaOH}$$

(b) Since 20 mol of NaOH combines with 20 mol of HCl, the problem now is to find how many grams of NaOH are in 20 mol of NaOH. The molecular mass of NaOH is

| Na | O | H | NaOH |

$$22.990 + 15.999 + 1.008 = 39.997$$

Therefore, the molar mass of NaOH is 39.997 g, and 20.00 mol has a mass of

$$20.00 \text{ mol} \times \frac{39.997 \text{ g}}{1 \text{ mol}} = 799.9 \text{ g}$$

Note: In Example 9.5 we found that 20 mol of HCl has a mass of 729.2 g. The result just obtained tells us that 800 g of NaOH combines with 729 g of HCl (approximately). You see how the mole concept helps us do this sort of calculation in a straightforward manner.

(c) A mole of HCl has a mass of about 36.5 g, and the molar mass of NaOH is approximately 40.0 g. Since 1 mol of HCl reacts with 1 mol of NaOH, we can find the proportion of HCl to NaOH in the above reaction (that is, the number of grams of HCl per gram of NaOH) by dividing the mass of 1 mol of HCl by the mass of 1 mol of NaOH:

$$\frac{\text{mass of HCl}}{\text{mass of NaOH}} = \frac{36.5 \text{ g}}{40.0 \text{ g}}$$

This fraction is a conversion factor between amounts of NaOH and amounts of

HCl. Given a value of 150 g for the mass of NaOH, we can solve the above equation for the mass of HCl:

$$\text{mass of HCl} = \frac{\text{mass of HCl}}{\text{mass of NaOH}} \times \text{mass of NaOH}$$

$$= \frac{36.5 \text{ g}}{40.0 \text{ g}} \times \text{mass of NaOH}$$

$$= \frac{36.5 \text{ g}}{40.0 \text{ g}} \times 150 \text{ g NaOH}$$

$$= 137 \text{ g HCl}$$

We see by inspection that this is a reasonable result—the fact that the amount of HCl is a little less than the amount of NaOH is consistent with the difference in the molecular weights.

Example 9.9 What is the mass of a single HCl molecule?

Solution The gram-molecular mass of HCl is 36.46 g. This amount of HCl contains 1 mol of HCl molecules, so we can write [as in Equation (9-1) above]

$$\text{mass of 1 mol of HCl} = N_A \times m(\text{HCl}) = 36.46 \text{ g}$$

where $m(\text{HCl})$ is the mass of a single HCl molecule. Solving for this mass, we have

$$m(\text{HCl}) = \frac{36.46 \text{ g}}{6.02 \times 10^{23} \text{ molecules}}$$

$$= 6.06 \times 10^{-23} \text{ g/molecule}$$

The same procedure can be used for any other molecule, atom, ion, or other particle.

Molar Volumes

We now come to an extremely important consequence of Avogadro's law. Recall what this law says: *Equal volumes of different gases at the same temperature and pressure contain the same number of molecules.* Reversing this statement leads to an equivalent rule: *A given number of molecules of any gas will always occupy the same volume* (at the same temperature and pressure).

As a result, a billion molecules of nitrogen occupy the same amount of space as a billion molecules of hydrogen. There are two reasons for this uniformity: (1) the volume of each molecule is much smaller than the space between molecules, and (2) the distance *between* molecules does not depend on the composition of the gas. The distance between molecules depends only on the temperature and pressure and does not depend on the kind of molecules making up the gas.

9.3 MOLAR MASS AND THE MOLE

Because 1 mol is one specific number of molecules, 1 mol of any gas occupies the same volume regardless of the kind of gas. The volume occupied by 1 mol is called the **molar volume**. The molar volume is a most interesting number, because once its value is known the molecular mass of any gas can be determined simply by weighing a known volume of the gas.

In order to make sure that differences in temperature and pressure do not confuse the results, we must make all gas measurements at one particular temperature and pressure (or at least make corrections for changes in temperature and pressure). You will recall that the conditions for gas measurements are called **STP** (standard temperature and pressure) and are defined to be a temperature of 0°C and a pressure of one atmosphere (101.3 kPa). (You will learn more about measuring pressures in Chapter 12.)

To find the molar volume of a gas, we need only measure the volume and mass (or weight) of a container of that gas. As shown in the example below, the volume of 1 mol of the gas can then be calculated.

When a container of gas is weighed, we must take into account the fact that the atmosphere pushes up on the container to some extent, reducing its weight. In weighing ordinary objects, we often ignore this buoyancy effect, but a chemist performing very precise measurements must correct for it. In dealing with very light materials such as gases, the chemist must be especially careful, for then the buoyancy effect becomes particularly important. After all, a container of hydrogen might be lighter than the air it displaces. A correction can be calculated, or the buoyancy effect can be eliminated by suspending the container in a vacuum chamber.

Example 9.10

A container with a volume of 2.50 L is filled with pure CO_2 at 0°C and 1 atm pressure. It is weighed on a balance, and the CO_2 is found to have a mass of 4.90 g after all necessary corrections have been made. What is the volume of 1 mol of CO_2?

Solution

The molar mass of CO_2 is 44.0 g/mol. Therefore our aim is to find the volume occupied by 44.0 g of CO_2. The first step is to calculate the gas density:

$$\text{density} = \frac{\text{mass}}{\text{volume}} = \frac{4.90 \text{ g}}{2.50 \text{ L}} = 1.96 \text{ g/L}$$

Then the above equation is solved for the volume:

$$\text{density} = \frac{\text{mass}}{\text{volume}}$$

$$\text{volume} = \frac{\text{mass}}{\text{density}}$$

We substitute into this equation the 44.0-g molar mass:

$$\text{volume} = \frac{44.0 \text{ g}}{1.96 \text{ g/L}}$$

$$= 22.4 \text{ L}$$

This is the molar volume of CO_2.

This molar volume we have calculated does not apply only to carbon dioxide; 1 mol of *any* gas will occupy the same volume at the same temperature and pressure. Therefore this molar volume (22.4 L/mol) is a *universal* number. It is a very useful quantity to know and has been measured with great precision. The presently accepted value for the molar volume of any gas at STP is 22.41383 L, with an uncertainty of 31 parts per million.

Example 9.11

What is the volume in gallons of 22.4 L?

Solution

A liter is 0.264 gal, so the volume of 22.4 L is

$$22.4 \text{ L} \times 0.264 \text{ gal/L} = 5.91 \text{ gal}$$

This is the volume of 1 mol of any gas at STP.

Example 9.12

What is the volume of 10.0 kg of helium at STP?

Solution

The molar mass of helium is 4.00 g/mol. (Since helium has only one atom per molecule, its molecular mass is the same as the atomic mass—a fact true of all the noble gases.) Thus 4.00 g of He occupies 22.4 L at STP. The density of helium is

$$\text{density} = \frac{\text{mass}}{\text{volume}} = \frac{4.00 \text{ g}}{22.4 \text{ L}}$$

$$= 0.179 \text{ g/L}$$

The volume of 10.0 kg of He is found by solving the above equation for volume and substituting 10.0 kg (10,000 g) for the mass:

$$\text{volume} = \frac{\text{mass}}{\text{density}} = \frac{10,000 \text{ g}}{0.179 \text{ g/L}}$$

$$= 55,900 \text{ L}$$

The volume in cubic meters is found by noting that $1 \text{ m}^3 = 1000 \text{ L}$, so

$$\frac{55,900 \text{ L}}{1000 \text{ L/m}^3} = 55.9 \text{ m}^3$$

55.9 m^3 is the volume of a sphere 4.74 m (15.6 ft) in diameter—a sizable balloon.

The above examples illustrate the important fact that 22.4 L of any gas at STP contains 1 mol of that gas. This fact provides us with a direct way of measuring the molecular mass of any gaseous substance: *by finding the mass (in grams) of 22.4 L of that gas at STP.*

Further, since the density of a gas (the mass per liter) is equal to its molar mass divided by 22.4 L, *the density of any gas (at STP) is directly proportional to its molar mass:*

$$\text{density} = \frac{\text{mass}}{\text{volume}}$$

$$= \frac{g}{L} = \frac{g/\text{mol}}{22.4 \text{ L/mol}}$$

$$= \frac{\text{molar mass}}{\text{molar volume}}$$

This fact is important to consider when the buoyancy of gases is being estimated. In order for a balloon to float in air, it must be filled with a gas less dense than air. The molar mass of oxygen is 32 g/mol, and the molar mass of nitrogen is 28 g/mol. Air is about 78% nitrogen by volume, so the average molar mass of air is about 29 g/mol. This means that any gas with a molar mass of less than 29 g/mol tends to float in air, whereas any gas with a molar mass of more than 29 g/mol is more dense than air and tends to sink.

Example 9.13 Ten liters of an unknown gas is weighed and is found to have a mass of 7.14 g. (**a**) What is the molecular mass of this gas? (**b**) Will this gas float in air?

Solution (**a**) First find the density of the gas:

$$\text{density} = \frac{\text{mass}}{\text{volume}} = \frac{7.14 \text{ g}}{10.0 \text{ L}}$$

$$= 0.714 \text{ g/L}$$

Solve the density equation for mass, and find the mass of 22.4 L of the unknown gas:

$$\text{mass} = \text{density} \times \text{volume}$$

$$= 0.714 \text{ g/L} \times 22.4 \text{ L}$$

$$= 16.0 \text{ g}$$

Since this is the mass of 1 mol of gas, the molecular mass must be 16.0 amu.
(**b**) Since the average molar mass of air is 29 g/mol, its density is given by

$$D = \frac{\text{mass}}{\text{volume}} = \frac{29.0 \text{ g/mol}}{22.4 \text{ L/mol}} = 1.29 \text{ g/L}$$

The air density is found to be greater than the density of the gas found above: 0.714 g/L. Therefore we would expect the gas to rise if released into the atmosphere. This result verifies the statement that a gas with a molar mass of less than 29 g/mol will tend to rise in air.

Glossary

Avogadro's hypothesis (Avogadro's law) A theorem developed by Amedeo Avogadro stating that, at the same temperature and pressure, equal volumes of different gases contain equal numbers of molecules.

Avogadro's number 6.02×10^{23}.

diffusion The mixing of the molecules of one gas (or liquid) with another due to the motion of the molecules.

gram-molecular mass The molecular mass of a compound or element expressed in gram units.

kilomole (kmol) A unit of 1000 moles.

law of combining volumes A law developed by J. L. Gay-Lussac stating that, under the same conditions of temperature and pressure, the volume ratios of gases in a chemical reaction will be simple whole numbers.

Loschmidt's number The number of molecules in 1 cm³ of a gas at STP: 2.71×10^{19}.

molar mass The mass of 1 mol of any substance.

molar volume The volume of 1 mol of gas at STP.

mole (mol) An Avogadro's number of any kind of object; a quantity of any substance containing Avogadro's number of molecules.

molecular mass The mass of a molecule relative to the mass of a ^{12}C atom.

STP Standard Conditions of Temperature and Pressure: 0°C and 1 atm.

Problems and Questions

9.1 Historical Review

9.1.Q State three concepts proposed by John Dalton that later turned out to be false.

9.2.Q Explain the fallacy in Dalton's hypothesis that water has the formula HO.

9.3.Q Devise two hypotheses that explain equally well the combining weights of H and O in water (8 parts O to 1 part H).

9.4.Q What is the best way to evaluate the truth or falsity of a hypothesis?

9.5.Q a. What distinguishes an ordinary guess from a scientific hypothesis?
   b. What can a scientist do to make a theory more than a hypothesis?

9.6.Q a. State the law of combining volumes.
   b. Give an example to illustrate its meaning.

9.7.Q State Avogadro's hypothesis. Why was it so important?

9.8.P Interpret the reaction

$$3H_2 + N_2 \longrightarrow 2NH_3$$

using Cannizzaro's synthesis of the law of combining volumes and Avogadro's hypothesis.

9.9.P In the reaction of Problem 9.8, what volume of NH_3 is produced if 16 cm³ of N_2 combines with sufficient H_2 to complete the reaction?

9.10.P Given the reaction

$$H_2 + Cl_2 \longrightarrow 2HCL$$

what volumes of H_2 and Cl_2 are required to produce 14.8 L of gaseous HCl?

9.11.P In the reaction

$$N_2 + 3F_2 \longrightarrow 2NF_3$$

we obtain 15 L of NF_3. What volumes of N_2 and F_2 are required for this reaction?

9.2 Avogadro's Number

9.12.Q What is diffusion?

9.13.Q Define Loschmidt's number.

9.14.Q a. What is a gram-atomic mass?
  b. How does a gram-atomic mass differ from an atomic mass?
  c. Is an atomic mass different from an atomic weight?

9.15.Q Define Avogadro's number in terms of the gram-atomic mass. What is the numerical value of Avogadro's number?

9.16.P How many molecules of nitrogen are there in 28 g of nitrogen (N_2) gas?

9.17.Q Describe two ways of measuring Avogadro's number.

9.18.P The atomic mass of sodium is 22.99.
  a. How many electrons are required to convert Na^{1+} ions into 22.99 g of Na atoms in an electrolysis experiment?
  b. How many electrons are required to convert Na^{1+} ions into 11.49 g of Na atoms?

9.19.P It is found that 6.02×10^{23} electrons are needed to produce 39.1 g of potassium by electrolysis.
  a. How many atoms of potassium are produced?
  b. How does 39.1 g compare to the atomic mass of potassium?

9.20.P It is found that 6.02×10^{23} electrons are required to convert a quantity of Mg^{2+} to Mg atoms by electrolysis. (Notice that two electrons are required for each atom.)
  a. How many atoms of Mg are formed?
  b. How many gram-atomic masses of Mg are produced?

9.21.Q What is the difference between Loschmidt's number and Avogadro's number?

9.22.Q Can you think of any manufactured products that are made in quantities of Avogadro's number? (That is, are 6×10^{23} things of any nature manufactured either during a year or over a decade?)

9.23.P If you had a factory capable of manufacturing 10,000 buttons per second, how long would it take to make an Avogadro's number of buttons? (Compare this answer with your answer to Question 9.22.)

9.24.P How tall a pile would be made by 6×10^{23} pinheads, each 1 mm in diameter, if you stacked them up over the city of Los Angeles? (Assume that each sphere occupies a cubical space when packed with the others, as shown in Figure 9-4.)

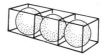

FIGURE 9-4 When spheres are packed solidly into a volume, each sphere occupies a cubical volume, so it is easy to calculate the volume of a large number of spheres.

9.25.P The world population is approximately 4.5×10^9. If you distributed an Avogadro's number of dollars among all the people in the world, how much money would each person get?

9.26.P An Avogadro's number of a certain molecule occupies 18 cm³.
  a. What is the volume of a single molecule?
  b. What is its diameter?

9.27.P An Avogadro's number of carbon atoms has a mass of 0.012 kg. What is the mass of each atom?

9.28.P What is the total mass of an Avogadro's number of atoms, each of which has a mass of 3.33×10^{-22} g?

9.29.P A sample of liquid containing an Avogadro's number of molecules has a volume of 14.82 cm³. What is the volume occupied by each molecule?

9.30.Q What is the difference between gram-atomic mass and gram-molecular mass?

9.31.Q In what kind of substance is the gram-atomic mass the same as the gram-molecular mass?

9.32.P How many molecules are there in 2 mol of nitrogen?

9.33.P How many grains of sand are there in half a mole of sand grains?

9.34.P Interpret each equation below in terms of moles of substances:

    a. $2H_2 + O_2 \longrightarrow 2H_2O$
    b. $2KClO_3 \longrightarrow 2KCl + 3O_2$
    c. $N_2 + 3H_2 \longrightarrow 2NH_3$

9.35.P Predict the number of moles of product substances produced in the reactions below from the given amounts of starting materials.

    a. $2Mg + O_2 \longrightarrow 2MgO$
        start with 2 mol of Mg
    b. $2Ca + O_2 \longrightarrow 2CaO$
        start with 1 mol of Ca
    c. $4K + O_2 \longrightarrow 2K_2O$
        start with 1 mol of K

9.36.P Give the number of molecules of each of the following substances:

    a. 32 g of O_2
    b. 35.5 g of Cl_2
    c. 10 g of Na

9.3 Molar Mass and the Mole

9.37.Q What information is needed to calculate the molecular mass of a compound?

9.38.Q a. How is the gram-molecular mass related to the molecular mass?
    b. How is the gram-molecular mass related to the molar mass?

9.39.P Calculate the molar mass of each of the following materials:

    a. Na   b. Cl_2
    c. KCl   d. H_2SO_4
    e. C_6H_6

9.40.P In the following equations, assume that the number in front of each compound represents the number of moles of that compound. Calculate how many grams of each compound take part in the reaction.

    a. $2H_2 + O_2 \longrightarrow 2H_2O$
    b. $H_2 + Cl_2 \longrightarrow 2HCl$
    c. $N_2 + 3H_2 \longrightarrow 2NH_3$
    d. $2KClO_3 \longrightarrow 2KCl + 3O_2$
    e. $2C_6H_6 + 15O_2 \longrightarrow 12CO_2 + 6H_2O$

9.41.P Calculate the molar mass of each of the following compounds:

    a. $Hg(NH_3)_2Br_2$
    b. $Ni(IO_3)_2$
    c. $Pr_2(MoO_4)_2$
    d. $Sr(ClO_4)_2$
    e. $Th(PO_3)_4$

9.42.P a. How many molecules are in a mole of H_2O?
    b. How many molecules are in a mole of HCl?

9.43.P Given the reaction

$$Mg + 2HCl \longrightarrow MgCl_2 + H_2$$

calculate the number of grams of H_2 formed from 10 g of Mg plus enough HCl to complete the reaction.

9.44.P Using the equation in Problem 9.43, calculate the number of moles of $MgCl_2$ formed from 3.5 mol of Mg plus excess HCl.

9.45.P Given the equation

$$N_2 + 3H_2 \longrightarrow 2NH_3$$

determine the following.
    a. How many grams of hydrogen are required to react with 3.5 kg of nitrogen?
    b. How many moles of nitrogen are required to form 300 g of NH_3? How many grams of nitrogen is that?

9.46.P Given the equation

$$2HCl + Ca(OH)_2 \longrightarrow CaCl_2 + 2H_2O$$

determine the following.
    a. If 60 g of HCl reacts with excess $Ca(OH)_2$, how many grams of $CaCl_2$ are formed?
    b. If 60 g of $Ca(OH)_2$ reacts with excess HCl, how many grams of $CaCl_2$ are formed?
    c. How many grams of HCl are required to form 60 g of $CaCl_2$?

9.47.Q State Avogadro's law.

9.48.Q What are the theoretical reasons behind Avogadro's law?

9.49.Q Define STP.

9.50.Q What is the relationship between the density and the molar mass of a gas?

9.51.P **a.** A gas at STP occupies 22,400 cm³ and has a molar mass of 71.0 g/mol. How many moles of the gas are present?
b. Another gas with a molecular mass of 28 amu occupies the same volume. How many moles of that gas are present?

9.52.P Find the density of a gas that occupies 16.4 L at STP and has a molar mass of 32.0 g/mol.

9.53.P What volume is occupied by 24.0 g of nitrogen at STP?

9.54.P A room has dimensions of 3.0 m × 4.0 m × 2.5 m.
a. What is the volume of the room in cubic meters?
b. What is the volume of the room in liters?
c. What is the mass of the air in the room, assuming an average molecular mass of 29 amu?

9.55.P What volume is occupied by 44.0 g of CO_2 at STP?

9.56.P What volume is occupied by 22.0 g of CO_2 at STP?

9.57.P What volume is occupied by 0.50 mol of CO_2 at STP?

9.58.P What volume is occupied by 10.0 mol of CO_2 at STP?

9.59.P **a.** A gas has a density of 1.97 g/L. What volume does 22.0 g of the gas occupy at STP?
b. What is the molecular mass of this gas?

9.60.P **a.** A gas has a density of 1.25 g/L at STP. What is the mass of an 11.2-L sample of this gas?
b. What is the molecular mass of this gas?

Self Test

1. Calculate the molar mass of each of the following compounds:
   a. $CaCl_2$
   b. $C_6H_{12}O_6$
   c. H_3PO_4

2. How many moles are represented by each of the following?
   a. 6.0 g of H_2
   b. 12.0 g of Na
   c. 80.0 kg of $MgBr_2$
   d. 90.0 g of AgCl

3. How many grams are there in each of the following?
   a. 2.0 mol of Cl_2
   b. 3.4 mol of Ba
   c. 6.7 kmol of $CoCl_2$
   d. 3.9 kmol of $Fe(NO_3)_2$

4. Calculate the mass of each product (in grams):
   a. $H_2 + Cl_2 \longrightarrow 2HCl$ (given 6 g of H_2)
   b. $2KClO_3 \longrightarrow 2KCl + 3O_2$ (given 61.2 g of $KClO_3$)

5. **a.** A gas has a density of 3.20 g/L at STP. What is the volume of 71.0 g of this gas?
   b. What is the molar mass of this gas?

6. **a.** Sixteen grams of a gas occupies a volume of 11,200 cm³. Find the density of this gas in grams per liter.
   b. What is the molar mass of this gas?

7. How many liters of product are formed in each of the following reactions?
   a. $N_2 + 3H_2 \longrightarrow 2NH_3$ (given 1.5 L of H_2)
   b. $H_2 + F_2 \longrightarrow 2HF$ (given 0.60 L of H_2)

8. Determine the following for 6.5 L of N_2 at STP.
   a. How many moles of N_2 are present?
   b. How many grams of N_2 are present?

9. Calculate the volume in liters of 10.0 g of O_2 at STP.

10. What is the density of methane (CH_4) at STP?

10
Chemical Equations

A worker wearing protective clothing samples pig iron from a blast furnace cast for spectrographic and chemical analysis.

Objectives After completing this chapter, you should be able to:
1. Demonstrate three ways of interpreting a chemical equation.
2. Demonstrate four types of equations used to describe common inorganic reactions: synthesis, decomposition, displacement, and double displacement.
3. Present several examples of each of the four types of equations.
4. Use the activity series to predict which elements will displace a given element from a compound.
5. Balance simple equations, using the trial-and-error method and the balance-sheet method.

10.1 The Meaning of a Chemical Equation

Describing a Reaction Qualitatively

There are many different ways to describe a chemical reaction. The simplest and most primitive way is to tell what we see when the reaction takes place.

For example, a spoonful of iron filings is put into a beaker of dilute hydrochloric acid. The filings sink to the bottom of the beaker while bubbles of gas rise from them. At the same time, the colorless acid begins to turn blue. The gas generated by the reaction is found to be flammable, burning in air or oxygen. Further investigation shows that water is given off by the burning gas. Since we know that water results from the reaction of hydrogen with oxygen, we are led to believe that the gas bubbles escaping from the hydrochloric acid consist of hydrogen.

If there is enough acid, all of the iron filings disappear, and nothing is left but a blue solution. When this solution is evaporated, a mass of blue crystals remains in the beaker. What has happened to the iron? What are these blue crystals?

We can make some guesses about what has happened, using our knowledge that hydrochloric acid contains hydrogen and chlorine. Since hydrogen has been released by the reaction, we guess that the chlorine was left behind to react with the iron. Therefore, the blue crystals are a compound of iron and chlorine.

We do not say that the iron is *dissolved* in the acid. Rather, we say that the iron has *reacted* with the acid, forming a compound that is soluble in water. If the iron were simply dissolved, you could get it back by evaporation, which is not the case in the present experiment. However, the word "dissolved" is often misused in common language to describe what happens when a metal reacts with an acid.

What we have done above is to describe the *phenomena* (plural of *phenomenon*) taking place in the beaker. A phenomenon is something that happens and that can be observed by the human senses, perhaps with the aid of instruments such as thermometers, microscopes, and cameras. Our description is a **phenomenological description**. It tells what we actually saw, without considering what happens to the atoms and molecules taking part in the reaction.

Such descriptions were the subject matter of chemistry prior to the time of Lavoisier. This was **qualitative chemistry**. However, once it became possible to make accurate measurements of quantities of chemicals taking part in a reaction, these measurements became a necessary part of any description. To describe a reaction completely, it became necessary to tell *how much* of the various ingredients went into the reaction. Adding measurements to the descriptions made the science into **quantitative chemistry**.

Describing a Reaction Quantitatively

The quantities of the substances entering into the Fe-HCl reaction just described may be measured in the following way: Weigh the iron before it is put into the

10.1 THE MEANING OF A CHEMICAL EQUATION

beaker, and then weigh the acid (or measure the acid volume). In order to ensure that all of the acid reacts with the iron, use enough iron so that some of it is left at the bottom of the beaker after the reaction is complete. Now filter out the remaining (unreacted) iron and weigh it. Then subtract the mass of the unreacted iron from the original mass of the iron to get the mass of reacted iron. Now evaporate the blue solution, and weigh the crystals of iron chloride left behind. (We have eliminated a lot of details from this description: the beaker and filter paper must be weighed, the crystals must be dried, and so on. These are techniques that chemists learn in the course of their work.)

From these measurements we find that 100 g of HCl reacts with 76.6 g of Fe, producing 173.9 g of iron chloride—Fe_xCl_y. (The use of x and y as subscripts indicates that we do not as yet know the number of Fe and Cl atoms in the compound.) These measurements give the proportion of pure HCl to Fe (by weight) in the reaction. (*Note*: The above numbers imply that the beaker contains 100 g of pure HCl. The calculations must take into account the fact that in actuality the HCl is dissolved in water. The weight of the water does not count, since it is not a reactant. Therefore the concentration of the acid must be known.)

These numbers by themselves are not very enlightening. Measurements taken even with the most accurate balances have no great meaning if they apply only to one particular sample of chemicals. More useful are *general relationships* between the quantities of Fe and HCl that react with each other.

What we would like to know is: how many molecules of HCl react with one atom of Fe? Or we could ask the equivalent question: how many moles of HCl react with 1 mol of Fe? (Remember that the mole is a fixed number of molecules.) With that kind of information, we can write down an equation to describe the reaction in detail.

The question is answered as follows: The measurements show that 100 g of HCl reacts with 76.6 g of Fe. The proportion (by mass) of HCl to Fe is

$$\frac{\text{mass(HCl)}}{\text{mass(Fe)}} = \frac{100 \text{ g HCl}}{76.6 \text{ g Fe}} = \frac{1.305 \text{ g HCl}}{1 \text{ g Fe}}$$

The number 1.305 is a conversion factor. It tells how many grams of HCl react with each gram of Fe. By solving the above equation for mass(HCl) and substituting the molar mass of Fe for mass(Fe), we can find how many grams of HCl react with 1 mol of Fe. Using the table of atomic masses, we find the molar mass of iron to be 55.9 g/mol. We then write

$$\text{mass (HCl)} = \frac{1.305 \text{ g HCl}}{1 \text{ g Fe}} \times \text{mass(Fe)}$$

$$= \frac{1.305 \text{ g HCl}}{1 \text{ g Fe}} \times 55.9 \text{ g Fe}$$

$$= 73.0 \text{ g HCl}$$

The number of moles of HCl is found by calculating the molar mass of HCl to be 36.5 g/mol and then saying

$$\text{no. of mol HCl} = \frac{1 \text{ mol}}{36.5 \text{ g}} \times 73.0 \text{ g} = 2.00 \text{ mol HCl}$$

This leads us to an important conclusion: 1 mol of Fe has combined with 2 mol of HCl. This conclusion helps us guess at the formula for iron chloride. We know that a mole contains a specific number (N_A) of molecules; 2 mol of HCl contains $2 \times N_A$ chlorine atoms, and 1 mol of Fe contains N_A Fe atoms. It follows, therefore, that in the reaction under discussion two atoms of chlorine combined with each atom of iron (see Figure 10-1). Accordingly, the formula for the blue iron chloride compound must be $FeCl_2$.

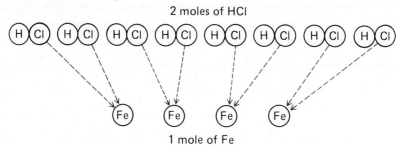

FIGURE 10-1 If 2 mol of HCl react with 1 mol of Fe, there are two chlorine atoms for every iron atom. (In this diagram we let 1 mol be represented by four molecules or atoms.)

This guess may be checked by considering the blue crystals. Since the reaction of 76.6 g of Fe with 100 g of HCl produced 173.9 g of iron chloride, the proportion of iron chloride to iron is (using the provisional formula $FeCl_2$)

$$\frac{\text{mass}(FeCl_2)}{\text{mass}(Fe)} = \frac{173.9 \text{ g FeCl}_2}{76.6 \text{ g Fe}} = \frac{2.27 \text{ g FeCl}_2}{1 \text{ g Fe}}$$

Solving for mass(Fe Cl_2), we get the amount of iron chloride obtained from a given mass of iron:

$$\text{mass}(FeCl_2) = \frac{2.27 \text{ g FeCl}_2}{1 \text{ g Fe}} \times \text{mass}(Fe)$$

Substitute into this equation the mass of 1 mol (55.9 g) of iron:

$$\text{mass}(FeCl_2) = \frac{2.27 \text{ g FeCl}_2}{1 \text{ g Fe}} \times 55.9 \text{ g Fe} = 126.9 \text{ g FeCl}_2$$

How many moles of iron chloride does this mass represent? To answer this question, we first calculate the molecular mass of $FeCl_2$:

$$\text{molecular mass of } FeCl_2 = 55.9 + 2 \times 35.5 = 126.9 \text{ amu}$$

Therefore, 1 mol of $FeCl_2$ weighs 126.9 g. But that is just the amount of iron chloride obtained from 1 mol of iron, proving that 1 mol of iron reacted with 2 mol of HCl to form 1 mol of $FeCl_2$. Accordingly, the equation for the reaction is

$$Fe + 2HCl \longrightarrow FeCl_2 + H_2$$

Let us review our logic. Because two Cl atoms enter into the reaction for each Fe atom, we deduce that one formula unit of $FeCl_2$ is formed in each reaction and one molecule of hydrogen (H_2) is released. This deduction is verified by weighing all the ingredients in the reaction—we find that 1 mol of Fe produces 1 mol of a compound with the same molecular mass as $FeCl_2$. (*Note*: Since $FeCl_2$ exists as an ionic crystal, the term *molecule* does not, strictly speaking, describe the smallest unit of $FeCl_2$. Therefore we use the term *formula unit* to represent the smallest applicable unit of a chemical substance.)

In general, any chemical equation tells us how many formula units of each ingredient enter into the reaction represented by the equation. It also tells the number of moles of each substance entering or produced by the reaction. So the above equation says that *one atom of iron combines with two molecules of hydrochloric acid to form one formula unit of iron chloride plus one molecule of hydrogen*. At the same time it says that 1 *mol of iron plus* 2 *mol of hydrochloric acid yields* 1 *mol of iron chloride plus* 1 *mol of hydrogen*.

Chapter 11 will continue this discussion of quantitative chemistry. In that chapter you will get more practice in doing problems of this nature, which will help you to master the many new concepts introduced.

The Several Meanings of the Equation

We have seen that a chemical equation can mean a great many different things, depending on the level of description. Let us review the various meanings of the equation.

1. *Qualitative meaning.* The equation tells us what compounds go into the reaction (reactants) and what compounds come out of the reaction (products).

2. *Quantitative meaning.* The equation tells us how much of each reactant compound takes part in the interaction and how much product to expect. Each term in the equation describes the number of moles of each compound that go into or come out of the reaction. Therefore, if we know the molecular mass of each ingredient, we can calculate the number of grams of each ingredient taking part in the reaction.

3. *Molecular meaning.* The equation tells us how many formula units of each compound go into the reaction. The equation describes what happens on a molecular level. For example, the equation $Fe + 2HCl \rightarrow FeCl_2 + H_2$ told us how many molecules of HCl reacted with each atom of Fe, and how many units of $FeCl_2$ and how many molecules of H_2 appeared as a result of the reaction. The same information is contained in any properly balanced equation. (Balancing of equations is an important skill that will be addressed in Section 10.3.)

10.2 Four Types of Equations

Many of the common chemical reactions can be classified into four basic types, which differ in the way their ingredients come together and in the kinds of products that are formed. We illustrate these reactions by letting A, B, C, and D represent four different elements, molecules, or groups. Then the four types of equations may be summarized as follows:

$$A + B \longrightarrow AB$$
$$AB \longrightarrow A + B$$
$$AB + C \longrightarrow BC + A$$
$$AB + CD \longrightarrow AD + CB$$

In these equations the combination A + B means that A combines with B. The combination AB means a compound formed from the two elements (or groups) A and B. (For example, if AB represents the compound $FeCl_2$, A stands for Fe and B represents Cl_2.)

The elements or compounds on the left side of each equation are the *reactants*; they are the ingredients entering into the reaction. The elements or compounds on the right side of each equation are the *products* of the reaction; they are the substances formed by the reaction.

As we have stated previously, chemical equations are not equations in the mathematical sense. This is why an arrow is used instead of an equal sign. The arrow means "yields," or "produces." Thus, the first of the above four expressions says "A plus B yields AB," or "A combines with B to produce the compound AB." In reading the fourth equation we say, "AB reacts with CD to yield AD and CB."

In the following sections, the four basic types of reactions will be discussed in detail.

Simple Combination, or Synthesis

The reaction A + B → AB is best described as the simple combination of A and B to form the molecule AB. Oxidation reactions are simple combinations. We have already encountered the combustion of carbon and hydrogen:

$$C + O_2 \longrightarrow CO_2$$
$$2H_2 + O_2 \longrightarrow 2H_2O$$

The formation of ferrous sulfide when iron filings are heated with sulfur is another reaction of this type:

$$Fe + S \longrightarrow FeS$$

This reaction describes the synthesis of FeS. In a **synthesis reaction** elements or compounds combine to form a new compound.

Chlorine and hydrogen react explosively when exposed to bright light:

$$H_2 + Cl_2 \longrightarrow 2HCl$$

The two ingredients on the left do not have to be single elements—they can be compounds. Consider, for example, the following reaction between phosphorus trichloride and chlorine:

$$PCl_3 + Cl_2 \longrightarrow PCl_5$$

Generally a simple combination is expected to be an **exothermic reaction** (one that gives off heat). This is because the product compound (AB) is more stable than the system made up of the separate elements, and a stable system is at a lower energy level than an unstable system. Therefore energy must be released when the compound is formed.

Another reaction fitting the format $A + B \to AB$ is the combination of water with a type of compound called an **acid anhydride**, resulting in the formation of an acid. (*Acid anhydride* means, literally, acid without water.) An example of such a reaction is the solution of carbon dioxide in water to form carbonic acid (the prime ingredient of soda water):

$$CO_2 + H_2O \longrightarrow H_2CO_3$$

The combination of sulfur trioxide with water plays an important role in the manufacture of sulfuric acid:

$$SO_3 + H_2O \longrightarrow H_2SO_4$$

Simple Decomposition

Reactions of the form $AB \to A + B$ are called **decomposition reactions**, since they involve breaking up the molecule AB to form the separate elements or compounds A and B. They generally require the application of energy to the molecule from the outside and so are mainly **endothermic reactions**, except for special cases in which the molecule is inherently unstable. One example of an endothermic decomposition reaction is the electrolysis of water by an electric current:

$$2H_2O \longrightarrow 2H_2 + O_2$$

Another famous decomposition reaction led to the discovery of oxygen by Joseph Priestley in 1774. This feat was accomplished by the heating and decomposition of red mercuric oxide. (The mercuric oxide was heated inside a closed glass container by concentrating sunlight onto it with a magnifying glass.) The reaction is described by the equation

$$2HgO \longrightarrow 2Hg + O_2$$

Another oxygen-producing reaction is the decomposition of potassium chlorate by heating:

$$2KClO_3 \longrightarrow 2KCl + 3O_2$$

An important industrial reaction is the decomposition of limestone by heating to form lime and carbon dioxide:

$$CaCO_3 \longrightarrow CaO + CO_2$$

The manufacture of aluminum is accomplished by the electrolysis of aluminum oxide dissolved in a mixture of molten NaF, CaF_2, and AlF_3. The reaction is described by the equation

$$2Al_2O_3 \longrightarrow 4Al + 3O_2$$

The production of aluminum requires a great deal of electrical energy, both to keep the mixture molten and to separate the aluminum from the oxygen.

The above reactions are endothermic, requiring an input of energy to make them go. There are also exothermic decompositions that take place when the molecule is unstable. (More properly, the molecule should be described as **metastable**, meaning that it is slightly stable but that a small disturbance may cause it to become unstable. The situation is something like that of a rock sitting at the edge of a cliff. It is stable for the time being, but give it a small push and it will come crashing down.)

Hydrogen peroxide (H_2O_2) is bought in pharmacies as a 3% solution for use as an antiseptic. It is a watery, bubbly liquid that, when poured into a cut, fizzes and releases oxygen. Concentrated hydrogen peroxide, on the other hand, is a bluish, syrupy liquid that is likely to decompose explosively with no apparent provocation. Dilute and concentrated H_2O_2 decomposes according to the equation:

$$2H_2O_2 \longrightarrow 2H_2O + O_2$$

The products of the reaction are water and oxygen.

Most explosives are by their very nature unstable compounds. One of the most treacherous is nitroglycerine. It is an oily liquid that decomposes upon being heated or jarred, forming large volumes of gases such as carbon monoxide, nitrogen, oxygen, and water vapor. This reaction shows that a decomposition reaction may be more complicated than the simple $AB \to A + B$. We can have $ABC \to A + B + C$, for example, as well as more complex reactions.

Displacement Reactions

In the reaction $AB + C \to BC + A$, the element C displaces element A from its position in the AB molecule. This will happen if element C is a more chemically active element than element A.

For example, if a piece of iron is placed in a $CuSO_4$ solution, some of the iron will go into solution and the displaced copper will precipitate out of solution. The reaction is

$$CuSO_4 + Fe \longrightarrow FeSO_4 + Cu$$

This reaction takes place because the iron is more active than the copper.

If sulfuric acid is used instead of copper sulfate, a similar reaction takes place. The only difference is that in sulfuric acid there are two hydrogen ions (instead of a copper ion) associated with the sulfate ion. Therefore the iron replaces the hydrogen in the sulfuric acid, forming $FeSO_4$ and releasing hydrogen gas into the air, according to the equation

$$H_2SO_4 + Fe \longrightarrow FeSO_4 + H_2$$

Reactions of this type afford a simple method of producing hydrogen (although commercially electrolysis of water is the preferred method). Many metals will displace hydrogen from an acid, producing the corresponding salt and freeing the hydrogen. For example, magnesium chloride is formed by the reaction of metallic magnesium with hydrochloric acid:

$$Mg + 2HCl \longrightarrow MgCl_2 + H_2$$

Not all metals react in this manner; mercury, copper, silver, gold, and the other noble metals do not displace hydrogen from an acid. How can we predict which metals will react with acids and which will not? We can make such a prediction if we arrange the metals into a series according to their activity. Such an **activity series** is presented in Table 10-1.

TABLE 10-1 Activity Series of Some Metals (Partial Listing)

lithium	hydrogen
potassium	copper
cesium	silver
barium	mercury
calcium	platinum
sodium	gold
magnesium	
aluminum	
zinc	
chromium	
iron	
cobalt	
nickel	
lead	

The most active metals, the alkali metals, are at the top of the table. As we go down the first column, the metals become less and less active, until lead is reached. The second column is a continuation of the first, starting with hydrogen. Here we find the least active metals, including silver, platinum, and gold. Hydrogen has a place in the series because it behaves like a metal (forms a positive ion) as far as this class of chemical reactions is concerned. (The primary difference between an acid and a salt is that in an acid hydrogen occupies the place that in a salt is filled by a metal.)

The table is interpreted as follows: Whenever one element is above another element in the series, the first (more active) element will displace the second (less active) element from a salt containing the second element. As we have seen, iron displaces copper from copper sulfate; this fact tells us that iron must be higher in the series than copper. The entire series was made up by observing which elements displace others from their salts and then placing them in the order of their activity.

The elements above hydrogen in the series will displace hydrogen from acids. We have seen that iron is "dissolved" by hydrochloric or sulfuric acid, forming iron chloride or iron sulfate. The higher in the series the metal is, the more rapid will be the reaction. Sodium and potassium react very violently, whereas lead is affected by acids rather slowly.

Another prediction that can be made from the series is that the metals below hydrogen will not displace hydrogen from an acid. Thus, copper, silver, and gold are relatively inert and are not attacked by dilute acids. (Copper and silver will react with nitric acid and other "oxidizing" acids, but there the reaction is more than a simple displacement reaction.)

The elements at the top of the series are so active that they will displace hydrogen from water. For example, when sodium is placed in water, hydrogen is generated (sometimes catching fire from the heat of the reaction) and sodium hydroxide is formed. The equation for this reaction is

$$2Na + 2H_2O \longrightarrow 2NaOH + H_2$$

Similar reactions occur with potassium and cesium.

In the case of metals less active than the alkali metals, the same kind of reaction may take place if the water is heated. Even iron will displace hydrogen from water if steam is passed over red-hot iron pellets:

$$3Fe + 4H_2O \longrightarrow Fe_3O_4 + 4H_2$$

Here the oxide of iron rather than the hydroxide is formed. This oxide is an interesting one. It occurs in nature as a red crystal with important magnetic properties and so has been given the name **magnetite**. It contains iron atoms with oxidation numbers of both 2 and 3; its formula may be written as $FeO \cdot Fe_2O_3$.

The above reaction can run backward as well as forward. It is one of many **reversible reactions**. The reverse reaction takes place when hydrogen gas is passed over heated magnetite:

$$Fe_3O_4 + 4H_2 \longrightarrow 3Fe + 4H_2O$$

If the reaction is able to go both ways, what determines its direction? What decides whether the iron replaces the hydrogen or the hydrogen replaces the iron?

The direction a reaction will take depends on how the reaction environment is set up. If hot steam is fed over the iron and the hydrogen is drawn away from it, the iron will displace the hydrogen from the water. If the hydrogen is passed into the chamber and the water vapor is pulled out, the reaction goes the other way. (These matters will be discussed in connection with the subject of equilibrium in Chapter 15.)

Displacement reactions take place among the nonmetals as well as among the metals. Iodine is manufactured from the ashes of kelp (a seaweed), which contain sodium and potassium iodide. When the ashes are treated with chlorine gas, the more active chlorine displaces the less active iodine, according to the reaction

$$2NaI + Cl_2 \longrightarrow 2NaCl + I_2$$

An activity series can be compiled for the nonmetals just as for the metals. At the head of this list is the gas fluorine, which is the most active of all elements. Fluorine, when exposed to water or to water vapor, will displace the oxygen from the water molecule, forming hydrogen fluoride:

$$H_2O + F_2 \longrightarrow 2HF + (O)$$

The released atomic oxygen immediately forms both molecular oxygen (O_2) and ozone (O_3). **Ozone** is an unstable form of oxygen with a characteristic odor. It is often detected near electrical discharges and is found naturally in the upper part of the earth's atmosphere.

Double Displacement Reactions

Reactions of the form $AB + CD \rightarrow AD + CB$ are called **double displacement reactions** (or *double decomposition reactions*). Such reactions simply involve the switching of partners by pairs of molecules.

One of the most common reactions is that of an acid neutralized by a base:

$$NaOH + HCl \longrightarrow NaCl + H_2O$$

Here the Na captures the Cl, while the H and the OH join to form water (H_2O, or HOH).

A neutralization reaction can sometimes run in the opposite direction. Magnesium chloride, for example, reacts with water to produce magnesium hydroxide and hydrochloric acid:

$$MgCl_2 + 2H_2O \longrightarrow Mg(OH)_2\downarrow + 2HCl$$

The vertical arrow in the above equation indicates that magnesium hydroxide is a solid with a very low solubility in water. Because of this property, the water cannot keep the $Mg(OH)_2$ in solution; as a result, the compound separates out of, or *precipitates from*, the solution in the form of a fine white powder.

It is the **precipitation** of the magnesium hydroxide that causes the reaction to go from left to right rather than in the reverse direction. The reason is as follows: When magnesium chloride is dissolved in water, all that exists at first is a mixture of Mg^{2+}, H^+, Cl^-, and OH^- ions. If magnesium hydroxide were soluble, all of these ions would remain in solution and nothing very interesting would happen. Since $Mg(OH)_2$ is not very soluble, however, it separates out of solution as a solid and is thus removed from the reacting system. See Figure 10-2. The reaction tends to go in the direction that forms the precipitate.

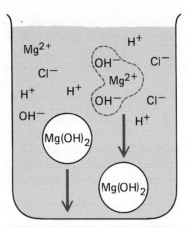

FIGURE 10-2 Insoluble magnesium hydroxide precipitates out of solution and so is removed from the system of reacting ions in the solution.

This reaction happens to be an important one; it is encountered when "hard water" (water containing high concentrations of dissolved minerals) is heated in hot-water tanks. One of the minerals that makes water hard is magnesium chloride. When the **hydrolysis reaction** just described takes place in the boiler, the magnesium hydroxide precipitates and forms scale on the inside of the tank, while the hydrochloric acid corrodes the tank's metal wall. (To avoid the acid reaction, many manufacturers line their hot-water tanks with glass.)

The same sort of action takes place in the combination of carbonic acid with calcium hydroxide (or barium hydroxide). If CO_2 is bubbled through a calcium hydroxide solution (lime water), first carbonic acid (H_2CO_3) is formed, as described in Section 10.1. Then the carbonic acid reacts with the calcium hydroxide to form a white precipitate of insoluble calcium carbonate:

$$H_2CO_3 + Ca(OH)_2 \longrightarrow CaCO_3 \downarrow + 2H_2O$$

This reaction can be used as a test to detect the presence of carbon dioxide.

Double displacement reactions are also common between pairs of salts. They are useful in situations in which one of the product salts is insoluble. The insoluble salt precipitates out of the solution and can be separated from the other ingredients by filtration. A common example is the reaction of sodium chloride with silver nitrate:

$$NaCl + AgNO_3 \longrightarrow NaNO_3 + AgCl \downarrow$$

Silver chloride is very insoluble in water and therefore precipitates out of the solution. Figure 10-3 demonstrates how the solid silver chloride can be filtered from the solution. Since a silver chloride precipitate is produced when any soluble silver salt is mixed with a soluble chloride, this reaction can be used as a test for silver (or for the chloride ion). It can also be used in the manufacture of silver chloride (or iodide, or bromide). These salts are important in the production of photographic film and paper.

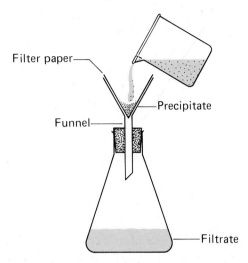

FIGURE 10-3 An insoluble compound precipitates out of a solution and is separated from the other compounds in the solution by filtering.

It must not be assumed that reactions such as these take place only in water solutions. There are some reactions that take place when the dry ingredients are mixed together and then melted. Such a reaction is called a **fusion**; the process may be considered a solution of one salt in the other (molten) salt. For example, barium sulfate is not very soluble in water but is highly soluble in molten sodium carbonate. Therefore, when a mixture of barium sulfate and sodium carbonate is melted, the following reaction takes place:

$$BaSO_4 + Na_2CO_3 \longrightarrow BaCO_3 \downarrow + Na_2SO_4$$

The barium carbonate is insoluble in the melt and precipitates out. It can then be separated from the sodium sulfate by dissolving the sulfate in water.

10.3 Balancing Equations

Methods of Balancing Equations

Getting the correct numbers of atoms and molecules on both sides of a chemical equation is called *balancing the equation*.

One fundamental rule governs the writing of a correct equation: *In ordinary chemical reactions, no atoms are created or destroyed.* As many atoms go into the reaction as come out of it. Therefore there must be just as many atoms on the left side of a chemical equation as on the right side. We specify *ordinary* chemical reactions here in order to rule out *nuclear* reactions, in which atoms are created, destroyed, or changed in form. A century ago this distinction would not have been necessary. In an ordinary chemical reaction, no atoms are changed from one kind to another. No elements are transmuted from one variety to another. Therefore the number of atoms of *each element* must be the same on the left side of a chemical equation as on the right side. An equation that has the same number of atoms of each element on each side is known as a **balanced equation**.

In this book we have already balanced equations. Consider the familiar equation describing the combustion of hydrogen. Would it be correct to write the equation in the following way?

$$H_2 + O_2 \longrightarrow H_2O$$

Clearly, this equation is incorrect, because there are two atoms of oxygen on the left and only one on the right. We equalize the number of oxygen atoms by putting two water molecules on the right side of the equation:

$$H_2 + O_2 \longrightarrow 2H_2O$$

Now the oxygen atoms are balanced, with two on each side of the equation. However, the hydrogen atoms are unbalanced, because there are two on the left and four on the right. This means that the number of hydrogens on the left must be doubled:

$$2H_2 + O_2 \longrightarrow 2H_2O$$

Of course, we could also get a balanced equation by writing

$$H_2 + \tfrac{1}{2}O_2 \longrightarrow H_2O$$

but a fractional molecule has little meaning for chemical reactions.

This method of balancing equations is basically a trial-and-error procedure, in which we start out with the simplest trial equation and then proceed to change the number of atoms or molecules of the various ingredients one at a time until a balance is found.

Before you can begin to balance an equation, you must know (1) the reactants and the products of the reaction and (2) the correct formulas for the compounds taking part in the reaction. In the course of balancing the equation you must not change the formulas for any of the compounds in the equation.

In the above equation, for example, you must know that the product of the reaction is water (H_2O) and not hydrogen peroxide (H_2O_2). Once these basic facts are known, balancing the equation is not difficult.

Some equations are already balanced when first written down and require no further manipulation. For example, in the reduction of copper oxide by hydrogen to form metallic copper and water, the trial equation is

$$CuO + H_2 \longrightarrow Cu + H_2O$$

Each side of the equation has one Cu, one O, and two Hs. The equation is thus balanced.

A more complicated case is the reaction of calcium sulfate and aluminum chloride to form aluminum sulfate and calcium chloride. We set up a trial equation, using our knowledge of oxidation numbers to write the formulas for the four compounds. The oxidation numbers of interest are $+2$ for Ca, $+3$ for Al, -2 for SO_4, and -1 for Cl.

The trial equation becomes

$$CaSO_4 + AlCl_3 \longrightarrow Al_2(SO_4)_3 + CaCl_2$$

This equation is obviously not balanced, because there are three chlorides on the left and two on the right. Also, the right side has three sulfates, whereas there is only one sulfate on the left. How can we get the chlorides to balance? We can't do it simply by doubling one of the molecules, the way we did in the water synthesis reaction. What we must do is to double the number of $AlCl_3$ molecules on the left, while multiplying the number of $CaCl_2$ molecules on the right by three. This action gives us six Cl atoms on each side of the equation. (We decided to do this because the number 6 can be divided by both 2 and 3. This procedure is common whenever there are two atoms of one component on one side of the equation and three atoms of the same component on the other side.) The resulting equation is

$$CaSO_4 + 2\,AlCl_3 \longrightarrow Al_2(SO_4)_3 + 3\,CaCl_2$$

The Cl and Al atoms are now balanced, but there are three sulfate groups on the right side and only one on the left. We then try having three calcium sulfate molecules on the left:

$$3\,CaSO_4 + 2\,AlCl_3 \longrightarrow Al_2(SO_4)_3 + 3\,CaCl_2$$

(*Note*: It would not be correct to write $Ca(SO_4)_3$ on the left, since Ca has an oxidation number of $+2$ and SO_4 has an oxidation number of -2.)

Now all the elements are balanced. Notice that we have not had to worry about the separate elements inside the sulfate group. Since that group does not change in the reaction, it can be treated as a whole unit for the sake of balancing the equation.

Consider the combustion of gasoline, one ingredient of which is the organic compound heptane (C_7H_{16}). Whenever an organic compound containing carbon and hydrogen burns in air, the carbon oxidizes to form CO_2 and the hydrogen forms H_2O. Thus the reaction is of the form

$$?C_7H_{16} + ?O_2 \longrightarrow ?CO_2 + ?H_2O$$

The question marks indicate unknown numbers of molecules. To balance the equation, we make a first trial with one molecule of heptane. Since there are seven carbon atoms in the heptane molecule, there must be seven CO_2 molecules on the right side of the equation to equalize the count of carbon atoms. In addition, eight H_2O molecules are required on the right to account for the sixteen hydrogen atoms in the C_7H_{16} on the left.

We now count the number of oxygen atoms on the right side of the equation. The seven CO_2 molecules contain fourteen oxygen atoms, and the eight H_2O molecules provide eight oxygen atoms. The total number of oxygen atoms on the right is $14 + 8 = 22$ oxygen atoms. Therefore, eleven O_2 molecules are needed on the left. The equation then becomes

$$C_7H_{16} + 11O_2 \longrightarrow 7CO_2 + 8H_2O$$

The equation is now balanced, proving that our initial guess of one heptane molecule was correct.

Some students find a *balance sheet approach* an easier and more systematic approach to balancing equations. We illustrate this method using the above equation for the oxidation of heptane. Begin by writing a trial equation and listing each element (or group) in the equation beneath the arrow.

$$C_7H_{16} + O_2 \longrightarrow CO_2 + H_2O$$
$$\begin{array}{c} C \\ H \\ O \end{array}$$

Next, list the number of atoms currently present on each side of the equation.

$$C_7H_{16} + O_2 \longrightarrow CO_2 + H_2O$$

7	C	1
16	H	2
2	O	3

The aim of this procedure is to make the numbers in the right-hand column and the numbers in the left-hand column match. First we see that seven Cs are needed on the right. But if there are seven Cs on the right, there will be $14 + 1 = 15$ Os. Enter these new numbers:

$$C_7H_{16} + O_2 \longrightarrow 7CO_2 + H_2O$$

7	C	~~1~~ 7
16	H	2
2	O	~~3~~ 15

Now we see that the Hs can be balanced by replacing the 2 with a 16 on the right. This requires eight H_2O molecules on the right side of the equation, raising the number of Os to $14 + 8 = 22$:

$$C_7H_{16} + O_2 \longrightarrow 7CO_2 + 8H_2O$$

7	C	~~1~~ 7
16	H	~~2~~ 16
2	O	~~3~~ ~~15~~ 22

The oxygen still remains to be balanced. This is done by replacing the 2 on the left with a 22. The twenty-two oxygen atoms require eleven O_2 molecules in the equation:

$$C_7H_{16} + 11O_2 \longrightarrow 7CO_2 + 8H_2O$$

7	C	~~1~~ 7
16	H	~~2~~ 16
22 ~~2~~	O	~~3~~ ~~15~~ 22

The "balance sheet" is now complete, because the numbers on the left match those on the right. This procedure gives the same balanced equation as obtained previously but in a more systematic manner.

Example 10.1

Balance the following equation:

$$?FeS_2 + ?O_2 \longrightarrow ?Fe_2O_3 + ?SO_2$$

Solution

$$?FeS_2 + ?O_2 \longrightarrow ?Fe_2O_3 + ?SO_2$$

1	Fe	2
2	S	1
2	O	5

$$2FeS_2 + ?O_2 \longrightarrow 1Fe_2O_3 + ?SO_2$$

2 ~~1~~	Fe	2
4 ~~2~~	S	1
2	O	5

$$2FeS_2 + ?O_2 \longrightarrow Fe_2O_3 + 4SO_2$$

2 ~~1~~	Fe	2
4 ~~2~~	S	~~1~~ 4
2	O	~~5~~ 11

$$2FeS_2 + 5.5O_2 \longrightarrow Fe_2O_3 + 4SO_2$$

2 ~~1~~	Fe	2
4 ~~2~~	S	~~1~~ 4
11 ~~2~~	O	~~5~~ 11

The equation is balanced as far as the number columns are concerned: the numbers on the right equal the numbers on the left. However, one more condition must be met: the coefficients in the equation itself must be whole numbers. To meet that condition, double every term in the equation:

$$4FeS_2 + 11O_2 \longrightarrow 2Fe_2O_3 + 8SO_2$$

The Meanings of the Balanced Equation

Let us review the several meanings of the balanced chemical equation. The equation

$$H_2 + I_2 \longrightarrow 2HI$$

can have any of the following interpretations:

1. One molecule of hydrogen reacts with one molecule of iodine to yield two molecules of hydrogen iodide.

2. Ten molecules of hydrogen reacts with 10 molecules of iodine to yield 20 molecules of hydrogen iodide.

3. N molecules of hydrogen reacts with N molecules of iodine to yield $2N$ molecules of HI.

4. 6.02×10^{23} molecules of hydrogen (one mole) reacts with 6.02×10^{23} molecules of iodine (one mole) to yield $2 \times 6.02 \times 10^{23}$ molecules (2 mol) of hydrogen iodide.

Glossary

acid anhydride A nonmetal oxide that reacts with water to form an acid. (*Anhydride* means, literally, without water. (*Hydro*, in this sense, is a root that means water, as in hydroplane, fire hydrant, etc.)

activity series A list of elements arranged in order of activity; also called the electromotive series. The first members in this series will replace the later members in their compounds.

balanced equation A symbolic statement of a chemical process in which there are the same number of atoms of each element on each side of the equation.

decomposition reactions Chemical reactions in which a single reactant (or compound) decomposes or separates into two or more products (AB → A + B).

displacement reactions Chemical reactions with two reactants, one a single material and the other a compound material. In this reaction the single material replaces (or displaces) one of the elements in the compound material (AB + C → BC + A).

double displacement reactions Chemical reactions in which two compounds exchange positive and negative partners (AB + CD → AD + CB); sometimes called *double decomposition* or *metathesis*.

endothermic reactions Chemical processes in which heat is absorbed from the surroundings.

exothermic reactions Chemical processes in which heat is given off to the surroundings.

fusion A reaction taking place in molten material, rather than in aqueous solution.

hydrolysis reaction The reaction of a compound with water, either forming a new compound or dissociating into ions.

magnetite An iron oxide ore strongly attracted by a magnet.

metastable molecule A molecule that decomposes easily, releasing energy in the process.

ozone An allotrope of oxygen (a form of the element with a different molecular structure) with the formula O_3.

phenomenological description A description of events observed to take place, without regard to underlying structure or theory.

precipitation A process in which an insoluble solid compound separates out from a solution, usually in the form of a powder.

qualitative chemistry That branch of chemistry which deals with the nature and properties of substances without making quantitative measurements.

quantitative chemistry That branch of chemistry which deals with measuring quantities of materials or quantities relating to the properties of materials.

reversible reactions Chemical reactions that can run in either direction—that is, the reactants can also be products and the products can be reactants. Usually the reaction reaches an equilibrium state in which there are as many molecules reacting in one direction as in the opposite direction within a short period of time.

synthesis reaction A reaction in which two elements or compounds combine to form a new compound: A + B → C.

Problems and Questions

10.1 The Meaning of a Chemical Equation

10.1.Q What makes quantitative chemistry different from qualitative chemistry?

10.2.Q What is a phenomenological description?

10.3.Q Give a phenomenological description of the burning of a piece of paper.

10.4.Q What is the difference between the dissolving of one substance in another and the reaction of one substance with another?

10.5.P Given that 18.25 g of HCl combines with 20.0 g of NaOH according to the reaction

$$HCl + NaOH \longrightarrow NaCl + H_2O$$

determine the following.
- a. How many moles of HCl and NaOH enter into the reaction?
- b. How many moles of products are formed?
- c. How many grams of NaCl and H_2O are formed?

10.6.P Given that 40.0 g of $KClO_3$ is heated so that it decomposes according to the equation

$$2KClO_3 \longrightarrow 2KCl + 3O_2$$

determine the following.
- a. How many moles of $KClO_3$ enter into the reaction?
- b. How many moles and how many grams of KCl will be formed; and how much oxygen will be formed?

10.7.Q What are the differences between the qualitative, quantitative, and molecular meanings of a chemical equation?

10.8.Q Read the following equation, giving three different interpretations:

$$Zn + 2HCl \longrightarrow ZnCl_2 + H_2$$

10.9.P Using the reaction of Question 10.8, answer the following:
- a. How many moles of hydrogen are formed for each mole of zinc?
- b. How many moles of HCl react with each mole of zinc?
- c. How many grams of H_2 are formed when 55 g of zinc reacts?

10.2 Four Types of Equations

10.10.Q List the four basic types of chemical reactions.

10.11.Q Give the general equation for a synthesis reaction.

10.12.Q Why would you expect a synthesis reaction to be exothermic?

10.13.Q Give an example of a synthesis reaction. Give the equation for the reaction. Does it have a name?

10.14.Q Give the general equation for a decomposition reaction.

10.15.Q Give two examples of decomposition reactions.

10.16.Q Why are decomposition reactions generally endothermic?

10.17.Q What kind of compound would be likely to have an exothermic decomposition reaction?

10.18.Q What is the general equation for a displacement reaction?

10.19.Q What kinds of chemical reactions are predicted by the activity series?

10.20.Q Which element in each of the following pairs is higher in the activity series for metals?
- a. sodium or calcium
- b. lead or gold

10.21.Q Which is higher in the activity series for nonmetals: chlorine or iodine?

10.22.Q Explain what factor decides which way the reaction goes when steam is passed over hot iron.

10.23.P Predict the products of the following displacement reactions:
- a. $HCl + Zn \rightarrow$
- b. $H_2SO_4 + Fe \rightarrow$
- c. $CuSO_4 + Fe \rightarrow$
- d. $K + H_2O \rightarrow$
- e. $NaI + Cl_2 \rightarrow$

10.24.P Predict the products of the following decompositions:
- a. $H_2O_2 \rightarrow$
- b. H_2O + electric current $\rightarrow$
- c. HgO + heat $\rightarrow$
- d. $KClO_3$ + heat $\rightarrow$
- e. $H_2CO_3 \rightarrow$

10.25.Q What is the general equation for a double displacement reaction?

10.26.Q What is a precipitate?

10.27.Q What is "hard water"?

10.28.Q Explain how precipitation causes a double displacement reaction to go in one direction or another.

10.29.P Complete the following double displacement reactions:
- **a.** $NaOH + HCl \rightarrow$
- **b.** $Ca(OH)_2 + H_2CO_3 \rightarrow$
- **c.** $KCl + AgNO_3 \rightarrow$
- **d.** $KOH + H_2SO_4 \rightarrow$
- **e.** $CaCO_3 + 2HCl \rightarrow$

10.3 Balancing Equations

10.30.P Balance the following equations:
- **a.** $?H_2SO_4 + ?NaOH \rightarrow ?Na_2SO_4 + ?H_2O$
- **b.** $?C_3H_8 + ?O_2 \rightarrow ?CO_2 + ?H_2O$
- **c.** $?Mg + ?HCl \rightarrow ?MgCl_2 + ?H_2$
- **d.** $?AgNO_3 + ?Cu \rightarrow ?Cu(NO_3)_2 + Ag$
- **e.** $?CO + ?Fe_3O_4 \rightarrow ?CO_2 + ?FeO$
- **f.** $?B_2H_6 + ?H_2O \rightarrow ?H_3BO_3 + ?H_2$

10.31.P In the following reactions, 3 mol of the first reactant is added to enough of the second reactant so that all of the reactants combine. Predict the number of moles of the first product listed on the right side of each equation:
- **a.** $?Na_2CO_3 + ?HCl \rightarrow ?NaCl + ?H_2O + ?CO_2$
- **b.** $?Ca + ?O_2 \rightarrow ?CaO$
- **c.** $?CuS + ?O_2 \rightarrow ?CuO + ?SO_2$
- **d.** $?O_2 + ?CH_3OH \rightarrow ?CO_2 + ?H_2O$
- **e.** $?C_6H_{12}O_6 \rightarrow ?C_2H_6O + ?CO_2$
- **f.** $?TiCl_3 + ?HCl \rightarrow ?TiCl_4 + H_2$
- **g.** $?NOCl \rightarrow ?Cl_2 + ?NO$
- **h.** $?C_8H_{18} + ?O_2 \rightarrow ?CO_2 + ?H_2O$
- **i.** $?NH_3 + ?O_2 \rightarrow ?H_2O + ?NO$
- **j.** $?H_2S + ?O_2 \rightarrow ?SO_2 + ?H_2O$

Self Test

1. Give a phenomenological description of the events that take place when a piece of iron is placed in a solution of copper sulfate.

2. Use the activity table to explain why hydrogen is released when zinc is placed in a hydrochloric acid solution.

3. Given the reaction

$$2NaOH + H_2SO_4 \longrightarrow Na_2SO_4 + 2H_2O$$

   determine the following.
   a. What kind of reaction is this?
   b. How many moles of sulfuric acid react with 1 mol of sodium hydroxide?
   c. How many grams of sulfuric acid are required to react with 24.6 g of sodium hydroxide?

4. a. Name the four general types of reactions studied in this chapter.
   b. Give an example of each type of reaction.

5. Why do we use an arrow instead of an equal sign in a chemical equation?

6. What reaction can be used as a test for the presence of carbon dioxide?

7. Explain how precipitation causes a reaction to go more in one direction than in another.

8. Balance each of the following equations, and state which of the four categories it belongs to.
   a. $?H_2 + ?CO \rightarrow ?CH_3OH$
   b. $?H_2 + ?N_2 \rightarrow ?NH_3$
   c. $?Fe_3O_4 + ?H_2 \rightarrow ?Fe + ?H_2O$
   d. $?B_2H_6 + ?H_2O \rightarrow ?H_3BO_3 + ?H_2$
   e. $?H_2SO_4 + ?NaOH \rightarrow ?Na_2SO_4 + ?H_2O$
   f. $?C_3H_8 + ?O_2 \rightarrow ?CO_2 + ?H_2O$
   g. $?Mg + ?HCl \rightarrow ?MgCl_2 + ?H_2$
   h. $?AgNO_3 + ?Cu \rightarrow ?Cu(NO_3)_2 + ?Ag$
   i. $?CO + ?Fe_3O_4 \rightarrow ?CO_2 + ?FeO$
   j. $?NaOH + ?Al \rightarrow ?H_2 + ?Na_3AlO_3$
   k. $?HCl + ?Na_2CO_3 \rightarrow ?CO_2 + ?H_2O + ?NaCl$
   l. $?CO_2 + ?H_2O \rightarrow ?C_6H_{12}O_6 + O_2$

11
Stoichiometry

Objectives After completing this chapter, you should be able to:
1. Calculate the mole ratios of the substances taking part in a given reaction.
2. Calculate the mass ratios of the substances taking part in a given reaction.
3. Calculate the mass of any substance taking part in a reaction, if the number of moles of one of the reactants or products is given.
4. Calculate the mass of any substance taking part in a reaction, if the mass of any one of the reactants or products is given.
5. Calculate the percentage composition of a substance if the masses, weights, or volumes of the ingredients are given.
6. Calculate the mass (or weight, or volume) of a substance if its percentage composition is given and the quantity of one of the ingredients is given.

11.1 Equations and Mole Ratios

Stoichiometry is the measurement or calculation of the quantities of the elements entering into a chemical reaction. The term was introduced by J. B. Richter in 1792; it comes from a Greek word meaning measuring elements. We have already been doing a considerable amount of stoichiometry without giving it a formal name.

The balanced equation is the key to chemical calculations. In Chapter 10 we showed that a chemical equation represents the number of atoms, molecules, or moles of the substances taking part in the reaction, so the same equation can be interpreted in a number of different ways. We now proceed further with that concept.

Let us return to the familiar equation for the combustion of hydrogen, writing it in several forms:

1. $2H_2 + O_2 \longrightarrow 2H_2O$
2. 2 molecules H_2 + 1 molecule $O_2 \longrightarrow$ 2 molecules H_2O
3. $2 \times 6 \times 10^{23}$ molecules H_2 + 6×10^{23} molecules $O_2 \longrightarrow$ $2 \times 6 \times 10^{23}$ molecules H_2O
4. 2 mol H_2 + 1 mol $O_2 \longrightarrow$ 2 mol H_2O

The equation is still valid if all the numbers on both sides are doubled:

$$4 \text{ mol } H_2 + 2 \text{ mol } O_2 \longrightarrow 4 \text{ mol } H_2O$$

The essential feature of the equation is that it always contains twice as many moles of hydrogen as of oxygen—an important piece of information. The ratio of the number of moles of one ingredient to the number of moles of another ingredient is called the **mole ratio**. The mole ratio of hydrogen to oxygen in the above equation is 2/1:

$$\frac{\text{no. of mol H}}{\text{no. of mol O}} = \frac{2}{1}$$

At the same time, the mole ratio of hydrogen to water is 1/1, meaning that the number of moles of water always equals the number of moles of hydrogen.

For most practical purposes, thinking about reactions of *moles* of atoms or molecules is more realistic than thinking of reactions in terms of single atoms or molecules. A single molecule is much too small for the chemist to examine in a test tube. Even if two molecules could be isolated in a test tube, the chance that their encounter would make an effective reaction is very slight indeed. It would be like the chance that two billiard balls would collide within a space as vast as the solar system. (Nevertheless, with special equipment, modern physicists and chemists can and do study collisions between isolated atoms and molecules.)

As another example of mole ratios, consider the decomposition of $KClO_3$ (potassium chlorate), a reaction that takes place when the dry substance is heated:

$$2KClO_3 \longrightarrow 2KCl + 3O_2$$

In terms of moles,

$$2 \text{ mol } KClO_3 \longrightarrow 2 \text{ mol } KCl + 3 \text{ mol } O_2$$

The mole ratio of $KClO_3$ to oxygen is

$$\text{mole ratio} = \frac{\text{no. of mol } KClO_3}{\text{no. of mol } O_2} = \frac{2}{3}$$

At the same time, the mole ratio of $KClO_3$ to KCl is 2/2, which is the same as 1/1, and the mole ratio of KCl to O_2 is 2/3.

Example 11.1

How many moles of oxygen are formed from the decomposition of 5.0 mol of $KClO_3$?

Solution

$$\text{mole ratio} = \frac{\text{no. of mol } O_2}{\text{no. of mol } KClO_3} = \frac{3 \text{ mol } O_2}{2 \text{ mol } KClO_3}$$

From the above equation we know that the mole ratio of oxygen to $KClO_3$ is 3/2; three moles of oxygen are formed from every two moles of potassium chlorate. Then we have

$$\text{no. of mol } O_2 = \frac{3 \text{ mol } O_2}{2 \text{ mol } KClO_3} \times \text{no. of mol } KClO_3$$

$$= \frac{3 \text{ mol } O_2}{2 \text{ mol } KClO_3} \times 5.0 \text{ mol } KClO_3$$

$$= 7.5 \text{ mol } O_2$$

11.2 Mass Problems

Mass Ratios

The proportions by mass of the reacting substances are easily found from the mole ratios for a given reaction. The first step is to determine the molar mass (the number of grams per mole) of each substance in the equation. Then, a **mass ratio** can be found using the mole ratio.

Consider two substances A and B in a reaction.

$$\text{mass of A} = \text{no. of mol A} \times \frac{\text{no. of g A}}{1 \text{ mol}}$$

$$= \text{no. of mol A} \times \text{molar mass of A}$$

Similarly,

$$\text{mass of B} = \text{no. of mol B} \times \text{molar mass of B}$$

Divide the equation for B by the equation for A to find the proportion of substance B to substance A:

$$\frac{\text{mass of B}}{\text{mass of A}} = \frac{\text{no. of mol B}}{\text{no. of mol A}} \times \frac{\text{no. of g B/mol}}{\text{no. of g A/mol}}$$

$$= \frac{\text{no. of mol B}}{\text{no. of mol A}} \times \frac{\text{molar mass of B}}{\text{molar mass of A}}$$

Consider, for example, the combustion of hydrogen:

$$2H_2 + O_2 \longrightarrow 2H_2O$$

What is the proportion of oxygen to hydrogen (by mass)? In Chapter 9 we found the molecular mass of hydrogen to be 2 (approximately), so the molar mass of H_2 is 2 g/mol. Similarly, the molar masses of O_2 and H_2O are 32 g/mol and 18 g/mol, respectively.

Using these figures, we can write the equation for the combustion of oxygen in the following equivalent ways:

$$2H_2 + O_2 \longrightarrow 2H_2O$$

or

$$2 \text{ mol } H_2 + 1 \text{ mol } O_2 \longrightarrow 2 \text{ mol } H_2O$$

So

$$2 \times 2 \text{ g } H_2 + 1 \times 32 \text{ g } O_2 \longrightarrow 2 \times 18 \text{ g } H_2O$$

or

$$4 \text{ g } H_2 + 32 \text{ g } O_2 \longrightarrow 36 \text{ g } H_2O$$

In this equation the mass ratio of oxygen to hydrogen is

$$\frac{\text{mass of oxygen}}{\text{mass of hydrogen}} = \frac{\text{no. of mol } O_2}{\text{no. of mol } H_2} \times \frac{\text{molar mass of } O_2}{\text{molar mass of } H_2}$$

$$= \frac{1 \text{ mol } O_2 \times 32 \text{ g/mol}}{2 \text{ mol } H_2 \times 2 \text{ g/mol}} = \frac{32 \text{ g } O_2}{4 \text{ g } H_2} = \frac{8}{1}$$

The calculation shows that in this reaction there are always eight times as many grams of oxygen as there are grams of hydrogen. The mass ratio of water to hydrogen is, on the other hand,

$$\frac{\text{mass of water}}{\text{mass of hydrogen}} = \frac{36 \text{ g } H_2O}{4 \text{ g } H_2} = \frac{9 \text{ g } H_2O}{1 \text{ g } H_2}$$

Example 11.2 In the production of oxygen by the decomposition of potassium chlorate, what is the mass ratio of oxygen to $KClO_3$? That is, how many grams of oxygen are formed from each gram of $KClO_3$?

Solution The reaction equation is

$$2KClO_3 \longrightarrow 2KCl + 3O_2$$

or

$$2 \text{ mol } KClO_3 \longrightarrow 2 \text{ mol } KCl + 3 \text{ mol } O_2$$

The molecular mass of $KClO_3$ is the sum of the atomic masses:

$$(1 \times K) + (1 \times Cl) + (3 \times O) = (1 \times 39.1) + (1 \times 35.5) + (3 \times 16.0)$$
$$= 123 \text{ (rounded off to 3 digits)}$$

Therefore the molar mass of $KClO_3$ is 123 g/mol. Similarly, the molar mass of O_2 is 32.0 g/mol. The ratio of O_2 to $KClO_3$ is

$$\frac{\text{mass of } O_2}{\text{mass of } KClO_3} = \frac{3 \text{ mol} \times 32.0 \text{ g/mol}}{2 \text{ mol} \times 123 \text{ g/mol}} = \frac{0.390 \text{ g}}{1.0 \text{ g}}$$

This result means that each gram of $KClO_3$ produces 0.390 g of O_2.
Note: Since we are dealing with ratios in the above problem, we can use any kind of units. A kilogram of $KClO_3$ yields 0.390 kg of O_2. A pound of $KClO_3$ yields 0.390 lb of O_2. The mass ratio is the same in any system of units.

Example 11.3 When sugar (sucrose) is heated, it first melts, turning into a brown caramel, and then gives off water vapor, leaving behind a black deposit of carbon. The equation for the decomposition reaction is

$$C_{12}H_{22}O_{11} \longrightarrow 12C + 11H_2O$$

(a) How many grams of carbon are formed for each gram of sugar decomposed?
(b) How many grams of water are formed for each gram of sugar decomposed?

Solution The molecular mass of sugar is

$$(12 \times 12.0) + (22 \times 1.00) + (11 \times 16.0) = 342$$

Therefore the molar mass of sugar is 342 g/mol.
(a) The proportion of carbon to sugar is

$$\frac{\text{mass of carbon}}{\text{mass of sugar}} = \frac{12 \text{ mol carbon} \times 12.0 \text{ g/mol}}{1 \text{ mol sugar} \times 342 \text{ g/mol}} = \frac{144 \text{ g carbon}}{342 \text{ g sugar}}$$

$$= \frac{0.421 \text{ g C}}{1 \text{ g sugar}}$$

(b) The proportion of water to sugar is

$$\frac{\text{mass of water}}{\text{mass of sugar}} = \frac{11 \text{ mol water} \times 18.0 \text{ g/mol}}{1 \text{ mol sugar} \times 342 \text{ g/mol}} = \frac{198 \text{ g water}}{342 \text{ g sugar}}$$

$$= \frac{0.579 \text{ g water}}{1 \text{ g sugar}}$$

Note: Adding up the masses of the products (right side of the equation) always gives the sum of the masses of the reactants (left side). Performing this addition is a good way to check your arithmetic. In this case,

$$0.421 \text{ g} + 0.579 \text{ g} = 1.000 \text{ g}$$

Thus the total mass of carbon and water produced from the decomposition of sugar equals the initial 1.000 g of sugar, as the law of conservation of mass requires.

Mole-Mass Problems

In most problems we are interested in calculating the masses of the compounds taking part in the reaction. A simple way of doing problems involving quantities of reactants is to break each problem into two steps. First calculate the number of moles of the compounds of interest, and then calculate the number of grams. A **mole-mass problem** is one in which the number of moles of one of the compounds is given and we are required to find the number of grams of the other compounds.

Consider the reaction in which potassium reacts with bromine:

$$2K + Br_2 \longrightarrow 2KBr$$

In terms of moles, the equation says:

$$2 \text{ mol K} + 1 \text{ mol Br}_2 \longrightarrow 2 \text{ mol KBr}$$

Suppose you are given 6 mol of potassium (K) and are asked the following questions:

1. How many grams of bromine (Br_2) are needed to complete the reaction?
2. How many grams of potassium bromide (KBr) are formed?
3. How many grams of potassium are used?

First, set up the mole ratio of Br_2 to K:

$$\frac{\text{no. of mol Br}_2}{\text{no. of mol K}} = \frac{1 \text{ mol Br}_2}{2 \text{ mol K}}$$

Then multiply both sides of the equation by no. of mol K to solve for no. of mol Br_2:

$$\text{no. of mol } Br_2 = \frac{1 \text{ mol } Br_2}{2 \text{ mol K}} \times \text{no. of mol K}$$

$$= \frac{1 \text{ mol } Br_2}{2 \text{ mol K}} \times 6 \text{ mol K} = 3 \text{ mol } Br_2$$

The number of moles of KBr is found similarly:

$$\frac{\text{no. of mol KBr}}{\text{no. of mol K}} = \frac{2 \text{ mol KBr}}{2 \text{ mol K}} = \frac{1 \text{ mol KBr}}{1 \text{ mol K}}$$

So

$$\text{no. of mol KBr} = \frac{1 \text{ mol KBr}}{1 \text{ mol K}} \times \text{no. of mol K}$$

$$= \frac{1 \text{ mol KBr}}{1 \text{ mol K}} \times 6 \text{ mol K} = 6 \text{ mol KBr}$$

We have found that if we start with 6 mol of K, 3 mol of Br_2 will produce 6 mol of KBr.

Now find the molar masses of Br_2 and KBr:

$$\text{molar mass of } Br_2 = 2 \times 79.9 \text{ g} = 159.8 \text{ g}$$

$$\text{molar mass of KBr} = 39.09 \text{ g} + 79.90 \text{ g} = 118.99 \text{ g}$$

To find the mass of compound A, we use the relationship

$$\text{mass of A} = \text{no. of mol A} \times \text{molar mass of A}$$

or

$$\text{no. of g A} = \text{no. of mol A} \times \frac{\text{no. of g A}}{1 \text{ mol A}}$$

Thus, the mass of bromine required is

$$\text{no. of g } Br_2 = \text{no. of mol } Br_2 \times \frac{\text{no. of g } Br_2}{1 \text{ mol } Br_2}$$

$$= 3 \text{ mol} \times 159.8 \frac{\text{g}}{\text{mol}} = 479.4 \text{ g}$$

Similarly, the number of grams of potassium bromide produced is

$$\text{no. of g KBr} = \text{no. of mol KBr} \times \frac{\text{no. of g KBr}}{1 \text{ mol KBr}}$$

$$= 6 \text{ mol KBr} \times 118.99 \frac{\text{g}}{\text{mol}} = 713.94 \text{ g KBr}$$

The amount of potassium used in the reaction is easily found. Since we started with 6 mol, the mass of potassium is

$$\text{no. of g K} = \text{no. of mol K} \times \frac{\text{no. of g K}}{1 \text{ mol K}}$$

$$= 6 \text{ mol K} \times 39.09 \frac{\text{g}}{\text{mol}} = 234.5 \text{ g K}$$

We check the result by making sure that the law of conservation of mass has not been violated. We should find that

$$\text{mass of K} + \text{mass of Br} = \text{mass of KBr}$$

$$234.5 \text{ g} + 479.4 \text{ g} = 713.9 \text{ g}$$

We see that the mass of KBr just obtained is the same as that calculated from the mole ratio, verifying the correctness of all the arithmetic.

Example 11.4 How many grams of oxygen are produced by the decomposition of 2.50 mol of potassium chlorate?

Solution The reaction is the same as that in Example 11.2:

$$2KClO_3 \longrightarrow 2KCl + 3O_2$$

Set up the mole ratio:

$$\frac{\text{no. of mol } O_2}{\text{no. of mol } KClO_3} = \frac{3 \text{ mol } O_2}{2 \text{ mol } KClO_3}$$

Solve for the number of moles of O_2:

$$\text{no. of mol } O_2 = \frac{3 \text{ mol } O_2}{2 \text{ mol } KClO_3} \times \text{no. mol } KClO_3$$

Substitute 2.50 mol for the number of moles of $KClO_3$:

$$\text{no. of mol } O_2 = \frac{3 \text{ mol } O_2}{2 \text{ mol } KClO_3} \times 2.50 \text{ mol } KClO_3$$

$$= 3.75 \text{ mol } O_2$$

Find the mass of O_2:

$$\text{no. of g } O_2 = \text{no. of mol } O_2 \times \frac{\text{no. of g } O_2}{1 \text{ mol } O_2}$$

$$= 3.75 \text{ mol } O_2 \times \frac{32.0 \text{ g } O_2}{1 \text{ mol } O_2}$$

$$= 120 \text{ g } O_2$$

Mass-Mass Problems

The most common problems are those in which the mass of one reactant or product is given and we aim to find the mass of another reactant or product. Such a problem is called a **mass-mass problem**.

This type of problem is illustrated by a reaction used in the smelting of iron ore: the reduction of iron oxide by carbon to obtain metallic iron. In this reaction iron ore is mixed with carbon and the mixture is heated to a very high temperature. The carbon pulls the oxygen from the iron oxide, forming carbon monoxide and allowing the molten iron to be poured off:

$$Fe_2O_3 + 3C \longrightarrow 2Fe + 3CO$$

In terms of moles,

$$1 \text{ mol } Fe_2O_3 + 3 \text{ mol C} \longrightarrow 2 \text{ mol Fe} + 3 \text{ mol CO}$$

Let us start with 100 kg of iron oxide and ask the following questions:

1. How many kilograms of carbon are needed for the iron oxide to react completely?

2. How many kilograms of iron are obtained by the end of the process?

We will present two methods of solving these problems. Both methods are equivalent, but each emphasizes a different feature of the logic.

Using Mass Ratios

Earlier in this section you learned to calculate the proportion (or ratio) of substance B to substance A. Once this proportion is known, it is a simple matter to calculate the quantity of substance B corresponding to a given quantity of A:

$$\frac{\text{mass of B (unknown)}}{\text{mass of A (given)}} = \frac{\text{mass of B}}{\text{mass of A}} \text{ (calculated ratio)}$$

Solving for the mass of B (unknown), we have

$$\text{mass of B (unknown)} = \frac{\text{mass of B}}{\text{mass of A}} \times \text{mass of A (given)}$$

Notice that we can use any units we like for the given mass, since the ratio has no units. Therefore the unknown mass always has the same units as the given mass.

The first step in solving any problem of this sort is to set up a table listing the molar masses of all the compounds involved.

Substance	Molecular Mass	Molar Mass
Fe	55.85	55.85 g/mol
C	12.00	12.00 g/mol
Fe_2O_3	$2 \times 55.85 + 3 \times 16.00 = 159.7$	159.7 g/mol

Next, find the proportion of carbon to iron oxide:

$$\frac{\text{mass of C}}{\text{mass of Fe}_2\text{O}_3} = \frac{\text{no. of mol C}}{\text{no. of mol Fe}_2\text{O}_3} \times \frac{\text{molar mass of C}}{\text{molar mass of Fe}_2\text{O}_3}$$

$$= \frac{3.00 \text{ mol C}}{1.00 \text{ mol Fe}_2\text{O}_3} \times \frac{12.0 \text{ g/mol}}{159.7 \text{ g/mol}}$$

$$= \frac{0.225 \text{ g C}}{1.00 \text{ g Fe}_2\text{O}_3}$$

Then the mass of carbon wanted is found by multiplying both sides of the equation by the given mass of Fe_2O_3:

$$\text{mass of C} = \frac{0.225 \text{ g C}}{1.00 \text{ g Fe}_2\text{O}_3} \times \text{mass of Fe}_2\text{O}_3$$

$$= \frac{0.225 \text{ g C}}{1.00 \text{ g Fe}_2\text{O}_3} \times 100 \text{ kg Fe}_2\text{O}_3$$

$$= 22.5 \text{ kg C}$$

The mass of iron produced is found in the same way. First find the proportion of iron to iron oxide:

$$\frac{\text{mass of Fe}}{\text{mass of Fe}_2\text{O}_3} = \frac{\text{no. of mol Fe}}{\text{no. of mol Fe}_2\text{O}_3} \times \frac{\text{molar mass of Fe}}{\text{molar mass of Fe}_2\text{O}_3}$$

$$= \frac{2.00 \text{ mol Fe}}{1.00 \text{ mol Fe}_2\text{O}_3} \times \frac{55.85 \text{ g/mol}}{159.7 \text{ g/mol}}$$

$$= \frac{0.699 \text{ g Fe}}{1.00 \text{ g Fe}_2\text{O}_3}$$

Then multiply both sides of the equation by the given mass of Fe_2O_3:

$$\text{mass of Fe} = \frac{0.699 \text{ g Fe}}{1.00 \text{ g Fe}_2\text{O}_3} \times \text{mass of Fe}_2\text{O}_3$$

$$= \frac{0.699 \text{ g Fe}}{1.00 \text{ g Fe}_2\text{O}_3} \times 100 \text{ kg Fe}_2\text{O}_3$$

$$= 69.9 \text{ kg Fe}$$

The calculation shows that 22.5 kg of carbon combines with 100 kg of Fe_2O_3 to yield 69.9 kg of metallic iron.

The Mole-Ratio Method An alternative method of doing the same kind of problem is to first convert the problem into a mole-mass problem and then proceed to the solution. The steps are as follows.

Step 1: Determine the mole ratio of the unknown (B) and given (A) substances from the balanced equation:

$$\frac{\text{no. of mol B}}{\text{no. of mol A}}$$

Step 2: Find the molar masses of the reactants and products of interest, using the table of atomic masses.

Step 3: Find the number of moles of the given substance (A). We know that

$$\text{no. of g A} = \text{no. of mol A} \times \frac{\text{no. of g A}}{1 \text{ mol A}}$$

$$= \text{no. of mol A} \times \text{molar mass of A}$$

Therefore,

$$\text{no. of mol A} = \frac{\text{no. of g A}}{\text{molar mass of A}}$$

Step 4: Find the unknown number of moles by multiplying the mole ratio by the given number of moles:

$$\text{no. of mol B} = \underbrace{\frac{\text{no. of mol B}}{\text{no. of mol A}}}_{\text{ratio from Step 1}} \times \text{no. of mol A}$$

Step 5: Find the mass of B (unknown) by multiplying the number of moles of B by the molar mass (mass per mole):

$$\text{no. of g B} = \text{no. of mol B} \times \frac{\text{no. of g B}}{1 \text{ mol B}}$$

This sequence seems complicated at first, but see how easily it is applied to the problem solved above by the mass-ratio method. The equation for the reduction of iron ore is

$$Fe_2O_3 + 3C \longrightarrow 2Fe + 3CO$$

Step 1: The mole ratios of interest are

$$\frac{\text{no. of mol C}}{\text{no. of mol Fe}_2O_3} = \frac{3 \text{ mol C}}{1 \text{ mol Fe}_2O_3}$$

$$\frac{\text{no. of mol Fe}}{\text{no. of mol Fe}_2O_3} = \frac{2 \text{ mol Fe}}{1 \text{ mol Fe}_2O_3}$$

Step 2: The molar masses of the compounds of interest are again,

Substance	Molecular Mass	Molar Mass
Fe	55.85	55.85 g/mol
C	12.00	12.00 g/mol
Fe_2O_3	$2 \times 55.85 + 3 \times 16.00 = 159.7$	159.7 g/mol

Step 3:

$$\text{no. of mol Fe}_2\text{O}_3 = \frac{\text{no. of g Fe}_2\text{O}_3}{\text{molar mass of Fe}_2\text{O}_3}$$

$$= \frac{100{,}000 \text{ g Fe}_2\text{O}_3}{159.7 \frac{\text{g}}{\text{mol}}}$$

$$= 626 \text{ mol Fe}_2\text{O}_3$$

Note: 100 kg = 100,000 g to three significant figures.

Step 4:

$$\text{no. of mol C} = \frac{3 \text{ mol C}}{1 \text{ mol Fe}_2\text{O}_3} \times 626 \text{ mol Fe}_2\text{O}_3$$

$$= 1880 \text{ mol C}$$

$$\text{no. of mol Fe} = \frac{2 \text{ mol Fe}}{1 \text{ mol Fe}_2\text{O}_3} \times 626 \text{ mol Fe}_2\text{O}_3$$

$$= 1250 \text{ mol Fe}$$

Step 5:

$$\text{mass of C} = 1880 \text{ mol} \times 12.0 \frac{\text{g}}{\text{mol}} = 22{,}600 \text{ g} = 22.6 \text{ kg}$$

$$\text{mass of Fe} = 1250 \text{ mol} \times 55.85 \frac{\text{g}}{\text{mol}} = 69{,}800 \text{ g} = 69.8 \text{ kg}$$

You see that, aside from small variations due to rounding-off errors, the two methods give the same results. Method 1, the mass-ratio method, is more direct. First calculate the mass of unknown substance obtained from one gram of given substance, and then multiply this number by the mass of given substance. Method 2, the mole-ratio method, breaks the procedure down into more detailed steps. First find the mole ratio of the unknown substance to the given substance, and then find the number of moles of the given substance. Next determine the number of moles of the unknown substance, and finally calculate the number of grams of the unknown substance.

In all problems of this nature, it is essential to keep track of the units, for forcing the units to cancel so as to give the desired result is the safest way to ensure that the equations are written correctly. It is not necessary to memorize the equations, because after you become familiar with the method the logic of the problem will tell you what equations to use.

Example 11.5   When a solution of sodium hydroxide reacts with a solution of ferric chloride, $FeCl_3$, a precipitate of ferric hydroxide, $Fe(OH)_3$, is formed:

$$3\,NaOH + FeCl_3 \longrightarrow 3\,NaCl + Fe(OH)_3$$

The $Fe(OH)_3$ may be filtered out of the solution, dried, and weighed. In one such procedure, 24.6 g of $Fe(OH)_3$ was obtained. How many grams of $FeCl_3$ were in the original solution? How many grams of NaOH were required to precipitate all of the $FeCl_3$? (*Note:* This is an example of measuring the quantity of a compound in a solution by forming and weighing a precipitate.)

Solution   Use the mole-ratio method.

Step 1: The mole ratios of the compounds of interest are

$$\frac{\text{no. of mol } FeCl_3}{\text{no. of mol } Fe(OH)_3} = \frac{1 \text{ mol } FeCl_3}{1 \text{ mol } Fe(OH)_3}$$

$$\frac{\text{no. of mol NaOH}}{\text{no. of mol } Fe(OH)_3} = \frac{3 \text{ mol NaOH}}{1 \text{ mol } Fe(OH)_3}$$

Step 2: Find the molar masses of the ingredients of interest.

$Fe(OH)_3$ $55.8 \text{ g} + 3 \times (16.0 \text{ g} + 1.0 \text{ g}) = 106.8 \text{ g}$

$FeCl_3$ $55.8 \text{ g} + 3 \times 35.5 \text{ g} = 162.3 \text{ g}$

NaOH $23.0 \text{ g} + 16.0 \text{ g} + 1.0 \text{ g} = 40.0 \text{ g}$

Step 3:

$$\text{no. of mol } Fe(OH)_3 = \frac{24.6 \text{ g}}{106.8 \frac{\text{g}}{\text{mol}}} = 0.230 \text{ mol } Fe(OH)_3$$

Step 4:

$$\text{no. of mol } FeCl_3 = \text{no. of mol } Fe(OH)_3 = 0.230 \text{ mol } FeCl_3$$

$$\text{no. of mol NaOH} = \frac{3 \text{ mol NaOH}}{1 \text{ mol } Fe(OH)_3} \times \text{no. of mol } Fe(OH)_3$$

$$= 3 \times 0.230 \text{ mol } Fe(OH)_3 = 0.690 \text{ mol NaOH}$$

Step 5:

$$\text{mass of } FeCl_3 = \text{no. of mol } FeCl_3 \times \text{molar mass}$$

$$= 0.230 \text{ mol} \times 162.3 \frac{\text{g}}{\text{mol}} = 37.3 \text{ g}$$

$$\text{mass of NaOH} = \text{no. of mol NaOH} \times \text{molar mass}$$

$$= 0.690 \text{ mol} \times 40.0 \frac{\text{g}}{\text{mol}} = 27.6 \text{ g}$$

Note: In this problem the given compound is one of the products on the right side of the equation. As you can see, it is as easy to calculate the amount of reactant on the left side as it is to calculate the amount of product on the right. Given the amount of any one of the compounds taking part in a reaction, you can calculate all the other quantities from the known mole ratios of all the compounds.

Example 11.6 Hydrogen combines with bromine to form hydrogen bromide:

$$H_2 + Br_2 \longrightarrow 2HBr$$

If 6.0 g of H_2 is used, (a) how many grams of Br_2 are needed to complete the reaction, and (b) how many grams of HBr are formed?

Solution Use the mass-ratio method.

Step 1: The molar masses of the ingredients are

H_2 $2 \times 1.0 = 2.0$ g/mol

Br_2 $2 \times 80 = 160$ g/mol

HBr $1.0 + 80 = 81$ g/mol

Note: The mass of hydrogen used in the reaction is given to two significant figures. Therefore the molar masses need be calculated only to two significant figures.

Step 2: The mass ratios are

$$\frac{\text{no. of g Br}_2}{\text{no. of g H}_2} = \frac{\text{no. of mol Br}_2}{\text{no. of mol H}_2} \times \frac{\text{molar mass of Br}_2}{\text{molar mass of H}_2}$$

$$= \frac{1 \text{ mol Br}_2}{1 \text{ mol H}_2} \times \frac{160 \text{ g/mol}}{2.0 \text{ g/mol}}$$

$$= \frac{80 \text{ g Br}_2}{1 \text{ g H}_2}$$

$$\frac{\text{no. of g HBr}}{\text{no. of g H}_2} = \frac{\text{no. of mol HBr}}{\text{no. of mol H}_2} \times \frac{\text{molar mass of HBr}}{\text{molar mass of H}_2}$$

$$= \frac{2 \text{ mol HBr}}{1 \text{ mol H}_2} \times \frac{81 \text{ g/mol}}{2.0 \text{ g/mol}}$$

$$= \frac{81 \text{ g HBr}}{1 \text{ g H}_2}$$

Step 3: The unknown masses are

$$\text{mass of Br}_2 = \frac{\text{no. of g Br}_2}{\text{no. of g H}_2} \times \text{no. of g H}_2 \text{ (given)}$$

$$= \frac{80 \text{ g Br}_2}{1 \text{ g H}_2} \times 6.0 \text{ g H}_2$$

$$= 480 \text{ g Br}_2$$

$$\text{mass of HBr} = \frac{\text{no. of g HBr}}{\text{no. of g H}_2} \times \text{no. of g H}_2 \text{ (given)}$$

$$= \frac{81 \text{ g HBr}}{1 \text{ g H}_2} \times 6.0 \text{ g H}_2$$

$$= 490 \text{ g HBr (rounded off to two digits)}$$

11.3 Percent Composition

A common problem is that of finding the **percent composition** of a group of things. Consider a classroom containing 100 students, of which 62 are male and 38 are female. We say the class consists of 62 percent (62%) male and 38 percent (38%) female students. (The term comes from the Latin *per centum*, meaning "by the hundred.")

How do we calculate percentages if the class has only 80 students—35 male and 45 female? We first find the *fraction* of students who are male (or female) and then multiply by 100.

$$\text{fraction of male students} = \frac{35}{80} = 0.44$$

Multiplying by 100 gives us the percentage:

$$0.44 \times 100 = 44\%$$

What this means is that if there were 100 students, 44 of them would be male. Similarly,

$$\text{fraction of female students} = \frac{45}{80} = 0.56$$

Multiplying by 100 gives 56%. Naturally, 44% + 56% = 100%, since the total number of students is 100% of the whole.

The rule, then, for finding the percentage of anything is as follows: Divide the part of interest by the whole quantity to get the fraction of the whole:

$$\text{fraction of whole} = \frac{\text{part}}{\text{whole}}$$

The percentage is the fraction of the whole multiplied by 100:

$$\text{percentage} = \frac{\text{part}}{\text{whole}} \times 100$$

Let us now apply the concept of percentage to chemistry. In chemistry a very common problem is to find the percentage composition of a mixture or compound. Often it is the percentage *by mass* (or by weight) that is wanted. However, many times (especially when dealing with gases or liquids) we are concerned with the percentage composition *by volume*.

A bottle of mouthwash, for example, is labeled 22% alcohol (by volume). This means that 22 cm³ of every 100 cm³ is alcohol and the rest is water and other ingredients. (This mouthwash has a higher alcohol content than most wines.)

Example 11.7   A bottle contains 32.5 cm³ of alcohol dissolved in enough water to make 157.5 cm³ of solution.
(a) What is the percentage of alcohol in the solution (by volume)?
(b) How much water must be added to dilute the existing solution into a 10% solution?

Solution   (a) The percentage of alcohol is given by

$$\frac{\text{part}}{\text{whole}} \times 100 = \frac{32.5 \text{ cm}^3}{157.5 \text{ cm}^3} \times 100 = 20.6\%$$

Note: Since 32.5 cm³ has three significant figures, the answer is rounded off to three significant figures (20.6%).

(b) Instead of solving directly for the volume of water to be added, first solve for the total amount of solution that would be present if the concentration of alcohol were 10%. The amount of alcohol is still the original 32.5 cm³. Then the equation is

$$\% \text{ solution} = \frac{\text{part}}{\text{whole}} \times 100$$

or

$$10.0\% = \frac{32.5 \text{ cm}^3}{\text{whole}} \times 100$$

where "whole" is now the total volume of solution.
Multiply both sides of the equation by "whole":

$$10.0 \times \text{whole} = 32.5 \text{ cm}^3 \times 100$$
$$= 3250 \text{ cm}^3$$

Divide both sides of the equation by 10.0:

$$\text{whole} = 325 \text{ cm}^3$$

This is the total volume. To find the amount of water that must be added, subtract the original volume:

$$\text{volume added} = \text{total volume} - \text{original volume}$$
$$= 325 \text{ cm}^3 - 158 \text{ cm}^3 = 167 \text{ cm}^3$$

Note: This problem could be solved directly for the volume of added water, but the solution is more complicated than the one given above. This example illustrates the fact that the problem-solving strategy may determine the difficulty of the solution. Choosing unknowns carefully can sometimes make a great difference.

Example 11.8 Brass is an alloy of copper and zinc. The combination of red-orange copper and white zinc produces the familiar yellow metal used for bed frames, belt buckles, and a variety of hardware items. There are many types of brass, and the percentage of zinc (by weight) may vary from 15% to 35%. Given a brass containing 31.5% zinc, determine the following.
(a) What is the percentage of copper in the alloy?
(b) How many grams of copper are in 819 g of brass?

Solution (a) In a mixture of two substances A and B, the percentage compositions of the two components must add up to exactly 100%:

$$\%(A) + \%(B) = 100\%$$

Therefore,

$$\%(B) = 100\% - \%(A)$$

For the present problem,

$$\%(\text{copper}) = 100\% - \%(\text{zinc})$$
$$= 100.0\% - 31.5\% = 68.5\%$$

(b) From the definition of percentage,

$$\%(\text{copper}) = \frac{\text{part}}{\text{whole}} \times 100\%$$

$$= \frac{\text{mass of copper}}{\text{mass of brass}} \times 100\% = 68.5\%$$

Therefore,

$$\text{mass of copper} \times 100\% = \text{mass of brass} \times 68.5\%$$

To find the mass of copper, divide both sides of the equation by 100%:

$$\text{mass of copper} = \text{mass of brass} \times 0.685$$

Finally,

$$\text{mass of copper} = 819 \text{ g} \times 0.685 = 561 \text{ g}$$

Example 11.9 How much brass having a composition of 28.0% zinc can be made with 15.5 kg of zinc?

Solution The definition of percentage is

$$\%(\text{zinc}) = \frac{\text{part}}{\text{whole}} \times 100\%$$

Therefore,

$$\%(\text{zinc}) = \frac{\text{mass of zinc}}{\text{mass of brass}} \times 100\% = 28.0\%$$

We want to find the mass of brass, so we multiply both sides of the equation by the mass of brass and then divide both sides by 28.0%:

$$\text{mass of zinc} \times \frac{100\%}{28.0\%} = \text{mass of brass}$$

Transpose:

$$\text{mass of brass} = \text{mass of zinc} \times \frac{100\%}{28.0\%}$$

$$= 15.5 \text{ kg} \times 3.57 = 55.4 \text{ kg}$$

The solution shows that for every kilogram of zinc we end up with 3.57 kg of brass.

Example 11.10 What is the percentage (by mass) of
(a) hydrogen and
(b) oxygen in a sample of water?

Solution

(a) In 1.0 mol of H_2O are 2.0 g of hydrogen and 16.0 g of oxygen. The percentage of hydrogen is given by

$$\%(H) = \frac{\text{part}}{\text{whole}} \times 100$$

$$= \frac{2.0 \text{ g}}{(2.0 + 16.0) \text{ g}} \times 100$$

$$= \frac{2.0}{18.0} \times 100 = 11.1\%$$

(b) The percentage of oxygen can be found in two ways. Using the method of part (a) above, we have

$$\%(O) = \frac{16.0 \text{ g}}{18.0 \text{ g}} \times 100$$

$$= 88.9\%$$

Or, more simply, since

$$\%(H) + \%(O) = 100\%$$

we know that

$$\%(O) = 100\% - 11.1\% = 88.9\%$$

Glossary

mass-mass problem A problem in which the mass of one of the compounds taking part in a reaction is given and one is required to find the mass of the other compounds.

mass ratio The ratio of the number of grams of one substance to the number of grams of another substance taking part in a reaction.

mole-mass problem A problem in which the number of moles of one of the compounds taking part in a reaction is given and one is required to find the mass of the other compounds.

mole ratio The ratio of the number of moles of one substance to the number of moles of another substance taking part in a reaction.

percent composition The ratio of the mass (or weight, or volume) of one ingredient in a sample to the mass (or weight, or volume) of the whole, multiplied by 100.

stoichiometry Calculating or measuring the quantitative relationships between the reactants and products in a chemical reaction.

Problems and Questions

11.1 Equations and Mole Ratios

11.1.Q What is a mole ratio?

11.2.P **a.** What is the mole ratio of HgO to O_2 in the reaction

$$2HgO \longrightarrow 2Hg + O_2$$

b. What is the mole ratio of NaOH to $Fe(OH)_3$ in the reaction

$$3NaOH + FeCl_3 \longrightarrow 3NaCl + Fe(OH)_3 \downarrow$$

c. What is the mole ratio of NaBr to $CaBr_2$ in the reaction

$$2Na_3PO_4 + 3CaBr_2 \longrightarrow Ca_3(PO_4)_2 + 6NaBr$$

11.3.P Balance the following equations, and calculate the number of moles of product compounds formed from the amount of reactant indicated.

a. $?Al + ?O_2 \longrightarrow ?Al_2O_3$
   1.5 mol

b. $?Zn + ?HCl \longrightarrow ?ZnCl_2 + ?H_2$
   0.3 mol

c. $?KClO_3 \longrightarrow ?KCl + ?O_2$
   1.5 mol

11.4.P Balance the following equations, and calculate how many moles of each reactant are required to form 2.5 mol of H_2O in each reaction.

a. $?CuO + ?H_2 \longrightarrow ?Cu + ?H_2O$
b. $?H_3PO_4 + ?Ca(OH)_2 \longrightarrow ?Ca_3(PO_4)_2 + ?H_2O$
c. $?CH_4 + ?O_2 \longrightarrow ?CO_2 + ?H_2O$

11.5.P Balance the following equations, and calculate how many moles of each product are formed from 1.35 mol of the first reactant.

a. $?H_2SO_4 + ?NaOH \longrightarrow ?Na_2SO_4 + ?H_2O$
b. $?H_2O + ?Na \longrightarrow ?NaOH + ?H_2$
c. $?NH_3 + ?HCl \longrightarrow ?NH_4Cl$

11.6.P Balance the following equations, and calculate how many moles of the second reactant are required to use up 0.5 mol of the first reactant.

a. $?H_2 + ?N_2 \longrightarrow ?NH_3$
b. $?CuSO_4 + ?H_2O \longrightarrow ?CuSO_4 \cdot 5H_2O$
c. $?Pb(OH)_2 + ?HNO_3 \longrightarrow ?Pb(NO_3)_2 + ?H_2O$

11.2 Mass Problems

11.7.P Oxygen is made by decomposing 2.54 g of potassium chlorate, according to the equation

$$2KClO_3 \longrightarrow 2KCl + 3O_2$$

a. How many grams of KCl are formed?
b. How many grams of O_2 are formed?

11.8.P Given that 12.5 g of O_2 is formed in the reaction

$$2PbO_2 \longrightarrow 2PbO + O_2$$

determine the following.
a. How many grams of PbO_2 are used?
b. How many grams of PbO are formed?

11.9.P Given that 25.0 g of MgO reacts with water to form the base magnesium hydroxide, according to the equation

$$MgO + H_2O \longrightarrow Mg(OH)_2$$

determine the following.
a. How many grams of $Mg(OH)_2$ are formed?
b. How many grams of H_2O are needed to complete the reaction?

11.10.P Given that 55.0 kg of H_2SO_4 is formed in the reaction

$$H_2O + SO_3 \longrightarrow H_2SO_4$$

determine the following. (This reaction is the final step in the manufacture of sulfuric acid.)
a. How many kilograms of H_2O are needed?
b. How many kilograms of SO_3 are needed?

11.11.P Epsom salt, $MgSO_4 \cdot 7H_2O$, decomposes when heated, yielding magnesium sulfate and liberating its water of crystallization, according to the equation

$$MgSO_4 \cdot 7H_2O \longrightarrow MgSO_4 + 7H_2O$$

a. How many grams of magnesium sulfate ($MgSO_4$) are formed from 525 g of Epsom salt?
b. How many moles of water result when 736 g of $MgSO_4$ is produced by this reaction?

11.12.P When sodium bicarbonate (baking soda) is heated, it decomposes, forming sodium carbonate (washing soda) and releasing water vapor and gaseous carbon dioxide, according to the equation

$$2NaHCO_3 \longrightarrow Na_2CO_3 + H_2O + CO_2$$

a. If 250 kg of $NaHCO_3$ is decomposed,
  (1) How many moles of CO_2 are produced?
  (2) How many grams of Na_2CO_3 are formed?
b. If 3.45 mol of H_2O is produced,
  (1) How many grams of $NaHCO_3$ were required?
  (2) How many grams of Na_2CO_3 are produced?

11.13.P Zinc replaces the silver from a solution of silver nitrate, so the silver precipitates out of the solution. The equation is

$$Zn + 2AgNO_3 \longrightarrow Zn(NO_3)_2 + 2Ag$$

a. If 25.0 g of Zn is used, how many moles of $AgNO_3$ are required?
b. If 25.0 g of Zn is used, how many grams of Ag are produced?
c. Consider the above reaction as a possible way of recovering silver from old photographic film. The silver on the film is first dissolved in nitric acid and then precipitated by the addition of powdered zinc. However, suppose Zn is worth $4.70/g and silver sells for $3.70/g. Would it pay to use this reaction?

11.14.P Calcium carbonate (limestone) is heated to form calcium oxide (lime), according to the equation

$$CaCO_3 \longrightarrow CaO + CO_2$$

If 10^6 kg of $CaCO_3$ is processed daily, how many kilograms of CaO are produced in a five-day week?

11.15.P The reaction between potassium permanganate and hydrochloric acid produces chlorine gas, according to the equation

$$2KMnO_4 + 16HCl \longrightarrow 2KCl + 2MnCl_2 + 5Cl_2 + 8H_2O$$

Given that 16.0 g of $KMnO_4$ is used, determine the following.

a. How many grams of HCl are required to complete the reaction?
b. How many moles of $MnCl_2$ are formed?
c. How many moles of Cl_2 are formed?

11.16.P Given that 0.50 mol of PCl_5 is used in the reaction

$$?PCl_5 + ?H_2O \longrightarrow ?H_3PO_4 + ?HCl$$

determine the following.

a. How many grams of H_2O are needed to complete the reaction?
b. How many moles of H_3PO_4 are formed?

11.17.P The equation below represents the combustion of benzene (C_6H_6):

$$2C_6H_6 + 15O_2 \longrightarrow 12CO_2 + 6H_2O$$

Given that 10.0 g of benzene is burned, determine the following.

a. How many moles of C_6H_6 enter into the reaction?
b. How many moles of H_2O are formed?

11.18.P Carbonyl chloride (phosgene) decomposes when heated:

$$COCl_2 \longrightarrow CO + Cl_2$$

Given that 150 g of Cl_2 is produced, determine the following.

a. How many moles of $COCl_2$ were decomposed?
b. How many grams of CO are produced?

11.19.P If 52.6 g of sodium chromate reacts with barium chloride according to the equation

$$Na_2CrO_4 + BaCl_2 \longrightarrow BaCrO_4\downarrow + 2NaCl$$

how many grams of barium chromate are formed?

11.20.P If 152.4 kg of sulfur is produced by the reaction

$$H_2S + Cl_2 \longrightarrow 2HCl + S$$

how many kilograms of H_2S are needed to complete the reaction?

11.21.P If 256.3 g of zinc sulfide burns in oxygen, according to the equation

$$2ZnS + 3O_2 \longrightarrow 2SO_2 + 2ZnO$$

how many grams of ZnO are formed?

11.22.P In the following reaction, 0.57 g of SO_3 is formed:

$$SO_2 + NO_2 \longrightarrow SO_3 + NO$$

How many grams of SO_2 and of NO_2 are needed?

11.23.P Ammonia will react spontaneously with oxygen according to the equation

$$4NH_3 + 3O_2 \longrightarrow 2N_2 + 6H_2O$$

If 75.2 g of ammonia reacts, how many grams of N_2 are formed?

11.24.P Phosphorus is produced in part by the reduction of P_4O_{10} by carbon:

$$P_4O_{10} + 10C \longrightarrow P_4 + 10CO$$

How much P_4O_{10} and how much carbon are required to produce 150 kg of phosphorus?

11.3 Percent Composition

11.25.P What is the percentage by weight of each element in each of the following compounds?
    a. $KMnO_4$
    b. $MgCl_2$
    c. H_2SO_4
    d. H_2O_2

11.26.P What is the volume of pure alcohol in 100 cm^3 of a 12.0% (by volume) alcohol solution?

11.27.P Household bleach is a liquid consisting of 5.25% sodium hypochlorite (by weight) dissolved in water. If you purchase 500 g of bleach, what weight of sodium hypochlorite are you paying for? What weight of water is in the bottle?

11.28.P Rubbing alcohol is 70% isopropyl alcohol by volume. What volume of water is present in a 0.5-L bottle?

11.29.P A 0.050-g vitamin tablet contains 0.40 mg of folic acid and 20.0 mg of niacin.
    a. What percent of the tablet is folic acid (by weight)?
    b. What percent of the tablet is niacin?
    c. What percent of the tablet is other ingredients?

11.30.P A 4-L bottle of wine claims "12% alcohol by volume." How many liters of alcohol are present in the bottle? How many liters of other substances are present?

11.31.P A 1-fl. oz (29.5-mL) bottle of flavor extract used in cooking contains 80% alcohol. What volume of alcohol is present?

11.32.P A container of 25 bouillon cubes weighs 92 g. Each cube contains 1 g of carbohydrate. What is the percentage of carbohydrate in each cube?

11.33.P A 1-oz (28-g) serving of oatmeal contains 55 mg of potassium. What is the percentage of potassium in each serving?

11.34.P A patent medicine contains 0.0020% of a certain preservative, by volume. What is the volume of preservative in 0.50 mL of this drug?

11.35.P A package of cheese has the following "nutritional information" on the label:

serving size	1 oz (28 g)
protein	1 g
percentage U.S. RDA protein	2%

(RDA is recommended daily allowance.)
    a. How many grams of protein are in the recommended daily allowance?
    b. How many ounces of that cheese would you need to eat daily in order to get all of the U.S. RDA of protein?

Self Test

1. Balance the following equations, and calculate the number of moles and the number of grams of each of the products.
   a. $?Cl_2 + ?NaBr \longrightarrow ?NaCl + ?Br_2$
      (2.32 mol)
   b. $?TiCl_4 + ?H_2O \longrightarrow ?TiO_2 + ?HCl$
      (56.2 g)
   c. $?CaO + ?SO_3 \longrightarrow ?CaSO_4$
      (3.521 kg)
   d. $?H_2O_2 \longrightarrow ?H_2O + ?O_2$
      (27.5 mg)
   e. $?C_6H_{12}O_6 + ?O_2 \longrightarrow ?CO_2 + ?H_2O$
      (150 g)

2. The reaction
   $$10K + 2KNO_3 \longrightarrow 6K_2O + N_2$$
   involves the oxidation of potassium by potassium nitrate. Given that 15 g of K is used in the reaction, determine the following.
   a. How many moles of KNO_3 are required?
   b. How many grams of N_2 are formed?

3. A solution of sodium hydroxide slowly attacks glass by means of the reaction
   $$2NaOH + SiO_2 \longrightarrow Na_2SiO_3 + H_2O$$
   (The product is sodium silicate, a soluble compound. The glass is essentially "dissolved.") Given that 0.50 g of NaOH reacts with glass, determine the following.
   a. How many grams of SiO_2 are transformed into Na_2SiO_3?
   b. How moles of SiO_2 react?

4. Sodium reacts with ammonia to form the compound sodium amide (sodamide) according to the equation
   $$2Na + 2NH_3 \longrightarrow 2NaNH_2 + H_2$$
   a. How many grams of sodium are required to form 13 mol of hydrogen?
   b. How many grams of sodium are required to form 24.8 kg of sodium amide?

5. Hydrofluoric acid rapidly etches glass with the reaction
   $$SiO_2 + 4HF \longrightarrow SiF_4 + 2H_2O$$
   How many grams of HF are required to form 6.0 g of SiF_4?

6. Limestone ($CaCO_3$) reacts with acid in rain according to the equation
   $$CaCO_3 + H_2SO_3 \longrightarrow CaSO_3 + H_2O + CO_2$$
   The $CaSO_3$ (calcium sulfite) is soluble in water, so the limestone is essentially dissolved away. How many grams of H_2SO_3 are required to eat away 22.4 kg of limestone?

7. A packet of instant broth weighs 6.0 g and contains 1.0 g of protein and 2.0 g of carbohydrates.
   a. What is the percent composition by weight of protein?
   b. What is the percent composition by weight of carbohydrates?
   c. What is the percent composition by weight of the other ingredients?

8. A box of cereal contains the statement that 1.0 oz of the cereal contains 3.0 g of protein, 23.0 g of carbohydrates, 1.0 g of fat, and 375 mg of sodium.
   a. What is the percentage by weight of protein?
   b. What is the percentage by weight of carbohydrates?
   c. What is the percentage by weight of fat?
   d. What is the percentage by weight of sodium?
   e. What is the percentage by weight of other ingredients?

9. A 15-mL (15 cm³) bottle of eye drops contains 0.012% naphazoline hydrochloride (by volume). (*Note:* Do not be put off by the unfamiliar name of the chemical compound. You do not have to know anything about naphazoline hydrochloride to do this problem.)
   a. How many milliliters of naphazoline hydrochloride are in the bottle?
   b. How many milliliters of other ingredients are in the bottle?

10. A 50-mL bottle of insect repellant contains 80% 2-ethyl-1,3 hexandiol (by volume) and 20% inert ingredients.
    a. What is the volume in milliliters of active ingredient?
    b. What is the volume in milliliters of inert ingredients?

12
The Gas Laws

Objectives After completing this chapter, you should be able to:
1. Describe a gas by its properties.
2. Describe at least two methods of measuring gas pressure.
3. Define the common pressure units: lb/in^2, $newton/m^2$, pascal, kilopascal, torr, bar, and atmosphere.
4. Convert gas pressures from one set of units to another.
5. Explain the operation of the barometer and the manometer.
6. Describe the difference between absolute pressure and gauge pressure.
7. Discuss the historical background of pressure measurements and the gas laws.
8. Perform calculations involving Boyle's law, Charles's law, and the universal gas law.
9. Perform calculations involving the law of partial pressures.
10. Describe qualitatively the relationship between the kinetic theory of gases and the universal gas law.
11. Describe the difference between an ideal gas and a real gas.

12.1 Measuring Gas Pressure

Introduction to Properties of Gases

A hundred years ago, scientists began to accept the idea that a **gas** was composed of tiny molecules dashing about at random and colliding with one another like a swarm of blind bees. Before that time there was no theoretical model with which to describe a gas on the microscopic level. There were only descriptions of the properties of gases and a few relationships among these properties. The relationships had been discovered purely by experiment, and their basic underlying causes were not understood. In this chapter you will learn about those properties of gases important to the study of chemistry, together with some of the explanations for these properties.

Properties that describe a particular gas and help us tell the difference between one gas and another are called **particular properties of gases.** Properties that apply to all gases are considered **general properties of gases.** Let us first examine some particular properties before going on to study the general properties of gases.

Particular Properties

One important property of a gas is its **density**—that is, the mass per unit volume. Density units of gases are usually given in grams per liter, and as we have already seen in Chapter 9, the density of a gas is directly proportional to its molecular mass.

Another property useful in describing a gas is its color. Most common gases (including oxygen, nitrogen, carbon dioxide, and all the noble gases) are colorless and hence invisible to the naked eye. This is fortunate, for it would be a nuisance if the atmosphere around us were colored and imparted a tint to the objects viewed through it. (Sunlight does tend to get reddish as it passes through the atmosphere, especially at sunset, but that is because the blue wavelengths of the sun's spectrum are scattered away from the line of vision, leaving the red wavelengths to be seen.)

Chlorine, on the other hand, is a greenish-yellow gas; bromine vapor is reddish-brown; iodine vapor is purple; and nitric oxide has a dark brown color. Colored gases should not be confused with the colored smoke obtained from flares and fireworks; smoke actually consists of a fine suspension of solid particles in the air.

Odor is an important property of gases and vapors. (A **vapor** is the gaseous state of a substance that is liquid at ordinary temperatures. Water vapor is what evaporates from liquid water. Physically, there is no difference between a gas and a vapor.) Fortunately, the common atmospheric gases are odorless as well as colorless. However, automobile fumes, for example, have a familiar pungency. An entire fragrance industry is based on the bottling of pleasing odors. A perfume may be a liquid, but it is the vapor that you smell. Anything smelled is actually in the gaseous state, for it is the molecules of vapor encountering nerve endings inside the nose that produce the sensations of smell. Foods generally emit odoriferous gases; this is why your sense of taste diminishes when your nose is blocked.

12.1 MEASURING GAS PRESSURE

General Properties of Gases

A general gas property is one that applies to all gases rather than defining a particular gas. *Temperature* is one general property. All gases exist at one temperature or another; the particular temperature depends on the circumstances or surroundings, not on the specific gas.

Temperature is measured by thermometers of many kinds. The mercury-filled thermometer is commonly used in the chemistry laboratory, but many more sophisticated types have also come into use in recent years. As we will see later in this chapter, the properties of a given volume of gas can be used to determine the temperature of that gas, so a container of gas can itself serve as a thermometer.

One of the most important properties of gases is pressure. The **pressure** of a gas is the force that the gas exerts on each unit area of the surfaces in contact with the gas. (*Unit area* means 1 in^2, 1 cm^2, etc.) We often think of air pressure as representing the weight of the air in a column reaching from the top of the atmosphere down to the surface on which it presses, as shown in Figure 12-1(a). However, it is important to note that the atmosphere also presses on vertical surfaces, and there is even upward pressure on the bottom surface of anything in contact with the air, as illustrated in Figure 12-1(b).

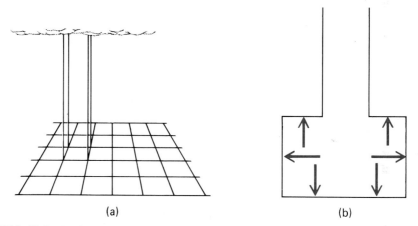

FIGURE 12-1 (a) The pressure at the bottom of a column of air is equal numerically to the weight of the air above a unit area (e.g., 1 in^2) at the bottom of the column. (b) The pressure is felt equally not only at the bottom of the air column but on the sides as well as on any bottom surfaces that there might be.

The ability of air to press in all directions is demonstrated by partially filling a balloon with air. When the balloon is closed tightly and carried to a high altitude, it expands into a spherical form, the air within pressing outward equally in all directions. (See Figure 12-2.) At high altitudes the pressure of the air inside the bag is no longer counterbalanced by that of the outside air pushing in. (In a later section you will learn why the air exerts pressure evenly in all directions.)

All fluids exert pressure on the walls of their containers. (Since a fluid is defined as *anything that flows*, gases as well as liquids are considered fluids.) This pressure is always measured in units of *force per unit area*. In the English system,

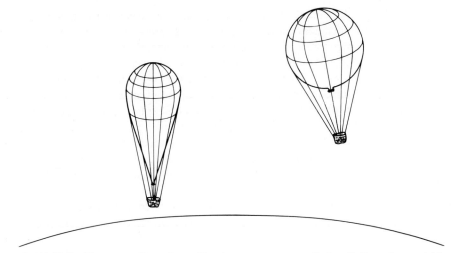

FIGURE 12-2 The expanding air applies its pressure equally in all directions within a rising balloon.

pounds per square inch is the unit most commonly used for measuring pressures of all kinds of fluids, both liquids and gases.

The concept of force per unit area is illustrated in Figure 12-3. Say the total force exerted by the air on the bottom of the container is 500 lb. For many purposes this total force is not a useful number. What we want to know is the pressure—the force exerted on each square inch of the container wall. If the area of the bottom wall is 10 in^2, the pressure is given by the equation

$$\text{pressure} = \frac{\text{force}}{\text{area}} = \frac{500 \text{ lb}}{10 \text{ in}^2} = 50 \text{ lb/in}^2$$

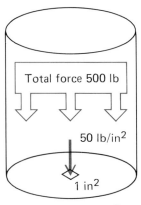

FIGURE 12-3 If the pressure on a surface is 50 lb/in^2, the total force applied by the gas on a 10-in^2 surface is 500 lb.

The reason we describe gases in terms of the pressure rather than the total force exerted on the wall of the container is that we don't want the gas property to vary with the size of the container. If one sample of a gas is in a large container and another sample of the same gas is in a small container and both are at the same pressure, the large sample exerts more total force on the walls of its container than does the small sample. But this difference is simply the result of the size of the containers. As we saw in our discussion of Avogadro's number (Chapter 9), equal volumes of gas at the same pressure contain equal numbers of molecules. Since the number of molecules per unit volume is a significant gas parameter, pressure is considered one of the fundamental gas properties.

Atmospheric pressure is the pressure exerted by the atmosphere on all surfaces. This pressure decreases gradually as altitude increases. At sea level, atmospheric pressure is about 14.7 pounds per square inch (or 14.7 lb/in^2). Thus, each square inch of your body supports 14.7 lb of air. At an altitude of 5 mi (at the top of Mount Everest in the Himalayas), the pressure is one-quarter its sea-level value. (In other words, three-quarters of the atmosphere lies below the 5-mi altitude.)

At a given altitude, the atmospheric pressure depends to some extent on climatic conditions. Therefore it is convenient to define a **standard atmospheric pressure:** 14.696 lb/in^2. Although the actual pressure at sea level may vary from day to day, this standard pressure is an arbitrary constant that allows us to compare measurements made at different times and under differing conditions. We will see later in this chapter that if the volume of a gas is measured at *any* pressure, a simple calculation will give the volume the same gas would have at *standard* pressure.

Before going into more detail about the meaning of standard pressure conditions, we make a detour to describe some of the more common instruments used to measure pressures.

Pressure-Measuring Instruments

The **barometer** is the instrument used to measure atmospheric pressure. The name comes from the Greek word *baros*, meaning "weight"; the suffix *meter* means "to measure." One way weather forecasters predict changes in the weather is by looking at a barometer to see whether the air pressure is going up or down.

The history of the barometer takes us back to the very beginning of modern science in the seventeenth century. Its invention marks a break from a medieval way of thinking and points the way to a modern, scientific logic.

The invention of the barometer was an indirect outgrowth of efforts to pump water from deep wells. (This is a prime example of how scientific research in one area can produce unexpected benefits in another area.) It had been known for a long time that when a piston is raised above a column of water, the water rises with the piston, as shown in Figure 12-4(a). The explanation given was based on the famous quotation from Aristotle: "Nature abhors a vacuum." (A **vacuum** is a volume of space without air or other gas in it.) If the water did not rise with the

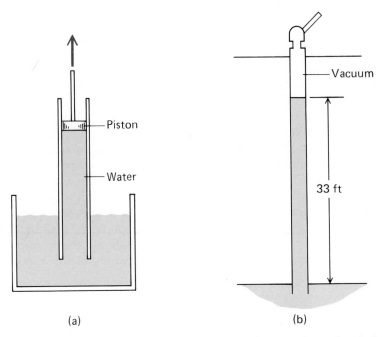

FIGURE 12-4 (a) A piston pulled up in a cylinder raises the water beneath it, behaving as though it did not want a vacuum to exist between the piston and the water. (b) However, if the piston rises more than 33 ft above the water level, a vacuum will appear to form in the space between the water and the piston. (Actually the space will be filled with water vapor.)

piston, a vacuum would be created in the space between the water and the piston; since nature hates a vacuum, the water has to rise. This argument was used to support the belief that the creation of a vacuum was impossible.

One observable fact was hard to reconcile with this explanation. It was that pumps, no matter how well made, could not raise water above a height of about 33 ft. The pump was always at the top of the well, however, as shown in Figure 12-4(b). Therefore, if the well was more than 33 ft deep, there had to be a vacuum above that column of water. How could this fact be explained?

Even the great scientist Galileo Galilei (1564–1642), who brought physics out of the Middle Ages, continued to believe in the idea that nature abhors a vacuum. But to explain why water could not be raised more than 33 ft, he speculated that perhaps there was some kind of limit to nature's feelings about vacuums. Galileo suggested to his assistant, Evangelista Torricelli (1608–1647), that he look into the matter more closely. In 1643 Torricelli began a series of experiments using liquid mercury instead of water. Since mercury is 13.6 times more dense than water, he reasoned that nature would hold up a column of mercury about 30 in high—much easier to handle than a column of water 33 ft high.

In his most famous experiment, Torricelli took a glass tube about 3 ft long, closed at one end and open at the other. He filled it with mercury, held his finger

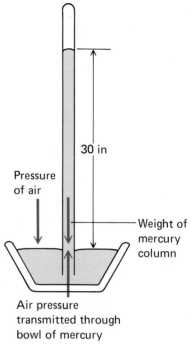

FIGURE 12-5 Torricelli's original mercury barometer. Normal atmospheric pressure will support a column of mercury about 30 in high. The space above the mercury is actually filled with mercury vapor, which exerts very little pressure.

over the open end, and then placed the tube upright in a bowl filled with mercury, with the open end of the tube beneath the surface. (See Figure 12-5.) When Torricelli took his finger away from the open end of the tube, the level of the mercury in the column dropped until there was just 30 in of mercury poised above the level of the mercury in the bowl. It seemed to Torricelli that the simplest way of explaining this phenomenon was to assume that the atmosphere pushes down on the top of the mercury in the bowl, and that this force, transmitted to the mercury in the tube, pushes it up. The mercury column is held steady at a height such that the weight of mercury per unit area in the tube just equals the weight of the atmosphere per unit area pressing down on the bowl. Then the system is in equilibrium.

But the weight of the mercury column is directly proportional to its height, and the weight of the atmosphere per unit area is just the atmospheric pressure on the mercury surface. This means that the atmospheric pressure can be determined simply by measuring the height of the mercury column. To this day we measure atmospheric pressure by means of mercury barometers operating on the same principle as Torricelli's mercury column.

Torricelli did not have time to pursue this matter very far before his untimely death in 1647. However, the eminent mathematician and physicist Blaise Pascal, who was also interested in the behavior of liquids, took up the question and devised a crucial experiment to test Torricelli's hypothesis. In 1646 Pascal had a

FIGURE 12-6 A modern mercury barometer. A thumb-screw keeps the level in the mercury pool adjusted so that it just reaches the bottom of the pointer. Then the scale at the top of the column accurately reads the difference in level between the mercury in the column at the top and the mercury in the pool at the bottom.

barometer taken up to the top of a mountain in central France, while a similar barometer remained at the foot of the mountain. In the elevated barometer the height of the mercury column dropped 8 cm for each 1000-m increase in altitude, while the other barometer remained relatively unchanged.

By demonstrating that atmospheric pressure decreases at higher elevations, this experiment suggested that there is just a certain amount of air above the surface of the earth, so the higher the altitude, the smaller the amount of air above. It also implied that outside this blanket of air covering the earth's surface is a vacuum. Far from abhorring a vacuum, nature is largely filled with vacuum—the space between the planets and stars.

The work of Torricelli and Pascal showed that explanations in science can be based on mechanical principles—how forces act on objects and how these objects move in response to such forces. To say that "nature abhors a vacuum" supposes that nature has a kind of "consciousness," for abhorrence is an emotion; it ascribes to nature the same kinds of feelings that humans have. Such *anthropomorphic explanations* have never been useful in predicting the workings of nature and are no longer considered part of science.

Modern barometers are of two general types: *mercury* and *aneroid*. Figure 12-6 shows a schematic view of a mercury barometer. The mercury at the bottom is kept in a cup, open to the air, with a flexible bottom. The thumbscrew at the bottom allows the user to adjust the level of the mercury in the cup so that it just reaches the little pointer (A). Then the height of the mercury column above that fixed pointer is read on the millimeter scale (B).

An aneroid barometer, shown in Figure 12-7, contains an airtight cylindrical box with a flexible diaphragm as its lid. The center of the diaphragm is connected by a series of levers and pulleys to a pointer. When the atmospheric pressure rises or falls, the diaphragm goes in or out, causing the pointer to move around a circular scale that can be calibrated in any convenient units. Such barometers are usually used in home weather stations. Professional meteorologists use recording barometers, which are aneroid barometers with a pen on the end of the rotating pointer. The pen makes a permanent record of the atmospheric pressure on a strip of paper.

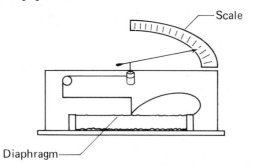

FIGURE 12-7 An aneroid barometer consists of a metal cylinder with a flexible, corrugated diaphragm that bends with changes in the air pressure. A series of gears and levers transmits this motion to a pointer which indicates the pressure on a circular scale.

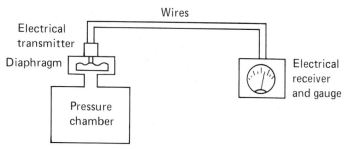

FIGURE 12-8 A manometer indicates the pressure within a closed chamber on an indicator located on a remote instrument board, or console.

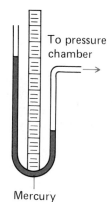

FIGURE 12-9 A mercury manometer. One end of the U-tube is connected to the pressure chamber. The difference between the levels of the mercury column indicates the amount of gas pressure within the chamber, over and above normal atmospheric pressure.

A **manometer** is an instrument used for measuring the pressure within a container of gas. The same principles used for measuring atmospheric pressures can be used for measuring any gas pressure. For example, an aneroid barometer can be converted into a manometer by putting it into a box connected by a tube to the gas container whose pressure is being measured. (See Figure 12-8.) Often the diaphragm is linked to an electrical device that sends a signal to a remote indicator, so that the pressure can be read at a control panel some distance away from the actual pressure gauge.

Mercury manometers are often found in the laboratory. One common type of mercury manometer is shown in Figure 12-9. It is simply a tube bent into a U shape, half filled with mercury. One end of the tube is connected to the container whose pressure is being measured; the other end is open to the air. If the gas pressure in the container is greater than the atmospheric pressure, the level of the mercury on the far side of the manometer will rise. The difference in the mercury level between the two sides of the tube reflects the difference between the gas pressure in the container and the air pressure outside. This difference in pressure between the inside and the outside is called the **gauge pressure**. The actual pressure inside the container—called the **absolute pressure**—is the gauge pressure plus the atmospheric pressure.

In both the English and the metric system, several different kinds of units are used to measure gas pressures. In the next section we will define these units and point out the relationships among them.

Units of Pressure

Since pressure is defined as a force per unit area, a unit of pressure must consist of a force divided by an area. In the English system, pressure is measured in *pounds per square inch*. This unit is commonly used by engineers, particularly in connection with steam pressure. (Much of our knowledge concerning gases was acquired during the eighteenth and nineteenth centuries, when the theory of the steam engine was an important area of research, the aim being to get the maximum work out of the energy stored in fuel.)

In the metric system, the basic unit of force is the newton and the unit of area is the square meter. Thus, pressure in the metric system is measured in newtons per square meter. The **newton (N)** is a rather small unit of force (about a quarter of a pound), and the meter is a large area. Therefore, 1 N/m² is a very small pressure. Converting from English to metric units, we find that

$$1 \text{ lb/in}^2 = 6891 \text{ N/m}^2$$

Recently the newton per square meter was given the name **pascal (Pa)** in honor of Blaise Pascal, who contributed much to our understanding of pressures in fluids. Since the pascal is such a small unit of pressure, the **kilopascal (kPa)** is preferred for ordinary purposes.

$$1 \text{ lb/in}^2 = 6891 \text{ N/m}^2 = 6891 \text{ Pa}$$
$$= 6.891 \text{ kPa}$$

The pascal and the kilopascal are the fundamental units of pressure in the SI system of measurement.

Since air pressure has long been measured by means of the mercury barometer, or manometer, it has been customary to use the height of the mercury column itself as a measure of the pressure. Thus, for example, atmospheric pressure is given as 30 in of mercury in the English system, or 76 cm (or 760 mm) of mercury in the metric system.

Standard atmospheric pressure (commonly called an **atmosphere**, or **atm**) is defined as the pressure measured by a mercury column exactly 76.00 cm (29.92 in) high, at sea level, at a temperature of 0°C. Sea level is specified because the weight of the mercury depends on its distance from the center of the earth. Since the density of mercury depends on the temperature, both location and temperature must be specified. In terms of SI pressure units,

$$1 \text{ atm} = 1.01325 \times 10^5 \text{ Pa}$$
$$= 101.325 \text{ kPa}$$

The factor

$$1 \text{ atm} = 14.7 \text{ lb/in}^2$$

is used to convert from atmospheres to English units.

The combination of *one standard atmospheric pressure* and *a temperature of 0°C* defines a set of conditions known as *standard temperature and pressure*—generally abbreviated *STP* (see Chapter 9). Many measurements in chemistry must be made at standard temperature and pressure in order to give meaningful and consistent results. For example, the statement that 1 mol of a gas occupies a volume of 22.4 L is true only under STP conditions. If the gas is actually at a different temperature and pressure, we can calculate what the volume would be at STP. In this way measurements made under a variety of conditions can be related to STP measurements.

Meteorologists define a pressure unit called the **bar**:

$$1 \text{ bar} = 10^5 \text{ N/m}^2 = 10^5 \text{ Pa}$$

Accordingly,

$$1 \text{ atm} = 1.01325 \text{ bar}$$

In order to avoid the use of the decimal point, meteorologists commonly use the **millibar (mb)** in weather reports. Since a millibar equals one-thousandth of a bar,

$$1 \text{ bar} = 1000 \text{ mb}$$

Accordingly,

$$1 \text{ atm} = 1013 \text{ mb}$$

Another unit that has come into common use (particularly among scientists working with low pressures) is the torr, named after Torricelli. One **torr** is the pressure measured by 1 mm of mercury in a manometer. Since 760 mm Hg is an atmosphere, 1 torr is 1/760 of an atmosphere. Thus a torr is a small unit of pressure and represents a partial vacuum.

Although the word *vacuum* is commonly used to mean an absence of air, in practice there are various grades of vacuums. An ordinary mechanical vacuum pump will only pump part of the air out of a vacuum chamber, down to a pressure of about 0.01 torr. More sophisticated pumps, such as the vapor diffusion pump, are needed to get down to lower pressures. Pressures of 10^{-6} torr are routinely obtained in the laboratory and in industry (for example, for the manufacture of TV picture tubes). With special procedures much lower pressures can be obtained, and in modern practice pressures of 10^{-12} torr are not unusual. However, never (even in outer space) is a *complete* absence of gas molecules found.

The modern trend is to try to use either the newton per square meter or the pascal as the unit of choice for all scientific work. However, the many units of pressure in common use for various purposes make for a great deal of confusion. Table 12-1 summarizes the relationships among the various units discussed above.

The unit factor method is used to convert from one pressure unit to another. Multiply the pressure expressed in one unit by the conversion factor that gives the desired cancellation of units. For example, we convert from atmospheres to pounds per square inch as follows:

$$x \frac{\text{lb}}{\text{in}^2} = A \text{ atm} \times 14.7 \frac{\text{lb/in}^2}{\text{atm}} = 14.7 A \frac{\text{lb}}{\text{in}^2}$$

Thus 10.0 atm is converted to pounds per square inch as follows:

$$x \frac{\text{lb}}{\text{in}^2} = 10.0 \text{ atm} \times 14.7 \frac{\text{lb/in}^2}{\text{atm}} = 147 \frac{\text{lb}}{\text{in}^2}$$

The examples given below show how this method is applied to a variety of pressure units.

12 THE GAS LAWS

TABLE 12-1 Pressure Unit Conversions

$$1 \text{ newton per square meter } (N/m^2) = 1.45 \times 10^{-4} \text{ lb/in}^2$$
$$= 1 \text{ Pa}$$
$$1 \text{ pascal (Pa)} = 1 \text{ N/m}^2$$
$$= 9.87 \times 10^{-6} \text{ atm}$$
$$1 \text{ kilopascal (kPa)} = 1000 \text{ Pa}$$
$$= 9.87 \times 10^{-3} \text{ atm}$$
$$1 \text{ pound per square inch (lb/in}^2) = 6891 \text{ Pa}$$
$$= 6.891 \text{ kPa}$$
$$1 \text{ atmosphere (atm)} = 76.0 \text{ cm Hg}$$
$$= 760 \text{ mm Hg}$$
$$= 760 \text{ torr}$$
$$= 29.92 \text{ in Hg}$$
$$= 1.01325 \times 10^5 \text{ N/m}^2$$
$$= 101.325 \text{ kPa}$$
$$= 14.7 \text{ lb/in}^2$$
$$= 1.01325 \text{ bar}$$
$$1 \text{ bar} = 1 \times 10^5 \text{ N/m}^2$$
$$= 100 \text{ kPa}$$
$$1 \text{ torr} = 1 \text{ mm Hg}$$
$$= 1/760 \text{ atm}$$
$$= 133.3 \text{ N/m}^2$$
$$= 133.3 \text{ Pa}$$
$$= 0.0193 \text{ lb/in}^2$$

Note: To convert from units in the left column to units in the right column, multiply the number to be converted by the conversion factor in the right column. For example, 110 Pa is converted to pounds per square inch as follows:

$$110 \text{ Pa} \times \frac{1.45 \times 10^{-4} \text{ lb/in}^2}{1.00 \text{ Pa}} = 160 \times 10^{-4} \text{ lb/in}^2$$
$$= 0.016 \text{ lb/in}^2$$

Example 12.1   A tank contains a gas at a pressure of 12.0 atm. State this pressure in the following units:
(a) inches of Hg
(b) centimeters of Hg
(c) pounds per square inch
(d) newtons per square meter
(e) bars
(f) torrs
(g) kilopascals

Solution

(a) One atmosphere is 29.92 in Hg. Therefore,

$$x \text{ in Hg} = 12.0 \text{ atm} \times 29.92 \frac{\text{in Hg}}{1 \text{ atm}}$$

$$= 359 \text{ in Hg}$$

Note: Round off to three significant digits.

(b) One atmosphere is 76.0 cm Hg. Therefore,

$$x \text{ cm Hg} = 12.0 \text{ atm} \times 76.0 \frac{\text{cm Hg}}{1 \text{ atm}}$$

$$= 912 \text{ cm Hg}$$

(c) One atmosphere is 14.7 lb/in^2. Therefore,

$$x \text{ lb/in}^2 = 12.0 \text{ atm} \times \frac{(14.7 \text{ lb/in}^2)}{1 \text{ atm}}$$

$$= 176 \text{ lb/in}^2$$

(d) One atmosphere is 1.01×10^5 N/m^2. Therefore,

$$x \text{ N/m}^2 = 12.0 \text{ atm} \times \frac{(1.01 \times 10^5 \text{ N/m}^2)}{1 \text{ atm}}$$

$$= 12.1 \times 10^5 \text{ N/m}^2$$

$$= 1.21 \times 10^6 \text{ N/m}^2$$

(e) One atmosphere is 1.01325 bar. Therefore,

$$x \text{ bar} = 12.0 \text{ atm} \times \frac{1.01325 \text{ bar}}{1 \text{ atm}}$$

$$= 12.2 \text{ bar}$$

(f) One atmosphere is 760 torr (760 mm Hg). Therefore,

$$x \text{ torr} = 12.0 \text{ atm} \times \frac{760 \text{ torr}}{1 \text{ atm}}$$

$$= 9120 \text{ torr}$$

(g) One atmosphere is 101.325 kPa. Therefore,

$$x \text{ kPa} = 12.0 \text{ atm} \times \frac{101.325 \text{ kPa}}{1 \text{ atm}}$$

$$= 1220 \text{ kPa}$$

Example 12.2 A vacuum chamber contains air at a pressure of 0.0150 torr. What is this pressure in the following units?
(a) pascals   (b) pounds per square inch

Solution (a) We see from Table 12-1 that 1 torr equals 133.3 Pa. Therefore,

$$x \text{ Pa} = 0.0150 \text{ torr} \times \frac{133.3 \text{ Pa}}{1 \text{ torr}}$$

$$= 2.00 \text{ Pa}$$

(b) Similarly, 1 torr equals 0.0193 lb/in². Therefore;

$$x \text{ lb/in}^2 = 0.0150 \text{ torr} \times \frac{(0.0193 \text{ lb/in}^2)}{1 \text{ torr}}$$

$$= 2.90 \times 10^{-4} \text{ lb/in}^2$$

12.2 Boyle's Law

Historical Introduction

Scientific investigation of gases and their properties began with the invention of the barometer by Torricelli (described in Section 12.1) and the invention of the air pump by Otto von Guericke in 1650. With his pump, von Guericke was able to evacuate the air from the interior of large containers and perform a variety of experiments, including one showing that the sound of a ringing bell will not pass through a vacuum. He also staged a spectacular demonstration in which air was pumped from a sphere 56 cm (22 in) in diameter, formed from a pair of hemispheres joined at their rims. When a vacuum existed inside this sphere, two teams of horses were unable to pull the hemispheres apart, but when air was let back into the sphere through a valve, the two halves easily separated. (See Figure 12-10.) This experiment, known as the Magdeburg Hemispheres after the city in which it took place, demonstrated in a graphic way the enormous force that the atmosphere can exert.

Hearing of von Guericke's work, in 1657 the Irish scientist Robert Boyle (1627–1691) devised his own air pump in order to perform his own experiments regarding the properties of gases. One particularly important experiment, done in 1662, was designed to determine how the volume of a fixed quantity of gas changes when the pressure on it is varied.

Boyle did the experiment with a glass tube shaped like a U, but with one leg longer than the other. The short leg was closed, and the long leg was open to the air. Boyle poured some mercury into the open end in such a way that there was an

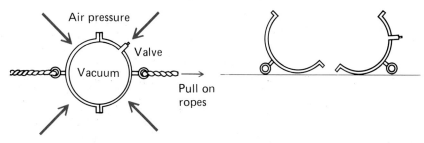

FIGURE 12-10 Otto von Guericke's famous experiment with the Magdeburg hemispheres. When there was a partial vacuum inside the sphere, the air pressure on the outside was so great that the two hemispheres could not be pulled apart. When air was allowed inside the sphere through a valve, the two hemispheres fell apart by themselves.

equal amount of mercury in each leg, as shown in Figure 12-11(a). The length of the trapped air column in the closed leg was then measured. Next, he poured more mercury into the open end until the air column in the closed end was reduced to half its original length. (For example, if the length of the air column in the closed leg was originally 10 cm, he would reduce it to 5 cm.) Measuring the height of the mercury column in the open leg, he found that it had risen 30 in above the top of the mercury in the closed leg, as shown in Figure 12-11(b).

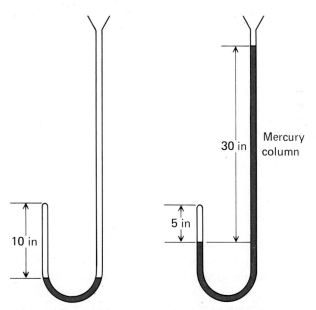

FIGURE 12-11 Demonstration of Boyle's law. With the mercury column at the same level in the two tubes, the height of the trapped air column in the tube on the left is 10 in (at 1 atm of pressure). With the mercury in the right-hand column at a height of 30 in, the trapped air column contracts to a height of 5 in and the total pressure inside the trapped column is 2 atm.

Boyle knew that the air pressure on the open end of the tube was normal atmospheric pressure—equivalent to 30 in of mercury. With the mercury in the two tubes at the same level, the air pressure in the closed end of the tube had to be the same: 30 in Hg. After he added mercury to a height of 30 in, the pressure in the closed end of the tube equaled the atmospheric pressure plus the added 30 in Hg, for a total of 60 in Hg, twice its original value. Boyle saw that doubling the pressure in the closed part of the tube caused the volume of the trapped air to become half what it had been.

Continuing to pour mercury into the long tube, he found that *increasing* the pressure by a factor of 3 caused the volume to *decrease* by a factor of 3. It was clear that he had found an important rule: *The volume of a fixed quantity of gas varies inversely with the pressure.*

This relationship is expressed mathematically by the proportionality statement

$$\text{volume} \propto \frac{1}{\text{pressure}}$$

The symbol $\propto$ (similar in form to the Greek letter alpha) means "is proportional to" or "varies as." This relationship is an inverse proportion, because as the pressure increases, the volume decreases. To convert this proportionality into an equation, we must introduce a *constant of proportionality*:

$$\text{volume} = \text{constant} \times \frac{1}{\text{pressure}}$$

This relationship is finally condensed into the simple equation

$$\text{volume} = \frac{\text{constant}}{\text{pressure}}$$

It is easy to see that this equation represents an inverse proportion: If the pressure is doubled, the volume is halved. If the pressure is tripled, the volume is divided by 3.

In symbols, the equation is written

$$V = \frac{K}{P}$$

The constant K is a number that must be determined in each situation and depends in part on the quantity of gas in the container—the greater the number of moles of gas, the greater the volume. (More about this constant later.)

The above relationship is now called **Boyle's law**. It is a law that has numerous applications, as we will see in the next section. (Note that the word *law* in this context means "a relationship determined by experiment and observation." *Laws of nature* are rules that determine how things in the world operate.) When Boyle did his experiments, he had no understanding of *why* gases behave the way they do. He was concerned mainly with finding out *how* they behave. The theoretical explanation for the behavior came much later, at the end of the nineteenth century.

In 1676, about fifteen years after Boyle's experiments, the French physicist Edmé Mariotte repeated Boyle's work. In his experiments, Mariotte emphasized one point that Boyle had neglected to mention. This was the fact that the volume of the gas trapped in the tube varies with the temperature. As the temperature goes up, the gas expands, and so if a precise relationship is to be found between pressure and volume, the experiment must be done in such a way that the temperature does not change. Therefore, the complete statement of Boyle's law (sometimes called **Mariotte's law**) is as follows: *The volume of a fixed quantity of gas at a constant temperature is inversely proportional to the pressure.*

Applications of Boyle's Law

In this section we will discuss some applications of Boyle's law and learn how to solve problems involving this law.

Boyle's law is seen in operation in the flight of gas-filled balloons (as opposed to hot-air balloons). A balloon containing hydrogen or helium may be only partly filled when on the ground, but as it rises into the upper atmosphere the gas expands, filling the balloon completely so it swells out into a spherical shape. The gas inside the balloon expands because of the decreased pressure surrounding it. This effect is in agreement with Boyle's law: as the external pressure on the gas decreases, the volume of the gas in the balloon increases proportionately.

Chemists are interested in Boyle's law because one method of determining the molecular mass of a gas is to measure the volume of a known mass of the gas and then apply the rule (developed in Chapter 9) that 1 mol of a gas at STP has a volume of 22.4 L. Since this rule is true only at standard temperature and pressure, if the measurement is made at another pressure, it is necessary to use Boyle's law to determine what the gas volume would be at exactly 1 atm pressure.

To see how this is done, consider the following problem: The air in a cylinder has a volume of 600 cm^3 when it is at a pressure of 3 atm. What would be its volume at a pressure of 1 atm?

The relationship between pressure and volume is

$$\text{volume} = \frac{\text{constant}}{\text{pressure}}$$

or
$$V = \frac{K}{P}$$

Multiplying both sides of the equation by P, we find that

$$PV = K$$

In other words, as the pressure varies, the volume varies in such a way that the pressure times the volume always equals the same number—the constant K.

The known quantities in the problem are the pressure and the volume under one set of conditions: $V = 600 \text{ cm}^3$ when $P = 3$ atm. Let us call these known values V_1 and P_1. Putting them into Boyle's law, we have

$$P_1 V_1 = K$$

We want to find the volume of the gas at a different pressure, which we will call P_2. If V_2 represents the gas volume to be found, the product of V_2 and P_2 equals the same constant K:

$$P_2 V_2 = K$$

Since the two products are both equal to the same constant, they must be equal to each other:

$$P_2 V_2 = K = P_1 V_1$$

or

$$P_2 V_2 = P_1 V_1$$

In other words, the pressure multiplied by the volume under one set of conditions equals the pressure multiplied by the volume under any other set of conditions (as long as the temperature is unchanged).

To find V_2, we divide both sides of the equation by P_2:

$$\frac{P_2 V_2}{P_2} = \frac{P_1 V_1}{P_2}$$

$$V_2 = \frac{P_1}{P_2} \times V_1$$

The unknown volume V_2 is found by substituting the known quantities into the equation:

$$V_2 = \frac{3 \text{ atm}}{1 \text{ atm}} \times 600 \text{ cm}^3$$

$$= 1800 \text{ cm}^3$$

Example 12.3 A balloon contains 5.0 m^3 of air on the ground, where the barometer shows 76.2 cm of mercury. What would the barometric pressure have to be for the balloon to expand to a volume of 14 m^3?

Solution Write the equation for Boyle's law:

$$P_2 V_2 = P_1 V_1$$

To find P_2, divide both sides of the equation by V_2:

$$\frac{P_2 V_2}{V_2} = \frac{P_1 V_1}{V_2}$$

Cancel:

$$P_2 = P_1 \times \frac{V_1}{V_2}$$

Then substitute:

$$P_2 = 76.2 \text{ cm} \times \frac{5 \text{ m}^3}{14 \text{ m}^3}$$

$$= 76.2 \text{ cm} \times 0.357$$

$$= 27 \text{ cm Hg} \quad \text{(rounding off to two digits)}$$

Note: Since the above equation contains only ratios, it makes no difference what kinds of units are used. When m³ is divided by m³, the units of volume cancel out. Then whatever unit of pressure is used for P_1 will be the unit for P_2. We could just as easily have started with 30.0 in of mercury for the initial (starting) pressure; the final pressure would then have been 10.7 in Hg.

Example 12.4 A pump cylinder contains 12.7 L of air at a pressure of 131 kPa (kilopascals). What is the volume in the cylinder when the piston forces the air to a pressure of 246 kPa? (Assume that the temperature does not change.)

Solution Solve the Boyle's law equation for V_2:

$$V_2 = \frac{P_1}{P_2} \times V_1$$

$$= \frac{131 \text{ kPa}}{246 \text{ kPa}} \times 12.7 \text{ L}$$

$$= 6.76 \text{ L}$$

Note: To get a better feeling for the pressure units in this problem, note that

$$131 \text{ kPa} = \frac{131 \text{ kPa}}{101 \text{ kPa/atm}} = 1.30 \text{ atm}$$

12.3 Charles's Law

Historical Introduction

As we have seen, Edmé Mariotte had noticed as early as 1676 that gases expand when warmed. However, it was many years before precise measurements could be made to determine *how much* a volume of gas expands for a given change of temperature. Such measurements awaited the development of better thermometers and the establishment of a proper temperature scale.

In 1699 Guillaume Amontons, a French scientist, studied several gases and found that each gas changed in volume by the same amount for a given change of temperature. After that, there was no further investigation of the subject until approximately 1787, when French physicist Jacques Alexandre Charles repeated Amontons's experiment. Making careful measurements of volume and temperature, Charles found the following important rule: *For each increase in temperature of 1°C, the volume of a gas increases by 1/273 of its volume at 0°C if the gas pressure is kept constant.*

Unfortunately, Charles failed to publish this discovery, and so it was not until 1802 that the rule became known through the work of the eminent French chemist Joseph Louis Gay-Lussac (1778-1850). This incident points up the importance of prompt publication of scientific discoveries in technical journals. Not only does it inform the world of the new knowledge, but it ensures that the proper person receives credit. Because of Charles's failure to publish, there was a long controversy over the proper credit for the discovery. Today the relationship between the temperature and the volume of a gas is frequently called **Charles's law** but is sometimes called **Gay-Lussac's law**.

As we investigate the relationships between the temperature, pressure, and volume of a gas in the following sections, we shall learn that Charles's/Gay-Lussac's law is just one facet of a more general law—the universal gas law.

Example 12.5

A container of gas has a volume of 4.50 L at a temperature of 0°C. What would be its volume at 50°C if kept at constant pressure?

Solution

According to Charles's law, the volume of a gas increases by 1/273 of its volume at 0°C for every degree increase of temperature. Therefore, for a 50°C change of temperature, the change in volume is

$$\text{change in volume} = \text{change in temp.} \times \frac{1}{273°C} \times \text{original volume}$$

$$= 50°C \times \frac{1}{273°C} \times \text{original volume}$$

$$= 50°C \times \frac{1}{273°C} \times 4.50 \text{ L}$$

$$= 0.82 \text{ L}$$

$$\text{final volume} = \text{original volume} + \text{change in volume}$$

$$= 4.50 \text{ L} + 0.82 \text{ L} = 5.32 \text{ L}$$

Example 12.6 A container of gas has a volume of 100 L at 0°C. What would be its volume if cooled down to a temperature of −273°C?

Solution A decrease in temperature causes a reduction in volume. Using the same mathematics as in the preceding example, we have

$$\text{change in volume} = -273°C \times \frac{1}{273°C} \times 100 \text{ L}$$

$$= -100 \text{ L}$$

$$\text{final volume} = 100 \text{ L} - 100 \text{ L} = 0 \text{ L}$$

The gas apparently shrinks to nothing at −273°. Actually, the gas will liquefy before it reaches that point, and the above equations will no longer apply. The temperature −273°C is absolute zero; we see from this example how Charles's law predicts the existence of such a point on the temperature scale.

Applications of Charles's Law

The meaning of Charles's law is demonstrated by repeating his experiment. The apparatus used is shown in Figure 12-12. It looks like a Boyle's law apparatus, but the tube with the bulb of trapped gas is in an oven that can be warmed or cooled, and the oven has a thermometer for measuring the gas temperature.

An important feature of this experiment is the rubber tube connecting the two vertical glass sections. This rubber tube allows the height of the mercury column to be adjusted so that the level of the mercury in the tall tube is always the same distance above its level in the other tube as the gas expands or contracts. This ensures that *the pressure in the gas bulb remains constant while the volume of the gas is changing.* (Thus only one variable at a time changes.)

As the temperature of the gas varies, we measure its change in volume by observing how the mercury level changes on the left-hand scale. We then draw a graph that plots the volume as a function of the temperature (see Figure 12-13). Say the original gas volume was 100 cm³ at a temperature of 0°C. If raising the temperature to 10°C increases the volume by 3.66 cm³, raising the temperature from 10°C to 20°C will increase the volume by another 3.66 cm³.

The first thing we notice is that each increase of 10°C changes the volume by the same amount. As a result, the graph in Figure 12-13 is a straight line. That is, there is a **linear relationship** between the gas volume and temperature.

The next observation we make is that the change in volume for each degree change in temperature is 0.366 cm³—just 1/273 of the original volume of 100 cm³. This change agrees with the measurements of Charles and Gay-Lussac.

Furthermore, we find that the same behavior takes place if we lower the temperature instead of increasing it. Each reduction of 1°C in temperature reduces

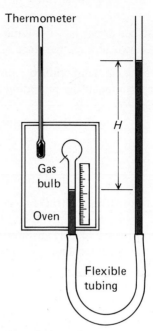

FIGURE 12-12 Demonstration of Charles's law. The flexible tubing is adjusted so that the height H of the mercury column is always the same. This ensures that the gas pressure inside the bulb does not change. Then the gas volume is measured as the temperature is varied, using the scale on the left-hand column.

the volume by 1/273 of the original volume at 0°C. If we extend the graph back far enough, we find a temperature at which the volume vanishes entirely (Figure 12-14). We saw in Example 12.6 that Charles's law predicts that this remarkable event will take place at a temperature of −273°C. As noted in that example, Charles's law does not hold at such low temperatures, so we do not expect the gas to disappear in actuality.

The dotted line in Figure 12-14 shows what happens to a real gas, such as nitrogen, at very low temperatures. The gas becomes a liquid at the **condensation point**, and the liquid becomes a solid at the **freezing point**. At each transition there is an abrupt change in volume. At absolute zero the volume of the substance is determined by the volume of its molecules, for at that temperature the molecules are as close together as they can get. The only substance that does not solidify at absolute zero (at normal atmospheric pressure) is helium, which remains liquid. However, at high pressures even helium solidifies.

In Chapter 2 a new temperature scale was defined—the Kelvin scale—in which the zero temperature is at −273.15°C (the point where the straight line in Figure 12-14 intersects the horizontal axis). The Kelvin scale is named in honor of the person who first suggested its importance to chemistry and physics: William Thomson, better known as Lord Kelvin (1824–1907).

To convert from Celsius degrees to kelvins, use this relationship:

$$\text{Kelvin temperature} = \text{Celsius temperature} + 273.15$$

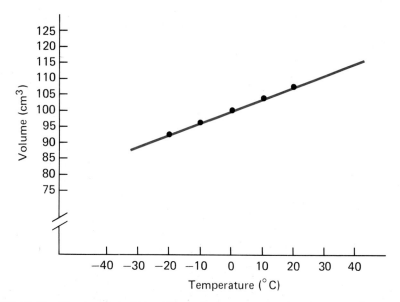

FIGURE 12-13 The results of the Charles's law experiment. As the temperature is varied, the gas volume changes in such a way that there is a linear (straight-line) relationship between the temperature and the volume. However, when the Celsius scale is used for temperature we cannot say that the volume is directly proportional to the temperature, because the volume is not zero when the temperature is zero.

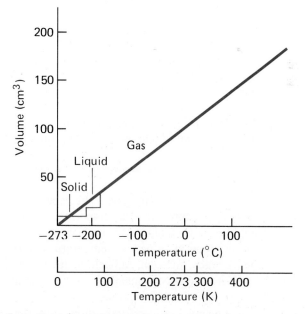

FIGURE 12-14 Extending the straight-line graph of temperature versus volume to the left, we find that the volume becomes zero (in theory) at a temperature of $-273°C$. This temperature is called **absolute zero** and is used as the zero of a new temperature scale called the Kelvin scale. A real gas does not follow the straight line down to zero volume.

(For many practical purposes, 273.15 may be rounded off to 273.) In symbols,

$$T_K = T_C + 273.15$$

Conversely,

$$\text{Celsius temperature} = \text{Kelvin temperature} - 273.15$$

or

$$T_C = T_K - 273.15$$

Example 12.7 Carbon dioxide solidifies at a temperature of $-78.5°C$ to form the white solid known as **dry ice**. What is the temperature of dry ice on the Kelvin scale?

Solution
$$T_K = T_C + 273.15$$
$$= -78.5°C + 273.15$$
$$= 194.7 \text{ K}$$

Note: We do not write "degrees Kelvin" or use a degree sign before the K because the unit is the kelvin. In the Celsius and Fahrenheit scales the unit is the degree.

Note on Temperature Scales

There is some confusion about the use of the Kelvin scale. A number of books use *degrees Kelvin* as the temperature unit on that scale. Thus, the melting point of ice is given as 273.15°K. Some other writings use *degrees Absolute* to mean the same thing, expressing the melting point of ice as 273.15°A. However, the Système International (SI), the current version of the metric system, recommends use of the *kelvin* as the unit of temperature; according to that scheme the melting point of ice is 273.15 K.

In all work dealing with the behavior of gases at various temperatures, it is necessary to express the gas temperature in Kelvin units. There are two reasons for using the Kelvin temperature scale in scientific work. First of all, a large amount of work is done at very low temperatures (cryogenic research)—temperatures at which oxygen, nitrogen, and helium are liquids. Since helium boils at a temperature of 4 K, it is most convenient to use the Kelvin scale to describe work done at the temperature of liquid helium.

Second, and most important, since the graph in Figure 12-14 goes to zero when the Kelvin temperature is zero, and since the volume of the gas increases in a straight line from that point, we conclude that *the volume of a given quantity of gas is directly proportional to its temperature measured on the Kelvin scale.* This statement is the modern version of Charles's law.

We express the above relationship by the equation

$$\text{volume} = \text{constant} \times \text{temperature (in kelvins)}$$

or

$$V = C \times T$$

where C is a constant of proportionality, whose value depends on the amount of gas and on the pressure, and T is the temperature *measured in kelvins*. (Notice that the constant C is not the same as the constant K introduced in Section 12.2.)

In many situations we know the volume of a gas at one temperature and want to determine what the volume of that gas will be at another temperature if the pressure is kept constant. Let V_1 be the known volume of gas at the temperature T_1. We would like to find V_2—the volume of the same amount of gas at another temperature, T_2. We know that

$$V_1 = C \times T_1 \quad \text{and} \quad V_2 = C \times T_2$$

We solve each of the two equations for the constant C:

$$\frac{V_1}{T_1} = C \quad \text{and} \quad \frac{V_2}{T_2} = C$$

Since the expressions on the left side of both equations are equal to the same constant C, they are equal to each other. We then write

$$\frac{V_2}{T_2} = \frac{V_1}{T_1}$$

Solve for V_2 by multiplying both sides of the equation by T_2:

$$V_2 = T_2 \times \frac{V_1}{T_1}$$

$$= \frac{T_2}{T_1} \times V_1$$

This equation tells us that the volume at temperature 2 is equal to the volume at temperature 1 multiplied by the ratio of temperature 2 to temperature 1.

It is essential to remember that the Kelvin temperature scale must be used in all of these gas problems. One of the most common errors that students make is to use the wrong temperature scale.

Example 12.8 A balloon has a volume of 5.00 m³ at a temperature of 20°C, which is a normal temperature for a cool day. What will be the balloon's volume on a very hot day, when the temperature rises to 35°C, assuming the pressure remains unchanged?

Solution

First convert Celsius temperatures to kelvins:

$$T_K = T_C + 273$$
$$T_1 = 20°C + 273 = 293 \text{ K}$$
$$T_2 = 35°C + 273 = 308 \text{ K}$$

Then set up the Charles's law equation:

$$V_2 = V_1 \times \frac{T_2}{T_1}$$
$$= 5.00 \text{ m}^3 \times \frac{308 \text{ K}}{293 \text{ K}}$$
$$= 5.26 \text{ m}^3$$

The balloon expands by about 5%:

$$\frac{5.26 - 5.00}{5.00} \times 100 = 5.2\%$$

Example 12.9 Given that a gas has a certain volume at 20°C, determine the temperature required to double the volume.

Solution Since the volume is directly proportional to the temperature, we can immediately say that the temperature must be doubled in order to double the volume. This can be seen from the Charles's law equation:

$$V_2 = \frac{T_2}{T_1} \times V_1$$

If $T_2/T_1 = 2$, then

$$V_2 = 2 \times V_1$$

However, do not jump to the conclusion that the temperature wanted is 2 times 20°C, or 40°C. As shown in the last example, this small change of temperature gives only a 5% increase in volume. Remember that in problems of this kind, the Kelvin temperature scale must be used. Therefore, the *Kelvin* temperature must be doubled.

Since the starting temperature is 273 K plus 20, or 293 K, the temperature required to double the volume is 2 times 293 K, or 586 K. Then, to find the corresponding Celsius temperature, subtract 273:

$$T_C = 586 \text{ K} - 273 = 313°C$$

Absolute Zero

Figure 12-14 shows that when a quantity of gas is cooled at constant pressure, the plot of volume versus temperature is a straight line that can be extrapolated back until it goes all the way down to zero. The same sort of thing happens if the gas volume is kept constant and its pressure is followed during the cooling. The pressure is seen to decrease linearly and extrapolate to zero at a temperature of $-273.15°C$.

Substances can be cooled down to 4 K by immersing them in liquid helium, and they can be cooled further by special techniques that remove from them all but a last bit of heat. An important fact revealed by such experiments is that the point where the straight line crosses zero is the *lowest* temperature that can be reached by any means. For this reason it is called **absolute zero**. Absolute zero is zero on the Kelvin temperature scale, corresponding to exactly $-273.15°$ Celsius. Absolute zero is a true zero of temperature. There are no negative temperatures on the Kelvin scale as there are on the Celsius and Fahrenheit scales.

Physically, an object at absolute zero has no thermal energy that can be extracted from it. It is not correct to say (as some do) that all molecular motion stops at absolute zero. There is still a very small "zero point" energy left inside the object, but there is no way of getting this energy out. Therefore at absolute zero an object has the smallest amount of internal energy that it can possibly have.

Interestingly, there is no way of cooling anything down to exactly absolute zero. We can approach it very closely—to within a thousandth of a degree—but can never quite make the final step to the bottom rung of the ladder.

There is a great amount of chemical and physical research done at temperatures near absolute zero. **Cryogenics**—the study of matter and energy at such low temperatures—is important because a great deal can be learned about the structure of matter when its atoms are not vibrating vigorously.

At very low temperatures, many materials behave in unfamiliar ways and display properties much different from those we expect. When cooled in liquid nitrogen, a banana becomes so hard that it can be used as a hammer, and a rubber ball becomes so brittle it will shatter like glass when dropped onto the floor.

One important practical application of cryogenics has been the development of **superconducting magnets**. When certain metals (alloys of niobium and tin, for example) are cooled to low temperatures, they gain the ability to conduct electric currents with no resistance at all. Therefore they are called **superconductors**. A superconducting ring can carry an electric current for thousands of years with no further source of energy, once the current is set up. This electric current produces a magnetic field, and so the coil carrying the current behaves just like a magnet. In the past, very strong magnets have used copper coils. Now, extremely powerful superconducting magnets can be made that use much less electric power than conventional copper coils.

Alloys of niobium and tin become superconductors at temperatures below 18 K. Other superconducting materials require lower temperatures to enter the superconducting state. Most metals never do exhibit superconductivity. One of the frontiers of research is the quest for materials that will become superconducting at higher temperatures, so that liquid nitrogen (boiling point 77 K) could be used for cooling instead of the liquid helium (boiling point 4 K) presently needed. Such a development would greatly reduce the difficulty and expense of superconductivity and would open the way for commercial developments of the greatest importance (for cxample, superconducting power lines).

12.4 Varying Both Temperature and Pressure

The Combined Gas Law

We have seen that the volume of a fixed quantity of gas is inversely proportional to the pressure if the temperature is held constant (Boyle's law). We have also seen that the same volume is directly proportional to the temperature if the pressure is held constant (Charles's law).

In real situations it is rare that one variable stays constant while the other two change. When the temperature changes, the pressure is likely to change at the same time. Therefore we now ask this question: What happens to the gas volume if both pressure and temperature change simultaneously?

The answer is that *the volume of a fixed quantity of gas is directly proportional to the temperature and inversely proportional to the pressure* even when both temperature and pressure change at the same time. Putting this relationship into equation form, we have

$$V = K \frac{T}{P}$$

You can see that if the pressure is held constant, the volume is directly proportional to the temperature. If the temperature is held constant, the volume is inversely proportional to the pressure. In this single equation, Boyle's law and Charles's law exist together; for that reason it is called the **combined gas law.**

The method of using the combined gas law is similar to the way we used Boyle's and Charles's laws. Usually we deal with problems in which the pressure, volume, and temperature (P_1, V_1, and T_1) under one set of conditions are known and we would like to find what happens to the gas under another set of conditions (P_2, V_2, T_2).

Variables under the first condition—P_1, V_1, and T_1—are substituted into the combined gas law, which is then rearranged (by dividing both sides of the equation by T_1 and multiplying by P_1) to give

12.4 VARYING BOTH TEMPERATURE AND PRESSURE

$$K = \frac{P_1 V_1}{T_1}$$

A similar equation can be written for the temperature, pressure, and volume under the second condition:

$$K = \frac{P_2 V_2}{T_2}$$

These equations tell us that under both sets of conditions the pressure multiplied by the volume and divided by the temperature is equal to the same constant K. Therefore the right sides of these two equations can be set equal to each other:

$$\frac{P_2 V_2}{T_2} = \frac{P_1 V_1}{T_1}$$

There are six quantities in this equation. If we know any five of these quantities, we can solve the equation for the sixth. For example, suppose we are given all three of the starting conditions—P_1, V_1, and T_1—so we know the state of the gas under the first condition. We then change the pressure and the temperature to new values P_2 and T_2 and solve for the new volume V_2.

We can solve the above equation for V_2 by multiplying both sides by T_2 and dividing by P_2, to obtain

$$V_2 = \frac{P_1}{P_2} \times \frac{T_2}{T_1} \times V_1$$

This appears at first to be a complicated equation, but it can be broken down into simple parts. What this equation tells us is that the final volume V_2 is equal to the initial volume V_1 multiplied by the ratio T_2/T_1 and then multiplied by the ratio P_1/P_2.

As always, the temperatures in these problems must be in kelvins.

Example 12.10   One liter of gas at room temperature (20°C) and a pressure of 1.0 atm is heated to a temperature of 200°C and is allowed to expand so that its final pressure is 0.50 atm. What is its final volume?

Solution   Write the equation

$$V_2 = \frac{P_1}{P_2} \times \frac{T_2}{T_1} \times V_1$$

Then convert temperatures from degrees Celsius to kelvins:

$$T_1 = 20°C + 273 \text{ K} = 293 \text{ K}$$

$$T_2 = 200°C + 273 \text{ K} = 473 \text{ K}$$

The given pressures are

$$P_1 = 1.0 \text{ atm}$$

$$P_2 = 0.50 \text{ atm}$$

It makes no difference what units are used for pressure, since we are taking the ratio of the two pressures. However, we must be sure that the pressures we use are *absolute* pressures rather than gauge pressures, which tell what the pressure is *above* atmospheric pressure.

Substituting into the equation, we have

$$V_2 = \frac{1.0 \text{ atm}}{0.50 \text{ atm}} \times \frac{473 \text{ K}}{293 \text{ K}} \times 1 \text{ L}$$

$$= 3.2 \text{ L}$$

Note: If we had merely changed the pressure, the final volume would have been 2.0 L. If we had changed the temperature alone, the final volume would have been 1.61 L. Since both were changed, the final volume is the product of the two numbers.

The Universal Gas Law

An equation that incorporates *all* the essential properties of a gas into a single relationship would be very useful. As we consider the combined gas law introduced in the last section—a relationship involving the pressure, the temperature, and the volume of a gas—we see that one important ingredient has been left out: the *amount* of gas.

Experiment shows that 2 g of hydrogen occupies twice as much volume as 1 g at the same temperature and pressure. Therefore we might think of including the *mass* of gas as a factor in the equation. However, 1 g of hydrogen occupies more volume than 1 g of oxygen. Therefore, if we used the mass of gas as a factor, we would need a different equation for each gas. Can we find one equation that will work for all gases?

Such an equation can be obtained by using the *number of moles of gas* to express the quantity of gas in the calculation, since a given number of moles of any gas always occupies the same volume under the same conditions of temperature and pressure. The resulting equation is called the **universal gas law**. It states that *the volume of a given amount of gas is directly proportional to the number of moles of gas, directly proportional to the temperature, and inversely proportional to the pressure.*

In symbols,

$$V = R\frac{nT}{P}$$

12.4 VARYING BOTH TEMPERATURE AND PRESSURE

Here, n is the number of moles of gas, T is the temperature, P is the pressure, and R is a constant of proportionality. What makes this equation so interesting is the fact that the constant R has the same numerical value for *all* gases. This can be shown by experiment. If you take a measured volume of a gas, find the number of moles by weighing, measure the temperature and the pressure, and substitute all these numbers into the equation

$$R = \frac{PV}{nT}$$

(the universal gas law solved for R), you will always find the same value for R, no matter what kind of gas you use for the experiment. For this reason R is called the *universal gas constant*. (Of course this result is a consequence of the fact that a mole of any gas always occupies the same volume at a given temperature and pressure.)

The actual numerical value of R will depend on the units used for measuring temperature and pressure (English or metric units), as shown in the following example.

Example 12.11 It is known from many measurements that 1 mol of any gas occupies a volume of 22.414 L at a temperature of 0°C (273.15 K) and at a standard atmospheric pressure of 101.32 kPa (76.0 cm Hg). These numbers can be used to find a numerical value for R by substituting into the equation

$$R = \frac{PV}{nT} = \frac{101.32 \text{ kPa} \times 22.414 \text{ L}}{1 \text{ mol} \times 273.15 \text{ K}}$$

$$= 8.314 \frac{\text{kPa-L}}{\text{mol-K}}$$

If pressure is measured in centimeters of mercury,

$$R = \frac{PV}{nT} = \frac{76.0 \text{ cm Hg} \times 22.414 \text{ L}}{1 \text{ mol} \times 273.15 \text{ K}}$$

$$= 6.236 \frac{\text{cm Hg-L}}{\text{mol-K}}$$

If V is measured in milliliters (mL) and pressure is measured in torrs (mm-Hg),

$$R = \frac{PV}{nT} = \frac{760 \text{ torr} \times 22414 \text{ mL}}{1 \text{ mol} \times 273.15 \text{ K}}$$

$$= 62{,}363 \frac{\text{torr-mL}}{\text{mol-K}}$$

You see that the unit for R is complex; it is a composite of all the units used in calculating the constant.

Notes About Units

In the SI system, if n is measured in gram-moles, V is measured in cubic meters, P in newtons per square meter (pascals), and T in kelvins, then

$$R = 8.31 \text{ N-m/mol-K}$$
$$= 8.31 \text{ Pa-m}^3/\text{mol-K}$$

Often the pressure is given in atmospheres and the volume is in liters. In that case,

$$R = 0.08206 \text{ atm-L/mol-K}$$

In doing practical problems, a good strategy is to convert the given units of pressure and volume into one of the above sets of units and then use the appropriate value of R. An alternative strategy is to remember that a gram-mole of any gas occupies 22.4 L at STP and then use the STP values in the given system of units to calculate a value for R.

The equation discussed above is called the *universal gas law* because it represents the condition of all gases, regardless of their composition. The word *universal* is used in the sense of "all encompassing." The universal gas law is a relationship involving the number of moles of a gas, its pressure, its temperature, and its volume. Given any three of these quantities, we can solve the equation for the fourth.

At this point it is necessary to emphasize that there are many exceptions to the universal gas law. It does not describe all gases equally well. At temperatures so low that the gases are close to becoming liquid the universal gas law is not obeyed at all. For such gases there is a more complicated but more exact equation that works better. For gases at reasonably high temperatures and low enough densities, however, the universal gas law is a good predictor of behavior.

A gas that obeys the universal gas law is called an **ideal gas** because it behaves the way a gas is ideally supposed to behave. For that reason the gas law is sometimes called the **ideal gas law**. In Section 12.6 we will explain the theoretical reasons for this ideal behavior. We will define just what a gas must be like in order to be an ideal gas, and we will see what there is about a real gas that makes it deviate from the predictions of the ideal gas law.

Example 12.12 What is the volume of 84.0 g of oxygen at a temperature of 20°C and a pressure of 176.0 cm of mercury?

Solution First find the number of moles of gas. The molar mass of oxygen is 32.0 g; the number of moles in 84.0 g is

$$n = \frac{84.0 \text{ g}}{32.0 \frac{\text{g}}{\text{mol}}} = 2.63 \text{ mol}$$

12.4 VARYING BOTH TEMPERATURE AND PRESSURE

The Kelvin temperature is

$$T_K = T_C + 273 = 20°C + 273 = 293 \text{ K}$$

Substitute these numbers into the universal gas law:

$$V = R\frac{nT}{P}$$

$$= 6.23 \frac{\text{cm Hg-L}}{\text{mol-K}} \times \frac{2.63 \text{ mol} \times 293 \text{ K}}{176.0 \text{ cm Hg}}$$

$$= 27.3 \text{ L}$$

Example 12.13 What is the volume of a container that contains 90.0 g of methane (CH_4) at a temperature of 40°C and a pressure of 75.0 kPa?

Solution The temperature is 40°C + 273, or 313 K. The molecular mass of methane is 12.0 + 4.0 = 16.0, so the number of moles of methane is

$$n = \frac{90.0 \text{ g}}{16.0 \frac{\text{g}}{\text{mol}}} = 5.63 \text{ mol}$$

Substituting into the universal gas law:

$$V = R\frac{nT}{V}$$

$$= 8.314 \frac{\text{kPa-L}}{\text{mol-K}} \times \frac{5.63 \text{ mol} \times 313 \text{ K}}{75.0 \text{ kPa}}$$

$$= 411 \text{ L}$$

Example 12.14 A tank with a volume of 100 L stores 230 mol of nitrogen at a temperature of 30°C. What is the pressure inside the tank?

Solution Solve the universal gas law for the pressure:

$$P = R\frac{nT}{V}$$

$$= 6.23 \frac{\text{cm Hg-L}}{\text{mol-K}} \times \frac{230 \text{ mol} \times 303 \text{ K}}{100 \text{ L}} = 4340 \text{ cm Hg}$$

$$= \frac{4340 \text{ cm Hg}}{76.0 \frac{\text{cm Hg}}{\text{atm}}} = 57 \text{ atm}$$

Note: We took care to convert 30°C to 303 K.

Example 12.15   A helium tank has a volume of 45.0 L. The gas inside is at a pressure of 12.3 atm (1250 kPa) and a temperature of 25°C. How many moles of helium are in the tank?

Solution   The temperature is 25°C + 273 = 298 K. Solve the universal gas law for the number of moles:

$$n = \frac{PV}{RT} = \frac{1250 \text{ kPa} \times 45.0 \text{ L}}{8.31 \frac{\text{kPa-L}}{\text{mol-K}} \times 298 \text{ K}} = 22.7 \text{ mol}$$

12.5   Dalton's Law of Partial Pressures

On the way toward formulating his atomic theory, John Dalton made an important contribution to the study of gases and their behavior. In 1801, Dalton was just beginning to visualize a gas as a collection of tiny particles. He knew that the atmosphere consists of a mixture of gases, chiefly nitrogen and oxygen. The question that came to mind was: What determines the pressure of a mixture of gases? In other words, how is the ideal gas law changed if there are two or more gases mixed in the same container?

The conclusion Dalton came to was this: *If a volume is occupied by a mixture of two or more different gases, the pressure of each gas is the same as the pressure it would exert if it were alone in the container, and the total pressure of all the gases is the sum of the individual pressures.* This statement is known as **Dalton's law of partial pressures**.

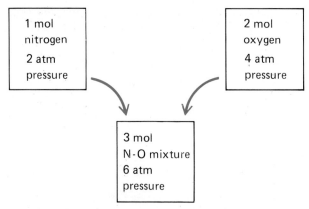

FIGURE 12-15 The law of partial pressures. The total pressure in a container of gas is equal to the sum of the partial pressures of each of the gases within the container.

Mathematically, the law says

$$P_T = P_1 + P_2 + P_3 + \cdots$$

where P_T is the total pressure of the gas mixture and P_1, P_2, etc., are the *partial pressures* of the individual gases. A partial pressure is the pressure that a gas would exert if it were by itself in the container at the same temperature.

The following example illustrates the meaning of Dalton's law. Suppose one container contains 1 mol (6.02×10^{23} molecules) of nitrogen at a pressure of 2 atm (see Figure 12-15), and another container of the same size at the same temperature contains 2 mol ($2 \times 6.02 \times 10^{23}$ molecules) of oxygen. The pressure in the second container will be 4 atm because there are twice as many moles (twice as many molecules) occupying the same volume. In other words, *each mole* of gas in this container contributes 2 atm to the total pressure.

Now suppose the content of the second container is pumped into the first so that the two gases are mixed together. The one container now contains a total of 3 mol ($3 \times 6.02 \times 10^{23}$ molecules) of gas. Since each mole contributes a pressure of 2 atm to the total pressure, the pressure of this mixture is 6 atm. In technical terms, the nitrogen is at a *partial pressure* of 2 atm and the oxygen is at a *partial pressure* of 4 atm, so the total pressure is 6 atm.

In terms of the above equation,

$$P_T = P_1 + P_2$$
$$= 2 \text{ atm} + 4 \text{ atm} = 6 \text{ atm}$$

FIGURE 12-16
Collecting a gas over water in a flask.

Example 12.16 A gas is bubbled through water and collected in an inverted flask, as shown in Figure 12-16. The result is a mixture of water vapor and the collected gas. If the water level inside the flask is adjusted to be equal to the level of the water in the pan outside the flask, the total pressure inside will equal the atmospheric pressure outside: 770.0 torr. The partial pressure of the water vapor in the flask is known to be 17.5 torr. What will be the pressure of the dry gas when the water vapor is removed?

Solution Let P_1 be the water vapor pressure and P_2 be the gas partial pressure. Then,

$$P_T = P_1 + P_2$$
$$P_2 = P_T - P_1$$
$$= 770.0 \text{ torr} - 17.5 \text{ torr} = 752.5 \text{ torr}$$

The partial pressure rule has an important consequence. Suppose two gases occupy the same volume. Both are at the same temperature, but they have different partial pressures, P_1 and P_2. There are n_1 mol of one gas and n_2 mol of the second gas in the container. The total pressure of the two gases when mixed together is

$$P_T = P_1 + P_2$$

where

$$P_1 = n_1 \frac{RT}{V} \quad \text{and} \quad P_2 = n_2 \frac{RT}{V}$$

Adding the two partial pressures gives

$$P_T = n_1 \frac{RT}{V} + n_2 \frac{RT}{V}$$

$$= (n_1 + n_2) \frac{RT}{V}$$

The V and T terms may be factored out, because both gases occupy the same volume and have the same temperature since they are mixed together. R is also factored out, since it is a constant. Finally, using the fact that the total number of moles n_T is equal to n_1 plus n_2, we have

$$P_T = n_T \frac{RT}{V}$$

This equation says that the total pressure is given by the universal gas law, where n is the *total number* of moles of gas, even though the container holds a mixture of several different kinds of gases. The *kind* of gas does not matter with the gas law, because the size of the gas molecules is insignificant compared to the distance between them. Therefore the nature of the individual molecules is not important in determining the relation between pressure and volume. In the next section we will present more details about the basic assumptions concerning the molecular theory of gases and will show how the universal gas law is a logical consequence of those assumptions.

12.6 Kinetic Theory and the Ideal Gas Law

Kinetic Theory

In this section we delve more deeply into the nature of gases. Our aim is to see how the behavior of gases results from the properties of the molecules that make up the gas. Boyle's law, Charles's law, and the universal gas law are *macroscopic laws*, describing the behavior of bulk matter. A look at the *microscopic world* inside a gas shows how this large-scale behavior results from the actions of the parts that make up the whole. The fundamental theory that explains the behavior of gases is called the **kinetic theory of gases.**

The word *kinetic* implies motion. Kinetic energy is the energy possessed by an object because of its motion. Kinetic theory is a theory of molecular motion. Scientists had to understand the nature of energy before they could invent kinetic theory, and so it did not become fully developed until the latter half of the nineteenth century. In particular, it was necessary for them to understand that heat is a form of energy, and this knowledge did not emerge until the very middle of the nineteenth century, when such men as Julius Robert Mayer and James Prescott Joule established our modern concept of energy.

An essential part of the energy concept is the idea that our perception of heat results from the motion of the molecules in the heated object: the faster the molecules are moving, the hotter the object is. The definition of temperature arises directly from this concept: *The temperature of any object is directly proportional to the average kinetic energy of its individual molecules.*

Once temperature had been defined in this fashion, the behavior of gases became easy to understand. The kinetic theory of gases became fully developed with the work of James Clerk Maxwell, a great Scottish mathematician and physicist, and Ludwig Boltzmann, an Austrian physicist. In the years following 1860, they worked out the details of the mathematical theory of gases. Much of the mathematics involved is exceedingly complex; here we shall speak only of the simplest parts of the theory.

The kinetic theory of ideal gases is based on the following assumptions:

1. An ideal gas consists of small molecules moving about in a disorganized (random) fashion. The distance between molecules is very large compared to the size of an individual molecule. Since *most of the gas is empty space*, the physical size of each molecule is unimportant in the simpler aspects of the theory.

2. As the molecules move about randomly, they collide with each other. These collisions are like billiard-ball collisions—that is, they are **elastic collisions** between infinitely small, hard spheres. The molecules do not tend to stick together, nor do they attract or repel each other except at the very instant of collision.

3. The molecules also collide with the walls of the container (or any surface in contact with the gas). The pressure of the gas is the cumulative effect of all these collisions. As each molecule rebounds from the wall, it pushes on the wall for an instant. (See Figure 12-17.) The sum of all these tiny pushes from all the billions of molecules adds up to the gas pressure. (It is like the force exerted by a stream from a fire hose. The water jet consists of millions of droplets; the total force is the sum of the tiny pushes from all the individual drops.)

4. The temperature of the gas is directly proportional to the average kinetic energy of the molecules, so the hotter the gas, the faster the molecules are moving. The average is used because the molecules do not all move at the same speed—some molecules move faster than others.

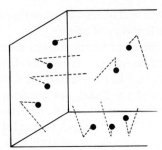

FIGURE 12-17 The pressure of a gas is the result of the numerous collisions between the molecules of the gas and the surface of the container in contact with the gas. Each collision imparts a tiny force to the wall, and the sum of all the little forces makes the totality of the force exerted by the gas on the wall.

Let us see how these basic assumptions lead to the universal gas law. (We will not go into the mathematics of the situation, but will describe qualitatively what happens.)

First of all, the gas pressure depends on only two major factors. The pressure depends on (1) the force with which each molecule hits the wall during each collision and (2) the number of collisions per unit area per unit time.

These two major factors are, in turn, determined by a number of minor factors. (1) The force imparted by each collision depends on the mass and the velocity of the molecule. (2) How frequently the molecules collide with the walls depends on (a) the number of molecules inside the container, (b) their velocities, and (c) the volume of the container. (The farther the molecules travel between encounters, the fewer encounters there will be per second.)

Both the mass and the velocity of each molecule enter into the kinetic energy of each molecule ($\frac{1}{2}mv^2$). Therefore, detailed mathematical calculations lead to the conclusion that *the pressure of a gas is directly proportional to the average kinetic energy of its molecules, which in turn is related to its temperature.* Thus, kinetic theory tells us that the pressure of a given volume of gas is directly proportional to its temperature.

The larger the container, the fewer the collisions between the gas molecules and the walls of the container per unit area. For example, suppose a 1-L tank contains enough gas so that one billion collisions occur per second in each square centimeter of inside surface. Now put the same amount of gas into a 2-L container. Two things happen. First, the molecules have farther to travel back and forth between collisions. Second, the collisions are spread out over a greater surface area. The net result is that only half a billion collisions occur per second per square centimeter, and the gas pressure is reduced to half what it was in the smaller volume.

In this way kinetic theory explains why, for a given temperature, the pressure of a gas is inversely proportional to its volume.

The same argument points out why the pressure is directly proportional to the number of moles of gas in the container; the more molecules there are, the more collisions per second there will be on each unit area of the container wall.

We have been able to explain the relationships among all the variables in the universal gas law equation

$$P = \frac{nRT}{V}$$

The only information the theory does not give is the numerical value of the constant R. The theory cannot predict the value of this number, for it is a quantity that can be found only by actual measurement.

Ideal and Real Gases

Earlier in this chapter we warned that not all gases obey the ideal (or universal) gas law. This law is, in fact, an approximation to the real gas law obeyed by a **real gas**. All gases are real, but some obey the ideal gas law with a high degree of accuracy and so behave like an "ideal" gas (a hypothetical gas that behaves in an ideal manner and obeys the ideal gas law).

What is the difference between an ideal gas and a real gas? To understand this difference, we must look again at the assumptions upon which the ideal gas law is based:

1. It is assumed that gas molecules are infinitely tiny, taking up no space at all. One result of this assumption is that when the temperature of the gas is reduced to zero, its volume vanishes. (This is what happens when the straight line of the volume-temperature graph is extrapolated back to absolute zero.) Of course, this cannot happen in reality because the molecules do have a definite (but small) volume. When the temperature of a real gas gets low enough, the molecules cluster so closely together that the size of the molecules determines the gas volume.

2. It is assumed that the molecules do not interact except to bounce off each other like very hard billiard balls. That is, it is assumed that there are no actions at a distance such as there are between electrically charged particles. However, when molecules are very close together, they attract each other by means of the weak, short-range van der Waals force, and this attraction pulls them even closer together. (The molecules behave as though they were a little bit sticky.) It is this attractive force that causes a gas to condense into a liquid. As a result, when a real gas is cooled, before it reaches absolute zero it first becomes a liquid and then (except in the case of helium) freezes into a solid.

At temperatures just above the liquefaction point, gases do not behave in the manner described by the ideal gas law. Instead, they obey a more exact (and more complex) equation called the **van der Waals equation**, which describes the behavior of a real gas better than does the ideal gas law. The closer it is to the liquefaction point, the further a real gas deviates from the properties of an ideal gas.

Gases such as oxygen and nitrogen are so far above their boiling points at room temperature that they obey the ideal gas law quite well. Carbon dioxide, on the other hand, deviates measurably from the ideal gas law because at room temperature it is not far above its freezing point. (At normal atmospheric pressure, CO_2 freezes directly into a solid without going through the liquid phase.) For example, at a pressure of 1 atm and a temperature of 0°C, a mole of CO_2 has a volume 0.5% below that predicted by the ideal gas law. Increasing the pressure to 10 atm brings the molecules closer together and increases the deviation from the ideal gas law to 8%.

In general, the higher the pressure and the lower the temperature, the more a gas will deviate from the ideal gas law.

Glossary

absolute pressure The actual pressure inside a container; the sum of the atmospheric pressure and the gauge pressure.

absolute zero The lowest possible temperature that can be reached; zero on the Kelvin scale.

atmosphere (atm) A unit of gas pressure assumed to be the standard pressure at sea level. By definition, it is equal to the pressure that will support 76.0 cm of mercury in a barometer.

bar A unit of gas pressure equal to 1×10^5 N/m^2, or 100 kPa.

barometer An instrument for measuring atmospheric pressure, invented by Torricelli.

Boyle's law A principle stating that the volume of a gas is inversely proportional to its pressure. (See also **Mariotte's law**.)

Charles's law A principle stating that the volume of a gas is directly proportional to the temperature (measured in kelvins) if the pressure is kept constant; also called Gay-Lussac's law.

combined gas law A principle stating that, for a given quantity of gas, the pressure multiplied by the volume and divided by the temperature (in kelvins) is equal to a constant.

condensation point The temperature at which a gas or vapor condenses into a liquid.

cryogenics The study of matter at very low temperatures.

Dalton's law of partial pressures A principle stating that when two or more gases are mixed together in one container, each gas exerts a partial pressure equal to the pressure it would exert if it were the only gas in the container, and the sum of the partial pressures is equal to the total pressure in the container.

density The mass of an object divided by its volume. For gases, density is usually given in grams per liter.

dry ice Solid carbon dioxide.

elastic collisions Collisions between two objects in which no internal energy is exchanged.

freezing point The temperature at which a liquid condenses into a solid.

gas A physical state of matter in which the molecules are far apart, are free to move, and have very little attraction for each other. A gas tends to fill any container it occupies.

gauge pressure The difference between the pressure inside a container and the atmospheric pressure outside.

Gay-Lussac's law See **Charles's law**.

general gas law See **combined gas law**.

general properties of gases Characteristics of gases that apply to all gases rather than a specific gas.

ideal gas A gas that obeys the ideal gas law; a model of a gas based on infinitely small, noninteracting molecules.

ideal gas law A simplified version of the van der Waals equation which assumes that all gases react in an identical, ideal manner: $PV = nRT$, where P is pressure, V is volume, n is the number of moles of gas, T is temperature, and R is a constant.

kilopascal (kPa) One thousand pascals or one thousand newtons per square meter; a unit of pressure in the SI system.

kinetic theory of gases An explanation of the properties of gases based on the idea that a gas is a cluster of molecules in motion, and that the average kinetic energy of the molecules is directly proportional to the gas temperature.

linear relationship A relationship between two variables of the form $y = kx + b$, where k and b are constants. When y is plotted as a function of x, a straight line results.

manometer A gauge that measures gas pressures within a container.

Mariotte's law Boyle's law, with the additional provision that the temperature remain constant as the pressure varies.

millibar One-thousandth of a bar.

newton (N) A basic unit of force in the metric system; the amount of force that will give a mass of 1 kg an acceleration of 1 m/sec^2 (named after Sir Isaac Newton, 1642–1727).

particular properties of gases Characteristics that relate to a specific gas and that allow one to tell the difference between that gas and another.

pascal (Pa) A unit of pressure equal to 1 N/m^2.

pressure The force exerted on a unit area of any surface. (The force may arise from either a gas, a liquid, or a solid.)

real gas A gas that obeys the van der Waals gas law rather than the ideal gas law.

standard atmospheric pressure The pressure that will support 76.0 cm of mercury in a barometer; nominally equal to atmospheric pressure at sea level, or 14.696 lb/in^2.

superconducting magnets Magnetic coils made of superconductors, which can carry an electric current for an indefinite period of time without a source of energy.

superconductors Substances that conduct electric current with no resistance. (To date, all known superconductors operate only at extremely low temperatures.)

torr A unit of pressure equal to 1 mm of mercury in a manometer.

universal gas law $PV = nRT$, where P is pressure, V is volume, n is the number of moles of gas, T is temperature, and R is the universal gas constant.

vacuum A volume of space in which the gas pressure is lower than the atmospheric pressure. (A hard vacuum, or high vacuum, is one in which the number of molecules per unit volume is less than 10^{-6} that of the atmosphere.)

van der Waals equation An equation developed by Johan Didrik van der Waals (1837–1923) to describe the gaseous state of matter:

$$\left(P + \frac{n^2 a}{V^2}\right)(V - nb) = RT$$

where P is pressure, V is volume, T is temperature, n is the number of moles of gas, a and b are constants related to specific properties of the gas, and R is the universal gas constant.

vapor The gaseous phase of a liquid or a solid.

Problems and Questions

12.1 Measuring Gas Pressure

12.1.Q Define or explain the following:
 a. a general property
 b. a particular property
 c. the density of a gas
 d. the pressure exerted by a gas

12.2.Q Which of the following statements refer to general properties of gases, and which refer to particular properties of gases? Explain your answers.
 a. The gas smells like rotten eggs.
 b. The gas is colorless.
 c. The gas is at a pressure of 42 lb/in^2.
 d. The gas has a volume of 22.4 L.
 e. The gas reacts with water explosively.

12.3.Q Why is it necessary to establish a standard atmospheric pressure?

12.4.Q Why is it more useful to discuss the pressure of a gas rather than the force it exerts on the walls of its container?

12.5.Q Two gas samples are brought to a laboratory for identification. Decide which of the following sets of samples are of the same gas or of different gases.
 a. temperature, pressure
 b. pressure, volume
 c. volume, temperature
 d. density, pressure
 e. color, temperature
 f. odor, volume

12.6.Q Compare the densities of gases and liquids in general and give the fundamental reason for the difference you find.

12.7.Q The expanding gas in the cylinder of an automobile engine exerts a force of 1120 lb on the piston. The piston area is 9.4 in^2. What is the gas pressure in pounds per square inch?

12.8.Q A hydraulic lift raises an automobile weighing 2500 lb. The piston operating the lift has an area of 20.5 in^2. What is the pressure of the fluid in the cylinder?

12.9.Q Discuss, define, or explain the following.
  a. the historical development of the barometer
  b. the quotation "nature abhors a vacuum"
  c. an anthropomorphic model of nature
  d. the mercury barometer
  e. the aneroid barometer
  f. the manometer
  g. gauge pressure
  h. absolute pressure

12.10.Q Why isn't the concept that "nature abhors a vacuum" accepted today?

12.11.Q Why is it important to know barometric pressure readings?

12.12.Q Describe Pascal's crucial experiment to test Torricelli's hypothesis, and explain how Pascal's observation demonstrated that the height of the mercury column measured the pressure of the atmosphere above it.

12.13.Q How is a manometer different from a barometer?

12.14.Q How would the reading on a barometer change if you took it to the bottom of a mine shaft?

12.15.Q Explain how a suction cup works.

12.16.Q Stage magicians show the following trick: The magician places a piece of paper over the open top of a tumbler filled to the brim with water. The tumbler is then turned upside down and the water does not spill out. Explain what happens in terms of the concepts presented in this chapter.

12.17.Q Could a barometer attached to the outside of an airplane serve as an altitude meter (an altimeter)? Explain your answer.

12.18.Q The daily weather report lists the atmospheric pressure in "inches." What is being measured in inches? Is it the length of the barometer? What changes when the pressure changes?

12.19.Q If air were present above the mercury in a barometer, what effect would it have on the barometer readings?

12.20.Q Define or explain the following terms:
  a. standard atmospheric pressure
  b. one atmosphere
  c. one bar
  d. one pascal
  e. vacuum

12.21.Q Why is it proper to say that when a light bulb breaks it implodes rather than explodes?

12.22.Q What does the weather forecaster mean by the statement "The mercury is falling"?

12.23.P a. Is 1350 mb more or less than 1 atm?
  b. Is 35 torr more or less than 1 atm?

12.24.P The TV weatherperson reports that the current barometric pressure is 28.76 in of mercury.
  a. Is this pressure higher or lower than standard atmospheric pressure?
  b. State the above pressure reading in the following units:
    (1) millimeters of Hg
    (2) torrs
    (3) atmospheres
    (4) newtons per square meter
    (5) millibars

12.25.P An undersea diver reports that the pressure on the diving bell is 4 atm. Give the pressure in
  a. pounds per square inch
  b. bars
  c. centimeters of Hg
  d. newtons per square meter

12.26.P The pressure between the double walls of a vacuum bottle (a Dewar flask or a "Thermos" bottle) is 0.1 torr. Give this pressure in
  a. millimeters of Hg
  b. millibars
  c. atmospheres
  d. pascals

12.27.P A TV picture tube is filled with gas at a pressure of 10^{-6} torr. Express this pressure in
  a. millimeters of Hg
  b. millibars
  c. atmospheres
  d. inches of Hg
  e. pascals

12.2 Boyle's Law

12.28.Q Define or explain the following terms:
  a. Boyle's law
  b. Mariotte's improvement on Boyle's law
  c. inversely proportional
  d. the Magdeburg hemispheres experiment

12.29.Q a. Describe Boyle's experiment with the mercury-filled glass tube.
  b. Describe Boyle's findings.

12.30.Q Give a theoretical explanation for Boyle's law in terms of the molecular model of gases.

12.31.Q As the piston in a bicycle tire pump moves up and down, what happens to the pressure of the air trapped in the pump?

12.32.Q A weather balloon is launched to record conditions high in the atmosphere. At first the balloon looks like a long sausage. What will it look like at a high altitude? Explain your answer in terms of Boyle's law.

12.33.Q At a football game, a number of helium balloons are released. They soar high into the air. A short time later they come falling back to earth, broken open. Explain what happened.

12.34.Q A balloon filled with helium is put into a freezer. Describe what happens to the volume of the balloon.

12.35.Q A sealed plastic bag of frozen vegetables is taken out of the freezer and dropped into boiling water. A few minutes later the bag bulges outward. Explain why.

12.36.P A scuba diver rises from the bottom of the water to the surface in an emergency ascent. The pressure at the bottom is 3 atm, and at the surface it is 1 atm. The volume of the diver's lungs at the start of the ascent is 2200 mL. To what volume would the diver's lungs expand if she did not exhale to reduce the amount of air contained within them?

12.37.P A jet plane uses 300 m^3 of air per min while it is on the runway, where the air pressure is 101 kPa. In the air it uses 1800 m^3/min. What is the atmospheric pressure at the flight altitude? (Assume that the engines must use the same number of moles of air per minute to support the fuel combustion.)

12.38.P A piston is moved outward so that the volume of gas in the cylinder expands from 2.5 L to 25 L. The gas is initially at a pressure of 650 torr. What is its final pressure?

12.39.P A sample of gas has a volume of 485 mL at 25°C and 760 mm Hg pressure. What volume will it occupy at 285 mm of pressure and a temperature of 25°C?

12.40.P A gas has a volume of 21.0 L at a pressure of 910 kPa and a temperature of 210 K. What is the pressure when its volume is changed to 6.9 L at 210 K?

12.41.P A balloon is filled with helium at 25°C and a pressure of 1 atm. The balloon rises until the pressure is 0.87 atm, the temperature is 25°C, and the volume is 850 L. What volume of He was put into the balloon on the ground?

12.42.P On the upstroke of a piston, the cylinder has a volume of 105 mL and a pressure of 1.1 atm. At the bottom of the stroke, the volume is 55.0 mL. What is the pressure at the bottom?

12.43.P Imagine that the inside of a TV picture tube is lined with a thin rubber balloon. The air pressure inside the tube is 10^{-6} torr, and the balloon has a volume of 3000 cm^3. The glass tube is cracked, so the balloon is now subject to an atmospheric pressure of 745 torr. What is the final volume of the balloon?

12.3 Charles's Law

12.44.Q Describe Charles's/Gay-Lussac's law (original version).

12.45.Q Describe an experiment you could do to verify Charles's law.

12.46.Q How many years elapsed between the work of Boyle and that of Charles? What was the reason for this great delay?

12.47.Q Define or explain the following:
  a. Charles's law (modern version)
  b. directly proportional
  c. condensation point
  d. freezing point
  e. Kelvin temperature scale
  f. cryogenics

12.48.Q Is it correct to say that the volume of a gas is proportional to the temperature measured on the Celsius scale? What kind of relationship is there between volume and temperature?

12.49.Q Why is it inappropriate to check the pressure of an auto tire after a long, high-speed drive? When is the best time to measure tire pressures?

12.50.Q Why do aerosol cans have instructions on the label that say, "Do not incinerate or place near a fire"?

12.51.P A gas at a temperature of 20°C with a volume of 2.5 L is warmed to 100°C. What is its new volume?

12.52.P A gas that has a volume of 2.5 L at a temperature of 20°C is heated at constant pressure until it expands to 10.0 L. What is its new temperature in degrees Celsius?

12.53.P A gas is warmed until it expands to three times its original volume. If the gas started at 10°C, what is its final temperature in degrees Celsius?

12.54.P If 1 L of helium is warmed from 0°C to 100°C, what is its new volume?

12.55.P A 5000-L volume of ammonia gas is cooled from a temperature of 1000°C to a temperature of 20°C, with the pressure remaining unchanged at 500 kPa. What is the final volume of the gas?

12.56.P A nitrogen-filled balloon has a volume of 40 cm^3 at standard temperature and pressure. What is the volume at a temperature of 200°C and 1 atm of pressure?

12.57.P A sample of gas is heated from 20°C to some higher temperature. While being heated (at constant pressure), it expands from a volume of 2.0 L to a volume of 6.0 L. Another sample of the same gas is heated from 60°C to a higher temperature, and it also expands from 2.0 L to 6.0 L. Is the change in temperature the same in both cases?

12.58.P If 2 L of gas is heated from 20°C to 40°C and an equal volume of the same gas is heated from 60°C to 80°C, are the final volumes the same or different?

12.59.P **a.** A 1.0-L volume of gas at a pressure of 760 torr and a temperature of 200 K is allowed to expand to 2.0 L without change of temperature. What is the new gas pressure?
  b. The gas in part a is now heated to 273 K, with the pressure kept constant. What is the volume after the second change? (*Note*: Use both Boyle's and Charles's laws.)

12.60.Q What is absolute zero?

12.61.Q What is the relationship between K, °K, and °A?

12.62.Q Describe some changes in properties of matter at low temperatures.

12.63.Q Describe the importance of the temperature −273.15°C.

12.64.Q Is it possible to reduce the internal energy of a substance to zero?

12.65.Q What happens to a gas when cooled toward 0 K?

12.66.Q What is a superconductor?

12.67.Q Why is it important to have a temperature scale that has an "absolute zero"?

12.68.Q What kind of problems require the use of the Kelvin temperature scale?

12.69.Q What is one important practical use of superconductors?

12.4 Varying Both Temperature and Pressure

12.70.Q Define the combined gas law and explain its meaning.

12.71.Q Explain why it is not permissible to use degrees Celsius in working problems with the gas law.

12.72.Q Explain how $P_1 V_1 / T_1$ can be set equal to $P_2 V_2 / T_2$.

12.73.P A gas has a volume of 1.00 L at a temperature of 0°C and a pressure of 760 torr. Its temperature is changed to 100°C and its pressure is changed to 2.00 atm. What is its new volume?

12.74.P A quantity of helium has a volume of 2 L at a temperature of 0°C and a pressure of 500 torr. It is warmed to a temperature of 50°C, and the pressure is allowed to stay at 500 torr.
  a. Is the gas volume larger or smaller after warming?
  b. What is the new volume in liters?

12.75.P A quantity of hydrogen has a volume of 35.0 mL at a temperature of 20°C and a pressure of 0.010 torr. It is heated to a temperature of 100°C while the volume is kept constant. What is the new gas pressure?

12.76.P A volume of gas occupies 12.4 L at a temperature of 5.8°C and a pressure of 3611 mb. It is allowed to expand to 24.2 L while the pressure decreases to 1509 mb. What is the new temperature of the gas in degrees Celsius?

12.77.P A quantity of ammonia occupies 1500 L at a pressure of 4.6 atm and a temperature of 300°C. It is stored at a pressure of 1.0 atm and a temperature of 12°C. What must the volume of the storage container be?

12.78.P A sample of gas weighs 25.5 g at STP. The temperature is raised to 100°C while the volume is kept constant.
   a. What is the final gas pressure?
   b. What is the change in pressure?

12.79.P The gas sample used in Problem 12.78 is heated from STP to 100°C, but this time the pressure is kept constant. What is the ratio of the final volume to the initial volume?

12.80.P A 15.5-L sample of nitrogen is generated at a temperature of 500 K and a pressure of 3.2 atm. It is stored at 20°C and a pressure of 101 kPa. What is its volume in the storage container?

12.81.P A gas is compressed in a cylinder so that its volume is reduced to one-tenth of its original value. At the same time, its Kelvin temperature doubles. What is the ratio of its final pressure to its original pressure?

12.82.P A weather balloon is at STP on the ground. It rises to a high altitude where the pressure is 152 torr and the temperature is −30°C. If the balloon has a volume of 51.7 L on the ground, what is its volume aloft?

12.83.Q a. Explain how we can assume that the volume (V) of a gas is directly proportional to the number of moles (n) of the gas.
   b. Under what conditions is the volume directly proportional to the number of moles of gas?

12.84.Q State and explain the universal gas law.

12.85.Q Does the universal gas law apply to all gases?

12.86.Q What important assumption do we make whenever we use the universal gas law to solve a problem?

12.87.Q If a given quantity of gas is in a container of fixed volume, and if the temperature is determined by the surroundings, what remaining property of the gas adjusts itself to all these conditions? (That is, what property can be calculated from the gas law?)

12.88.P What is the volume of 2.5 mol of an ideal gas at STP?

12.89.P A tank contains 11.2 L of N_2 at a pressure of 1×10^3 kPa and a temperature of 20°C.
   a. How many moles of N_2 are in the tank?
   b. How many grams of N_2 are present?
   c. If the temperature were changed to 40°C and all other factors except the pressure were kept constant, how many grams of N_2 would there be in the tank?

12.90.P Consider 11.2 g of O_2 at a pressure of 1520 torr in a 3.0-L volume.
   a. What is the temperature of the gas in degrees Celsius?
   b. If 22.4 g of O_2 is added to the container and the temperature is raised to 50°C while the volume is kept constant, what is the new gas pressure?

12.91.P A 10-g sample of gas is in a 9.0-L container at a pressure of 450 torr and a temperature of 220 K. What is the molecular weight of this gas?

12.92.P What is the mass of 5 L of CO_2 stored at a temperature of 273 K and a pressure of 1 atm?

12.93.P What is the volume of 75 g of He stored at a temperature of 20°C and a pressure of 120 kPa?

12.94.P Suppose the gas in Problem 12.93 is moved to a new tank at a pressure of 1 atm and a temperature of 10°C.
   a. What will its new volume be?
   b. What will its mass (in grams) be under these new conditions?
   c. How many moles of gas are present under the old and the new conditions?

12.95.P Calculate the value of R using cubic meters as the unit of volume and atmospheres as the unit of pressure.

12.96.P Calculate the density of helium at STP in units of grams per liter.

12.97.P A sample of a gas has a mass of 12.1 g and occupies a volume of 10.0 L at 16°C and a pressure of 780 torr. What is the molecular weight of this gas?

12.98.P An experiment is carried out in a pressure chamber at a pressure of 4.2 atm and a temperature of 1000°C. The gas produced has a volume of 3.45 L. Find the volume that the gas would occupy at STP.

12.99.P A plant is grown at a low pressure, 40 torr, and a temperature of 20°C. It produces 0.25 mol of CO_2 in a certain amount of time.
   a. How many liters of CO_2 are produced under the experimental conditions?
   b. How many liters of CO_2 does this represent at STP?

(*Hint*: First do part b; then do part a.)

12.100.P a. What is the volume of 10 mol of an ideal gas at STP?
  b. What is the volume of the gas in part a at a pressure of 3.5 atm and a temperature of 400 K?

12.5 Dalton's Law of Partial Pressures

12.101.Q State Dalton's law of partial pressures, and explain its meaning.

12.102.Q Explain the relationship between Dalton's law of partial pressures and the universal gas law.

12.103.Q What is assumed to be true about the two or more gases referred to in Dalton's law of partial pressures?

12.104.P Two gases are mixed together. One has a partial pressure of 400 torr and the other has a partial pressure of 600 torr. What is the combined pressure, in torrs and in atmospheres?

12.105.P Two gases are mixed together in a 2.5-L container at a pressure of 70 cm and a temperature of 20°C. One gas originally had a volume of 1 L at a pressure of 50 cm and a temperature of 20°C.
  a. What is the partial pressure of that gas after being mixed?
  b. What is the partial pressure of the other gas after being mixed?

12.106.P An emergency life raft consists of a compressed gas cylinder and an empty rubber tube. If the raft has a gas volume of 41.5 L at STP, what volume of compressed gas would be needed in the cylinder at a pressure of 4500 kPa and a temperature of 20°C?

12.107.P a. What volume will be occupied by 0.50 g of H_2 at STP?
  b. If 0.50 g of H_2 is mixed with 0.50 g of O_2 at STP, what volume will the mixture occupy?

12.108.P A mixture of CO and CO_2 contains 1.5 mol of CO and 3.0 mol of CO_2. The partial pressure of the CO is 75 kPa.
  a. What is the partial pressure of the CO_2?
  b. What is the total pressure of the mixture?

12.109.P Equal masses of He and Xe are mixed together. The partial pressure of the He is 101 kPa. What is the partial pressure of the Xe?

12.110.P A gas is mixed with water vapor. The gas has a volume of 1.4 L at a temperature of 25°C and a pressure of 1.0 atm. The partial pressure of the water vapor at 25°C is 23.8 torr. What is the total pressure of the gas–water vapor system?

12.111.P Two liters of N_2 at a pressure of 700 torr and a temperature of 10°C is mixed with 6.75 L of O_2 at a pressure of 800 torr and a temperature of 10°C. The final volume is 6.75 L. What is the pressure of the gas mixture?

12.112.P A 0.5-L can is pressurized to 150 kPa at a temperature of 0°C. What would be the pressure if an equal number of moles of a second gas were added at the same temperature?

12.113.P Air is a mixture of approximately 80% N_2 and 20% O_2. What is the partial pressure of the N_2 at STP?

12.6 Kinetic Theory and the Ideal Gas Law

12.114.Q a. What is kinetic theory?
  b. What is kinetic energy?

12.115.Q What is the meaning of the word *kinetic* in Question 12.114?

12.116.Q What is the relationship between temperature and kinetic energy?

12.117.Q What is the relationship between heat and energy?

12.118.Q State the basic assumptions of the kinetic theory of gases.

12.119.Q What is meant by an *elastic collision*?

12.120.Q What causes the pressure of a gas on the walls of its container?

12.121.Q Explain why the pressure of a gas is proportional to its temperature.

12.122.Q Explain why the number of molecules in a container of gas determines its pressure.

12.123.Q Explain how the volume of a container of gas determines its pressure.

12.124.Q Can the constant R be calculated from kinetic theory alone? How is it obtained?

12.125.Q What is an ideal gas?

12.126.Q What are the two essential features of a real gas that make it behave differently from an ideal gas?

12.127.Q What are van der Waals forces?

12.128.Q What is the van der Waals equation?

12.129.Q Under what conditions should you expect measurable deviations from the ideal gas law?

12.130.Q Compare the amount of molecular motion in two gas systems, identical except for the fact that one gas is at a lower temperature than the other.

12.131.Q Compare the size of a molecule of liquid H_2O with that of a molecule of H_2O vapor. Compare the distances between molecules in the liquid and the gaseous H_2O.

12.132.Q Which gas would you expect to have larger deviations from the ideal gas law—a gas consisting of polar molecules or one consisting of nonpolar molecules?

12.133.Q Compare equal volumes of a real gas and an ideal gas at the same temperature. Which would you expect to have the greater pressure?

12.134.Q One tank contains 2 mol of a gas. Another tank contains 10 mol of the same gas at the same temperature.
   a. Is it possible for the average kinetic the molecules in the two tanks.
   b. Compare the total kinetic energy of all the molecules in the two tanks. (This is the total thermal energy.)

12.135.Q Some of the molecules in a gas move faster than others. As they collide, they exchange energy.
   a. It is possible for the average kinetic energy of the molecules to change if the gas is isolated from the outside world?
   b. If energy enters the gas container from the outside world, what happens to the average kinetic energy of the molecules?
   c. In part b of this question, what happens to the temperature of the gas?

Self Test

1. A container of gas at a given temperature, pressure, and volume (PVT) is heated.
   a. What happens to the pressure if the volume is kept constant?
   b. What happens to the volume if the pressure is kept constant?
   c. Explain the changes in parts a and b in terms of the kinetic theory of gases.

2. A container of gas is at a given PVT. Gas is added until the number of moles has doubled.
   a. What happens to the pressure if V and T are kept constant?
   b. What happens to the volume if P and T are kept constant?
   c. What must happen to the temperature if both P and V are to stay unchanged?
   d. Can the temperature change in part c of this question occur by itself, or must there be some help from the outside?

3. One mole of a gas at STP is cooled to $-30°C$ and is compressed to a volume of 11.2 L. What is its new pressure in pascals?

4. A gas occupies a cylinder at a temperature of 20°C. If we want to double its volume (keeping P and n constant), what change must we make in its temperature? (*Hint*: First find the change in temperature on the Kelvin scale; then convert to degrees Celsius.)

5. A 10-g sample of neon occupies a volume of 15.5 L at a temperature of 0°C. What is the pressure, in atmospheres, of this gas?

6. A gas-filled tube at a pressure of 790 torr is found to contain a mixture of oxygen and water vapor. The partial pressure of the water vapor is 13 torr. What is the pressure of the pure oxygen?

7. a. Describe how the pressure of a gas on the walls of its container is related to the motion of the gas molecules.
   b. Explain in terms of kinetic theory how the pressure is related to the gas temperature.

8. The molecules of gas in a container are continually colliding with the walls of the container. Explain how the number of collisions per second is affected by each of the following changes.
   a. lowering the temperature
   b. reducing the pressure
   c. increasing the volume
   d. adding more molecules

9. Describe the differences between an ideal gas and a real gas.

10. What would be the volume and pressure of a sample of an ideal gas at 0 K? What happens to a real gas as it approaches that temperature?

13
Solutions

Objectives After completing this chapter, you should be able to:
1. Give examples of the different types of solutions (solid-liquid, solid-solid, etc.).
2. Express concentrations in several different ways.
3. Solve problems involving concentration expressed in grams per liter, moles per liter, and percentage composition.
4. Explain the concepts of dilute and concentrated solutions.
5. Explain the concept of solubility.
6. Explain how the dipole-dipole interaction (hydrogen bond) leads to the solubility of polar and ionic compounds in water.
7. Explain why nonpolar compounds are insoluble in water.
8. Explain the meaning of molarity and solve problems involving molarity of a solution.
9. Explain the meaning of normality and solve problems involving the normality of a solution.

13.1 Properties of Solutions

Types of Solutions

The possible solvents and solutes divide into three broad classes—solids, liquids, and gases. Thus we can expect nine different types of **solutions**, one for each solvent-solute combination. Some examples of these combinations are given in Table 13-1.

Solid-solid solutions include the class of **alloys**. Common alloys are 14-carat gold (copper and gold), brass (zinc and copper), and steel (carbon and iron, usually with other metals such as chromium and iridium). They are made by mixing together the molten metals and then allowing them to cool.

Liquids dissolved in solids are not uncommon. Metals saturated with oil are used to make bearings with permanent lubrication. In living matter, liquids pass through solid cell membranes. During this process, called **osmosis**, the liquid is dissolved in the solid.

Gases dissolved in solids are more common than most people realize. All solids absorb gases from the atmosphere, so a vessel being evacuated to a very low pressure must be "outgassed" by heating in order to drive out unwanted gases from the container walls. Hydrogen, in particular, dissolves readily in metals. Because hydrogen tends to diffuse directly through the walls of containers, it is difficult to make a hydrogen container that does not leak to some extent.

The solutions most often discussed are solids dissolved in liquids (usually water). A large proportion of inorganic reactions take place in aqueous (water) solutions. However, in organic chemistry a wide variety of other solvents are used.

Liquid-liquid solutions are also very common. Alcohol diluted with water and ether dissolved in alcohol are examples.

Gases dissolved in liquids comprise an important class of solutions. The carbon dioxide that dissolves from the atmosphere into the ocean offsets to some

TABLE 13-1 *Examples of Solute-Solvent Combinations*

Solvent	Solute		
	Solid	Liquid	Gas
Solid	copper in gold (gold alloy)	oil in bearing metal (permanent lubrication)	CO in charcoal filter
Liquid	sugar in water (tea)	isopropyl alcohol in water (rubbing alcohol)	CO_2 in water (soda)
Gas	smog	water vapor in air	oxygen in nitrogen (air)

Note: The definitions in this table are somewhat arbitrary. The term *water vapor in air* is a reference to the ability of water (a liquid under ordinary conditions) to disperse its molecules in vapor form so that they can mix with air without settling out. Water vapor is not to be confused with a *fog*, which consists of a *suspension* of water droplets in air.

extent the increase in atmospheric carbon dioxide due to the combustion of coal and oil. It is atmospheric oxygen dissolved in water that allows fish to sustain life.

We do not usually think of a solid or a liquid as dissolving in a gas. However, tiny particles of solids are formed, for example, in exhaust emissions, and if the particles are small enough to remain suspended without settling out they are called a **colloidal suspension**. Particles consisting of only a few molecules may be considered a solution of the vaporized solid in air.

Gas-gas solutions consist of any mixture of gases. Here the term *solution* is used in its broadest sense.

Concentration

The **concentration** of a solution is the amount of solute in a unit volume of solution. Concentration may be expressed both qualitatively and quantitatively.

A solution that has a *small* amount of solute in a given amount of solution is said to be **dilute**. In contrast, a solution with a *large* quantity of solute in a given volume of solution is said to be **concentrated**. There is, however, no official definition of small or large. So although these qualitative terms have some descriptive use, there is need for quantitative measures that tell *how much* solute there is for a given amount of solvent.

There are four standard quantitative measures:

1. Grams of solute per liter (or per 100 mL) of solution.

2. Percentage concentration by volume. (This was treated in Chapter 11.)

3. Percentage concentration by mass. This form is appropriate for alloys.

4. Number of moles of solute per liter of solution. This is the form most often used in quantitative inorganic chemistry. (We will deal with this topic separately in Section 13.3.)

In general, concentration is calculated by use of the formula

$$\text{concentration} = \frac{\text{quantity of solute}}{\text{volume of solution}}$$

Concentration may also be defined in terms of quantity of solute per unit quantity of solvent. Since the volume of solution is not necessarily the same as the volume of solvent, these two definitions are not the same. For example, if 20 mL of sulfuric acid is mixed with 50 mL of water, the volume of solvent is 50 mL, but the volume of solution is (approximately) 70 mL.

Furthermore, in some systems the volume of solution is less than the combined volumes of solute and solvent. This occurs when the solute molecules can locate themselves in spaces between the solvent molecules, making the total volume less than might be expected. Therefore we must be careful in labeling solutions to distinguish between *solvent* and *solution* volumes.

Most often it is the volume of solution that is measured, so the definition of concentration in terms of quantity of *solution* is applicable.

Example 13.1 A solution of $CuCl_2$ contains 65.3 g of $CuCl_2$ in 75.0 mL of solution. What is the concentration in grams per liter?

Solution
$$\text{no. of liters} = 75.0 \text{ mL} \times \frac{0.001 \text{ L}}{1 \text{ mL}}$$
$$= 0.0750 \text{ L}$$
$$\text{concentration} = \frac{\text{mass of solute}}{\text{volume of solution}}$$
$$= \frac{65.3 \text{ g CuCl}_2}{0.0750 \text{ L}}$$
$$= 871 \text{ g/L}$$

Example 13.2 A solution of NaOH has a concentration of 25.0 g/L. How many liters of solution contain 4.30 g NaOH?

Solution
$$\text{concentration} = \frac{\text{mass of solute}}{\text{volume of solution}}$$

Solve for volume of solution:
$$\text{volume of solution} = \frac{\text{mass of solute}}{\text{concentration}}$$
$$= \frac{4.30 \text{ g NaOH}}{25.0 \frac{\text{g NaOH}}{\text{L}}}$$
$$= \frac{4.30}{25.0} \text{ g NaOH} \times \frac{\text{L}}{\text{g NaOH}}$$
$$= 0.172 \text{ L}$$

Example 13.3 Gold is generally alloyed with copper and silver to increase its hardness. In the goldsmith's system of rating the purity of gold, 24 carats is equivalent to 100% gold. Common jewelry is rated at 14 carats. What is the percentage of gold in a ring made of 14-carat alloy? (See Section 11.3 for percentage problems.)

Solution A rating of 14 carats means that out of every 24 parts of alloy, 14 parts are gold. The "parts" can be measured in any convenient unit. Therefore,
$$\text{percentage of gold} = \frac{\text{part}}{\text{whole}} \times 100\%$$
$$= \frac{14.0 \text{ g}}{24.0 \text{ g}} \times 100\%$$
$$= 0.583 \times 100\% = 58.3\%$$

Solubility

The **solubility** of a substance is the maximum quantity of that substance that can dissolve in a given solvent. The *Handbook of Chemistry and Physics* lists the solubility of the most common compounds in terms of grams of solute per 100 mL of water. A solution containing the maximum possible amount of a given solute is said to be **saturated**.

The solubility of most compounds depends on the temperature of the system. For most solids dissolved in water, solubility increases with temperature. That is, the hotter the water, the greater the solubility of the compound. Figure 13-1 plots the solubility of a number of salts as a function of temperature. This graph shows that each compound behaves differently. As the temperature increases from 0°C to 100°C, the solubility of NaCl changes by a relatively small amount, whereas the solubility of potassium nitrate (KNO_3) changes from 10 g per 100 mL of water to over 160 g per 100 mL of water.

With gases dissolved in liquids, the trend is reversed. That is, heating the solution generally decreases the solubility of the gas. This effect is observed when water is heated. Just before it begins to boil, small bubbles appear in the water. These bubbles consist of dissolved air coming out of solution.

An interesting phenomenon occurs sometimes with compounds whose solubility increases with temperature (such as those in Figure 13-1). Suppose a saturated solution of such a compound is formed in hot water and then is cooled to a lower temperature. Since the solubility of the compound is reduced at the lower temperature, there is now more of the compound in solution than can stay dissolved. What happens?

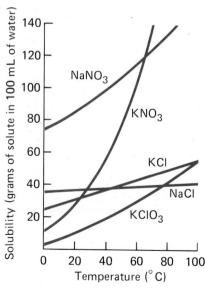

FIGURE 13-1 Curves of solubility versus temperature for a number of common salts

We would expect the excess compound to separate out of solution—to *precipitate*. However, sometimes this does not happen. If such a saturated solution is cooled carefully so that nothing disturbs it, the excess compound remains in solution, forming a **supersaturated solution**—a solution containing more solute than is normally allowed. A supersaturated solution is unstable. It can remain supersaturated for a period of time, but if a single crystal of the solute is introduced into the solution, this "seed" forms a nucleus around which the dissolved material will solidify. Crystals will grow in the water until all of the surplus solute has come out of solution. (Crystallization from supersaturated solution is readily demonstrated with sodium thiosulfate, the common "hypo" used for fixing developed photographs.)

The formation of solutions is important in chemistry because many chemical reactions take place only in solution. When two dry chemicals are mixed so that they sit in contact with each other, often nothing happens. But when chemicals are in solution, their molecules are free to intermingle and collide with one another, so that reactions of interest take place.

Example 13.4 The solubility of NaCl is 35.7 g per 100 mL of water at 20°C. How many grams of NaCl are needed to make a saturated NaCl solution in 1.50 L of water?

Solution Convert the solubility into units of grams per liter:

$$\frac{35.7 \text{ g}}{100 \text{ mL}} \times \frac{1000 \text{ mL}}{1 \text{ L}} = \frac{357 \text{ g}}{1 \text{ L}}$$

By definition,

$$\text{solubility} = \frac{\text{mass of solute}}{\text{volume of solvent}} = \frac{\text{grams of NaCl}}{\text{liters of H}_2\text{O}}$$

Solve for grams of NaCl:

$$\text{grams of NaCl} = \text{solubility} \times \text{liters of H}_2\text{O}$$

$$= \frac{357 \text{ g}}{1 \text{ L}} \times 1.50 \text{ L} = 536 \text{ g}$$

13.2 Water and the Formation of Solutions

The Hydrogen Bond

Water is a remarkable substance, with certain properties that make it unique among compounds. One of the characteristics of water is its ability to dissolve a great many elements and compounds. This property makes it the carrier of life through the bloodstream, for blood is a suspension of cells and a solution of compounds carrying both nourishment and chemical signals through the body.

The basis of water's properties is the fact that the water molecule is highly polar. You will recall (from Section 7.3) that the water molecule is V-shaped, with an angle of 104° between the two covalent bonds linking the hydrogen atoms and the oxygen atom. Since the shared electrons responsible for the bonds spend more time near the oxygen atom than near the hydrogen atoms, the molecule is an electric dipole, with a negative charge at the oxygen end and a positive charge at the hydrogen end.

When two water molecules are close to each other, there is an attraction between the negative end of one molecule and the positive end of the other molecule. There is also a repulsion between the two negative ends and the two positive ends, but the attraction is stronger than the repulsion because the attracting atoms are closer together than are the repelling atoms. (See Figure 13-2.) The same kind of attraction also takes place between two bar magnets.

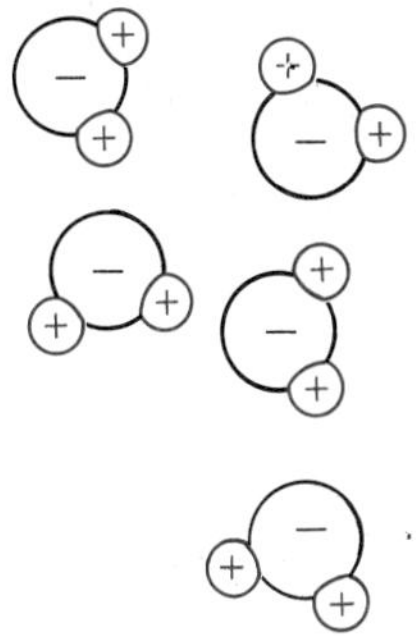

In general such an interaction between polar molecules is called a *dipole-dipole interaction*. In the particular case of a hydrogen atom, the bond is called a **hydrogen bond**. Hydrogen bonds occur when hydrogen exists in a compound together with highly electronegative atoms such as oxygen, fluorine, and nitrogen atoms. Hydrogen bonds occur partly because the small size of the hydrogen atom allows the hydrogen atom in one molecule to get close to the O, F, or N atom in the other molecule.

Hydrogen bonds in water cause the water molecules to cling to one another, making the water more than just a cluster of separate molecules bouncing around. To observe the strength of these bonds, fill a tumbler carefully with water until the surface of the water curves above the lip of the container. The **surface tension** of the water, produced by the hydrogen bond, prevents the water from spilling over the edge. Because of surface tension, a dry needle will float on the water's surface.

Without the attractive bonds, water molecules would behave like the molecules of a gas: they would fly about independently, only interacting when they collided with each other. This is what happens when water is heated to a temperature high enough to vaporize it. At room temperature, however, the hydrogen bonds pull the water molecules together, and the result is a liquid rather than a gas. The hydrogen bond is responsible for the fact that water has a much higher boiling point than either the hydrogen or the oxygen of which it is composed.

Since the hydrogen bond is relatively weak, the water molecules are not bound with great rigidity. Fortunately for the presence of life in this world, the hydrogen bond is just strong enough so that at room temperature water is a liquid instead of a gas or a solid.

FIGURE 13-2 Water molecules attracted to each other by means of the hydrogen (dipole-dipole) bond. The end of the molecule opposite the hydrogen atoms has a net negative charge, and the hydrogen end has a positive charge. The negative and positive ends attract each other.

Polar Covalent Solute-Solvent Systems

In the last section we saw that water is a liquid because its molecules are electric dipoles and the negative end of one dipole attracts the positive end of a neighboring dipole. As a result of this dipole-dipole interaction, a small attraction exists between water molecules, even though each molecule has no net electric charge. The dipole-dipole interaction also explains why polar covalent molecules in general are soluble in water.

The organic compound methanol (CH_3OH) has the form of a tetrahedron, with the carbon atom at the center. (See Figure 13-3.) The face containing the three hydrogen atoms is positively charged, and the other end of the molecule is negatively charged. When methanol is dissolved in a container of water, the attraction of the negative end of the methanol molecule to the positive end of a water molecule creates a weak bond between them.

The bond between the methanol molecule and the water molecule makes the methanol an integral part of the water. Without this bonding of individual methanol molecules to the water molecules, the methanol would tend to separate out and rise to the top of the container (since methanol is less dense than water). It is the dipole-dipole interaction that helps makes the methanol soluble in water.

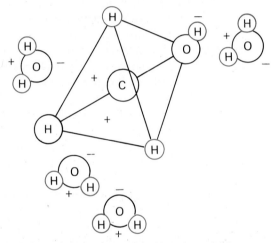

FIGURE 13-3 The polar-covalent compound methanol is soluble in water because of the attraction between the positive end of the methanol molecule and the negative end of the water molecule (and vice versa). In this diagram a tetrahedral model is used for the methanol molecule.

Ionic Solute–Polar Solvent Systems

In an ionic compound such as sodium chloride, the outer electron of the sodium atom (the valence electron) is completely captured by the chlorine atom, and the sodium chloride crystal consists of ions held together by electrostatic attraction. When sodium chloride dissolves in water, the attraction of the sodium ion to the chloride ion is overcome by the attraction of encircling water molecules, which orient themselves so that their negative ends (the oxygen ends) point toward the positive sodium ion. (See Figure 13-4.) This encirclement by the water molecules

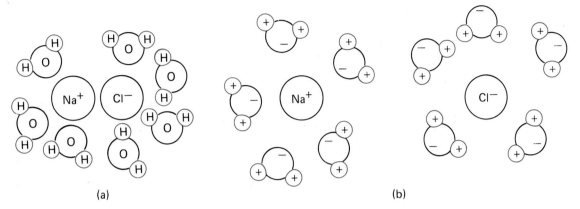

FIGURE 13-4 (a) In a solution of sodium chloride, an NaCl pair becomes surrounded by polar water molecules attracted by the charges on the Na^+ and Cl^- ions. (b) The ions become separated and are surrounded by clusters of water molecules.

causes the sodium ions and the chloride ions to separate from each other; the resulting solution consists of positive and negative ions, each surrounded by a shell of polarized water molecules.

Energy is required to break the bonds between the Na^+ and Cl^- ions in the solid. This energy is supplied by the attraction between the ions and the polarized water molecules. Two processes take place, one endothermic and one exothermic. The separation of the Na^+ and Cl^- ions (breakup of the crystal lattice) requires an input of energy and hence is an endothermic reaction. The joining of the Na^+ and Cl^- ions to the water molecules results in the emission of energy and hence is an exothermic reaction.

If more energy is given off than is absorbed, the net result is an exothermic reaction, and the solution ends up warmer than were the separate solute and solvent. For example, dissolving concentrated sulfuric acid in water produces a great amount of heat—enough to cause dangerous boiling and spattering if the mixing is not done carefully. (Always pour acid into water slowly to reduce spattering.)

The **heat of solvation** is the energy needed to liberate a given ion from its initial structure (crystal lattice or ionic molecule). When water is the solvent, this energy is called the **heat of hydration**.

A more commonly used quantity is the **heat of solution**: the amount of heat given off (or absorbed) when a specific quantity of a substance is dissolved in a large quantity of water. The heat of solution may be either positive or negative. When it is positive, heat is absorbed by the solute from the surroundings; consequently, the solution becomes colder. A negative heat of solution means that heat is given off by the solute and the solution becomes hotter. Although this may seem backwards, remember that these energy shifts are changes in the internal potential energy of the molecular system. When the potential energy of the system goes down as a result of a reaction, this means that energy has been given off.

Nonpolar Solutes and Polar Solvents

The carbon tetrachloride molecule (CCl_4) has a tetrahedral shape, with the carbon atom at the center of the four chlorine atoms. (See Figure 13-5.) Because of its symmetry, the molecule is not polarized. There is some attraction between CCl_4 molecules because of the van der Waals force; it is this attraction that allows CCl_4 to be a liquid at room temperature. However, there is little attraction between a CCl_4 molecule and a water molecule. Therefore, when carbon tetrachloride is mixed with water, no solution is formed. Instead, the carbon tetrachloride remains in a separate layer, with the water floating on top since the water is less dense than the CCl_4.

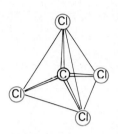

FIGURE 13-5
The carbon tetrachloride molecule is symmetrical and so does not have a net electric charge at one end or the other.

As a result of our investigation into the effect of bonding on the formation of solutions, we can make some general rules:

1. Ionic compounds usually dissolve easily in water because of the ion-dipole attraction.

2. Polar covalent compounds dissolve in polar covalent solvents such as water because of the dipole-dipole interaction. The solubility depends on the amount of polarity.

3. Nonpolar compounds do not dissolve in polar covalent solvents because the solute molecules are not attracted to the solvent molecules.

13.3 Molarity and Normality

Molarity

It is often useful to express the concentration of a solution in terms of the number of moles of solute per liter of solution. This quantity is called the **molarity** of the solution. For example, in the reaction

$$H_2SO_4 + 2NaOH \longrightarrow Na_2SO_4 + 2H_2O$$

1 mol of sulfuric acid neutralizes 2 mol of sodium hydroxide. (**Neutralization** is a reaction between an acid and a base involving an equal number of H^+ and OH^- ions.) If we know that the H_2SO_4 and the NaOH solutions have concentrations of 1 mol/L, then we can say immediately that 1 L of the H_2SO_4 solution will react with 2 L of the NaOH solution.

The molarity of a solution is denoted by the italic capital letter M and is defined as follows:

$$\text{molarity} = M = \frac{\text{no. of moles of solute}}{\text{no. of liters of solution}}$$

13.3 MOLARITY AND NORMALITY

Note that the definition uses the amount of *solution* and not the amount of *solvent*.

A 1.00-*M* solution contains 1 mol/L and is called a *one molar solution*. A 0.25-*M* solution contains 0.25 mol/L and is called a *point-two-five molar solution*.

Example 13.5 How many grams of $CaCl_2$ are needed to make 1.00 L of 1.00-*M* solution?

Solution The molar mass of $CaCl_2$ is

$$1 \times Ca \text{ atomic mass} + 2 \times Cl \text{ atomic mass} = 1 \times 40.1 \text{ g} + 2 \times 35.5 \text{ g}$$
$$= 111.1 \text{ g}$$

From the definition of molarity,

$$\text{molarity} = \frac{\text{no. of moles of solute}}{\text{no. of liters of solution}}$$

Therefore,

$$\text{no. of moles of solute} = \text{molarity} \times \text{no. of liters of solution}$$
$$= \frac{1.00 \text{ mol}}{1.00 \text{ L}} \times 1.00 \text{ L}$$
$$= 1.00 \text{ mol} = 111 \text{ g}$$

Without using the equation, you can say immediately that a mole of any solute is needed to make 1 L of a 1-*M* solution of that solute.

Note: To prepare this solution, dissolve the $CaCl_2$ in somewhat less than a liter of water; then add enough water in a calibrated flask to make up exactly 1 L of solution.

Example 13.6 How many grams of $CaCl_2$ are required to make 300 mL of 0.250-*M* solution?

Solution As in the above example,

$$\text{no. of moles of solute} = \text{molarity} \times \text{no. of liters of solution}$$

In this case,

$$\text{no. of liters of solution} = 300 \text{ mL} \times \frac{1.00 \text{ L}}{1000 \text{ mL}} = 0.300 \text{ L}$$

and

$$0.250 M = \frac{0.250 \text{ mol}}{1 \text{ L}}$$

Substitute into the equation:

$$\text{no. of moles of CaCl}_2 = \frac{0.250 \text{ mol}}{1 \text{ L}} \times 0.300 \text{ L}$$

$$= 0.0750 \text{ mol CaCl}_2$$

$$\text{no. of grams of CaCl}_2 = \text{no. of moles} \times \frac{\text{no. of grams}}{1 \text{ mol}}$$

$$= 0.0750 \text{ mol} \times \frac{111 \text{ g}}{1 \text{ mol}}$$

$$= 8.33 \text{ g CaCl}_2$$

Example 13.7 Calculate the molarity of a solution containing 75.5 g of KOH in 540 mL of solution.

Solution The molar mass of KOH is

$$39.1 + 16.0 + 1.0 = 56.1 \text{ g/mol}$$

The number of moles of KOH is

$$\text{no. of moles} = \text{no. of grams of KOH} \times \frac{1 \text{ mol}}{\text{no. of grams}}$$

$$= 75.5 \text{ g KOH} \times \frac{1 \text{ mol}}{56.1 \text{ g}} = 1.35 \text{ mol KOH}$$

The number of liters of solution is

$$540 \text{ mL} \times \frac{1 \text{ L}}{1000 \text{ mL}} = 0.540 \text{ L}$$

The molarity of the solution is

$$M = \frac{\text{no. of mol}}{\text{no. of liters}} = \frac{1.35 \text{ mol KOH}}{0.540 \text{ L}}$$

$$= \frac{2.49 \text{ mol}}{1 \text{ L}} = 2.49 M$$

Normality

Measuring concentrations by molarity has one drawback, which can be illustrated by the following example involving the neutralization of phosphoric acid with potassium hydroxide:

$$H_3PO_4 + 3\,KOH \longrightarrow K_3PO_4 + 3\,H_2O$$

13.3 MOLARITY AND NORMALITY

In terms of moles,

$$1 \text{ mol } H_3PO_4 + 3 \text{ mol } KOH \longrightarrow 1 \text{ mol } K_3PO_4 + 3 \text{ mol } H_2O$$

If both acid and base have a concentration of 1.00 M, then 1 L of each solution contains 1 mol of solute. Therefore, in terms of liters,

$$1 \text{ L } H_3PO_4 + 3 \text{ L } KOH \longrightarrow 1 \text{ L } K_3PO_4 + 3 \text{ L } H_2O$$

It would be convenient if the concentrations were such that 1 L of acid would neutralize 1 L of base. For that reason a system of normal solutions has been developed. Solutions of two ionic reactants have the same **normality** when equal volumes of solutions have just the right amounts of the various reactants to combine with no excess left over.

Returning to the above example, let us define a normal solution. A KOH solution has one OH^- ion for each K^+ ion. Therefore each KOH molecule contributes one OH^- ion to the reaction; a mole of KOH contributes a mole of OH^- ions. Accordingly, a 1.00-N solution of KOH (a normal solution) contributes 1.00 mol of OH^- per liter of solution. The H_3PO_4 molecule may contribute three H^+ ions to the reaction. A mole of H_3PO_4 may thus contribute 3 mol of H^+ ions, which could combine with the OH^- ions from 3 mol of KOH.

We want to arrange matters so that 1 L of H_3PO_4 solution will combine with 1 L of the 1.00-N KOH solution. We reason as follows: Since 1 mol of H_3PO_4 combines with 3 mol of KOH, $\frac{1}{3}$ mol of H_3PO_4 combines with 1 mol of KOH. Therefore, 1 L of solution prepared with $\frac{1}{3}$ mol of H_3PO_4 solution will combine with 1 L of the 1.00-N KOH solution. A solution with $\frac{1}{3}$ mol of H_3PO_4 per liter of is a 1.00-N solution. Thus, the normality is three times the number of moles of H_3PO_4 per liter of solution.

Generalizing this argument results in the following rule: The normality of an acid, base, or salt solution is n times the number of moles of acid, base, or salt per liter, where n is

1. the number of H^+ ions contributed to the reaction by each acid molecule, or

2. the number of OH^- ions contributed to the reaction by each base molecule, or

3. the number of either positive or negative charges contributed to the reaction by each salt molecule.

Note that the normality of a solution is defined only in terms of a specific reaction and may vary from one reaction to another.

The following formula expresses the normality of a solution:

$$\text{normality} = N = \frac{n \times \text{no. of moles of solute}}{\text{no. of liters of solution}}$$

where n is as defined above.

The following examples illustrate the use of these definitions.

Example 13.8 Given the reaction

$$2\,\text{NaOH} + \text{H}_2\text{SO}_4 \longrightarrow \text{Na}_2\text{SO}_4 + 2\text{H}_2\text{O}$$

determine the following.
(a) How many moles of NaOH are in 500 mL of a 0.10-N solution?
(b) How many moles of H_2SO_4 are in 500 mL of a 0.10-N solution?
(c) How many milliliters of a 0.10-N solution of NaOH are required for 500 mL of a 0.10-N solution of H_2SO_4 to combine completely?

Solution (a) Since

$$N = \frac{n \times \text{no. of moles}}{\text{no. of liters}}$$

it follows that

$$\text{no. of moles} = \frac{1}{n} \times N \times \text{no. of liters}$$

In this case $n = 1$, since 1 mol of OH^- is released per mole of NaOH.

$$\text{no. of moles} = 0.10\,\frac{\text{mol}}{\text{L}} \times 0.50\,\text{L} = 0.050\,\text{mol NaOH}$$

(b) In this case $n = 2$, since 2 mol of H^+ is released per mole of H_2SO_4.

$$\text{no. of moles} = \frac{1}{2} \times N \times \text{no. of liters}$$

$$= \frac{1}{2} \times 0.10\,\frac{\text{mol}}{\text{L}} \times 0.50\,\text{L} = 0.025\,\text{mol H}_2\text{SO}_4$$

(c) We know that 1 mol of H_2SO_4 combines with 2 mol NaOH, so 0.025 mol of H_2SO_4 combines with 0.050 mol NaOH. In the given problem, these are the numbers of moles present in 0.50 L of H_2SO_4 solution and NaOH solution, respectively. Therefore, 0.50 L of 0.1-N H_2SO_4 combines fully with 0.5 L of 0.1-N NaOH. The solution to the problem verifies the fact that complete neutralization takes place when equal volumes of acids and bases having the same normality are mixed together.

Example 13.9 How many milliliters of a 0.50-N solution of KOH are required to neutralize 150 mL of a 0.10-N solution of H_3PO_4?

Solution If the KOH solution had the same normality as the H_3PO_4, it would take 150 mL of KOH solution to neutralize the 150 mL of acid. But the KOH solution has five times the normality of the H_3PO_4 (0.50N compared with 0.10N). Therefore the volume of KOH needed is one-fifth the volume of H_3PO_4, or 30 mL.

Note: You see that it is not necessary to know the molecular masses of the compounds taking part in the reaction to use the concept of normality.

Example 13.9 illustrates how normality is used in chemical analysis to find the quantity of acidity or alkalinity in a solution (quantitative analysis). Suppose, for example, you are given 100 mL of some acid solution. The exact acid does not matter. You can determine how acidic it is by the following procedure.

First prepare a basic solution of known normality, say a 1.00-N solution of NaOH. This is done by dissolving 1 mol of NaOH (40 g) in water and diluting the solution to exactly 1.00 L. Then put into the acid solution a small quantity of an **indicator**—a compound that can distinguish acidic and basic solutions by a change of color. A commonly used indicator is the organic compound phenolphthalein. In acidic solutions phenolphthalein is colorless, and in basic solutions it is red. This compound is one of a number of indicator compounds used to determine, among other things, when a neutralization process has been completed.

After preparing the NaOH solution, drop it slowly into the acid solution by means of a **buret**—an apparatus with a valve and a calibrated tube that allows a precise measurement of the volume of solution mixed with the acid. (See Figure 13-6.) Add the NaOH solution to the acid drop by drop until the solution just

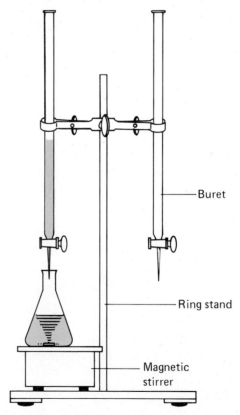

FIGURE 13-6 A buret is used to drop measured quantities of one solution into another. The solution in the flask contains an indicator that changes color when the reaction is completed. The amount of reactant in the flask can then be measured.

begins to turn pale pink. At this point the acid is neutralized; close the valve on the buret and note the amount of normalized solution used. This procedure is called **titration**. The normality of the acid can be determined from the measurement of the volume of normalized solution used, as shown in the example below.

Example 13.10 It requires 45.3 mL of 1.00-N NaOH solution to neutralize 100.0 mL of H_2SO_4.
(a) What is the normality of the acid?
(b) What is the acid concentration in moles per liter?

Solution (a) By definition, the normality of the acid solution is

$$N_a = n_a \times \frac{\text{no. of acid moles}}{\text{acid volume}}$$

where n_a is the number of H^+ ions per acid molecule or the number of moles of H^+ per mole of acid molecules. Therefore,

$$N_a \times \text{acid volume} = n_a \times \text{no. of acid moles} = \text{no. of } H^+ \text{ ions (in moles)}$$

Similarly, using N_b for the normality of the base solution, we have

$$N_b \times \text{base volume} = n_b \times \text{no. of base moles} = \text{no. of } OH^- \text{ ions (in moles)}$$

The acid-base mixture is neutralized when the number of H^+ ions equals the number of OH^- ions. Therefore the following relationship is true:

$$N_a \times \text{acid volume} = N_b \times \text{base volume}$$

Thus,

$$N_a = N_b \times \frac{\text{base volume}}{\text{acid volume}}$$

$$= 1.00N \times \frac{45.3 \text{ mL}}{100 \text{ mL}}$$

$$= 0.453N$$

(b) From the definition of normality,

$$\frac{\text{no. of moles}}{\text{acid volume}} = \frac{N}{n} = \frac{0.453 \text{ mol}}{2 \text{ L}} = 0.227 \text{ mol/L} = 0.227M$$

13.4 Colligative Properties of Solutions

Certain properties of solutions do not depend on the nature of the dissolved substance; they depend only on the concentration of the molecules or ions of solute. Such characteristics, called **colligative properties**, are quite common and are put to use in many practical ways.

For example, when the weather is very cold and water has frozen on the sidewalks, a sprinkling of salt helps to melt the ice. This procedure works because the freezing (melting) point of a solution of salt dissolved in water is lower than the freezing point of pure water.

Sprinkling salt on the ice results in the following sequence of events (see Figure 13-7): A little of the salt in contact with the ice dissolves in the moisture on the surface of the ice. If the freezing point of the resulting salt solution is lower than the surrounding temperature, the solution will not freeze. The salt solution is not in equilibrium with the ice, so the ice in contact with the solution melts faster than it refreezes. As a result, more of the ice melts and more of the salt is dissolved. The process continues until the freezing point of the diluted salt solution is the same as the temperature of the surroundings.

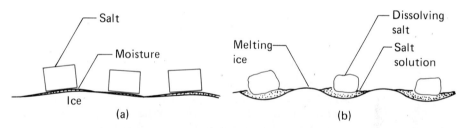

FIGURE 13-7 (a) Salt is able to melt ice because when the salt in contact with the ice begins to dissolve, it forms a salt solution whose freezing point is lower than the temperature of the surroundings. The salt solution therefore remains liquid even though its temperature may be lower than the freezing point of pure water. (b) The ice in contact with the salt solution tends to change from the solid to the liquid state, and so more salt solution is formed.

The melting of the ice results from a general phenomenon: *The freezing point of water is reduced when a nonvolatile compound is dissolved in it.*

The same effect is at work whenever antifreeze is added to the water in a car radiator. The antifreeze (usually ethylene glycol) lowers the freezing point of the radiator water. The higher the antifreeze concentration, the lower the freezing point of the solution. At the same time, the antifreeze raises the **boiling point** of the water, so it does not evaporate away as rapidly as it might otherwise.

In general, the formation of a solution changes both the boiling point and the freezing point of the solution. The amount of change depends on the kind of solvent, the kind of solute, and the concentration of the solution. Let us examine the mechanism underlying these changes as it affects solutions in which the solute evaporates *less* easily than the solvent and therefore has a higher boiling point.

Consider what takes place at the surface of a container of water. (See Figure 13-8.) The water molecules are in a constant state of motion. Most move too slowly to be released from the water, but a few move rapidly enough to escape from the forces binding them to the other water molecules (dipole-dipole forces). These fast-moving molecules evaporate from the surface of the water and form water vapor just above the surface. The partial pressure of this water vapor is called its **vapor pressure**.

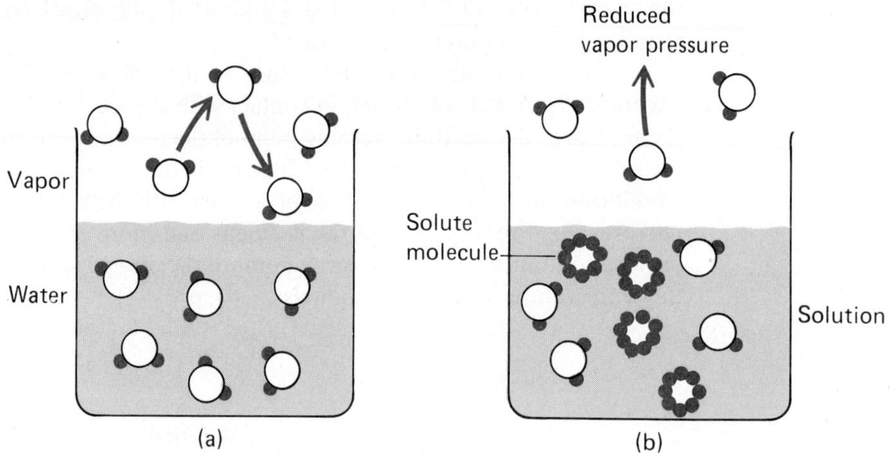

FIGURE 13-8 (a) Some water molecules are escaping from the surface of water while other water molecules are returning to the water. An equilibrium is reached in which there is some water vapor above the water surface. The partial pressure of the water vapor is called the vapor pressure. (b) When a foreign substance is dissolved in water, the molecules of this substance interfere with the evaporation of water molecules. As a result, there are fewer water molecules per unit volume in the space above the water, and the vapor pressure of the water is reduced.

While some molecules are escaping from the water, other water molecules are returning to the liquid from the atmosphere. For this reason the vapor pressure does not rise above a certain value, which depends on the temperature. An increase in temperature results in an increase in vapor pressure, for at a higher temperature more of the water molecules have enough energy to escape from the liquid into the vapor.

Figure 13-9 shows a plot of vapor pressure versus temperature. The curve does not go beyond 100°C because at that point the vapor pressure of the water equals the atmospheric pressure and boiling takes place. Think of the atmosphere as being like a lid on top of the water. As long as the lid can hold in the vapor, the water remains liquid. But when the vapor pressure above the liquid becomes equal to or greater than the atmospheric pressure, the lid blows off and the water boils.

There is a sudden dip in vapor pressure at the freezing point. The vapor pressure below 0°C is that of ice. Even though ice is a solid, it continually evaporates; this process is known as **sublimation**.

13.4 COLLIGATIVE PROPERTIES OF SOLUTIONS

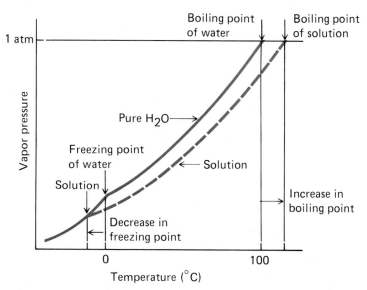

FIGURE 13-9 Graph of vapor pressure versus temperature. The boiling point is the temperature at which the vapor pressure equals the atmospheric pressure. The vapor pressure curve for a nonvolatile substance dissolved in water is lower than that for pure water. As a result, the solution's boiling point is higher and its freezing point is lower.

What happens when another substance is dissolved in the water? Figure 13-8(b) shows how the molecules of this other substance take up space at the surface of the solution and interfere with the evaporation of the water. As a result, the vapor pressure of the solution is less than that of the pure water.

In Figure 13-9, where vapor pressure is plotted against temperature, the curve for the solution is parallel to but below that for pure water. We see that the solution must be raised to a temperature *higher* than the boiling point of pure water before its vapor pressure equals the atmospheric pressure. Thus the boiling point of the solution is higher than that of pure water.

At the other end of the temperature scale, the molecules or ions of the solute interfere with the freezing of the water. It is important to understand that when a solution of salt in water freezes, the substance that solidifies is pure water, not frozen salt solution. (This is one way of separating the salt from, or desalinizing, ocean water.)

The water in a solution freezes when the vapor pressure of the solution equals the vapor pressure of the pure ice. Examination of the vapor pressure curve in Figure 13-9 shows that addition of a solute lowers the freezing point below that of pure water.

Since the amount of change in the freezing or boiling point depends on the number of dissolved particles (molecules or ions) per unit quantity of solution, measurement of a solution's freezing or boiling point can be used to determine the molecular mass of compounds. If we know the number of grams of solute in a given solution, we can measure the depression of the freezing point to find the number of moles of solute. Then we can figure that

$$\text{molar mass} = \frac{\text{no. of grams of solute}}{\text{no. of moles of solute}}$$

This method is not reliable for compounds that ionize in water solution. Ions tend to clump together in solution, so the number of effective particles in an ionized solution is not the same as the number of ions. As we saw in Figure 7-3, the sodium ions in a NaCl crystal are surrounded by chloride ions, and the chloride ions are surrounded by sodium ions. The same effect is seen to some extent in an NaCl solution. This ion clustering makes the theory of ionic solutions very complex.

However, when nonionic compounds dissolve, the number of effective particles is the same as the number of solute molecules. It has been found experimentally that the change in either the freezing point or the boiling point is directly proportional to the number of moles of solute per kilogram of solvent. (Notice that here the amount of *solvent* is important, not the amount of *solution*.)

Writing this relationship as a word equation, we have

$$\text{change in freezing point} = K_f \times \frac{\text{no. of moles of solute}}{\text{no. of kilograms of solvent}}$$

$$\text{change in boiling point} = K_b \times \frac{\text{no. of moles of solute}}{\text{no. of kilograms of solvent}}$$

In these equations K_f and K_b are constants of proportionality, determined experimentally. For a given solvent, these constants do not depend on the kind of solute; the change in the freezing point or the boiling point depends only on the concentration of the solution. The constants do depend on the solvent, however. Table 13-2 lists values of these constants for a few common solvents. It is useful to have a variety of solvents available, for many organic compounds are insoluble in water but are soluble in such compounds as benzene or carbon tetrachloride.

TABLE 13-2 *Freezing-Point and Boiling-Point Constants*

Solvent	Formula	T_f (°C)	K_f (kg°C/mol)	T_b	K_b (kg°C/mol)
benzene	C_6H_6	5.48	4.90	80.15	2.53
carbon tetrachloride	CCl_4	−22.8	29.8	76.8	5.03
phenol	C_6H_5OH	43.0	7.4	181.8	3.56
water	H_2O	0.0	1.86	100.0	0.51

Example 13.11 Common antifreeze is usually ethylene glycol, $C_2H_4(OH)_2$, whose molar mass is 62.07 g/mol. It is a liquid with a density of 1.11 g/cm³. Suppose 1.00 L of antifreeze is mixed with 3.00 L of water.
(a) What is the freezing point of the solution?
(b) What is the boiling point of the solution?

Solution **(a)** Find the mass of antifreeze:

$$\text{density} = \frac{\text{mass}}{\text{volume}}$$

$$\text{mass} = \text{density} \times \text{volume}$$

$$= 1.11 \frac{\text{g}}{\text{cm}^3} \times 1000 \text{ cm}^3$$

$$= 1110 \text{ g}$$

Find the number of moles of antifreeze:

$$\text{no. of moles} = \frac{\text{mass}}{\text{molar mass}}$$

$$= \frac{1110 \text{ g antifreeze}}{62.07 \ \frac{\text{g}}{\text{mol}}}$$

$$= 17.88 \text{ mol}$$

Find the mass of the solvent:

$$\text{mass} = \text{density} \times \text{volume}$$

$$= 1.00 \frac{\text{g}}{\text{cm}^3} \times 3.00 \text{ L} \times 1000 \frac{\text{cm}^3}{\text{L}}$$

$$= 3000 \text{ g}$$

$$= 3.00 \text{ kg}$$

Find the reduction in the freezing point:

$$\text{change in freezing point} = K_f \times \frac{\text{no. of moles of antifreeze}}{\text{no. of kilograms of water}}$$

$$= 1.86 \frac{\text{kg}°\text{C}}{\text{mol}} \times \frac{17.88 \text{ mol antifreeze}}{3.00 \text{ kg water}}$$

$$= 11.1°\text{C}$$

The freezing point of the solution is 11.1°C lower than the freezing point of pure water; it is −11.1°C.

(b) Find the increase in boiling point:

$$\text{change in the boiling point} = K_b \times \frac{\text{no. of moles of antifreeze}}{\text{no. of kilograms of water}}$$

$$= 0.51 \text{ kg°C} \times \frac{17.88 \text{ mol antifreeze}}{3 \text{ kg water}} = 3.04°C$$

Therefore the boiling point of the solution is

$$100°C + 3.04°C = 103.04°C$$

Glossary

alloy A solid solution composed of two or more metals.

boiling point The temperature at which the vapor pressure of a liquid is equal to the atmospheric pressure.

buret A cylinder open at one end and closed at the other with a stopcock. The side of the cylinder is marked in volume measurements such as millimeters.

colligative properties Characteristics of a solution that depend only on the concentration of the solute.

colloidal suspension A suspension of particles in a fluid, in which the particles are too small to settle to the bottom but too large to be considered part of a solution.

concentrated A term used to describe solutions in which a large amount of solute is used relative to the amount of solvent.

concentration The amount of solute per unit volume of solvent or solution.

dilute A term used to describe solutions in which a small amount of solute is used relative to the amount of solvent.

heat of hydration The heat of solvation of a water solution.

heat of solution The amount of heat given off (or absorbed) when a specific quantity of a substance is dissolved in a large amount of water.

heat of solvation The energy required to free a given ion from its initial position in a crystal lattice or ionic molecule.

hydrogen bond The attraction of two polar molecules to each other through a dipole-dipole interaction involving a hydrogen atom.

indicator A chemical that changes color with changes in the concentration of hydrogen ions in the solution in which it is placed.

molarity The number of moles of solute per liter of solution.

neutralization A reaction between an acid and a base involving an equal number of H^+ and OH^- ions.

normality The number of moles of reactant per liter of solution multiplied by the number of ions contributed to the reaction by each molecule of reactant.

osmosis A process wherein a semipermeable membrane which separates a solvent and a solution allows the solvent to flow into the solution and at the same time blocks the solute from flowing into the pure solvent.

saturated solution A solution containing the maximum amount of solute that can dissolve in a given amount of solvent.

solubility The maximum amount of solute that can dissolve in a given volume of solvent.

solution A mixture of solvent and solute that cannot be separated by gross mechanical means.

sublimation The conversion of a solid to a vapor without passage through the liquid phase.

supersaturated solution A solution in which the concentration of solute is greater than the saturated amount. Such a solution is in an unstable condition.

surface tension The attraction of molecules at the surface of a liquid to the molecules in the body of the liquid, causing the surface to resist being stretched.

titration Measuring the quantity of an acid that neutralizes a given volume of base (or vice versa) for the purpose of calculating the acid (or base) concentration.

vapor pressure A pressure exerted by a vapor that is over a solvent and in equilibrium with it at a given temperature.

Problems and Questions

13.1 Properties of Solutions

13.1.Q Define and illustrate the following:
   a. solution
   b. solubility
   c. saturated solution
   d. precipitate
   e. mixture
   f. supersaturated solution

13.2.P A solution of lithium sulfate contains 24.5 g of solute per 100 mL of solution. What is the concentration in grams per milliliter?

13.3.P A solution of magnesium acetate contains 1.09 g of solute in 32.4 mL of solution. What is the concentration in grams per liter?

13.4.P A saturated solution of mercurous carbonate contains 4.5×10^{-5} g of solute per liter of solution. What is the mass of solute in 75 mL of solution?

13.5.P A solution of potassium acetate contains 18.0 g of solute per 100 mL of solution. What volume of solution contains 99.0 g of potassium acetate?

13.6.P The saturation concentration of silver nitrate is 952 g/100 mL of water at 100°C. How much silver nitrate is required to make a saturated solution at 100°C in 375 mL of water?

13.7.P A liter of saturated strontium chromate contains 1.2 g of solute at 15°C. What is the solubility in grams per 100 mL?

13.8.P A saturated solution of zinc sulfate is made by dissolving 663.6 g of solute in 100 mL of water at 100°C. The solution is cooled to 20°C, at which temperature its saturation concentration is 96.5 g/100 mL.
   a. How much zinc sulfate remains in this solution at 20°C?
   b. How much zinc sulfate will precipitate out of solution when it is cooled?
   c. How much zinc sulfate could be in solution at 20°C if the volume were increased by adding 100 mL of water to the original 100 mL?

13.9.P Bromobenzene has the following saturated solution data:

0.045 g/100 mL in water
10.4 g/100 mL in alcohol
71.3 g/100 mL in ether

   a. Which solvent can dissolve the most bromobenzene per milliliter?
   b. How much bromobenzene is required to make a saturated solution with 1 L of alcohol?
   c. How much water would have to be used to hold the same amount of solute as 100 mL of alcohol would hold?

13.10.P A solution of $CaCl_2$ is saturated with 59.5 g in 100 mL of water at 0°C, and NaCl is saturated with 35.7 g/100 mL at 0°C.
   a. Which solution contains more solute per milliliter of solvent?
   b. How much water is required to make a saturated solution with 59.5 g NaCl?
   c. How much water is required to make a saturated solution with 35.7 g $CaCl_2$?

13.11.P A mole of sodium bromide (NaBr) has a mass of 102.9 g. The solubility of NaBr at 0°C is 79.5 g/100 mL water. How many moles of NaBr can be dissolved in 200 mL of water?

13.2 Water and the Formation of Solutions

13.12.Q Define the following:
   a. hydrogen bond
   b. heat of hydration
   c. heat of solution

13.13.Q Explain how it is possible to float a razor blade on water even though the density of steel is greater than the density of water.

13.14.Q In each of the following sets, which is the solvent and which the solute?
   a. sugar in water
   b. water with dissolved salt
   c. silver in mercury (the amalgam used for tooth fillings)
   d. copper in gold (14-carat gold)
   e. water with dissolved carbon dioxide (soda water)
   f. isopropyl alcohol in water (rubbing alcohol)

13.15.Q Explain why ionic compounds tend to be soluble in water.

13.16.Q Explain why polar-covalent compounds tend to be soluble in water.

13.17.Q Explain why symmetrical covalent compounds tend to be insoluble in water.

13.18.Q Which of the following combinations will form solutions?
    a. $CHCl_3$ and water
    b. NO_2 and air
    c. CCl_4 and water
    d. NaCl and water
    e. $CuSO_4$ and water
    f. talcum powder and air
    g. flour and water

13.19.Q Give a reason for each of your answers to Question 13.18 in terms of solute-solvent classifications and interactions.

13.3 Molarity and Normality

13.20.Q Define or explain the following terms:
    a. molarity
    b. neutralization
    c. normality
    d. titration

13.21.P Calculate the molarity of each of the following aqueous solutions:
    a. 40 g NaOH in 1000 mL of solution
    b. 200 g $CaCl_2$ in 500 mL of solution
    c. 4.0 g NaCl in 50 mL of solution
    d. 60 mg KNO_3 in 10 mL of solution
    e. 0.5 mol HBr in 0.25 L of solution
    f. 6.0 mmol H_3PO_4 in 1.5 L of solution

13.22.P Calculate the number of grams of solute in each of the following solutions:
    a. 2 mol KOH in 250 mL of solution
    b. 0.25 mol $NaNO_2$ in 0.025 L of solution
    c. 6.5 L of 3.0-M H_2SO_4
    d. 100 mL of 0.011-M $NaHCO_3$
    e. 3.25 mmol $NaC_2H_3O_2$ in 0.05 mL of solution
    f. 9.5 mg $Ca(OH)_2$ in 10.5 mL of solution

13.23.P Calculate the volume in milliliters of each of the following solutions:
    a. 3.5 mol $MgSO_4$ in a 3.5-M solution
    b. 1.0 mol HNO_3 in a 15-M solution
    c. 26 g $MgBr_2$ in an 8.4-M solution
    d. 7.1 g NH_4OH in a 10-M solution
    e. 65 mg Na_3PO_4 in a 6.0-M solution
    f. 15.8 mmol $KHSO_3$ in a 2.1-M solution

13.24.P Calculate the normality of each of the following solutions. (Assume that all H^+ or OH^- ions are released for reaction.)
    a. 40 g NaOH in 500 mL of solution
    b. 196 mg H_2SO_4 in 100 mL of solution
    c. 49 g H_2SO_4 in 250 mL of solution
    d. 98 g H_3PO_4 in 1500 mL of solution
    e. 2 mol $Ca(OH)_2$ in 600 mL of solution
    f. 0.01 mol HNO_3 in 1.0 mL of solution

13.25.P A solution of HCl contains 72 g of solute in 2500 mL of solution.
    a. Calculate the molarity of this solution.
    b. Calculate the normality of this solution.

13.26.P A solution of $Ca(OH)_2$ contains 1.0 g of solute in 1500 mL of solution.
    a. Calculate the molarity of this solution.
    b. Calculate the normality of this solution.

13.27.P An 800-mL solution of NaOH is labeled 0.5 M.
    a. What is the normality of this solution?
    b. How many grams of NaOH are in the solution?

13.28.P A 200-mL solution of H_2SO_4 is labeled 0.4N.
    a. What is the molarity of this solution?
    b. How many grams of H_2SO_4 are in this solution?

13.29.P What volume of NaOH is required to neutralize the acid when 16 mL of 0.2-N HCl is titrated with 0.4-N NaOH in the reaction

$$HCl + NaOH \longrightarrow NaCl + H_2O$$

13.30.P What volume of 0.6-M $HC_2H_3O_2$ is required to neutralize 40 mL of 0.01-M $Ca(OH)_2$ in the reaction

$$2HC_2H_3O_2 + Ca(OH)_2 \longrightarrow 2H_2O + Ca(C_2H_3O_2)_2$$

13.4 Colligative Properties of Solutions

13.31.P Fifty grams of ethanol (C_2H_6O) is mixed with 500 g of water. Calculate the effect of the alcohol on the boiling point and the freezing point of the water.

13.32.P Fifty grams of glycerol ($C_3H_8O_3$) is mixed with 500 g of water. Calculate the effect on the boiling point and the freezing point of the water.

13.33.P Compare the answers to Problems 13.31 and 13.32. Which compound has the greater effect on the boiling and freezing points of water? Explain your answer.

13.34.P Fifty grams of ethanol is added to 500 g of phenol. What effect does the alcohol have on the freezing and boiling points of the phenol?

13.35.P Using the answers to Problems 13.31–13.34, determine which solvent—water or phenol—has its freezing point changed more upon the addition of a given amount of ethanol. Would the answer to this question be different if glycerol were used as a solute instead of ethanol?

Self Test

1. Define the following terms and give illustrations of each.
   a. solution
   b. alloy
   c. solubility
   d. saturated solution
   e. surface tension
   f. molarity
   g. normality

2. Explain why ethanol is soluble in water. (The formula for ethanol is C_2H_5OH.)

3. Describe what happens when the white solid NaCl mixes with water and seems to disappear.

4. A solution contains 10 g NaOH in 925 mL of solution.
   a. Calculate the molarity of the solution.
   b. Calculate the normality of the solution.

5. A solution contains 25 g of H_2SO_4 in 1800 mL of solution.
   a. Calculate the molarity of the solution.
   b. Calculate the normality of the solution.

6. H_2SO_4 in a 0.10-N solution reacts with 150 mL of 0.05-N $Ca(OH)_2$ solution in the reaction

$$H_2SO_4 + Ca(OH)_2 \longrightarrow CaSO_4 + 2H_2O$$

   a. Calculate the volume of H_2SO_4 required for neutralization.
   b. Calculate the number of moles of each of the reactants.

7. A beaker contains 20 g of HCl in 400 mL of solution. This acid neutralizes 250 mL of NaOH solution in the reaction

$$NaOH + HCl \longrightarrow NaCl + H_2O$$

   a. What mass of NaOH is required to complete the reaction?
   b. What is the normality of the HCl solution?
   c. What is the normality of the NaOH solution?

8. A solution consists of 10% ethanol (alcohol) and 90% water (by volume).
   a. What is the freezing point of the solution?
   b. What is the boiling point of the solution?
   (*Hints:* Assume any convenient amount of solution. The density of ethanol is 0.79 g/cm³. Give the answer in degrees Celsius.)

14
Acids, Bases, and Salts

An artist etches a plate in an acid bath.

Objectives After completing this chapter, you should be able to:
1. Explain the behavior of acids, bases, and salts with the Arrhenius model.
2. Explain what is meant by electrolytic dissociation.
3. Explain the difference between a strong acid and a weak acid in terms of ionization.
4. Explain the behavior of acids and bases with the aid of the Bronsted-Lowry model.
5. Describe the formation of a hydronium ion when an acid dissolves in water.
6. Define acids and bases in terms of proton transfer and in terms of electron transfer.
7. Describe degree of acidity in terms of hydrogen or hydronium ion concentration.
8. Calculate the percentage ionization in a given solution.
9. Describe the relationship between hydronium ion and hydroxide ion concentration in a solution.
10. Describe the relationship between pH and hydronium ion concentration.
11. Calculate the pH of a solution, given the hydronium ion concentration.
12. Calculate the hydronium ion concentration of a solution, given its pH.
13. State the pH values of a variety of common solutions.

14.1 Models of Acids, Bases, and Salts

In Chapter 5 we discussed a number of atomic models that help explain the observed behavior of matter. In this chapter we will consider three models developed to explain the behavior of acids, bases, and salts. One of our aims is to understand why these compounds form ions when dissolved in water.

The Arrhenius Model

Svante August Arrhenius (1859–1927), a Swedish chemist, was one of the first scientists to advance the idea that atoms are composed of electric charges. He was intrigued by the fact that when certain substances (for example, sodium chloride) are dissolved in water, the resulting solution is an excellent conductor of electric current. Such substances are called **electrolytes**, and the solutions they form are called **electrolytic solutions**. Other compounds, such as sugar (sucrose), are **nonelectrolytes**, for their water solutions do not conduct electric current.

Arrhenius developed the theory that when an electrolyte dissolves in water it breaks up into two parts, one with a positive electric charge and the other with a negative charge. This breaking up of the molecule is called **electrolytic dissociation**. It is the same process already discussed under the name *ionization*; the two charged parts of the molecule are the positive and negative ions.

The theory of Arrhenius was highly imaginative: it was the first scientific analysis of ionic solutions, and it was published before anyone knew much about the nature of an electric charge. Indeed, the electron had not yet been discovered. Therefore, when Arrhenius submitted this idea in 1884 as the basis of his doctoral dissertation, he met with a great deal of opposition and disbelief, for without knowing that atoms consist of electric charges, nobody could understand how atoms and molecules could break into positive and negative ions.

During the following two decades, however, both the electron and the proton were discovered, making it possible to formulate a model describing Arrhenius's ions in some detail. Eventually the opposition to Arrhenius's theory disappeared, and in 1903 he was awarded the Nobel prize in chemistry. Ironically, the dissertation that won the prize was the same one that had only reluctantly been approved by his Ph.D. committee two decades earlier.

Acids

An important part of Arrhenius's theory was his definition of acids, bases, and salts. We already encountered this definition in Chapter 8. An **Arrhenius acid** is a compound that dissociates into an H^+ ion plus a negative ion when dissolved in water. The strength of the acid is related to the number of ionized molecules. If most of the solute (acid) molecules become ionized when the compound is dissolved, it is a **strong acid**. If only a small fraction of the molecules ionize, it is a **weak acid**.

An H^+ ion is simply a hydrogen atom with the electron missing; it is a bare proton. Thus the Arrhenius model pictures an acid in solution as a mixture of protons and negative ions floating freely among the water molecules. The presence of the hydrogen ions makes the solution an acid; the *type* of negative ion determines the *kind* of acid. For example, sulfuric acid in water dissociates into two hydrogen ions and one sulfate ion:

$$H_2SO_4 \longrightarrow 2H^+ + SO_4^{2-}$$

Two electrons that are normally part of the hydrogen atoms have transferred over to the sulfate ion, giving it two extra negative charges.

One flaw in the Arrhenius model is that it does not explain why the acid molecule dissociates when the acid is mixed with water. A good model should present a mechanism for the process.

Another set of facts not explained by the Arrhenius model is the behavior of electrolytes in different solvents. Hydrochloric acid, for example, ionizes almost completely in water, but when the same acid dissolves in carbon tetrachloride, only a small fraction of the HCl molecules dissociate. Other compounds show the opposite behavior, ionizing weakly in water and strongly in other solvents. To be complete, a model of solutions must explain why the nature of the solvent determines the strength of the ionization.

In the next section we will consider a model that deals with these problems in some detail.

Bases

An **Arrhenius base** is defined as a compound that dissociates into a positive ion plus a negative hydroxide ion (OH^-) when it dissolves in water. A typical example is the dissociation of sodium hydroxide:

$$NaOH \longrightarrow Na^+ + OH^-$$

The positive ion may be polyatomic, as in the case of ammonium hydroxide:

$$NH_4OH \longrightarrow NH_4^+ + OH^-$$

These equations are incomplete, since they show only the ionization of the molecule and not the role played by the water molecule in the reaction.

Salts

Salts are defined as compounds that dissociate into positive and negative ions but do not yield either H^+ or OH^- ions. Potassium nitrate, for example, dissociates into potassium and nitrate ions:

$$KNO_3 \longrightarrow K^+ + NO_3^-$$

Some books define a salt as a metallic positive ion combined with a nonmetallic negative ion. This definition can be confusing. In the salt ammonium chromate [$(NH_4)_2CrO_4$], the positive ion (NH_4^+) contains no metals, and the negative ion (CrO_4^{2-}) contains the metal chromium. However, the ammonium ion (NH_4^+) behaves in solution like a metallic ion because of its positive charge, and the chromate ion (CrO_4^{2-}) behaves like the nonmetallic nitrate ion (NO_3^-).

The terms *metallic* and *nonmetallic* have more to do with the behavior of the ions than with their metallic content. The ions should be labeled "positive" or "negative," rather than "metallic" or "nonmetallic."

The Bronsted-Lowry Model

In 1923 Johannes Bronsted in Denmark and Thomas Lowry in England independently proposed a new model of acid-base-salt behavior that answered many of the objections to the Arrhenius theory. The new theory emphasized the crucial role played by solvent molecules in the dissociation of electrolytes. It also explained the variations in the behavior of the same compound when dissolved in different solvents.

One major difference between the Arrhenius model and the Bronsted-Lowry model is that the first focuses attention on the electrons, whereas the second concentrates on the protons. In the Arrhenius model, for example, the ionization of HCl in water is regarded simply as the transfer of an electron from the hydrogen atom to the chlorine atom. This is followed, somehow, by the separation of the hydrogen and chloride ions.

According to the Bronsted-Lowry model, on the other hand, the hydrogen ion separates from the HCl and attaches itself to a water molecule. (See Figure 14-1. Notice that in this diagram the x's represent electrons contributed by the oxygen or chlorine atoms and the o's represent electrons contributed by the hydrogen atoms.)

$$HCl + H_2O \longrightarrow H_3O^+ + Cl^-$$

Acid   Base   Hydronium   Chloride

FIGURE 14-1 Proton transfer in a water solution of hydrochloric acid. A positively charged hydronium ion is formed when a proton detaches itself from the HCl and attaches itself to the water molecule. (In this and similar diagrams, the x's represent electrons contributed by the chlorine atom, and the o's represent electrons contributed by the hydrogen atom.)

Why does the hydrogen ion migrate to the water molecule? The reason is that the water molecule is a highly polar molecule; one end (where the two hydrogen atoms are clustered) is positively charged, and the other end is negatively charged. Therefore, there is a strong attraction between the positive hydrogen ion (a bare proton) and the negative end of the water molecule. This attraction is stronger than the affinity between the hydrogen and chloride ions in the HCl molecule. When HCl is mixed with water, the HCl molecules bounce around among the H_2O molecules; during the collisions, a proton (an H^+ ion) is transferred from each HCl molecule to an H_2O molecule.

The water molecule with an attached H^+ ion is called a **hydronium ion**; it may be thought of as a hydrated hydrogen ion. (*Hydrated* simply means "with one or more water molecules attached.") Its formula is written as H_3O^+, or $H_2O \cdot H^+$.

In the Bronsted-Lowry model, the water solution of HCl is looked upon as a mixture of hydronium ions and chloride ions rather than as the mixture of hydrogen ions and chloride ions visualized by Arrhenius.

The Bronsted-Lowry model presents a mechanism for the dissociation of the acid molecule when it dissolves in water. It is based on a microscopic description of acid molecules colliding with water molecules; during the collision, the electrical attraction between the acid's proton and the polar water molecule causes the proton to separate from the negative ion. In addition, the model gives new definitions of acids and bases, built on the concept of **proton transfer**.

The new definitions are as follows: (1) A **Bronsted-Lowry acid** is a compound that *donates* a proton to a base during an acid-base reaction. The acid is called a proton donor. (2) A **Bronsted-Lowry base** is a compound that *accepts* a proton from an acid during an acid-base reaction. The base is called a proton acceptor.

These definitions can be understood by referring to Figure 14-1. There the HCl is seen donating a proton to the H_2O. According to the definitions above, the HCl is an acid and the H_2O is a base. (Notice that the Cl^- ion plays no part in the definition of an acid or a base.)

Up to now, water has not been thought of as either an acid or a base. The new definition broadens the concept of acid and base. We will see that at times H_2O acts as an acid and at other times it acts as a base.

Consider Figure 14-2, which shows what happens when ammonia (NH_3) dissolves in water. A proton transfers from the H_2O molecule to the NH_3 molecule, forming the ammonium ion (NH_4^+) and the hydroxide ion (OH^-). The water is now the proton donor and thus an acid; the ammonia is a proton acceptor and thus a base. Note that in the reaction of Figure 14-1 the water acted as a base, but in the reaction of Figure 14-2 the water behaves as an acid.

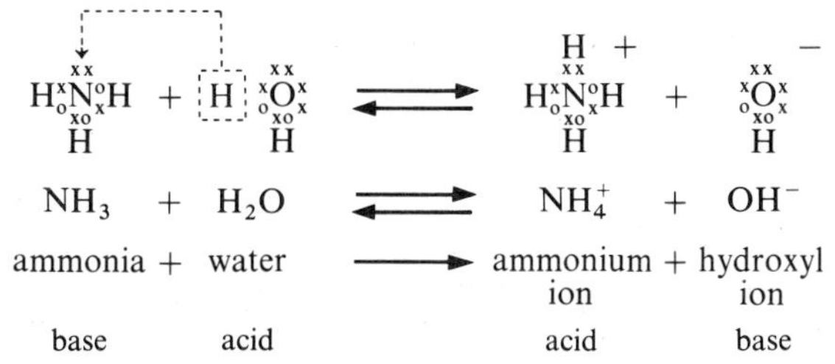

NH_3	+	H_2O	⇌	NH_4^+	+	OH^-
ammonia	+	water	⟶	ammonium ion	+	hydroxyl ion
base		acid		acid		base

FIGURE 14-2 Ammonia dissolving in water. A proton transfers from a water molecule to an ammonia molecule, forming an ammonium ion (+) and a hydroxide ion (−).

What determines how the water behaves? The determining factor is the strength of the attraction between the proton and the various molecules in the solution. In the dissociation of HCl, the proton is attracted more strongly to the water molecule than to the chloride ion. Therefore the reaction goes in the direction shown by the arrow in Figure 14-1.

There is nothing to prevent a proton from going from the hydronium ion to the chloride ion, causing the reaction to go in the reverse direction. This may happen in a few collisions. However, the great majority of the reactions go from left to right, and so we say that the HCl is a stronger acid than the hydronium ion. A reaction will always go in such a direction as to replace a stronger acid with a weaker one, because the stronger acid has a greater tendency to dissociate than does the weaker one. (In the reaction discussed above, the HCl is replaced by the hydronium ion.)

The equation in Figure 14-2 has arrows going in both directions. This is because there is a good chance that in a mixture of ammonia molecules, water molecules, ammonium ions, and hydroxide ions, a proton will transfer from an ammonium ion to a hydroxide ion, making a reaction that goes from right to left, as shown in Figure 14-3. In fact, the reaction is more likely to go from right to left than from left to right. This means that the solution will have more un-ionized ammonia molecules than ammonium ions. As a result, the solution is only partially ionized; accordingly, the ammonium hydroxide solution is a relatively weak base.

Another way of explaining the situation is to say that the hydroxide ion is a stronger base than is the ammonia molecule, so the hydroxide ion has a greater tendency to accept a proton than does the ammonia molecule. For this reason the reaction goes to the left more often than it goes to the right.

In Figure 14-3 the ammonia molecule is labeled $Base_1$ and the ammonium ion is labeled $Acid_1$. The water molecule is labeled $Acid_2$, and the hydroxide ion is labeled $Base_2$. Each acid is paired off with a base, and this pair is called a **conjugate acid-base pair**. In the present example, NH_3 and NH_4^+ make one pair, and H_2O and OH^- comprise another pair.

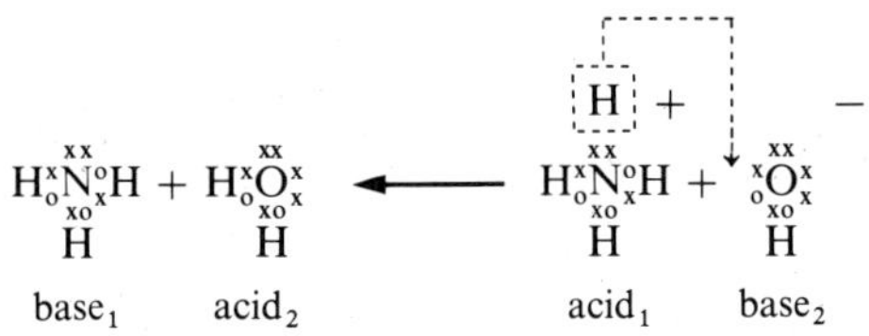

$$H_o^xN_x^oH + H_o^xO_{xo}^x \; \longleftarrow \; H_o^xN_x^oH + {}^xO_{xo}^x$$

| base$_1$ | acid$_2$ | | acid$_1$ | base$_2$ |

FIGURE 14-3   A proton transfers from the ammonium ion to the hydroxide ion, causing the reaction to go from right to left.

In the dissociation of HCl, HCl and Cl^- make one acid-base pair, and H_3O^+ and H_2O comprise the other pair. When HCl (an acid) dissociates, the H_2O is a base. On the other hand, when NH_3 dissolves, the H_2O is an acid. This example demonstrates how a given molecule behaves sometimes as an acid and sometimes as a base, depending on the other compounds in the reaction. For this reason an acid or a base cannot always be identified simply by looking at the compound's formula.

The Lewis Acid-Base Model

In 1923, G. N. Lewis proposed a third model of acid-base behavior that extended the concept of acid to include covalent compounds as well as ionic compounds. His picture was based on the transfer of electrons to form covalent bonds, in contrast to the proton transfer of the Bronsted-Lowry model. It is a broader concept and includes classes of compounds not treated by the Bronsted-Lowry model.

A **Lewis acid** is defined as a substance whose molecule *accepts* a pair of electrons to form a covalent bond. Conversely, a **Lewis base** is a substance whose molecule *donates* an electron pair. Consider the combination of calcium oxide (a base) with sulfur trioxide (an acid) to form calcium sulfate:

$$Ca^{2+}[:\ddot{\underset{..}{O}}:]^{2-} + \underset{:\ddot{\underset{..}{O}}:}{\overset{:\ddot{O}:}{\ddot{S}:\ddot{O}:}} \longrightarrow Ca^{2+}\left[\underset{:\ddot{\underset{..}{O}}:}{\overset{:\ddot{O}:}{:\ddot{O}:\ddot{S}:\ddot{O}:}}\right]^{2-}$$

donor   acceptor

As the above equation shows, the trioxide molecule accepts a pair of electrons donated by the calcium oxide. This electron pair is then shared with the sulfur atom to form a covalent bond in the SO_4^{2-} ion. By definition, then, the CaO molecule is a base, and the SO_3 molecule is an acid.

Acid-Base Neutralization

Neutralization of an acid by means of a base may be analyzed in terms of the transfer of a proton between acid-base pairs. Consider what happens when HCl and NaOH are each dissolved in water and the two solutions are mixed.

We have already seen that the HCl dissociates by transferring a proton to a water molecule:

$$HCl + H_2O \longrightarrow H_3O^+ + Cl^-$$

The NaOH, an ionic compound, also dissociates in the water:

$$NaOH \longrightarrow Na^+ + OH^-$$

When the two solutions are mixed, the hydronium ion donates a proton to the hydroxide ion:

$$H_3O^+ + Cl^- + Na^+ + OH^- \rightleftharpoons Na^+ + Cl^- + H_2O + H_2O$$

If the reaction ran all the way to the right, and if equal numbers of NaOH and HCl molecules had been mixed, the result would be a solution containing nothing but sodium ions, chloride ions, and water molecules. However, a small number of reactions run from left to right, leaving a few H_3O^+ and OH^- ions (in equal numbers). At this point the solution is neutralized.

In general we can think of acid-base reactions in the following way. Let HA be an acid, where A is any negative group, and H represents the hydrogen ion attached to it. BOH is a base, where B represents any positive ion. (For example, if HCl is the acid, A = Cl, and if NaOH is the base, B = Na.) The acid-base reaction involves transfer of the proton H^+ to the base:

$$HA + BOH \longrightarrow A^- + BHOH^+ \longrightarrow B^+ + A^- + HOH$$

$$HCl + NaOH \longrightarrow Cl^- + NaHOH^+ \longrightarrow Na^+ + Cl^- + HOH$$

The final step shows the water molecule splitting off, leaving the ionized salt in solution.

Autoionization

In certain situations water by itself can behave as both acid and base simultaneously. To see how this works, first recall the equation for the dissociation of HCl:

$$HCl + H_2O \rightleftharpoons H_3O^+ + Cl^-$$

Now substitute an OH^- ion for the Cl^- in this equation:

$$HOH + H_2O \rightleftharpoons H_3O^+ + OH^-$$

In this reaction one of the water molecules has transferred a proton to the other water molecule, forming a hydronium ion. You will notice that this reaction is just the reverse of the one that takes place during acid-base neutralization.

The reaction just described is called **autoionization** and is responsible for the fact that water always contains a certain number of ions. The neutralization reaction is much more probable than the autoionization reaction, so the number of ions found in pure water is relatively small. (That is, in the equation above, the reaction is more likely to go from right to left than from left to right.) The ionization of water will be discussed in more detail in Section 14.2.

14.2 Acid-Base Strength

Measures of Acidity

Since the beginning of chemistry, it has been known that some acids are more "acidic" than others. Some acids are more sour-tasting than others; some dissolve metals more readily than others. The acids with these properties are said to be *stronger* than other acids.

A classical test for **acid strength** is the **litmus test**. Litmus, a dye obtained from certain varieties of lichens (moss), has a useful property: in an acidic solution it has a red color and in an alkaline or basic solution it has a blue color. An easy method of using litmus is by means of litmus paper—paper that has been soaked in the dye. By dipping this paper into the unknown solution and observing whether the paper turns red or blue, we can determine whether the solution is acidic or basic. There are now many other indicators that allow the chemist to determine the acidity of a solution, but litmus has long been one of the most common.

In some acids, such as HCl, litmus turns bright red, although equal concentrations of other solutions, such as acetic acid, only turn the litmus a pale pink. This difference in shade indicates that HCl is a strong acid and acetic acid is a weak acid.

Another important indication of acid strength is the ability of the solution to conduct an electric current. If the solution is a good conductor (that is, carries electric current easily, with little resistance), the acid is a strong acid. If the solution is a poor conductor, the acid is weak.

Strong bases (such as NaOH) have intense reactions and fast reaction rates and turn litmus a strong blue color; their solutions are good conductors of electric current. Other bases (such as ammonium hydroxide) are weaker and are not good electrical conductors.

Why does a strong acid or base conduct an electric current easily? We will see in the next section that the ability to conduct electricity has to do with the degree of ionization of the compound in solution. If there are many ions in the solution, the solution is a good conductor; if there are few ions, the solution is a poor conductor.

Degree of Ionization

Suppose we prepare solutions of NaOH (sodium hydroxide) and of NH_4OH (ammonium hydroxide), measuring the compounds so that in each solution there is 1 mole of solute in 1 L of solution (a molar solution). Now let us measure the electrical conductivity of the two solutions. We do this by immersing a pair of electrodes in each solution and measuring the electric current that flows when the electrodes are connected to a battery. (See Figure 14-4.) By definition, the conductivity is proportional to the amount of current that flows, for a constant voltage. The higher the current, the greater the conductivity.

What result might we expect from this experiment? The natural expectation is that the two solutions will behave identically. We have put the same number of

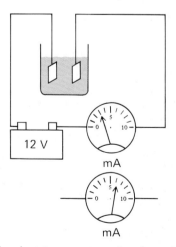

FIGURE 14-4 Measuring the electric current passing through a solution. If one solution passes twice as much current as the other when the voltage is kept constant, that solution has twice the conductivity of the other.

molecules of solute into each solution, and we know that the amount of electric current depends (among other things) on the number of ions present in the solution, since it is the ions moving through the solution from one electrode to the other that make up the electric current. (Remember that an electric current is a stream of electric charges, and each ion carries an electric charge, either positive or negative. Positive ions move toward the negative electrode, and negative ions move in the opposite direction toward the anode.)

However, the experimental results are surprising: the NaOH solution has a higher conductivity than the NH_4OH solution. The only way to understand these results is to hypothesize that more NaOH molecules than NH_4OH molecules have dissociated into ions. Consequently, there are more ions per cubic centimeter of solution in the NaOH solution than in the NH_4OH solution. The more ions there are, the more current the solution carries for a given applied voltage, since the amount of electric current is proportional to the ion concentration.

By measuring the current we can calculate how many ions are in the solution. We define the **ion concentration** to be the number of ions per unit volume of solution (for example, the number of ions per liter). We can then compare the number of ions present in the solution to the number of solute molecules.

When the ion concentrations of the NaOH and NH_4OH solutions are measured, essentially 100% of the NaOH are found to be ionized, whereas only a small percentage of the NH_4OH molecules are ionized. Defining the **degree of ionization** to be the percentage ionization, we see that the NaOH has a large degree of ionization and the NH_4OH has a small degree of ionization.

Example 14.1 A 0.10-M solution of NaOH is found to contain about 6×10^{19} $(OH)^-$ ions per milliliter. What is the percentage ionization of this solution?

Solution The 0.10-M solution contains 0.10 mol NaOH (4.0 g) in 1.0 L of solution. The number of molecules per liter of solution is

$$\frac{\text{no. of molecules}}{1 \text{ L}} = \frac{6.02 \times 10^{23} \text{ molecules}}{1 \text{ mol}} \times \frac{0.10 \text{ mol}}{1.0 \text{ L}}$$

$$= \frac{6.02 \times 10^{22} \text{ molecules}}{1 \text{ L}}$$

The number of molecules in 1 mL is

$$\frac{\text{no. of molecules}}{1 \text{ mL}} = \frac{6.02 \times 10^{22} \text{ molecules}}{1 \text{ L}} \times \frac{1 \text{ L}}{1000 \text{ mL}}$$

$$= \frac{6.02 \times 10^{19} \text{ molecules}}{1 \text{ mL}}$$

$$= \frac{6.0 \times 10^{19} \text{ molecules}}{1 \text{ mL}} \quad \text{(to two digits)}$$

14.2 ACID-BASE STRENGTH

The percentage ionization is

$$\% \text{ ionization} = \frac{\text{no. of ions/mL}}{\text{no. of molecules/mL}} \times 100\%$$

$$= \frac{6.0 \times 10^{19} \text{ ions/mL}}{6.0 \times 10^{19} \text{ molecules/mL}} \times 100\%$$

$$= 100\% \text{ ionization} \quad \text{(approximately)}$$

This solution is totally ionized.

Example 14.2 In a water solution of ammonium hydroxide, 1.34% of the NH_4OH molecules are ionized (that is, the degree of ionization of NH_4OH is 1.34%). How many OH^- ions are found in 1.0 L of a 0.10-M solution of NH_4OH?

Solution The dissociation reaction is represented by the equation

$$NH_4OH \longrightarrow NH_4^+ + OH^-$$

Each dissociated molecule of ammonium hydroxide produces one OH^- ion. Therefore we must find out how many NH_4OH molecules dissociate. First calculate how many molecules are in the 1.0-L volume. There is 0.10 mol NH_4OH in the volume, so

$$\frac{\text{no. of molecules}}{1 \text{ L}} = \frac{0.10 \text{ mol}}{1 \text{ L}} \times \frac{6.02 \times 10^{23} \text{ molecules}}{1 \text{ mol}}$$

$$= \frac{6.02 \times 10^{22} \text{ molecules}}{1.0 \text{ L}}$$

From the definition of percentage,

$$\% = \frac{\text{part}}{\text{whole}} \times 100\% = 1.34\%$$

where "part" is the number of ions (the number of molecules ionized), "whole" is the original number of molecules, and % is percent ionization.

Solve for "part," and substitute the % value given:

$$\text{part} = \frac{1.34}{100} \times \text{whole}$$

$$= 0.0134 \times \text{whole}$$

The total number of NH_4OH molecules in the liter (the "whole") is 6.02×10^{22}. The number of ions (the "part") is

$$\text{no. of molecules ionized} = 0.0134 \times \text{total number}$$

$$= 0.0134 \times 6.02 \times 10^{22} \text{ molecules}$$

$$= 8.07 \times 10^{20} \text{ molecules ionized}$$
$$= 8.1 \times 10^{20} \text{ (OH)}^- \text{ ions/L}$$
(rounded to two digits)

The number of ions per milliliter is

$$\frac{8.1 \times 10^{20} \text{ ions}}{1 \text{ L}} \times \frac{1 \text{ L}}{1000 \text{ mL}} = 8.1 \times 10^{17} \frac{\text{ions}}{\text{mL}}$$

Note: It is of interest to compare this number with the 6.0×10^{19} (OH)$^-$ ions per milliliter found in Example 14.1. Calculate the ratio of the number of ions in the NaOH solution to the number of ions in the NH$_4$OH solution:

$$\frac{\text{no. of ions in NaOH}}{\text{no. of ions in NH}_4\text{OH}} = \frac{6.0 \times 10^{19}}{8.1 \times 10^{17}}$$

$$= 75 \quad \text{(approximately)}$$

The concentration of (OH)$^-$ ions in the strongly ionized NaOH solution is about seventy-five times greater than in the weakly ionized NH$_4$OH solution.

The Ionization of Water

We tend to think of water as a neutral compound, neither acid nor base; but pure water has a definite, although small, electrical conductivity. Since water conducts electric current, even though to a very small degree, it must be ionized to some extent.

The fact that water is ionized should not be too surprising, for in Section 14.1 we demonstrated that pure water undergoes autoionization according to the equation

$$H_2O + H_2O \rightleftharpoons H_3O^+ + OH^-$$

Both hydronium ions and hydroxide ions are present in equal number. Because of this equality of the hydronium ion and hydroxide ion concentrations, the water behaves in a neutral fashion. It has no taste, and it does not cause litmus paper to change color.

Measurements of the conductivity of water lead to the conclusion that there are roughly 6×10^{16} hydronium ions per liter of water. This sounds like a very great number of ions, but to understand the meaning of this number we must compare it with the number of molecules in that same liter of water. After all, if only one out of every million molecules is ionized, the degree of ionization is very small.

To calculate the number of molecules in the liter, recall that a liter of water has a mass of 1000 g and that the molar mass of water is 18.0 g/mol. Therefore the number of moles of water in a liter is

14.2 ACID-BASE STRENGTH

$$\frac{\text{no. of moles}}{1\text{ L}} = \frac{\text{no. of grams}}{1\text{ L}} \times \frac{1\text{ mol}}{\text{no. of grams}}$$

$$= \frac{1000\text{ g}}{1\text{ L}} \times \frac{1\text{ mol}}{18.0\text{ g}}$$

$$= \frac{55.5\text{ mol}}{1\text{ L}}$$

Now find the number of water molecules in the liter:

$$\frac{\text{no. of molecules}}{1\text{ L}} = \frac{\text{no. of moles}}{1\text{ L}} \times \frac{\text{no. of molecules}}{1\text{ mol}}$$

$$= \frac{55.5\text{ mol}}{1\text{ L}} \times \frac{6.02 \times 10^{23}\text{ molecules}}{1\text{ mol}}$$

$$= \frac{3.34 \times 10^{25}\text{ molecules}}{1\text{ L}}$$

The 6×10^{16} hydronium ions per liter pales in comparison to the 3.34×10^{25} molecules of water per liter.

It is of interest to calculate the *fraction* of molecules that are ionized: the number of hydronium ions in the solution divided by the number of water molecules. This is the degree of ionization. Using round numbers, we have

$$\text{fractional ionization} = \frac{\text{no. of ions}}{\text{no. of molecules}}$$

$$= \frac{6 \times 10^{16}}{3 \times 10^{25}} = 2 \times 10^{-9} \quad \text{(approximately)}$$

This means that roughly two out of every billion water molecules are ionized. Although the degree of ionization is very low, it is readily measurable.

Another important measure of ionization is the **degree of acidity** of a solution: the number of *moles* of H_3O^+ ions per liter of solution. It is a number easily calculated if the number of ions per liter is known.

$$\frac{\text{no. of moles}}{1\text{ L}} = \frac{\text{no. of ions}}{1\text{ L}} \times \frac{1\text{ mol of ions}}{\text{no. of ions}}$$

$$= \frac{\text{no. of ions}}{1\text{ L}} \times \frac{1\text{ mol}}{6.02 \times 10^{23}\text{ ions}}$$

That is, the number of moles per liter is the number of ions per liter divided by 6×10^{23} ions per mole. The "no. of ions" terms cancel and we are left with moles of ions per liter.

For pure water,

$$\frac{\text{no. of moles of ions}}{1 \text{ L}} = \frac{6 \times 10^{16} \text{ ions}}{1 \text{ L}} \times \frac{1 \text{ mol}}{6.02 \times 10^{23} \text{ ions}}$$

$$= 1 \times 10^{-7} \frac{\text{mol}}{\text{L}} \quad \text{(approximately)}$$

The concentration of H_3O^+ ions is seen to be about 1×10^{-7} mol/L. This number is known as the **hydronium ion concentration** or the **hydrogen ion concentration**. [Often we ignore the H_2O part of the hydronium ion (H_3O^+) and concentrate on the H^+ part of it.]

The hydronium ion concentration is represented by the symbol $[H_3O^+]$, where the bracket symbol [ ] means "moles per liter" and the formula inside the square brackets represents the particular substance under discussion. Thus, $[H_3O^+]$ stands for *the number of moles of hydronium ions per liter of solution*. Similarly, the symbol $[OH^-]$ is used to describe the **hydroxide ion concentration** in moles per liter. We have seen that in water

$$[H_3O^+] = 1 \times 10^{-7} \text{ mol/L} \quad \text{(approximately)}$$

In pure water, the hydroxide ion concentration has the same value:

$$[OH^-] = 1 \times 10^{-7} \text{ mol/L}$$

Just as the degree of acidity of a solution is measured by the hydronium ion concentration (the number of moles of hydronium ions per liter of solution), the **degree of alkalinity** of a solution is measured by the number of moles of hydroxide ions (OH^-) per liter of solution.

Ion Concentrations in Acids and Bases

What happens when an acid is dissolved in water? The acid dissociates and ionizes. As a result, the hydronium ion concentration increases. The amount of increase depends on the strength of the acid. Dissolving a very weak acid in water might increase the hydronium ion concentration by a factor of 10, so in solution

$$[H_3O^+] = 10 \times 1 \times 10^{-7} \frac{\text{mol}}{\text{L}} = 1 \times 10^{-6} \frac{\text{mol}}{\text{L}}$$

What happens to the OH^- concentration when the H_3O^+ concentration increases? This is an important question. To see what happens, let us look again at the equation for the autoionization of water:

$$H_2O + H_2O \rightleftharpoons H_2O \cdot H^+ + OH^-$$

This is a reversible reaction; it goes in both directions. Think of the water as a crowd of molecules pushing and jostling one another. Occasionally a molecule of water separates into H^+ and OH^- ions, and the H^+ ion becomes attached to a water molecule to form H_3O^+. The H_3O^+ and OH^- ions float freely in the solution. As the ions move about and collide with one another, there is a chance that the H⁺ may leave the hydronium ion and rejoin one of the OH⁻ ions, producing the right-to-left reaction in the above equation.

When the system is in equilibrium, as many reactions go to the right (per second) as go to the left. (We will go into more detail about equilibrium in Chapter 15. For the time being, think of a system in equilibrium as a balanced system—one in which concentrations do not change over time. Therefore, in equilibrium there are just as many reactions going in the forward direction as going in the reverse direction during a short period of time.)

As we have seen, in pure water the value of [H$_3$O$^+$] equals the value of [OH$^-$]. However, if the number of H$_3$O$^+$ ions in the solution is raised by the addition of acid, there is a greater chance of collision between the H$_3$O$^+$ ions and the OH$^-$ ions. Some of the additional H$_3$O$^+$ ions combine with some of the OH$^-$ ions, forming H$_2$O molecules. This increases the number of reactions going from right to left in the above equation, and at the same time it reduces the number of OH$^-$ ions in the solution. (Somewhat analogously, if you increase the number of cats in a field containing a quantity of cats and mice, the number of mice will soon decrease.)

Mathematical analysis of the situation has shown that the number of OH$^-$ ions in the solution is represented by this very simple equation:

$$[H_3O^+] \times [OH^-] = K_w$$

In other words, the hydronium ion concentration multiplied by the hydroxide ion concentration equals a numerical constant K_w, called the **ion product constant for water**.

If we divide both sides of the equation by [H$_3$O$^+$], we have

$$[OH^-] = \frac{K_w}{[H_3O^+]}$$

This equation tells us that the hydroxide ion concentration is *inversely proportional* to the hydronium ion concentration. This means that as more acid is put into the solution (raising the hydronium ion concentration), the OH$^-$ concentration must go down. *The greater the hydronium ion concentration, the lower the hydroxide ion concentration*, and vice versa.

Before numbers can be put into the equation, we must know the value of the constant K_w. That value is found easily, since we know the values of [H$_3$O$^+$] and [OH$^-$] for pure water. We simply multiply those two numbers together to get the value of K_w. The value of K_w will stay the same for all equations involving water solutions.

Putting the values for the ion concentrations of water into the equation for the constant K_w, we obtain

$$K_w = [H_3O^+] \times [OH^-]$$

$$K_w = 1 \times 10^{-7} \frac{\text{mol}}{\text{L}} \times 1 \times 10^{-7} \frac{\text{mol}}{\text{L}}$$

$$= 1 \times 10^{-14} \, (\text{mol/L})^2$$

(K_w varies with temperature and is equal to the above value only at 24°C. At other temperatures the number is approximately correct.)

Knowing the value of K_w, we can find the hydroxide ion concentration given the hydronium ion concentration, or vice versa.

Example 14.3 In a weak acid solution, the hydronium ion concentration is found to be 1×10^{-3} mol/L. What is the hydroxide ion concentration?

Solution
$$[OH^-] = \frac{K_w}{[H_3O^+]}$$
$$= \frac{1 \times 10^{-14} \, (mol/L)^2}{1 \times 10^{-3} \, mol/L}$$
$$= 1 \times 10^{-11} \, mol/L$$

Note: The hydronium ion concentration in this solution is 10^4 times (10,000 times) greater than it is in pure water, whereas the hydroxide ion concentration is 10^4 times smaller than it is in pure water. The numbers in this problem are given only to one significant digit because we are only interested in an *order-of-magnitude* answer.

14.3 Ionization Concentrations—pH

The Concept of pH

Writing ion concentrations in terms of moles per liter is cumbersome, because the numbers are usually so small that they must be written as powers of ten. Therefore it has become customary to use only the power to which the ten is raised—the number in the exponent—to describe the concentration. Furthermore, so as to eliminate the minus sign, the *negative* of the exponent is used. (Since the exponent itself is a negative number, its negative is a positive number.)

For example, instead of using the entire number 1×10^{-7} to describe the hydrogen ion concentration of water, we take the exponent and change the minus sign to a plus. The number +7 is then used to represent the hydrogen (or hydronium) ion concentration of the water.

A special symbol used to represent hydrogen ion concentration is *pH*, which stands for *hydrogen potential*. It is always spelled with a small p and a capital H and is pronounced exactly the way it is spelled: "p-H." (The two letters are pronounced separately.)

To describe the hydrogen ion concentration of water, we simply say, "The pH of water is seven." The pH is universally used as a convenient measure of the acidity of any kind of solution. (Actually, the pH of water is not exactly 7.00 except at one particular temperature. However, for many purposes, an order-of-magnitude value—good to one significant digit—will suffice.)

The term *pH* has become common in many areas where measuring acidity is important. The farmer or gardener measures the pH of the soil to find out whether it is acid or alkaline. The environmentalist measures the pH of rainwater to determine how much acid is raining from the sky because of atmospheric pollution.

In most industrial processes dealing with the manufacture of chemical and pharmaceutical products, it is essential to keep the acidity of the ingredients within the proper range for the reactions taking place. Therefore electronic instruments constantly monitor the pH of solutions to ensure that the reactions take the right course. In a similar manner, the human body constantly keeps track of its blood pH to make sure that it stays at a level of 7.2. Otherwise the machinery of the body would not function properly.

Definition and Range of pH

The formal definition of pH is as follows: The pH of a solution is the negative of the number found in the exponent of the power-of-ten term when the hydrogen ion concentration of that solution (in moles per liter) is written in the form

$$[H^+] = 1 \times 10^{-pH} \text{ mol/L}$$

(Those familiar with the subject of logarithms will recognize that an alternative definition is

$$pH = -\log[H^+]$$

We will have more to say about logarithms in the next section.)

For example, if a solution has a hydrogen ion concentration of 1×10^{-3} mol/L, the pH of this solution is the negative of the number in the exponent:

$$pH = -(-3) = +3$$

Conversely, if a solution has a pH of 5, its hydrogen ion concentration is

$$[H^+] = 1 \times 10^{-5} \text{ mol/L}$$

Comparing the two solutions in the following table, we discover an important consequence of the definition of pH.

$[H^+]$	Magnitude	pH	Magnitude
1×10^{-3}	greater	3	smaller
1×10^{-5}	smaller	5	greater

The solution that has the greater hydrogen ion concentration—and is therefore more acidic—has the smaller pH value, and vice versa. In general, *the smaller the pH, the more acid the solution.*

A very strong acid might have a hydrogen ion concentration of 1 mol/L. We know that $1 = 10^0$, so the pH of this strong acid is 0 (zero). We have already seen that the pH of pure water is 7. Water is neutral, and its hydrogen ion concentration is the smallest that an acid can have. Any solution with a smaller hydrogen ion concentration is an alkaline (or basic) solution.

Thus we see that the pH values of acids range from 0 (approximately) to 7. In a similar manner, the pH values of basic solutions range from 7 to 14. (We will discuss basic solutions in more detail in a later section.) The range of values can be visualized with the help of the diagram in Figure 14-5.

Notice that while the pH values of all possible solutions range from 0 to 14, the corresponding hydrogen ion concentrations go from 1 to 1×10^{-14}—a ratio of one hundred million million to one. This fact illustrates the value of the pH scale. The numbers from 0 to 14 on the pH scale allow us to express a huge range of actual ion concentrations, a range too great to be comprehended directly.

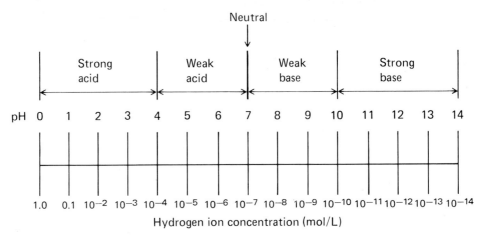

FIGURE 14-5 The pH value increases from 0 to 14 as the hydrogen ion concentration in the solution decreases from 1 to 10^{-14} mol/L.

Logarithms and the Definition of pH

We now generalize the concept of pH so that it can be used in all circumstances. (In this section the Arrhenius model is used for simplicity, and acidity is related to hydrogen ion concentration, or $[H^+]$.) According to our original definition of pH, if the hydrogen ion concentration of any solution is written in the form

$$[H^+] = 1 \times 10^{-pH} \text{ mol/L}$$

the pH of the solution is the number in the exponent, without the minus sign. (In this chapter the integer 1 means *exactly* 1.) So far we have considered only simple situations in which the pH is an integer. How do we treat problems with noninteger values of pH?

As an example, consider an acid with a hydrogen ion concentration of 5.0×10^{-3} mol/L. The pH is not simply 3, for that would imply that the hydrogen ion concentration was 1×10^{-3} mol/L. The problem is that the ion concentration is given in the form

$$[H^+] = 5.0 \times 10^{-3} \text{ mol/L}$$

and it must be rewritten in the form

$$[H^+] = 1 \times 10^{-pH} \text{ mol/L}$$

The number 5.0×10^{-3} lies between 1×10^{-3} and 1×10^{-2}. Therefore, $-pH$, the number in the exponent, must fall between -2 and -3; accordingly, the pH is between 2 and 3. How can we interpolate between 2 and 3 to find a more exact value of pH?

The solution is based on the concept of the logarithm. By definition, the **logarithm** of a number N (to the base 10) is the power to which 10 must be raised in order to get the number N. That is,

$$N = 10^{\log N}$$

The definition of pH given at the beginning of this section is

$$[H^+] = 10^{-pH}$$

Using the definition of the logarithm, we turn this equation around to obtain

$$pH = -\log [H^+]$$

where $\log [H^+]$ stands for *the logarithm of the hydrogen ion concentration.*

In the past the logarithms of numbers were calculated laboriously by methods described in books of mathematical analysis and then were published in standard numerical tables, where they could be looked up for purposes of computation. Logarithms were frequently used to shorten arithmetic calculations—two numbers can be multiplied by adding their logarithms together or divided by subtracting one logarithm from the other. However, the advent of electronic calculators has made the table of logarithms, as well as the use of logarithms for arithmetic calculations, completely obsolete.

However, there are still at least three terms that incorporate the *concept* of the logarithm: (1) the electronic term *decibel* (for example, the db scale on your stereo amplifier), (2) the geological term *Richter scale* used to describe the strength of earthquakes, and (3) the chemical term *pH*, now under discussion.

There are three ways of finding the logarithm of a number. The hard way is to compute the logarithm by means of certain well-known mathematical formulas. An easier (but obsolete) way is to look up the number in a published table of logarithms. The modern way is to use a calculator with a log key.

To solve the above problem, it is necessary to find the log of 5.0×10^{-3}. You simply punch this number into your calculator and press the log key. The calculator's reply is -2.3. This means that

$$5.0 \times 10^{-3} = 10^{-2.3}$$

Applying this result to the pH problem, we have

$$pH = -\log[H^+]$$
$$= -\log(5 \times 10^{-3}) = -(-2.3) = 2.3$$

This result agrees with our prediction that the pH would fall between 2 and 3.

Example 14.4 The hydrogen ion concentration in a particular acid solution is

$$[H^+] = 4.20 \times 10^{-1} \text{ mol/L}$$

Find the pH value.

Solution Key in the number 4.20×10^{-1} (or 0.42), and then press the log key. The answer is -0.377. By definition, the pH is the negative of the logarithm of the hydrogen ion concentration. Therefore, the pH of this acid is $-(-0.377)$, or $+0.377$.

Example 14.5 An acid solution has a pH of 2.6. What is the hydrogen ion concentration?

Solution We know that

$$[H^+] = 1 \times 10^{-pH}$$

In this example,

$$[H^+] = 1 \times 10^{-2.6} \text{ mol/L}$$

Using a calculator with the capacity to compute exponentials, find 10 to the -2.6 power (following the instructions for the particular calculator being used). The solution is obtained directly:

$$[H^+] = 10^{-2.6} = 2.5 \times 10^{-3} \text{ mol/L}$$

The procedure given here exactly reverses the procedure for finding the pH given in Example 14.4.

One final caution: Although the intelligent use of the calculator saves a great amount of arithmetic as well as the bother of looking up numbers in a table of logarithms, it is still necessary to be aware of decimal points and significant digits in order to make sure that the answer makes sense!

The pH of Bases

Although we tend to think of hydrogen ions in connection with acids, even solutions of bases have a certain number of hydrogen ions floating about. In Section 14.2 the concentrations of H^+ and OH^- ions in a solution were found to be related to each other by the equation

14.3 IONIZATION CONCENTRATIONS—pH

$$[H^+] \times [OH^-] = 1 \times 10^{-14} = K_w$$

Given the hydroxide ion concentration, $[OH^-]$, we can find the hydrogen ion concentration by substituting into the equation

$$[H^+] = \frac{K_w}{[OH^-]}$$

This equation states that the hydrogen ion concentration is inversely proportional to the hydroxide ion concentration: the greater the hydroxide ion concentration, the smaller the hydrogen ion concentration.

In a base, the hydroxide ion concentration is greater than 10^{-7} mol/L. Consider a weak base in which $[OH^-]$ equals 10^{-5} mol/L. The hydrogen ion concentration is

$$[H^+] = \frac{10^{-14} \, (mol/L)^2}{10^{-5} \, mol/L} = 10^{-9} \, mol/L$$

$[H^+]$ is 100 times smaller than the neutral value of 10^{-7} mol/L, and $[OH^-]$ is 100 times greater than that neutral value. The pH of this weak base is 9.

If the base is a very strong one in which $[OH^-]$ equals 1 mol/L, the hydrogen ion concentration is

$$[H^+] = \frac{10^{-14} \, (mol/L)^2}{1 \, mol/L} = 10^{-14} \, mol/L$$

The pH of this solution is 14.

TABLE 14-1 Hydrogen Ion Concentrations and pH Values for Acids and Bases

Description	$[H^+]$ (in moles/liter)	pH
strongest acid (HCl, H_2SO_4)	$10^0 = 1$	0
	10^{-1}	1
	10^{-2}	2
	10^{-3}	3
weak acid (acetic acid)	10^{-4}	4
	10^{-5}	5
	10^{-6}	6
neutral (water)	10^{-7}	7
	10^{-8}	8
weak base (ammonium hydroxide)	10^{-9}	9
	10^{-10}	10
	10^{-11}	11
	10^{-12}	12
strong base (sodium hydroxide)	10^{-13}	13
	10^{-14}	14

TABLE 14-2 Common Liquids and Their Approximate pH Values

Name	[H$^+$]	pH
gastric juice	1×10^{-1}	1
orange juice	1×10^{-3}	3
urine	1×10^{-5}	5
milk	1×10^{-6}	6
water	1×10^{-7}	7
blood	1×10^{-7}	7
milk of magnesia	1×10^{-10}	10
lye solution	1×10^{-13}	13

These examples demonstrate that in basic solutions the pH varies from 7 to 14. The hydrogen ion concentrations and pH values for acids and bases are summarized in Table 14-1; Table 14-2 gives the pH values for a few common liquids that are often found in the home.

You can see from Table 14-2 that gastric juice (found in the stomach) is quite a strong acid; this is because hydrochloric acid is one of its main components. This acidity of the stomach is completely normal, and efforts to neutralize the HCl by taking antacids can possibly do more harm than good. Milk is a very weak acid because of the lactic acid found in it, and orange juice is distinctly acid as a result of its citric acid. Milk of magnesia (magnesium hydroxide), on the other hand, is a weak base, as is lime water (calcium hydroxide solution).

Example 14.6 In a moderately strong base, the hydroxide ion concentration is

$$[OH^-] = 10^{-3} \text{ mol/L}$$

Determine the hydrogen ion concentration and the pH.

Solution The hydrogen ion concentration is given by

$$[H^+] = \frac{10^{-14} \text{ (mol/L)}^2}{10^{-3} \text{ mol/L}}$$

$$= 10^{-11} \text{ mol/L}$$

The pH of the solution is seen by inspection to be 11.

Glossary

acid strength A qualitative way of comparing acids. A strong acid has a high hydrogen ion concentration.

Arrhenius acid A compound that dissociates into a hydrogen ion plus a negative ion in solution.

Arrhenius base A compound that dissociates into a hydroxide ion plus a positive ion in solution.

autoionization A reaction in which two identical covalent molecules interact in solution and are both ionized.

Bronsted-Lowry acid A compound that donates a proton to a base when in solution.

Bronsted-Lowry base A compound that receives a hydrogen ion from an acid in solution.

conjugate acid-base pair A Bronsted-Lowry acid and a Bronsted-Lowry base that interact with each other in solution.

degree of acidity The concentration of hydrogen ions; the number of moles of H_3O^+ ions per liter of solution.

degree of alkalinity The concentration of hydroxide ions; the number of moles of OH^- ions per liter of solution.

degree of ionization The number of moles of dissociated molecules compared to the number of moles of solute molecules in the solution; fractional or percentage ionization.

electrolytes Substances that dissociate into ions in solution or when fused, thereby becoming electrically conducting.

electrolytic dissociation A process in which a substance breaks up into ions upon forming a solution.

electrolytic solutions Solutions that contain ions and so conduct electric current.

hydronium ion The ion $(H_3O)^+$, consisting of water with an extra attached proton.

hydronium ion concentration The number of moles of hydronium ions per liter of solution, represented by the symbol $[H_3O^+]$; equivalent to *hydrogen ion concentration*.

hydroxide ion concentration The number of moles of OH^- ions per liter of solution; represented by the symbol $[OH^-]$.

ion concentration The number of ions per unit volume of solution.

ion product constant for water A constant number, K_w, found by multiplying the concentration of hydroxide ions by the hydronium ion concentration in water; 1×10^{-14} $(mol/L)^2$.

Lewis acid A substance whose molecule accepts a pair of electrons to form a covalent bond.

Lewis base A substance whose molecule donates an electron pair to another molecule to form a covalent bond.

litmus test A test in which the chemical indicator litmus is used to determine the amount of acidity of a solution. Litmus appears red in the presence of acids and blue in the presence of alkaline solutions.

logarithm The exponent indicating the power to which a fixed number (the base) must be raised in order to produce a given number. For example, since $100 = 10^2$, the logarithm of 100 to the base 10 is 2, or $\log_{10} 100 = 2$.

nonelectrolyte Substances that do not form ions in solution and so do not conduct electric current.

pH The negative of the logarithm of the hydronium ion concentration in a solution.

proton transfer Transfer of a proton from an acid molecule to a water molecule or from a water molecule to a base molecule during dissociation of a compound in water solution.

strong acid An acid that has a high degree of ionization.

strong base A base that has a high degree of ionization.

weak acid An acid with a low degree of ionization.

weak base A base with a low degree of ionization.

Problems and Questions

14.1 Models of Acids, Bases, and Salts

14.1.Q Define acid and base according to the Arrhenius model.

14.2.Q Define a salt according to the Arrhenius model.

14.3.Q State two important facts that the Arrhenius model does not explain.

14.4.Q What is the difference between a strong acid and a weak acid?

14.5.Q Write the equations for the dissociation of the following compounds:
   a. H_3PO_4
   b. H_2SO_4
   c. HCl
   d. $NaOH$
   e. $Ca(OH)_2$

14.6.Q When an acid ionizes in water to form one or more H^+ ions, what happens to the electrons that were part of the hydrogen atoms?

14.7.Q Name a "metallic" positive ion that has no metal in it and a "nonmetallic" negative ion that has metal atoms in it.

14.8.Q a. What is a hydroxide ion?
   b. What is a hydronium ion?

14.9.Q Describe the dissociation of an acid in aqueous solution in terms of the Bronsted-Lowry theory.

14.10.Q Define acid and base in terms of the Bronsted-Lowry model.

14.11.Q a. What is a proton donor (acid or base)?
   b. What is a proton acceptor (acid or base)?

14.12.Q Describe one important advantage of the Bronsted-Lowry model over the Arrhenius model.

14.13.Q Explain neutralization in terms of Bronsted-Lowry acid-base theory.

14.14.Q Using the Bronsted-Lowry acid-base model, describe the following reactions (label the conjugate acid-base pairs):

   a. $HCl + NaOH \rightleftarrows H \cdot OH + Na^+ + Cl^-$
   b. $HC_2H_3O_2 + NH_4OH \rightleftarrows NH_4C_2H_3O_2 + H_2O$
   c. $NH_3 + H_2O \rightleftarrows NH_4^+ + OH^-$
   d. $H_2O + H_2O \rightleftarrows H_3O^+ + OH^-$
   e. $NH_3 + NH_3 \rightleftarrows NH_4^+ + NH_2^-$

14.15.Q a. Write the equation for the autoionization of water.
   b. What reaction is the reverse of autoionization of water?
   c. Which of the reactions listed in Problem 14.14 are autoionization reactions?

14.16.Q a. In which direction is the reaction of Problem 14.14.c more likely to go—to the right or to the left?
   b. What determines this direction?

14.17.P Identify the Lewis acid and the Lewis base in each of the following reactions. (*Note:* Draw the electron-dot diagram first.)
   a. $HCl + H_2O \longrightarrow H_3O^+ + Cl^-$
   b. $NH_3 + H_2O \longrightarrow NH_4^+ + OH^-$
   c. $NH_3 + BF_3 \longrightarrow NH_3BF_3$

14.2 Acid-Base Strength

14.18.Q What is a chemical acid-base indicator?

14.19.Q Describe the litmus test.

14.20.Q a. Describe two tests of acid strength.
   b. Describe the results of these two tests when applied to hydrochloric acid and acetic acid, respectively.

14.21.Q a. Describe two tests of base strength.
   b. Describe the results of these two tests when applied to sodium hydroxide and ammonium hydroxide, respectively.

14.22.P The degree of ionization of acetic acid is 1.34%. Six hundred grams of acetic acid ($HC_2H_3O_2$) is dissolved in water to make 2.0 L of solution. Calculate the concentration of H^+ ions per cubic centimeter. The ionization equation is

$$HC_2H_3O_2 \longrightarrow H^+ + C_2H_3O_2^-$$

14.23.P Assuming 100% ionization, how many grams of potassium hydroxide (KOH) would it take to form a solution containing 1 mol of hydroxide ions per liter of solution?

14.24.P A solution of NH_4OH is 1.50% ionized. If 100 g of NH_4OH is dissolved in 500 mL of solution, what is the concentration of OH^- in number of ions per liter?

14.25.P A solution of NH_4Cl is labeled $0.025M$, and the Cl^- concentration is found to be 1.0×10^{-10} mol/L. What is the degree of ionization?

14.26.P An acid solution contains 55 mol of solute per liter of solution. The H^+ concentration is found to be 1.0×10^{-7} mol/L. What is the degree of ionization?

14.27.Q Explain how pure water can be a conductor of electricity.

14.28.Q Describe the mechanism by which pure water is ionized.

14.29.Q Write the equation for an autoionization reaction involving the reactants $NH_3 + NH_3$.

14.30.P We find that approximately 8.1×10^{20} out of 6.02×10^{22} NH_4OH molecules become ionized in 2 L of solution.
  a. Calculate the degree of alkalinity of NH_4OH.
  b. Calculate the degree of ionization of the same compound.

14.31.P a. What is the degree of ionization of water?
  b. What is the degree of acidity of water?

14.32.Q a. What is the meaning of the bracket symbol [ ]?
  b. What is the meaning of the expression $[H_3O^+]$?

14.33.Q What is the relationship between hydronium ion concentration and hydroxide ion concentration in a water solution?

14.34.Q How do we find the value of the ion product constant, K_w?

14.35.Q Give a physical explanation for the fact that increasing the hydronium ion concentration decreases the hydroxide ion concentration in an aqueous solution.

14.36.P Given a water solution in which $[H_3O^+]$ equals 1×10^{-4} mol/L, calculate the hydroxide ion concentration, $[OH^-]$.

14.37.P Given a water solution in which $[OH^-]$ equals 1×10^{-3} mol/L, calculate the hydronium ion concentration, $[H_3O^+]$.

14.38.P If $[H_3O^+]$ equals 1×10^{-9} mol/L, what is the hydroxide ion concentration, $[OH^-]$?

14.39.P If $[H_3O^+]$ equals 1×10^{-10} mol/L, can $[OH^-]$ equal 1×10^{-6} mol/L? If not, what is the value of $[OH^-]$?

14.40.P Calculate $[OH^-]$ if $[H_3O^+]$ equals 1×10^{-1} mol/L.

14.3 Ionization Concentrations—pH

14.41.Q What does pH measure?

14.42.Q What is the rule for converting hydrogen ion concentration into pH?

14.43.Q a. What is the hydrogen ion concentration of water?
  b. What is the pH of water?

14.44.Q Describe two uses of pH measurements.

14.45.P Determine the pH corresponding to each of the following hydrogen ion concentrations:
  a. 1×10^{-4} mol/L
  b. 1×10^{-2} mol/L
  c. 1×10^{-8} mol/L

14.46.P Determine the hydrogen ion concentration corresponding to each of the following pH values:
  a. 6   b. 1
  c. 10

14.47.Q a. What is the range of pH values for acids?
  b. What is the range of pH values for bases?

14.48.P Determine the hydrogen ion concentration corresponding to each of the following pH values:
  a. 5   b. 12
  c. 1   d. 14

14.49.P Determine the pH corresponding to each of the following hydrogen ion concentrations:
  a. 1×10^{-10} mol/L   b. 1×10^{-2} mol/L
  c. 1×10^{-8} mol/L   d. 1×10^{-6} mol/L

14.50.Q Classify the solutions given in Problems 14.45 and 14.46 as acid or base and strong or weak.

14.51.P A strong acid has a pH of 1. It is diluted until it has a pH of 2.
  a. Determine the hydrogen ion concentration in the acid before and after dilution.
  b. What is the ratio of these two hydrogen ion concentrations?

14.52.P In general, when the pH of a solution is changed by adding or subtracting 1 from the pH value, what is the corresponding change in the hydrogen ion concentration expressed in moles per liter?

14.53.Q What is the relationship among pH, $[H^+]$, and logarithms?

14.54.Q a. What is the range of pH values for acids?
b. What is the corresponding range of hydrogen ion concentrations?

14.55.Q If you increase the pH of a solution by 1 (say, from 4 to 5), what effect does that have on the hydrogen ion concentration?

14.56.Q If you increase the pH of a solution by 3 (say, from 2 to 5), what effect does that have on the hydrogen ion concentration?

14.57.P Calculate the pH of a solution whose $[H^+]$ equals 1×10^{-7} mol/L.

14.58.P Calculate the pH of a solution whose $[H^+]$ equals 2.7×10^{-4} mol/L.

14.59.P Calculate the pH of a solution whose $[OH^-]$ equals 3.5×10^{-5} mol/L.

14.60.P Calculate the $[H^+]$ of a solution whose pH is 3.0.

14.61.P Calculate the $[H^+]$ of a solution whose pH is 7.5.

14.62.P Calculate the $[H^+]$ of a solution whose pH is 11.3.

14.63.P Calculate the pH of a solution that contains 6×10^{-3} mol of H^+ ions in 200 mL of solution.

14.64.P Calculate the number of moles of H^+ ions in 700 mL of a solution with a pH of 9.2.

14.65.P Calculate the $[H^+]$ of a solution whose pH is 2.7. If the $[H^+]$ is doubled, what is the new pH value?

14.66.P Calculate the number of moles of H^+ ions in 3.5 L of a solution with a pH of 8.8.

14.67.Q What is the relationship between $[H^+]$ and $[OH^-]$?

14.68.Q What is the relationship between pH and $[OH^-]$?

14.69.Q What is the range of pH values for a fairly strong acid?

14.70.Q What is the range of pH values for a fairly strong base?

14.71.Q What is the range of pH values for a weak acid?

14.72.Q What is the range of pH values for a weak base?

14.73.P What is the pH of a solution when its $[OH^-]$ equals 2.5×10^{-4} mol/L?

14.74.P What is the pH of a solution when its $[OH^-]$ equals 7.8×10^{-9} mol/L?

14.75.P What is the $[OH^-]$ of a solution whose pH is 6.7?

14.76.P What is the $[OH^-]$ of a solution whose pH is 10.3?

14.77.P Determine the $[OH^-]$ and the $[H^+]$ of a solution whose pH is 1.9.

14.78.P If the $[H^+]$ of a solution is 1.5×10^{-5} mol/L, what is the $[OH^-]$?

14.79.P If the $[OH^-]$ of a solution is 6.6×10^{-10} mol/L, what is the $[H^+]$?

14.80.P If the $[H^+]$ of a solution is 4.9×10^{-2} mol/L, what is the $[OH^-]$?

14.81.P a. If the $[H^+]$ of the solution in Problem 14.80 is doubled, what happens to the value of $[OH^-]$?
b. Generalize the answer of part a to any solution.

14.82.P If the $[OH^-]$ of a solution is reduced from 9.9×10^{-11} to 9.9×10^{-12}, what change occurs in the value of $[H^+]$?

14.83.P Given that the $[H^+]$ of a solution is 7.5×10^{-8}, determine its $[OH^-]$ and its pH.

Self Test

1. Define the following terms, and give illustrations for each.
   a. Arrhenius acid
   b. Bronsted-Lowry base
   c. conjugate acid-base pair
   d. strong acid versus weak acid
   e. neutralization, as described by both the Arrhenius model and the Bronsted-Lowry model
   f. degree of ionization
   g. autoionization
   h. K_w

2. Some manufacturers of frozen French fries add an alkali (base) to potatoes to make it easier to remove the skins. After peeling and before freezing, a weak acid such as citric acid is often applied. Suggest a reason for the addition of citric acid.

3. Label the conjugate acid-base pairs in each of the following reactions:
   a. $HCl + H_2O \rightleftharpoons H_3O^+ + Cl^-$
   b. $HCH_3CO_2 + H_2O \rightleftharpoons H_3O^+ + CH_3CO_2^-$
   c. $H_2SO_4 + H_2SO_4 \rightleftharpoons H_3SO_4^+ + HSO_4^-$

4. Calculate the pH values of each of the following solutions:
   a. distilled water, with $[H^+]$ equal to 1×10^{-7} mol/L
   b. urine, with $[H^+]$ equal to 2.0×10^{-5} mol/L
   c. sodium hydroxide solution, with $[OH^-]$ equal to 3.1×10^{-1} mol/L

5. Calculate the values of $[H^+]$ and $[OH^-]$ for each of the following solutions:
   a. orange juice, with a pH of 2.8
   b. gastric juice, with a pH of 0.9
   c. milk of magnesia, with a pH of 10.1

6. Calculate the degree of ionization of each of the following solutions:
   a. a 1.00-M solution of NH_4OH, with a pH of 9.0
   b. a 1.00-M solution of $HC_2H_3O_2$, with a pH of 4.8.
   c. a 0.10-M solution of H_2SO_4, with a pH of 1.0

7. Acetic acid ($HC_2H_3O_2$) is sometimes added to canned vegetables and fruits to make the product more acidic. Explain how the addition of $HC_2H_3O_2$ increases the acidity of the food product. If the pH changes to 6.1, what happens to the hydroxide ion concentration?

15
Reaction Dynamics

Objectives After completing this chapter, you should be able to:
1. Explain chemical reactions in terms of collisions between molecules and ions.
2. Explain the factors that determine whether or not a reaction will take place.
3. Explain the factors that determine the rate of a reaction.
4. Explain the difference between the total collision rate and the effective collision rate.
5. Explain the effect of the concentration of the reactants on the reaction rate, using the rate equation.
6. Write a rate equation for simple reactions.
7. Understand the role of catalysts in chemical reactions.
8. Explain the concept of equilibrium in reversible reactions.
9. Calculate the equilibrium constant for simple reactions, given the concentrations of reactants.
10. Calculate the concentrations of reactants at equilibrium, given the equilibrium constant for the reaction.
11. Use Le Chatelier's principle to predict the course of a reaction if conditions such as temperature, pressure, or concentration are changed.
12. Explain the influence of temperature changes on the equilibrium of a chemical reaction.
13. Use the enthalpy of a reaction to predict whether the reaction is exothermic or endothermic.
14. Predict how to increase the yield of a reaction by changing the concentration of the reactants or the temperature of the system.

15.1 Reactions and Molecular Collisions

In this chapter we delve further into some important questions concerning chemical reactions. We ask why some reactions take place and others do not, and why some reactions take place more rapidly than others. A question of interest to the practicing chemist is: What can we do to speed up a slow reaction, or slow down a fast reaction, so that it will be more useful in a manufacturing process? In order to get answers to these questions, we must return to the molecular model and determine what happens when molecules collide with one another.

Molecular Collisions and Reaction Probabilities

Chemical reactions occur because of the collisions of molecules or ions with one another whenever their paths happen to intersect. Imagine a dozen blindfolded players running about in random directions on a football field. Because of the large distance between them, there is not much chance that they will run into one another. Yet, given enough time, some chance collisions will occur among them sooner or later. The situation is similar in a gas: distances between molecules are relatively large, and one molecule may pass many others before engaging in an encounter close enough to cause a reaction. (See Figure 15-1.)

In a water solution, the reactant molecules are surrounded by water molecules and must make their way through the throng. It is as though the runners were in the middle of a dense crowd and had to battle their way through the horde in order to make contact. Although the presence of water molecules may impede the reaction, the advantage of the water solution is that it allows a great many

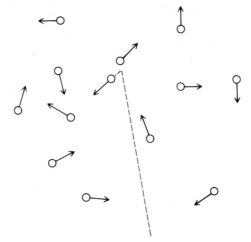

FIGURE 15-1 In a gas the distance between molecules is great compared with the diameters of the molecules, so a given molecule may pass by many other molecules before making a collision.

reacting molecules to be packed into a smaller space so it is easier for them to find each other. Furthermore, when a compound ionizes in an aqueous solution, the electrical attraction between the oppositely charged ions increases the probability that collisions will take place.

One of three things may happen when two molecules come close to each other (Figure 15-2):

1. The two molecules may simply ignore each other if their paths do not bring them close enough.

2. They may bounce off each other in an elastic collision—a collision in which there is no change in the internal states (or energies) of the molecules.

3. The molecules may stick together, forming an **intermediate state**.

(All of the comments in this section apply to single atoms as well as to molecules.)
In the collision depicted in Figure 15-2(c), if A and B are single atoms forming an intermediate state AB, one of two things may happen:

1. AB may split apart into the original atoms, A and B. The result is the same as if no reaction had taken place.

2. Alternatively, A and B may stick together to form a compound.

As an example of the latter case, let us consider the formation of iron sulfide by the reaction of iron and sulfur:

$$A + B \longrightarrow AB$$
$$Fe + S \longrightarrow FeS$$

On the other hand, if A and B are molecules, then three possibilities exist:

1. A and B may stick together permanently to form a new compound.

2. Intermediate state AB may hold together for less than a microsecond and then split apart into original molecules A and B.

3. Or else a different pair of molecules, C and D, may emerge from AB after an internal rearrangement of atoms. The types of events depicted in Figure 15-2(c) are known as **inelastic collisions**, in which there is a transfer of energy between the interacting molecules or atoms, as well as possibly a change in their internal structures.

For example, in the reaction of ammonia with hydrochloric acid, the following sequence of events may take place (possibility 1):

$$A + B \longrightarrow AB$$
$$NH_3 + HCl \longrightarrow NH_4Cl$$

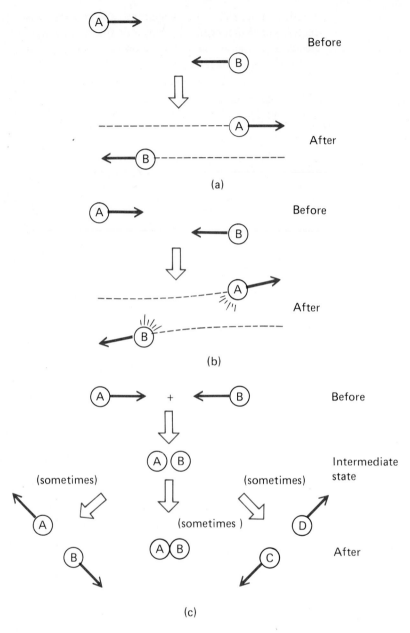

FIGURE 15-2 (a) Molecules A and B may pass each other but find so little attraction that they simply ignore each other. (b) Molecules A and B may bounce away from each other in an elastic collision, in which the internal states of the molecules do not change and so there is no change in internal energy. (c) A and B may combine to form an intermediate molecule AB. Sometimes this intermediate molecule is unstable and breaks up into two molecules, A and B, sometimes it remains as AB, and sometimes it forms different molecules, C and D.

On the other hand, the combustion of methane in oxygen gives us possibility 3:

$$A + B \longrightarrow AB \longrightarrow C + D$$

$$CH_4 + O_2 \longrightarrow CO_2 + 2H_2O$$

The intermediate state in this reaction is very complex and involves the *dissociation* (splitting) of the O_2 molecule into individual atoms.

In aqueous solutions, the situation is even more complicated. Encounters between the reactant and water molecules cause the formation of ions. The mixture of ions and water molecules represents an intermediate state in the reaction. When a freely floating ion encounters an ion of the opposite charge, they may join together to form a new compound. For example, in the reaction of lead chloride with silver sulfate, the intermediate state is a conglomerate of Pb^{2+}, Ag^+, Cl^-, and SO_4^{2-} ions. When Ag^+ and Cl^- ions combine, the result is AgCl, which is insoluble in water and so precipitates out. The effect is a rearrangement of the molecules—a double displacement reaction.

$$A + B \longrightarrow AB \longrightarrow C + D$$

$$PbCl_2 + Ag_2SO_4 \longrightarrow Pb^{2+} + 2Cl^- + 2Ag^+ + SO_4^{2-} \longrightarrow PbSO_4 + 2AgCl$$

An important question is: How hard must the molecules hit each other for a reaction like this to take place? The answer depends on which molecules are under discussion. Neon atoms can collide with great force, and nothing but elastic collisions will take place. On the other hand, a mixture of hydrogen and oxygen molecules will (under the right conditions) combine vigorously and explosively. The detailed theory of atomic and molecular collisions predicts that the *probability* (the chance) that hydrogen and oxygen atoms will stick together is very high, whereas the probability that neon atoms will stick together is extremely low.

Looking at this situation from a macroscopic point of view, we observe that *some chemical reactions are more probable than others*. Reactions do not happen simply at random. Collisions between molecules are random, but whether or not a reaction takes place during a collision depends on the properties of the molecules involved and the way they interact with each other. The reactions by which complex organic molecules are built up from simpler molecules follow paths of high probability.

Factors Determining the Likelihood of a Reaction

Several factors determine whether or not a reaction takes place.

1. The reacting molecules must find each other. That is, they must come close enough to react. If only two molecules are bouncing about in a large box, there is little probability that they will come close enough to have a collision of any kind.

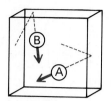

FIGURE 15-3
Two molecules bouncing around in a box have a small probability of colliding.

(See Figure 15-3.) However, chemistry rarely deals with only two molecules at a time. At least a reasonable fraction of a mole generally is involved. In a mole of any substance, there are 6.02×10^{23} molecules moving about, and as a result millions of collisions take place every second. Chemistry then becomes a statistical science. Although we cannot follow the path of any individual molecule in this enormous crowd, we can calculate the number of reactions that are *likely* to take place in any given period of time.

Clearly, the number of reactions depends on the concentrations of the molecules involved (that is, on the number of molecules per unit volume). If the number of A molecules in a given volume is doubled, the number of collisions per second will double. If the concentration of A is fixed and the concentration of B is doubled, again the number of collisions per second will double. Extending this argument, *the number of collisions per second depends on the product of the concentrations of A and B*. For example, if the number of A molecules *and* the number of B molecules are both doubled, the number of reactions per second is multiplied by four.

2. The chance of a reaction depends on the orientation of the colliding molecules. *Orientation* refers to the angle that one makes with respect to the other. (See Figure 15-4.) In the case of long and thin molecules, the chance of a reaction is greater when they meet each other broadside than when they approach each other head on.

3. The energy of the collision must be higher than a certain minimum amount in order for a reaction to take place. The major reason for this requirement is that a molecule consisting of two or more atoms must be broken apart before a reaction can take place. For example, in the reaction

$$H_2 + Cl_2 \longrightarrow 2HCl$$

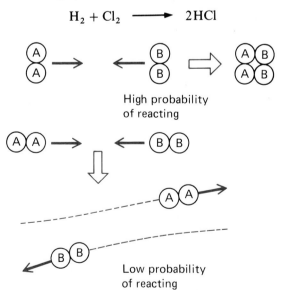

FIGURE 15-4 Two long molecules colliding broadside have a higher probability of reacting than do the same two molecules colliding head on.

the hydrogen and the chlorine molecules must be dissociated (split into separate atoms) before the H atoms and the Cl atoms can join together. (The intermediate state is H + H + Cl + Cl. Atomic hydrogen and chlorine are very unstable substances.)

To break the molecular bond takes a certain amount of energy. If the molecules are traveling too slowly, they will simply bounce away from each other upon colliding. But if they are traveling fast enough, the collision will split each molecule apart. (The same sort of thing happens when two cars collide on the highway. In a very low-energy collision, the two cars will bounce apart elastically. In a high-energy collision, the cars will be split into smaller pieces.) After the molecule is split, the individual fragments can rearrange themselves into a more stable form.

Because energy is needed to break the molecular bonds, we find that in order for a reaction to take place, the **kinetic energy** of the reacting molecules must be greater than a certain minimum amount (an amount that is different for each kind of reaction).

This minimum amount of energy the molecules must have before they can react is called the **activation energy**. The higher the activation energy, the more difficult it is for a given reaction to take place. It is as though each molecule must climb over an energy barrier before it can effectively collide with the other molecule. The height of this "potential barrier" plays a large role in deciding how easily and how rapidly a reaction takes place. (See Figure 15-5.)

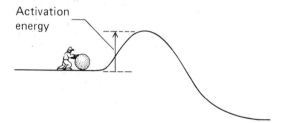

FIGURE 15-5 In order for two molecules to have an effective collision and enter into a reaction, they must have a certain amount of kinetic energy—the activation energy. The situation is analogous to that of a big ball that must overcome a hill before rolling down into the valley. To get the ball over the hill, you have to give it enough of a push so that it can coast over the top and get to the downward slope.

15.2 Reaction Rates

The **reaction rate** is a number that describes the speed of a given reaction—how rapidly or how slowly it occurs. It can be defined as the number of atomic, molecular, or ionic reactions taking place in a unit volume each second. For any given reaction, this rate depends on a number of factors, and in this section we survey the most important of these factors.

Total Collision Rate

One factor of importance in determining the reaction rate is the **total collision rate**: the number of collisions (of all kinds) per unit time between the interacting particles. (During the remainder of the discussion, you should remember that by *particle* we mean atom, molecule, or ion.) As indicated, not all collisions produce reactions of interest. In some reactions most of the collisions are elastic—particles bouncing off each other with no permanent effect. Those collisions between two particles that result in a reaction are called **effective collisions**. Our attention will be focused on the **effective collision rate**: the number of collisions per second that result in reactions of interest.

It can be shown that for a given reaction the effective collision rate is a fixed fraction of the total collision rate (at a given temperature). Therefore an increase in the total collision rate will produce an increase in the effective collision rate. The final result will be an increase in the rate of the chemical reaction.

Before considering what determines the effective collision rate, let us first examine a number of factors that affect the total collision rate.

The Concentration of the Reactants

Clearly, the more reacting particles there are in a given volume, the more collisions per second there will be. As shown in Section 14.1, the collision rate is proportional to the product of the concentrations of the two reactants (in a binary reaction). The concentration can be changed in two ways, as shown in Figure 15-6.

One way is to put more (or less) of the reacting particles into the same container. The concentration of a gas is increased by increasing the amount of gas, which in turn involves increasing the pressure. The concentration of a water solution is increased by adding more solute to a given volume of water.

The other way to change the concentration, in the case of a gas, is to change the volume of the container without changing the amount of gas. Decreasing the volume, which increases the number of molecules per unit volume, requires an increase in pressure.

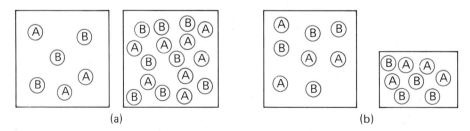

FIGURE 15-6 There are two ways of increasing the concentration of molecules in a container: (a) increasing the number of molecules in the container and (b) decreasing the size of the container without changing the number of molecules.

The Temperature

The higher the temperature, the faster the particles of a substance move, on the average. The faster they move, the farther they travel per unit time; as a result there are more collisions in that time. Therefore, increasing the temperature increases the total collision rate.

Surface Area

In reactions between a solid and a liquid (or a solid and a gas), the surface area of the solid plays an important role. The more area there is, the more collisions there will be between the particles of the liquid and the particles of the solid. (See Figure 15-7.) Therefore a rough surface increases the collision rate; if the solid is ground into fine powder, the collision rate is increased even more. (Flammable substances ground into very fine powders can be extremely explosive. The dust that collects in grain elevators is a major cause of the spontaneous explosions that sometimes take place inside these structures.)

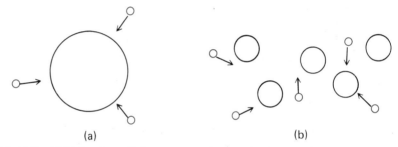

FIGURE 15-7 (a) Molecules collide at a certain rate with a large object. (b) If this object is broken into many small pieces, there is a greater surface area and therefore the same number of molecules will undergo more collisions per unit time.

Effective Collision Rate

Since there is a minimum activation energy that must be reached before particles can engage in reactions, changing the temperature of the reactants can have a dramatic impact on the effective collision rate. To understand the reason for this, let us look at the way kinetic energy is distributed among the particles of a liquid or a gas.

At a given temperature, the particles do not all move at the same speed. Some move faster and some slower than the average speed. This fact is illustrated by an **energy distribution curve**, as shown in Figure 15-8. The curve shows how many particles in the sample have a given amount of energy. We see that most of the particles cluster about some **average energy** that is proportional to the temperature of the substance. However, there is a long tail at the end of the curve, representing a few particles that have much more than the average energy. (Because of this long tail, the average energy is a bit to the right of the peak of the curve.)

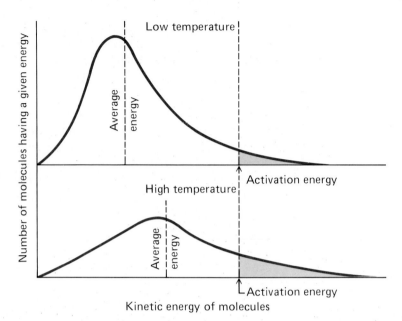

FIGURE 15-8 The energy distribution curve of molecules in a liquid or gas. The curve shows the number of molecules having a given energy. If the temperature is raised, the peak of the curve moves in a horizontal direction. The number of molecules having energies greater than a given value (the activation energy) is proportional to the shaded area under the curve. The higher the temperature, the greater the area.

Consider a hypothetical case in which the activation energy of the reaction is two or three times greater than the average energy of the particles. The number of particles having energies greater than the activation energy is represented by the amount of shaded area under the curve. These are the particles that will engage in effective collisions.

When the temperature of the system is raised, the curve broadens out. The peak moves down and to the right; also, the average energy moves to the right. In addition, the tail extends farther out to regions of higher energy. The activation energy, of course, does not change; it is independent of the temperature, depending only on the internal energies of the reacting particles. As a result, the number of particles in the shaded area under the curve (those with energies greater than the activation energy) is much greater at the higher temperature than it is at the lower temperature. This means that a moderate increase in temperature has caused a very large increase in the effective collision rate, because those particles with energies greater than the activation energy are the ones able to have effective collisions.

There is a useful rule of thumb for making a good approximation in many situations: *The rate of a reaction doubles for every 10°C increase in temperature.* This effect is mainly caused by the increase in the effective collision rate, as described above. For reactions with a very high activation energy, there may be an even greater change in reaction rate for a 10°C temperature change.

The Rate Equation

The factors affecting the reaction rate of a binary reaction between reactants A and B are summarized in the following rate equation:

$$r = k[A][B]$$

In this equation r is the reaction rate; it is a quantity that is proportional to the number of reactions per unit volume per second. [A] and [B] represent the concentrations of the two reacting substances, usually expressed in moles per liter. The constant of proportionality k is a number that represents the probability that a collision will produce a reaction. It is a constant only under one given set of conditions; its value depends on the temperature, the activation energy, and other factors.

This equation meets all the conditions set forth in the above paragraphs. The fact that r depends on the product of [A] and [B] means that the reaction rate is proportional to the concentration of each of the two reactants. The number k expresses the way the reaction rate varies with temperature. In a fast reaction the value of k will be large.

The above rate equation applies to reactions in which the collision initiating the reaction involves two molecules—one molecule of A combines with one molecule of B. Consider, however, the combustion of hydrogen:

$$2H_2 + O_2 \longrightarrow 2H_2O$$

Here *three* molecules enter into the reaction: two molecules of hydrogen and one of oxygen. A triple collision is necessary to produce this event. As a result, the rate equation for this reaction is written as

$$r = k[H_2][H_2][O_2]$$
$$= k[H_2]^2[O_2]$$

You see that the reaction rate is proportional to the *square* of the hydrogen concentration. The hydrogen concentration enters into the equation twice because the rate depends on the collision of each of the two hydrogen molecules with the oxygen.

In general, for a reaction of the form

$$aA + bB \longrightarrow cC + dD$$

where a, b, c, and d are the number of molecules (or moles) of ingredients A, B, C, and D, respectively, entering into the elementary reaction, the rate of the left-to-right reaction is given by the equation

$$r = k[A]^a[B]^b$$

The reaction rate is proportional to the concentration of A to the ath power multiplied by the concentration of B to the bth power. (We pronounce "a-th" and "b-th" just as we say "sixth" and "seventh.")

Fast and Slow Reactions

The fact that reactions go at various rates is demonstrated by many of the common phenomena we observe around us. Even among fast reactions such as combustions and explosions there exist many variations in speed. A mixture of hydrogen and oxygen can explode violently, and the controlled combustion of hydrogen in pure oxygen is used to propel large rockets. However, the explosion of nitroglycerine is a much faster reaction. It generates less energy per gram of reactants than does the H-O reaction, but the energy is released so rapidly that the expanding gases create a destructive shock wave; it is this sudden detonation that makes nitroglycerine useful in the construction of tunnels for breaking up rocks. (Dynamite is nitroglycerine absorbed in an inert material such as sawdust; this combination is safer to use than the pure nitroglycerine, since it is not set off so easily by accident.)

A familiar slow reaction is the rusting (oxidation) of iron. In a dry atmosphere rusting may take a long time, but the presence of moisture speeds up the reaction—a few days' exposure to water may thoroughly ruin an iron or steel tool.

Another highly important reaction is the photosynthesis reaction, which converts carbon dioxide and water into hydrocarbons within the leaves of growing plants. This application of solar energy is a key step in the world's food cycle. Photosynthesis consists of a complex series of reactions, and anyone who has watched a plant grow knows that it is a particularly slow process.

The reactions used in photography vary substantially in speed. An image is produced when light from the object being photographed passes through the lens and strikes the film. The response of the film to light takes between a millionth of a second and a few tenths of a second, depending on the amount of light available and the speed of the film. (In the early days of photography, several minutes were required for the film to react sufficiently to make a good image.) Developing the image to make it visible is another chemical reaction that takes anywhere from a few seconds to a minute.

Catalysts

In this section we discuss a method frequently employed by industry to speed up the chemical reactions used in manufacturing processes. The method goes back at least to 1817, when Sir Humphrey Davy discovered that the rate of oxidation of alcohol by air was increased when a hot platinum wire was present. In 1831 it was found that adding a red-hot platinum wire to a mixture of sulfur dioxide and air speeds up the oxidation of the sulfur dioxide to form sulfuric acid. (This principle is still used in the manufacture of sulfuric acid.)

In 1835 the great Swedish chemist Berzelius coined the word **catalyst**, used today to describe a material that when added to a reaction increases the rate of the reaction but is not itself changed by the reaction. The process of speeding up a reaction by the use of a catalyst is called **catalysis**.

A catalyst is not consumed during the reaction, so it can be used over and over again. The reaction itself, the products of the reaction, and the final concentrations of those products are not changed by the presence of the catalyst. The only change is that the reaction goes faster with the catalyst present than it does without the catalyst.

A common modern use of catalysis is in the catalytic converter found in the exhaust systems of automobiles. The purpose of the catalytic converter is to reduce the amounts of carbon monoxide (CO) and hydrocarbons such as methane (CH_4) that are emitted by the car engine. The catalyst in this case is finely divided platinum, which helps the CO and the hydrocarbons to oxidize to CO_2 and H_2O as they pass through the converter. The final result is the reduction of atmospheric pollution and the lessening of the smog that occurs when sunlight acts upon automobile exhaust fumes. Since the platinum catalyst is not depleted in the reaction, it lasts for a long time. However, if lead is present in the exhaust, the catalyst soon becomes "poisoned" and is no longer effective. For this reason, gasoline containing lead cannot be used in cars with catalytic converters.

Although a catalyst does not appear to be affected by the reaction that it aids, it does enter into the reaction by forming an intermediate product. Consider this reaction:

$$A + B \longrightarrow X + Y$$

With the catalyst the reaction is

$$A + CAT \longrightarrow X + INT$$
$$B + INT \longrightarrow Y + CAT$$

Here A reacts with the catalyst (CAT) to produce X and an intermediate product (INT). B, in turn, reacts with INT to produce Y, giving back the catalyst, which is now available to react over again. The reason this reaction goes faster with the catalyst than without is that the reactions A + CAT and B + INT have lower activation energies than does reaction A + B. As shown in Figure 15-9, reducing the activation energy increases the number of *effective* collisions and so increases the rate of the reaction. Thus the catalyst allows the reaction to go through a series of intermediate steps that occur more easily than would the original reaction without the catalyst.

Among common examples of catalysts are vitamins that aid the processes of digestion, enzymes that speed up fermentation of sugar into alcohol, and catalysts that aid the conversion (or *cracking*) of crude oil into gasoline.

One environmental problem that may relate to the introduction of a catalyst into the atmosphere is the potential depletion of the ozone (O_3) layer. Ozone exists in the upper atmosphere (at an altitude of 10 to 30 miles) because of the action of ultraviolet radiation from the sun on the oxygen of the air. The ozone layer is beneficial to life on the earth, since it absorbs most of the short-wavelength ultraviolet light reaching earth from the sun. An increase in the amount of this radiation reaching the surface of the earth would result in a higher incidence of blindness and skin cancer.

Ozone is unstable and slowly breaks down into ordinary oxygen (O_2). The

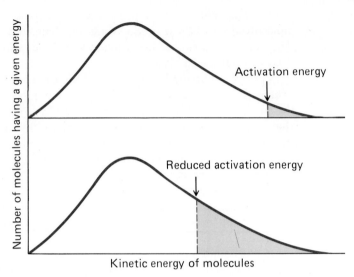

FIGURE 15-9 If the activation energy is reduced, the number of molecules having energies in the shaded area under the curve is increased, even though the temperature stays constant.

amount of ozone in the upper atmosphere is determined by a balance between the rate at which it is created and the rate at which it disappears. There has been some concern that nitrous oxide (NO) in the air acts as a catalyst to speed up the decomposition of the atmospheric ozone. If the amount of NO increases as a result of emissions from high-flying jet planes, then the ozone layer could be in danger. The consequence of a reduction in the ozone layer would be a decrease in the amount of protection afforded us against short-wavelength ultraviolet light.

Inhibitors, also called poisons or negative catalysts, are materials that slow down reactions. We have seen that lead in an automobile catalytic converter acts as a poison. Manganese or vanadium added to auto tires slows down oxidation of the rubber and lengthens the tire life.

15.3 Equilibrium

The Equilibrium Constant

There is a tendency to think of a chemical reaction as something that goes only in one direction. We write the equation

$$2H_2 + O_2 \longrightarrow 2H_2O$$

with an arrow going to the right. This equation accurately describes the burning of hydrogen in oxygen: the hydrogen continues to burn until it is all oxidized (assuming enough oxygen is present). We don't expect the water to break up spontaneously into hydrogen and oxygen.

However, it is not impossible for the reaction to go in the reverse direction—toward the left—under the proper conditions. This is just what happens when water is electrolyzed into hydrogen and oxygen. The electrolysis process supplies energy to the system from the outside in order to make the reaction reverse direction and go from right to left.

We find from experience that any given reaction has a *tendency* to go either to the right or to the left, depending on external conditions. We emphasize the word *tendency* because in theory the reaction can go in both directions; it is a reversible reaction. In practice the reaction goes in the direction dictated by the controlling conditions.

Under normal conditions, the hydrogen-oxygen reaction tends to go from left to right. However, at very high temperatures the reaction reverses, going from right to left. At even higher temperatures, the water molecules can no longer exist as water molecules; they dissociate into molecular hydrogen and oxygen. If the temperature is high enough, the molecular hydrogen and oxygen dissociate into atomic hydrogen and oxygen.

A somewhat analogous situation is that of a car on the side of a hill. The car can roll either up or down. By itself it will roll downhill, following the general rule stating that *when left to itself a system will move in the direction that will reduce its potential energy*. However, if energy is added to the car (either by giving it a push from the outside or by burning fuel in its engine), then the car can go uphill.

In many situations a reaction goes both to the right and to the left at the same time. In this case, the reaction does not go all the way in either direction—it does not reach a state of completion. (**Completion** means that all the reactants on one side of the equation are converted into products on the other side.)

This effect is observed when hydrogen and nitrogen are mixed at a temperature of 400°C. The reaction that takes place produces ammonia according to the equation

$$N_2 + 3H_2 \rightleftarrows 2NH_3$$

Suppose we start out with three parts of hydrogen (by volume) to one part of nitrogen. When we first introduce the gases into the chamber, they start reacting, and ammonia begins to appear. But as soon as the ammonia is formed, it starts breaking up into hydrogen and nitrogen according to the reverse reaction, indicated by the right-to-left arrow.

This process continues; the combination of H_2 and N_2 goes more and more slowly because there are fewer and fewer of these molecules in the chamber, while the breakup of NH_3 goes faster and faster because there are more of these molecules present. (Remember that, as shown in Section 15.2, reaction rates are proportional to the concentrations of the molecules involved.) There comes a time when the rate of the reaction to the right becomes equal to the rate of the reaction to the left. At this point the number of ammonia molecules produced per second equals the number of ammonia molecules breaking up per second. Once this condition is reached, the number of moles of each of the reactants (H_2, N_2, and NH_3) inside the chamber will stay constant. There will be no further concentration changes.

We call this a condition of **equilibrium**. It is akin to what happens at a large, crowded party. When people see an empty room, they make their way there, gradually filling it up. Eventually that room becomes uncomfortable, and guests start leaving it. When the number of people going in each minute equals the number of people going out each minute, the number of people in the room stops changing. The system has reached an equilibrium.

A system in equilibrium is *not* a system in which nothing is happening. Reactions continue to go on in both directions, but *in equilibrium the quantities of reactants present do not change.*

If the amounts of hydrogen, nitrogen, and ammonia in the reaction chamber are measured after equilibrium is reached, the following proportions are found (by volume): 67.35% hydrogen, 22.45% nitrogen, and 10.20% ammonia. We started out with three parts of hydrogen to one part of nitrogen, so the starting proportions were 75% hydrogen and 25% nitrogen. Only a small fraction of these gas molecules have combined to give ammonia.

Under the condition of equilibrium, the reaction rate for the production of ammonia (left to right) equals the reaction rate for the decomposition of ammonia (right to left). From our discussion in the last section, recall that the left-to-right reaction rate is given by

$$r_1 = k_1[N_2][H_2]^3$$

where k_1 is the rate constant for that left-to-right reaction and $[N_2]$ and $[H_2]$ are nitrogen and hydrogen concentrations, respectively, in moles per liter.

The reverse reaction rate is given by

$$r_2 = k_2[NH_3]^2$$

When the two rates are equal, we can set r_1 equal to r_2:

$$k_1[N_2][H_2]^3 = k_2[NH_3]^2$$

From this equation it follows that

$$\frac{k_1}{k_2} = \frac{[NH_3]^2}{[N_2][H_2]^3} = K$$

The ratio of the two rate constants k_1 and k_2 is itself a constant called K, the **equilibrium constant**.

The above equation represents an important relationship: *When a reaction is in equilibrium, the concentration of the products (the substances on the right side of the reaction equation, raised to the proper power) divided by the concentration of the reactants (the substances on the left, raised to the proper power) is equal to a constant.* This mathematical statement is known as the **law of mass action**.

To illustrate, we will calculate the equilibrium constant for the following reaction, in which carbon dioxide is reduced to carbon monoxide by the action of hydrogen:

$$CO_2 + H_2 \rightleftharpoons CO + H_2O$$

15.3 EQUILIBRIUM

At the equilibrium point ($r_1 = r_2$), the concentrations of the reactants and products are as follows:

CO_2 0.50 mol/L

H_2 0.70 mol/L

CO 1.20 mol/L

H_2O 0.060 mol/L

To calculate the equilibrium constant K, we write

$$\frac{[CO][H_2O]}{[CO_2][H_2]} = K$$

$$\frac{(1.2)(0.060)}{(0.50)(0.70)} = 0.21$$

(Note that K is given without units and the exponents are all ones.)

Given the value of K at a given temperature, and given the concentration of the reactants, we can calculate the concentration of the products at equilibrium. Thus this equation can be useful in the design of manufacturing processes. In the next section we will consider some of the details of this type of problem.

The meaning of the equilibrium constant can be understood more thoroughly by examining the equation that defines it. Think of a general reaction such as

$$a A + b B \rightleftharpoons c C + d D$$

where A, B, C, and D represent the reacting molecules and a, b, c, and d are their respective mole proportions. The equilibrium constant is the ratio of the product of the concentrations, raised to the appropriate power:

$$K = \frac{[C]^c \times [D]^d}{[A]^a \times [B]^b}$$

We note that if [C] and [D] are large, a large value of constant K is obtained. Therefore, if the value of the equilibrium constant is greater than 1, the reaction goes to the right and more of the product compounds C and D are formed. On the other hand, if the value of K is less than 1, the equilibrium position of the reaction is toward the left side of the equation, and so less of C and D are formed.

An important issue is how the position of a reaction's equilibrium point can be controlled by changing the conditions of the reaction. This topic will be the subject of Section 15.4.

Equilibrium Problems with Weakly Ionizing Compounds

Highly ionized solutions present complicated problems. For simplicity our discussion of aqueous equilibrium reactions will focus on weakly ionizing substances.

As we have seen, strong acids are almost completely ionized in a water solution. In the reaction

$$HCl + H_2O \longrightarrow H_3O^+ + Cl^-$$

virtually all the HCl dissociates into ions. Since relatively few reactions go toward the left, the arrow is drawn only to the right.

On the other hand, in a solution of acetic acid ($HC_2H_3O_2$), only a small fraction of the acetic acid molecules dissociate into ions:

$$HC_2H_3O_2 + H_2O \rightleftharpoons H_3O^+ + C_2H_3O_2^-$$

The reaction is written with double arrows; the equilibrium point is such that acetic acid molecules, hydronium ions, and acetate ions exist together in the solution.

For the above reaction, the law of mass action yields

$$K = \frac{[H_3O^+][C_2H_3O_2^-]}{[H_2O][HC_2H_3O_2]}$$

Since only a few acetic acid molecules dissociate (typically 1%-5%), only a small fraction of the water molecules react to form H_3O^+. Therefore the number of water molecules does not change very much. For this reason the concentration of water, $[H_2O]$, can be taken to be the number of moles per liter of pure water, a constant. (It is as though a few dozen people left the stadium during the seventh inning of the World Series—they would hardly be missed.)

We can rewrite the above equation to read

$$K[H_2O] = \frac{[H_3O^+][C_2H_3O_2^-]}{[HC_2H_3O_2]} = K_i$$

The quantity K_i is the equilibrium constant for weakly ionizing substances and equals the constant K multiplied by the concentration of pure water.

Example 15.1

Calculate **(a)** the percentage ionization and **(b)** the equilibrium constant K_i for the ionization of acetic acid. The concentrations at equilibrium are known to be

$$[H_3O^+] = [C_2H_3O^-] = 7.3 \times 10^{-4} \text{ mol/L}$$

$$[HC_2H_3O_2] = 3.0 \times 10^{-2} \text{ mol/L}$$

Note: The positive and negative ion concentration are equal, because for every hydronium ion formed in solution there is an acetate ion. These two ion concentrations are equal only when acetic acid is the only acid in the solution. If another acid is present, the positive and negative ion concentrations will differ. (This point will be discussed more fully in the next section.)

Solution **(a)** The percent ionization is given by

$$\% \text{ ionization} = \frac{[H_3O^+]}{[HC_2H_3O_2]} \times 100\%$$

$$= \frac{7.3 \times 10^{-4} \text{ mol/L}}{3.0 \times 10^{-2} \text{ mol/L}} \times 100\% = 2.4\%$$

This result agrees with our initial statement that acetic acid is weakly ionizing.

(b)
$$K_i = \frac{[H_3O^+][C_2H_3O_2^-]}{[HC_2H_3O_2]}$$

$$= \frac{(7.3 \times 10^{-4})(7.3 \times 10^{-4})}{(3.0 \times 10^{-2})}$$

$$= 1.8 \times 10^{-5}$$

Note: The fact that K_i is a small number shows that acetic acid is weakly ionizing and that the solution contains more neutral molecules than acetate ions.

Example 15.2 What is the concentration of un-ionized acetic acid in a solution containing H_3O^+ and $C_2H_3O_2^-$ with a concentration of 1.2×10^{-3} mol/L?

Solution From the result of Example 15.1 we know that K_i equals 1.8×10^{-5}.
We write the equation for the equilibrium constant:

$$K_i = \frac{[H_3O^+][C_2H_3O_2^-]}{[HC_2H_3O_2]}$$

We then put the known quantities into the equation and solve for $[HC_2H_3O_2]$:

$$K_i = \frac{(1.2 \times 10^{-3})(1.2 \times 10^{-3})}{[HC_2H_3O_2]} = 1.8 \times 10^{-5}$$

$$[HC_2H_3O_2] = \frac{(1.2 \times 10^{-3})(1.2 \times 10^{-3})}{1.8 \times 10^{-5}} = 8.0 \times 10^{-2} \text{ mol/L}$$

The units must be moles per liter, since those are the appropriate units for any quantity written inside square brackets.

Example 15.3 In Chapter 14 we discussed the ionization of water solutions. In such solutions the product of the H_3O^+ and OH^- concentrations equals the ionization constant K_w:

$$[H_3O^+][OH^-] = K_w = 1.0 \times 10^{-14}$$

> This equation is a type of equilibrium equation; that is, it is true under conditions of equilibrium. How is the above equation related to the type of equilibrium equation discussed in this section?

Solution

> Let us start by writing the equation for the ionization of water:
>
> $$H_2O + H_2O \rightleftharpoons H_3O^+ + OH^-$$
>
> The equilibrium equation for this reaction is
>
> $$\frac{[H_3O^+][OH^-]}{[H_2O][H_2O]} = K$$
>
> The concentration of un-ionized water, $[H_2O]$, is found from the known density of water (1 g/mL = 1000 g/L):
>
> $$[H_2O] = \frac{1000 \text{ g}}{1 \text{ L}} \times \frac{1 \text{ mol}}{18.0 \text{ g}} = 55.6 \frac{\text{mol}}{\text{L}}$$
>
> (The amount of ionization is so small that it does not appreciably affect the value of $[H_2O]$.)
>
> The equilibrium equation then becomes
>
> $$\frac{(1 \times 10^{-7})(1 \times 10^{-7})}{(55.6)(55.6)} = K = 3.2 \times 10^{-18}$$
>
> The only difference between the equilibrium constant K and the ionization constant K_w is that one includes the concentration of the un-ionized water and the other does not. Chemists use K_w more often than K.

In this section we have considered only the ionization reaction itself—that is, the reaction between a solute and the water in which it is dissolved. Reactions between two or more reactants make up a much more complicated system and are beyond the scope of this book. We can see in a general way why this is so. If A and B are two substances dissolved in water, one equilibrium equation is required to describe the ionization of A, another to describe the ionization of B, and then a third to describe the reaction between the A^+ and B^- ions.

There are some other general observations to be made about the equilibrium equation:

1. There is no clear distinction between the terms reactants and products. Since the arrows go in both directions, the compounds on the right side of the equation could just as well be called reactants and the compounds on the left could be called products.

2. There are concentrations of all the reactants (on both sides of the equation) present in the equilibrium solution. In some reactions all the ingredients are present in comparable amounts. In reactions where one of the products precipitates out of solution, the concentration of that product in the solution will be very small, but not zero. All substances, no matter how "insoluble," have a certain amount of solubility; the amount of a substance remaining in solution is determined by another equilibrium constant, called the *solubility product*.

3. Catalysts have no effect on the equilibrium of the reaction. A catalyst makes the reaction go faster, but it does not change the concentrations of the reactants at the equilibrium point. However, the rate of the reaction is an important matter to consider. A reaction that takes a year to reach equilibrium is not of much practical use, even if it gives a large yield of products. Therefore the catalyst plays a large role in the economic feasibility of a reaction for industrial purposes.

15.4 Le Chatelier's Principle

Stress and Equilibrium

Henri Louis Le Chatelier (pronounced *luh shah-tuhl-yay'*) was a French chemist who investigated the connection between heat and chemical processes. In 1888 he stated the rule that has become known as **Le Chatelier's principle**: *Every change of one of the factors of an equilibrium brings about a rearrangement of the system in such a way as to minimize the original change.*

For example, consider a system including one or more gases that is at equilibrium. If we *increase* the pressure on these gases, changes will take place in the system that tend to *reduce* the gas pressure. The end result is that the pressure will not increase as much as it would have without the readjustment of the system.

Le Chatelier's principle is an extremely general concept and applies to systems in many areas other than that of chemical reactions. A mundane example is that of a group of people at a party. During the course of the evening, more people come to the party, making the room crowded, hot, and noisy. As a result, some of the partygoers leave early. Therefore the room does not become as crowded as it would have if everybody had stayed. The new people coming into the room have produced a change in the system away from the original equilibrium. The system has responded by rearranging itself (losing some people) so as to minimize the change.

Le Chatelier's principle is often spoken of in terms of stress. In the example of the noisy party, the entrance of the newcomers causes a stress on the system in the form of noise, crowding, heat, etc. The system responds in such a way as to reduce the amount of stress, in this case by reducing the number of bodies (lowering the pressure).

In the modern terminology of systems theory, the effects under discussion come under the heading of negative feedback. An interesting example is seen in the events of the recent energy crisis. Industry and the public responded to the stress of higher energy prices by using less energy. As a result there has been a surplus of oil, and the prices have not risen as high as they would have if everybody had continued to burn oil at the original rate. This phenomenon is clearly an example of Le Chatelier's principle.

In the following sections we will discuss how chemical systems respond to various kinds of stresses. We will see that they all follow this general rule: *When a system in stable equilibrium is disturbed, it tries to get back to equilibrium, and it does this by reducing the stress that caused the original disturbance.*

Effect of Concentration Changes

Look again at a reaction previously considered:

$$HC_2H_3O_2 + H_2O \rightleftharpoons H_3O^+ + C_2H_3O_2^-$$

At equilibrium all four reactants are present in the solution, in proportions determined by the equilibrium constant. What happens if we now change the concentration of one of the ingredients?

As an example, let us increase the concentration of H_3O^+ ions in the solution without adding more acetic acid. This can be done by adding a quantity of some other acid, which ionizes, producing more H_3O^+ ions that are available to combine with $C_2H_3O_2^-$ in the right-to-left reaction. The right-to-left reaction now goes faster than the left-to-right reaction, and the system is no longer in equilibrium. The right-to-left reaction proceeds, reducing the H_3O^+ concentration until the system gets back into equilibrium. Therefore the *addition* of H_3O^+ to the system results in the production of *more* $HC_2H_3O_2$. Since relatively few acetic acid molecules were ionized to begin with, there is little change in $[HC_2H_3O_2]$. The major effect of adding H_3O^+ ions is to decrease the concentration of acetate ions until a new equilibrium position is reached.

We can see that this same conclusion follows from the equilibrium equation

$$\frac{[H_3O^+][C_2H_3O_2^-]}{[HC_2H_3O_2]} = K_i$$

The numerator divided by the denominator always equals the constant K_i. We have already noted that the denominator is approximately constant because the acetic acid is weakly ionizing and its concentration is not materially changed by the small percentage of molecules undergoing ionization. Therefore, if the value of $[H_3O^+]$ in the numerator is increased, the value of $[C_2H_3O_2^-]$ must become proportionately smaller so that the constant value of K_i is maintained.

We have already noticed (in Chapter 14) a similar effect in which the OH^- ion concentration is reduced when acid is added to water. The equilibrium reaction is

$$H_2O + H_2O \rightleftharpoons H_3O^+ + OH^-$$

From Example 15.3 the corresponding equilibrium equation is

$$\frac{[H_3O^+][OH^-]}{[H_2O]^2} = K$$

15.4 LE CHATELIER'S PRINCIPLE

Absorbing the $[H_2O]^2$ term in the constant (that is, multiplying both sides of the equation by $[H_2O]^2$), we obtain the familiar relationship between $[H_3O^+]$ and $[OH^-]$ in water solutions:

$$[H_3O^+][OH^-] = K_w = 1 \times 10^{-14}$$

Accordingly, if the H_3O^+ concentration is increased by adding acid, the OH^- concentration must decrease in response in order to keep the product constant.

An important class of reactions taking place in aqueous solution includes those in which at least one of the products has a very low solubility, so its concentration in the solution is extremely small. The result is the precipitation of that product out of the solution. Consider, for example, the reaction

$$Na_2SO_4 + BaCl_2 \longrightarrow BaSO_4\downarrow + 2NaCl$$

The barium sulfate is essentially insoluble, so it removes itself from the reacting system by forming a solid precipitate. As a result, the reaction tends to run all the way to the right.

Such observations suggest this general rule: *Whenever the conditions are such that one of the products of a reaction is removed from the system, the reaction tends to go to completion.* In other words, the reaction tries to produce *more* of the product to make up for what is being removed from the system, and it continues to do so as long as there are reactants left in the system.

Another class of reactions described by an equilibrium equation includes those reactions involving only gases. For example,

$$H_2 + I_2 \rightleftharpoons 2HI$$

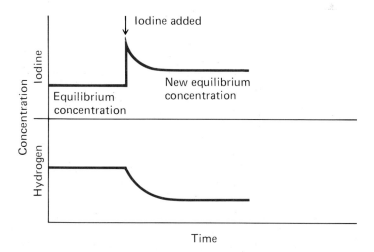

FIGURE 15-10 When iodine is added to a mixture of H_2 and I_2 in equilibrium, the concentration of iodine suddenly increases and then gradually decreases to a value somewhat higher than its original one.

This reaction takes place in a closed chamber into which hydrogen and iodine are introduced. These gases gradually combine and reach an equilibrium with the hydrogen iodide. At that point the reaction goes with equal speed in both directions. Now suppose we add more iodine vapor to the chamber, increasing the concentration of I_2. Some of the iodine will combine with the hydrogen, producing more HI, until equilibrium is again attained. The reaction will be driven toward the right.

The I_2 concentration over time is plotted in Figure 15-10. When the iodine vapor is first added, the I_2 concentration quickly jumps to a higher value. As the reaction continues, however, the I_2 concentration decreases until the equilibrium is once more reached, but at a higher value than before. At the same time, the hydrogen concentration falls to a lower level, since it reacts with the added iodine to form more HI.

Effect of Pressure Changes on a System

If at least one of the components of a reaction is a gas, the gas pressure plays an important role in the system equilibrium, because the pressure is related to the concentration of the gas molecules. As the pressure is increased, the molecules are squeezed closer together. Therefore, when we vary the pressure we also vary the concentration.

The gas law introduced in Chapter 12 shows how the pressure is related to the concentration:

$$PV = nRT$$

where P is the gas pressure, V is the volume, n is the number of moles of gas, R is the universal gas constant, and T is the temperature in kelvins.

This equation can be rearranged to read

$$\frac{n}{V} = \frac{P}{RT}$$

Since n/V is the concentration in moles per liter, the above expression tells us that the concentration is directly proportional to the pressure, for a given temperature.

There are two ways to vary the pressure. One is to change the number of moles by pumping more gas into the same container (or by pumping gas out). The other is to change the volume while keeping the number of moles of gas constant. Compressing a gas in a cylinder by pushing down on a piston reduces the volume and increases the pressure, thus increasing the concentration.

To see the chemical result of a pressure change, consider the reaction in which limestone decomposes to form lime and carbon dioxide:

$$CaCO_3 \rightleftharpoons CaO + CO_2$$
$$\text{limestone} \qquad \text{lime}$$

This reaction, which is used in the commercial manufacture of lime, is generally carried out at a high temperature—about 1000°C. To obtain a good yield of lime, one must continuously remove the CO_2 from the system. This forces the reaction to go to the right, as described in the previous section.

On the other hand, adding more CO_2 to the system, and thereby increasing its pressure, drives the reaction to the left. This is because the greater concentration of CO_2 in the system speeds up the absorption of gas by the lime, causing more limestone to form.

In the last section we saw that for the reaction

$$H_2 + I_2 \rightleftarrows 2HI$$

increasing the pressure by adding more iodine vapor to the reaction chamber drives the equilibrium point to the right, producing more HI. On the other hand, if the pressure is increased simply by decreasing the volume (as by squeezing the gas in a cylinder), the reaction is speeded up equally in both directions, because all of the reactants have their concentrations increased by the same amount. (Remember that the HI, as well as the H_2 and the I_2, is a gas.)

We can see how this result comes about by examining the equilibrium equation for this reaction:

$$\frac{[HI]^2}{[H_2][I_2]} = K$$

If the concentration of each of the ingredients is doubled, the numerator increases just as much as the denominator, so the equilibrium is unchanged. The basic reason for this result is that in this reaction the number of molecules is the same on both sides of the reaction equation.

On the other hand, in the reaction for the manufacture of ammonia,

$$N_2 + 3H_2 \rightleftarrows 2NH_3$$

the equilibrium equation is

$$\frac{[NH_3]^2}{[N_2][H_2]^3} = K$$

Doubling all the concentrations increases the denominator more than it increases the numerator, because there are more molecules on the left side of the reaction than on the right side. Therefore, when the pressure is increased, the concentration of NH_3 increases just enough to keep the above fraction a constant. For this reason ammonia is produced at high pressures (and in the presence of catalysts).

We can interpret these observations in terms of Le Chatelier's principle. Increasing the pressure on a system applies a stress to the system. If the system is able to respond by reducing the pressure of the system, it will do so and thus relieve the stress. When the ammonia reaction goes to the right, it reduces the number of molecules in the chamber and so reduces the pressure. The final pressure is therefore less than it would have been without the change in equilibrium caused by the original increase in pressure.

15.5 Heat, Temperature, and Equilibrium

Enthalpy

In Chapter 3 the concepts of exothermic and endothermic reactions were introduced. Now we will review these ideas and introduce two new topics: (1) the concept of enthalpy and (2) the influence of temperature change on the equilibrium of a chemical reaction.

An exothermic reaction is a reaction in which heat is evolved (or given off). Heat is given off because the molecules entering into the reaction contain more internal energy than do the products formed by the reaction. In other words, the system has more energy before the reaction than after. It gets rid of this extra internal energy by speeding up the molecules emerging from the reaction, transforming internal energy into kinetic energy of molecular motion. We see this transformation as an increase of temperature, and we say that heat has been evolved from the reaction.

The most obvious exothermic reactions are combustions in which a great amount of heat is given off: the burning of wood and oil, the metabolism of carbohydrates in the body to provide energy for motion and for maintaining the body's temperature in all weather. Other characteristic exothermic reactions are those between hydrogen and the halogens; for example, when a mole of hydrogen combines with a mole of chlorine, a considerable amount of heat is released:

$$H_2 + Cl_2 \longrightarrow 2HCl + 184{,}000 \text{ J} \quad (44{,}000 \text{ cal})$$

Indeed, most reactions in which elements combine to form compounds are exothermic. This is why compounds exist in the first place. When bound inside a molecule, atoms are in a state of lower energy than they are when in the free state. It is a general principle that systems tend to move in the direction of lower energy, because in that direction lies greater stability. Therefore, the formation of compounds from the elements is, in the majority of cases, accompanied by the release of energy.

A simple way of visualizing energy relationships in a chemical reaction is with a diagram showing the energy of the molecules before, during, and after the reaction, as in Figure 15-11. The horizontal axis does not represent any particular distance or scale; it merely refers to the course of the reaction before, during, and after. The vertical scale represents the **internal energy** of the molecule—the sum of the potential energy and the kinetic energy of the particles within the molecule.

The hill in the center of the graph in Figure 15-11 represents the activation energy discussed earlier in this chapter—basically the energy needed to dissociate (split apart) the reactant molecules so that their atoms can combine in different arrangements. This dissociation occurs when the molecules collide with each other so energetically that their internal atomic attractions are overcome and they break apart—that is, when their kinetic energy is greater than the activation

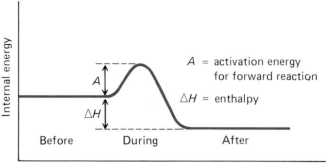

FIGURE 15-11 Energy relationships in an exothermic reaction. The system has less energy after the reaction than it did before.

energy. The molecular fragments then tend to come together in whatever way is most likely, in those combinations that have the strongest attractions. These new arrangements have less potential energy than the old ones.

The unneeded, left-over energy must be discarded by the molecules before they can settle down into their new forms. This energy, labeled ΔH in the diagram, is called the **enthalpy** of the reaction. It is the difference between the internal energy of the system after the reaction and that before the reaction:

$$\Delta H = E_{after} - E_{before}$$

In an exothermic reaction, E_{after} is smaller than E_{before}, so the enthalpy, ΔH, is a negative quantity. In the example given earlier of the combination of hydrogen and chlorine to form hydrogen chloride, the enthalpy of the reaction is $-92\,kJ$ ($-22\,kcal$) per mole of HCl (the 184 kJ, or 44 kcal, was for 2 mol of HCl).

In an endothermic reaction, energy must be absorbed by the system in order for the reaction to go. In endothermic reactions, the energy need not always come from heat; it may come from light or electrical energy. The photochemical reactions used by plant chlorophyll to synthesize carbohydrates from water and carbon dioxide are endothermic reactions in which the energy comes directly from sunlight. In electrolysis reactions (such as production of oxygen from water, or aluminum from bauxite), the energy comes from an electric current.

An important set of endothermic reactions is the combination of oxygen and nitrogen to form the various oxides of nitrogen—NO, NO_2, N_2O, etc. *The fact that the formation of these oxides is endothermic is what makes possible the existence of a breathable atmosphere.* Consider the fact that the atmosphere contains about 80% nitrogen and 20% oxygen. If nitrogen behaved as hydrogen does, burning in oxygen with an exothermic reaction, the first lightning bolt would start a flame that would continue until all the oxygen in the air had been consumed.

The *Handbook of Chemistry and Physics* and other references have tables giving the enthalpies of a number of specific reactions. The basic principles are as follows:

1. In an exothermic reaction, heat is evolved (given off); the enthalpy of such a reaction (the change in internal energy) is negative ($-\Delta H$).

2. In an endothermic reaction, heat is absorbed from the surroundings; the enthalpy of such a reaction is positive ($+\Delta H$).

Effects of Temperature Changes on Equilibrium

We have seen that increasing the temperature can speed up a reaction. This effect is useful in helping the reaction to reach its equilibrium point quickly. An important question is What does the increase of temperature do to the equilibrium point of the reaction?

The answer depends on whether the reaction is exothermic or endothermic. Consider, for example, the exothermic reaction for the manufacture of ammonia, discussed earlier in this chapter:

$$N_2 + 3H_3 \rightleftarrows 2NH_3 + 92{,}380 \text{ J} \quad (22{,}080 \text{ cal})$$

The forward reaction—the production of ammonia—evolves 92.38 kJ of thermal energy for each 2 mol of ammonia; that is, it has an enthalpy of -46.2 kJ/mol of ammonia (-11.0 kcal/mol). The reverse reaction—the dissociation of ammonia into hydrogen and nitrogen—is an endothermic reaction. Each mole of ammonia must absorb 46.2 kJ of energy to separate into its elements.

When this reaction is in equilibrium (in a closed chamber), the forward reaction absorbs exactly as much heat as is evolved by the reverse reaction, since the two reactions are going at the same rate. Therefore the temperature of the system does not change. However, if heat is removed from the system, the equilibrium point is shifted over to the right, so that more heat is produced. As a result, the forward reaction is able to go further toward completion, until a new equilibrium is established. (This shift of the equilibrium point is caused by the fact that removing heat from the system reduces its temperature, which slows down the reverse reaction more than it does the forward reaction.)

If, on the other hand, heat is added to the system from the outside, the equilibrium point will change so as to permit the system to *absorb* heat. Therefore the reaction will go toward the left.

We see here an application of Le Chatelier's principle. The change of temperature is thought of as a stress applied to the system. If a system in equilibrium is stressed by a change of temperature (by heat being added or taken away), the equilibrium point tends to shift in such a direction as to reduce that stress. Therefore, if heat is *added*, the equilibrium will shift in the direction of the *endothermic reaction*. Such a shift will absorb the extra heat and so prevent the temperature from rising as much as it would have otherwise. If heat is *removed*, the equilibrium will shift in the direction of the *exothermic reaction*, so that heat will be evolved and the reduction of temperature will be resisted. As always, *the system resists any changes being forced upon it.*

The above rule concerning change of equilibrium due to temperature change is often called **van't Hoff's law**. It is a special case of the more general Le

Chatelier's principle. (*Note:* In this context a *law* is a statement that describes a specific set of observed facts, whereas a *principle* is a much broader theoretical statement covering a wide range of phenomena.)

A knowledge of Le Chatelier's principle is of great importance in the design of industrial processes, since it enables planners to choose conditions that will make a given reaction go most effectively in the desired direction.

Glossary

activation energy The minimum amount of kinetic energy that a molecule must have in order to have an effective collision with another molecule.

average energy The result obtained by adding all the energies of the molecules in a system and then dividing the sum by the number of molecules present.

catalysis The process of speeding up a reaction with the help of a catalyst.

catalyst A material that increases the rate of a reaction without itself being consumed in the reaction.

completion A reaction state in which all reactants have been transformed into products.

effective collision A collision between two molecules that results in a reaction (as opposed to an elastic collision, in which no reaction takes place).

effective collision rate The rate at which molecules undergo effective collisions (the number of effective collisions per unit volume per second).

energy distribution curve A graph showing the number of molecules in a substance having kinetic energies lying within a given range. From this curve we can find the number of molecules having the minimum energy necessary for an effective collision.

enthalpy The gain or loss of internal energy during a chemical process, expressed in calories per mole (or joules per mole).

equilibrium A condition in which a forward and a reverse reaction occur at the same rate, so there is no net change in the amount of reactants on either side of the equation.

equilibrium constant The constant that expresses the ratio of product concentrations to reactant concentrations in the law of mass action equation.

inelastic collision A collision between two objects resulting in a change of internal energy, usually accompanied by a rearrangement of the particles within the colliding objects.

inhibitor A material that slows a chemical reaction, acting as a kind of poison to the reaction; the opposite of a catalyst; a negative catalyst.

intermediate state One or more temporary states that occur during the course of a reaction. In the reaction between molecules A and B, the intermediate state may consist of the combination AB.

internal energy The sum of the potential and kinetic energies of the particles within a molecule.

kinetic energy The energy possessed by an object because of its motion ($KE = \frac{1}{2}mv^2$). The kinetic energy of a pair of colliding molecules determines whether or not the collision will be effective in initiating a reaction.

Le Chatelier's principle The rule stating that when a stress is imposed on a system in equilibrium, the equilibrium position will shift in such a way as to reduce the stress.

mass action, law of The rule stating that when a reaction is in equilibrium, the concentration of the materials on the right (raised to the proper power) divided by the concentration of the materials on the left (raised to the proper power) is equal to a constant.

reaction rate The number of molecular reactions taking place per unit volume per second, measured by the change in concentration of the reactants per unit time.

total collision rate The number of collisions per unit time between atoms or molecules in a chemical system.

van't Hoff's law The rule stating that when heat is added to a system in equilibrium, the equilibrium point will shift in such a direction that heat will be absorbed by the system, and, conversely, when heat is removed from a system, the equilibrium point will shift so that heat will be given off.

Problems and Questions

15.1 Reactions and Molecular Collisions

15.1.Q What is the difference between an elastic and an inelastic collision?

15.2.Q Does a chemical reaction result from an elastic collision between molecules?

15.3.Q Describe the factors that determine whether or not a reaction will occur when two molecules collide.

15.4.Q What is dissociation?

15.5.Q What is activation energy?

15.6.Q What is the relationship between dissociation and activation energy?

15.7.Q Molecules A and B are reacting in a chamber. The reaction rate is such that 2×10^{20} molecules per second combine with each other.
  a. If we doubled the number of A molecules in the same chamber, what would happen to the reaction rate?
  b. If we doubled the number of A molecules and tripled the number of B molecules in the same chamber, what would happen to the reaction rate?

15.8.Q Predict the possible results of collisions between the following atoms or molecules:
  a. $H_2 + Ca$
  b. $H_2 + Br_2$
  c. $HCl + NH_3$

15.2 Reaction Rates

15.9.Q What is meant by reaction rate?

15.10.Q Discuss the effect on the reaction rate of changing the concentration.

15.11.Q Does the activation energy of a reaction depend on the temperature?

15.12.Q a. How does the number of molecules having energies greater than the activation energy depend on the temperature?
  b. How does increasing the temperature change the number of molecules having energies greater than the activation energy?

15.13.Q Discuss the effect on the reaction rate of changing the temperature. How is the energy distribution curve related to this effect?

15.14.Q What is the rate equation?

15.15.Q a. What factors enter into the rate equation?
  b. Which factor in the rate equation depends on the temperature?

15.16.Q Write a rate equation for each of the following reactions:
  a. $H_2 + I_2 \longrightarrow 2HI$
  b. $2HI \longrightarrow H_2 + I_2$
  c. $N_2 + 3H_2 \longrightarrow 2NH_3$
  d. $2H_2 + O_2 \longrightarrow 2H_2O$

15.17.Q What happens to the reaction rate of each of the reactions in Question 15.16 if the concentration of each of the reactants is doubled? (Double both reactants if there are two on the left side of the equation.)

15.18.Q What is meant by a fast reaction? How does it differ from a slow reaction?

15.19.Q Give two examples of slow reactions and two examples of fast reactions.

15.20.Q a. What is a catalyst?
  b. How does a catalyst affect the products of a reaction?
  c. How does a catalyst affect the final concentrations of the reaction products?
  d. How does a catalyst affect the rate of the reaction?

15.21.Q Give three examples of catalysts encountered in everyday life.

15.22.Q Explain the action of a catalyst in terms of intermediate reactions and activation energies.

15.23.Q In the freezing of fruit, citric acid is added to reduce the rate of oxidation of the fruit. Explain what happens in terms of catalysis.

15.24.Q a. Why is it not a good idea to store a car in a garage for the winter after the car has been coated with road salt?
  b. Why do we refrigerate many foods rather than keep them at room temperature?
  c. Why can photographic film (and flashlight batteries) be stored for long periods of time in a freezer?

15.3 Equilibrium

15.25.Q Define and explain the following:
a. condition of equilibrium
b. equilibrium constant
c. equilibrium equation

15.26.P Calculate the equilibrium constant for the following reaction:

$$2NO_2 \rightleftharpoons N_2 + 2O_2$$

At equilibrium the concentrations are

$$[N_2] = 1.3 \times 10^{-4} \text{ mol/L}$$
$$[O_2] = 2.6 \times 10^{-4} \text{ mol/L}$$
$$[NO_2] = 0.1 \text{ mol/L}$$

15.27.P The equilibrium constant K equals 0.59 for the reaction

$$H_2 + I_2 \rightleftharpoons 2HI$$

and at equilibrium the concentrations of HI and I_2 are

$$[HI] = 0.078 \text{ mol/L}$$
$$[I_2] = 0.10 \text{ mol/L}$$

What is the concentration of H_2?

15.28.P The equilibrium constant K equals 8.2×10^2 for the reaction

$$N_2 + 3H_2 \rightleftharpoons 2NH_3$$

and at equilibrium the concentrations of H_2 and NH_3 are

$$[H_2] = 0.02 \text{ mol/L}$$
$$[NH_3] = 0.01 \text{ mol/L}$$

What is the concentration of N_2?

15.29.P a. What is the concentration of $[NH_3]$ in the reaction

$$NH_3 + H_2O \rightleftharpoons NH_4^+ + OH^-$$

The ionization constant and the known equilibrium concentrations are

$$K_i = 1.79 \times 10^{-5}$$
$$[OH^-] = [NH_4^+] = 1.34 \times 10^{-3} \text{ mol/L}$$

b. What percentage of NH_3 molecules produce OH^- ions?

15.30.P For the reaction

$$HCN + H_2O \rightleftharpoons H_3O^+ + CN^-$$
$$K_i = 4.93 \times 10^{-10}$$
$$[HCN] = 0.01 \text{ mol/L}$$
$$[CN^-] = 1.4 \times 10^{-9} \text{ mol/L}$$

(HCN is hydrogen cyanide, and its solution is hydrocyanic acid.)
a. Calculate the concentration of $[H_3O^+]$.
b. Calculate the pH of this solution.
(*Hint:* It is not necessary for $[H_3O^+]$ to equal $[CN^-]$ if other acids are present.)

15.31.P For the reaction $H_2S + H_2O \rightleftharpoons H_3O^+ + HS^-$,

$$K_i = 9.1 \times 10^{-8}$$
$$[H_2S] = 0.10 \text{ mol/L}$$
$$[H_3O^+] = 1 \times 10^{-4} \text{ mol/L}$$

a. Calculate $[HS^-]$.
b. Find the pH of the solution.

15.32.P Compare the strengths of the two acids in the following two reactions. One involves the ionization of chloric acid:

$$HClO_3 + H_2O \rightleftharpoons H_3O^+ + ClO_3^-$$

with the ionization constant $K_i = 1 \times 10^3$. The other concerns the ionization of nitrous acid:

$$HNO_2 + H_2O \rightleftharpoons H_3O^+ + NO_2^-$$

with the ionization constant $K_i = 4.6 \times 10^{-4}$.

15.4 Le Chatelier's Principle

15.33.Q State Le Chatelier's principle in two different ways—in the original way and in terms of stress.

15.34.Q Give an example showing how changing one of the conditions of a system is equivalent to applying a stress to the system. Describe how Le Chatelier's principle applies to this situation.

15.35.Q A teacher assigns a student 50 pages of writing, which must be finished by the end of the semester. The student calculates that if he writes two pages per school day he can finish the assignment. Suddenly another teacher gives the student another 45 pages of writing to do. The student knows he cannot finish both assignments by the end of the semester at the rate he has been going. As a result he feels a certain amount of stress. Analyze this situation in terms of Le Chatelier's principle, and decide what the student can do to relieve the stress.

15.36.Q What is the effect of concentration changes on the reaction equilibrium of a chemical system?

15.37.Q Describe the effect of increasing or decreasing the pressure on the equilibrium of each of the following reactions, assuming the temperature is high enough so that all the reactants are gases:

  a. $H_2 + I_2 \rightleftharpoons 2HI$
  b. $N_2 + 3H_2 \rightleftharpoons 2NH_3$
  c. $2H_2 + O_2 \rightleftharpoons 2H_2O$
  d. $2SO_2 + O_2 \rightleftharpoons 2SO_3$

15.38.Q What effect does precipitation have on the equilibrium of a reaction?

15.39.Q When a gas is reacting with a solid to produce another solid, how can we make the reaction go one way or the other (to the right or to the left)? Explain the result in terms of Le Chatelier's principle.

15.40.Q In the reaction

$$SrSO_{3(s)} \rightleftharpoons SrO_{(s)} + SO_{2(g)}$$

the subscript (s) stands for solid and the subscript (g) stands for gas. What is the effect of each of the following conditions on the equilibrium of the reaction?

  a. SO_2 is allowed to escape.
  b. The SO_2 pressure is increased by squeezing the gas in a cylinder.
  c. More SO_2 is added to the sealed system.

15.5 Heat, Temperature, and Equilibrium

15.41.Q Define and/or explain the following:
  a. exothermic reaction
  b. endothermic reaction
  c. enthalpy
  d. Le Chatelier's principle
  e. van't Hoff's law

15.42.Q Which of the reactions in Table 15-1 are exothermic, and which are endothermic?

15.43.Q For each of the reactions in Table 15-1, determine which action—increasing the temperature or decreasing the temperature—would increase the yield of the product on the right side of the equation.

15.44.P Which of the reactions in Table 15-1 yields the most thermal energy? (Use the number of moles of reactants shown in the equations.)

15.45.P Which of the reactions in Table 15-1 yields the most energy per gram of product formed? (*Hint:* Convert moles to grams.)

15.46.P Which of the reactions in Table 15-1 yields the most energy per gram of reactant? (Use the total mass of all the reactants; this is the number that would be of interest in designing a rocket fuel.)

15.47.Q According to Le Chatelier's principle, what would happen to the equilibrium of the H_2-Cl_2 reaction in Table 15-1, if the temperature were increased? How does this differ from what happens to the reaction between H_2 and I_2?

15.48.Q Is the oxidation of nitrogen exothermic or endothermic? How does the existence of life on earth depend on the nature of this reaction?

TABLE 15-1 Some Typical Enthalpy Changes

	kJ/mol	kcal/mol
$N_2 + 3H_2 \longrightarrow 2NH_3$	-46.2	-11.0
$Ba + Cl_2 \longrightarrow BaCl_2$	-858	-205
$H_2 + Se \longrightarrow H_2Se$	$+85.7$	$+20.5$
$N_2 + 2O_2 \longrightarrow N_2O_4$	$+9.66$	$+2.31$
$2Ag + S \longrightarrow Ag_2S$	-31.8	-7.6
$2H_2 + O_2 \longrightarrow 2H_2O$	-286	-68.3
$H_2 + I_2 \longrightarrow 2HI$	$+26.4$	$+6.3$
$H_2 + Cl_2 \longrightarrow 2HCl$	-92.3	-22.1
$H_2 + F_2 \longrightarrow 2HF$	-268	-64.1

Note: Enthalpy changes have units of energy per mole of product compound.

Self Test

1. Define the following terms:
   a. activation energy
   b. effective collision rate
   c. energy distribution curve
   d. law of mass action
   e. catalysis
   f. equilibrium
   g. Le Chatelier's principle
   h. van't Hoff's law
   i. enthalpy

2. Write the rate equation for each of these reactions:
   a. $HCOOH + NH_4OH \rightleftharpoons NH_4OOCH + H_2O$
   b. $HCl + KOH \rightleftharpoons KCl + H_2O$
   c. $AgNO_3 + NaCl \rightleftharpoons AgCl\downarrow + NaNO_3$
   d. $N_2 + 3H_2 \rightleftharpoons 2NH_3$

3. Explain why raising the temperature increases the rate of a reaction.

4. Film developers have found that the temperature of the developing solutions has a critical effect on the developing times. Explain why this relationship between temperature and time is to be expected.

5. The rate of a reaction carried out in an ice-water slush bath (0°C) is measured. What would happen to the rate if the reaction were carried out at 10°C? At 50°C?

6. Write the K_i equation for each of these reactions:
   a. $CH_3COOH + H_2O \rightleftharpoons H_3O^+ + CH_3COO^-$
   b. $NH_3 + H_2O \rightleftharpoons NH_4^+ + OH^-$

7. a. Calculate the concentration of OH^- in the reaction
   $$NH_3 + H_2O \rightleftharpoons NH_4^+ + OH^-$$
   given that $K_i = 1.79 \times 10^{-5}$, $[NH_3] = 0.01$ mol/L, and $[NH_4^+] = 2.2 \times 10^{-3}$ mol/L.

   b. For the reaction in part a, calculate $[OH^-]$ if $[NH_4^+]$ is changed to 1.1×10^{-4} mol/L.
   c. Compare the $[NH_4^+]$ and $[OH^-]$ values in parts a and b of this question.
   d. Explain what could cause the changes in concentration seen in Problem 7.b.

8. According to Le Chatelier's principle, in which direction would increasing the pressure drive a reaction in which a solid plus a gas produces a solid? Explain your answer.

9. a. Calculate the number of kilojoules of heat produced when 200 g of hydrogen is burned in oxygen. (See Table 15-1 for enthalpy values.)
   b. Calculate the mass of oxygen used in the reaction in part a.

10. a. Calculate the number of kilojoules of heat produced when 200 g of hydrogen is burned in fluorine.
    b. Calculate the mass of fluorine used in the reaction in part a.
    c. Which would be the more effective oxidizer for the hydrogen fuel in a rocket engine—hydrogen or fluorine? Take into account the fact that the mass of the oxidizer must be carried along with the fuel. (*Hint:* Calculate the amount of heat per unit mass of the total fuel, including oxidizer.)

11. What happens to an exothermic reaction at equilibrium when the temperature is increased? Explain in terms of Le Chatelier's principle.

12. The process of dissolving salt (NaCl) in water is endothermic. Explain why someone making ice cream by hand would prefer a water-ice-salt slush to a simple water-ice slush.

16
Redox Reactions

Rusting iron—an example of a slow oxidation reaction.

Objectives After completing this chapter, you should be able to:
1. Define oxidation and reduction in terms of electron transfer.
2. Calculate the change in oxidation number of each of the elements taking part in a given reaction.
3. State the difference between a charge number and an oxidation number.
4. Write the partial equations of a redox reaction.
5. Balance redox equations using the oxidation number method.

16.1 The Concept of Redox

Oxidation Numbers

Chapter 7 introduced the concept of oxidation and reduction, and Chapter 10 presented a number of rules for assigning oxidation numbers to each element so that formulas for their compounds could be written correctly. This chapter will focus on chemical reactions involving oxidation and reduction processes. Such reactions are called **redox reactions**.

Let us first review the definitions of oxidation and reduction. **Oxidation** is a process in which one or more electrons are transferred from element A to element B. The oxidation number (or oxidation state) of each element is the number of electrons given up or received by that element. When element A gives up a number of electrons, its oxidation number is *increased*. Element A is then said to be *oxidized*, or in an *oxidized state*.

At the same time as element A gives up electrons, element B gains electrons. Its oxidation number *decreases*, and the process is known as **reduction**. Therefore element B is said to be *reduced*, or in a *reduced state*.

Clearly, both oxidation and reduction must take place at the same time in the same reaction; for this reason the combined and abbreviated name redox reaction is used.

Example 16.1   List the oxidation number of each atom in the following reaction:

$$2Na + 2H_2O \longrightarrow 2NaOH + H_2$$

This is an exothermic reaction that takes place when metallic sodium is placed in water. There is violent bubbling as hydrogen gas is given off. Sometimes the heat evolved by the reaction causes the hydrogen to catch fire, and the sodium seems to be burning in the water. In this reaction the affinity of the sodium for the hydroxyl group is so great that the water molecule is split and free hydrogen is released.

Solution   Rewrite the equation, listing beneath the symbols the oxidation number of each atom:

$$\begin{array}{ccccc} 2Na + & 2H & O & H & \longrightarrow & 2Na & O & H & + 2H \\ 0 & +1 & -2 & +1 & & +1 & -2 & +1 & 0 \end{array}$$

To the left of the arrow, each sodium atom has an oxidation number of zero (0). Each of the hydrogen atoms in the water molecule shares one electron with the oxygen atom. The H—O bond is a polar covalent bond—the electron spends most of its time near the oxygen atom. Although the hydrogen atom is not ionized, its electron behaves as though it has been partially transferred to the oxygen atom. For this reason there is a net positive charge at the end of the molecule where the hydrogen atoms reside. Thus we give each hydrogen atom an oxidation number of $+1$, meaning that it has given up one electron. The oxygen atom, having received two electrons, has an oxidation number of -2.

On the right side of the equation, one electron is transferred from each sodium atom to the oxygen atom. Since the sodium atom becomes more positive, we say its oxidation number is increased from 0 to $+1$. Accordingly, the sodium atom has been oxidized. On the other hand, two of the four hydrogen atoms have gone from an oxidation state of $+1$ to 0. These atoms have been reduced.

In this reaction the hydrogen atom is the **oxidizing agent** and the sodium atom is the reducing agent. (The atom that becomes reduced is always the oxidizing agent; the atom that becomes oxidized is always the reducing agent.)

Note: The hydrogen atoms in the OH groups do not change their oxidation state during this reaction. Their oxidation numbers are $+1$ on both sides of the equation. Similarly, the oxygen atoms retain their oxidation number of -2. Since the OH group does not change its identity during the reaction, we can write it in the equation as a unit (OH) with an oxidation number of -1.

The above example points out the difference between **charge number** and oxidation number. In NaOH the sodium is ionized, with a charge number of $1+$. Its oxidation number is $+1$, so in this case the charge number and the oxidation number have the same value (except for the convention that the plus sign comes *first* in the oxidation number and *last* in the charge number). However, in H_2O the water is not ionized. The bond is a polar covalent bond, so we would *not* say that the hydrogen atom has a *charge* of $1+$. However, its oxidation number is $+1$ because its electron is located closer to the oxygen atom than to the hydrogen atom, on the average.

Not all reactions are redox reactions. An example of a non-redox reaction is the neutralization of hydrochloric acid with sodium hydroxide to produce sodium chloride and water:

$$\underset{+1\ -1}{H\ Cl} + \underset{+1\ -1}{Na(OH)} \longrightarrow \underset{+1\ -1}{Na\ Cl} + \underset{+1\ -1}{H\ (OH)}$$

In this reaction the oxidation numbers of all the ingredients are the same on both sides of the equation. Since there is no change in oxidation number, we conclude that this is not a redox reaction.

Partial Equations

A useful device for analyzing redox reactions is the **partial equation**. We illustrate the use of the partial equation with the following example:

$$Fe + CuSO_4 \longrightarrow FeSO_4 + Cu$$

This reaction takes place when a piece of iron is put into a copper sulfate solution.

The iron slowly becomes covered with a red-brown copper coating. If there is sufficient $CuSO_4$ in the solution, the iron completely disappears, and a mass of copper granules remains at the bottom of the container.

A careful analysis of the reaction shows a number of processes taking place. First, the $CuSO_4$, when dissolved in water, breaks up into Cu^{2+} and SO_4^{2-} ions. Each iron atom introduced into the solution transfers two electrons to a copper ion, which then becomes a neutral copper atom. We write two equations—one describing electrons being transferred away from the iron and one showing electrons being transferred to the copper. The equation describing each of these *electron transfer reactions* is called a partial equation:

$$Fe_s - 2e^- \longrightarrow Fe^{2+}_{aq}$$
Oxidation No.:   0   +2

$$Cu^{2+}_{aq} + 2e^- \longrightarrow Cu_s$$
Oxidation No.:   +2   0

The first equation shows solid Fe (subscript s) losing two electrons to become an Fe^{2+} ion in aqueous solution (subscript aq). The second equation shows a Cu^{2+} ion in aqueous solution gaining two electrons to become solid Cu.

The oxidation number of the Fe increases from 0 to +2, indicating that the Fe has become oxidized during the reaction. At the same time, the oxidation number of the Cu decreases from +2 to 0, indicating that the Cu has been reduced. In this reaction the Cu is the oxidizing agent and the Fe is the reducing agent.

Adding the two partial equations yields the net redox equation:

$$Fe_s - 2e^- \longrightarrow Fe^{2+}_{aq}$$
$$Cu^{2+}_{aq} + 2e^- \longrightarrow Cu_s$$
$$\overline{Fe_s + Cu^{2+}_{aq} \qquad \qquad Fe^{2+}_{aq} + Cu_s}$$

The bottom equation shows only those parts of the complete reaction that take part in the redox process. It leaves out the $(SO_4)^{2-}$ group, which does not change its oxidation state in the reaction. The construction of partial equations and the net equation is useful as a preliminary step in balancing a complicated redox equation.

The example above is a simple one because the number of Fe and Cu atoms on the right equals the number of Fe and Cu atoms on the left and the number of electrons added to each Cu ion equals the number of electrons taken from each Fe atom. Thus, the initial equation is balanced as written. However, reactions are not always this simple, and in the next section we will see what happens when we try to balance equations of a more complex nature.

16.2 Balancing Redox Equations

Chapter 10 introduced techniques for balancing chemical equations, based on the concept that the number of atoms of each species must be the same on both sides of the equation. To balance an equation, we adjusted the number of molecules on each side of the equation until the atom counts were balanced.

There is another consideration involved in balancing redox reactions: for every element oxidized, another element is reduced. Accordingly, *the rise in oxidation number of one species must equal the decrease in oxidation number of another species.* Or, for ionic compounds, the gain in ionic charge of one species must equal the loss in ionic charge of another species. This means that *the number of electrons gained by one element must equal the number of electrons lost by another element.*

In this section we will use the oxidation number method of balancing equations. To illustrate this method, let us return to the first equation of this chapter. The question marks indicate that it is not yet balanced.

$$?Na + ?H_2O \longrightarrow ?NaOH + ?H_2$$

We first treat the H_2O molecule as a combination of H and OH:

$$H(OH) \longrightarrow H^{+1} + (OH)^{-1}$$

Next we write the two partial equations representing the oxidation and reduction processes:

$$Na^0 - e^- \longrightarrow Na^{+1} \quad \text{(oxidation)}$$
$$H^{+1} + e^- \longrightarrow H^0 \quad \text{(reduction)}$$

The second partial equation must be modified to show that the hydrogen molecule contains two atoms:

$$Na^0 - e^- \longrightarrow Na^{+1} \quad \text{(gain of one oxidation number)}$$
$$2H^{+1} + 2e^- \longrightarrow (H^0)_2 \quad \text{(loss of two oxidation numbers)}$$

In the second equation, each atom loses one oxidation number, so the total loss is two oxidation numbers (a gain of two electrons). At this point the gains and losses of oxidation numbers in the two equations are unequal. We remedy this situation by multiplying both sides of the first partial equation by 2:

$$2(Na^0 - e^- \longrightarrow Na^{+1})$$

We then have

$$2Na^0 - 2e^- \longrightarrow 2Na^{+1} \quad \text{(gain of two oxidation numbers)}$$
$$2H^{+1} + 2e^- \longrightarrow (H^0)_2 \quad \text{(loss of two oxidation numbers)}$$

The two partial equations are now balanced: the atom counts are balanced and the gain of two oxidation numbers in one equation is balanced by the loss of two oxidation numbers in the other equation. When the two partial equations are added, we find that the electrons cancel and thus do not appear in the resulting equation:

$$2Na^0 - 2e + 2H^{+1} + 2e \longrightarrow 2Na^{+1} + (H^0)_2$$

Finally, we have

$$2\,\text{Na}^0 + 2\,\text{H}^{+1} \longrightarrow 2\,\text{Na}^{+1} + (\text{H}^0)_2$$

The last equation is important because it gives the proportions of the elements entering into the oxidation-reduction part of the reaction. We can now put these proportions into the original equation, knowing that there must be two Na atoms on each side and one H_2 molecule on the right:

$$2\,\text{Na} + \text{H}_2\text{O} \longrightarrow 2\,\text{NaOH} + \text{H}_2$$

This equation is not yet balanced, for there are two O's on the right and only one on the left. Also, there are four H's on the right and only two on the left. The reason for this imbalance is that we have neglected the nonredoxed species, the OH groups. They do not appear in the partial equations because the oxidation number of the OH group does not change during the reaction.

Since there are two OH's on the right, we need two H_2O ($H \cdot OH$) molecules on the left, as in the equation

$$2\,\text{H(OH)} \longrightarrow 2\,\text{H}^{+1} + 2(\text{OH})^{-1}$$

During the redox reaction, the first two H's on the right of the above equation are reduced to yield the neutral H_2 molecule. The two OH's remain unchanged in oxidation number when they form the two NaOH molecules. Therefore, in the final equation we must put two water molecules on the left to obtain a completely balanced equation:

$$2\,\text{Na} + 2\,\text{H}_2\text{O} \longrightarrow 2\,\text{NaOH} + \text{H}_2$$

The above equation can also be balanced using the simpler method of equating numbers of atoms on both sides. Counting oxidation numbers seems to be an unnecessary complication. However, keeping track of the oxidation numbers helps to clarify certain important details about the reaction, such as the fact that in the example above some of the hydrogen atoms are reduced and some remain unchanged.

Furthermore, in very complicated equations, accounting for oxidation numbers makes the bookkeeping more orderly. As the following examples show, many redox equations do not follow the simple format of the four basic equation types discussed in Chapter 10. We will encounter equations with three or more reactants on each side. Such equations are easier to handle with the balancing method described in this section.

Example 16.2

Solution

$$?\text{K}_2\text{Cr}_2\text{O}_7 + ?\text{H}_2\text{S} + ?\text{H}_2\text{SO}_4 \longrightarrow ?\text{Cr}_2(\text{SO}_4)_3 + ?\text{K}_2\text{SO}_4 + ?\text{S} + ?\text{H}_2\text{O}$$

In this complex reaction, the chromium in potassium dichromate ($K_2Cr_2O_7$) oxidizes the sulfur in hydrogen sulfide (H_2S). The reaction takes place when hydrogen sulfide gas is bubbled through potassium dichromate dissolved in dilute sulfuric acid. A yellow precipitate of sulfur results.

Notice that the sulfur in the $(SO_4)^{-2}$ group is unchanged as far as oxidation number is concerned. Only the sulfur in the H_2S enters into the redox part of the reaction. Therefore the $(SO_4)^{-2}$ group does not enter into these partial equations:

$$Cr^{+6} + 3e^- \longrightarrow Cr^{+3} \quad \text{(reduction: loss of three oxidation numbers)}$$
$$S^{-2} - 2e^- \longrightarrow S^0 \quad \text{(oxidation: gain of two oxidation numbers)}$$

Multiply the first equation by 2 and the second equation by 3 to equalize the number of electrons exchanged (as well as the change in oxidation number):

$$2(Cr^{+6} + 3e^- \longrightarrow Cr^{+3})$$
$$3(S^{-2} - 2e^- \longrightarrow S^0)$$

Multiplying through and eliminating the parentheses, we have

$$2Cr^{+6} + 6e^- \longrightarrow 2Cr^{+3}$$
$$3S^{-2} - 6e^- \longrightarrow 3S^0$$

The Cr's gain six electrons, and the S's lose six electrons. Add the two partial equations to get

$$2Cr^{+6} + 6e^- + 3S^{-2} - 6e^- \longrightarrow 2Cr^{+3} + 3S^0$$

Canceling the six electrons, we have

$$2Cr^{+6} + 3S^{-2} \longrightarrow 2Cr^{+3} + 3S^0$$

In this reaction two Cr's are needed for three S's. As a first trial, we enter these numbers into the original equation, noticing that the potassium dichromate and chromium sulfate molecules already have two Cr atoms in them:

$$K_2Cr_2O_7 + 3H_2S + ?H_2SO_4 \longrightarrow Cr_2(SO_4)_3 + K_2SO_4 + 3S + ?H_2O$$

The H's, the K's, the sulfate groups, and the O's still need to be balanced. The seven O's in the $K_2Cr_2O_7$ require seven H_2O molecules on the right. These seven H_2O molecules, in turn, require fourteen H's on the left. Six of them can come from the three H_2S molecules. The remaining eight can be accounted for if we have four H_2SO_4 molecules. The SO_4 groups and the K's are then found to be balanced:

$$K_2Cr_2O_7 + 3H_2S + 4H_2SO_4 \longrightarrow Cr_2(SO_4)_3 + K_2SO_4 + 3S + 7H_2O$$

Using the simple bookkeeping method suggested in Section 10.3 might be helpful for the nonredox atom balancing.

Example 16.3

Solution

$$?P + ?HNO_3 + ?H_2O \longrightarrow ?H_3PO_4 + ?NO$$

In this example phosphorus is oxidized to form phosphoric acid, and the nitrogen in the nitric acid is reduced to the nitrogen in nitrous oxide. We write the partial equations

$$P^0 - 5e^- \longrightarrow P^{+5} \quad \text{(oxidation: loss of five electrons)}$$
$$N^{+5} + 3e^- \longrightarrow N^{+2} \quad \text{(reduction: gain of three electrons)}$$

Multiply the first equation by 3 and the second equation by 5 to equalize the gain and loss of electrons:

$$3(P^0 - 5e^- \longrightarrow P^{+5})$$
$$5(N^{+5} + 3e^- \longrightarrow N^{+2})$$

Multiply through:

$$3P^0 - 15e^- \longrightarrow 3P^{+5} \quad \text{(loss of 15 electrons)}$$
$$5N^{+5} + 15e^- \longrightarrow 5N^{+2} \quad \text{(gain of 15 electrons)}$$

Add the partial equations:

$$3P^0 - 15e^- + 5N^{+5} + 15e^- \longrightarrow 3P^{+5} + 5N^{+2}$$

Cancel the electrons:

$$3P^0 + 5N^{+5} \longrightarrow 3P^{+5} + 5N^{+2}$$

Then enter these proportions into the original equation:

$$3P + 5HNO_3 + ?H_2O \longrightarrow 3H_3PO_4 + 5NO$$

If we use two H_2O molecules on the left to balance the H's, we find that the oxygen atoms are, as a result, automatically balanced:

$$3P + 5HNO_3 + 2H_2O \longrightarrow 3H_3PO_4 + 5NO$$

Note: Once the proportions of P and N atoms have been found from the partial equations and entered into the final equation, the only other quantities needing adjustment are the numbers of molecules that do not take part in the partial equations—in this case, the H_2O molecules.

A preliminary balancing of the oxidation numbers (or transferred electrons) in the partial equations helps to eliminate a great deal of trial and error in balancing the final equation. Other techniques are described in the literature for balancing redox equations, but the oxidation number method presented here is one of the simplest and most convenient.

Glossary

charge number The number of excess electric charges, either positive or negative, on an ion.

oxidation A reaction in which an atom transfers electrons to another atom and, as a result, increases in oxidation number.

oxidizing agent A substance that accepts electrons in a reaction, and so causes another substance to become oxidized.

partial equation A chemical equation that isolates the oxidation or reduction of a particular species in a chemical process.

redox reactions Reduction-oxidation reactions.

reduction A reaction in which an atom accepts electrons, decreasing its oxidation number; the opposite of oxidation.

Problems and Questions

16.1 The Concept of Redox

16.1.Q Define or explain each of the terms below:
    a. oxidation number
    b. charge number
    c. oxidation
    d. reduction
    e. oxidizing agent
    f. reducing agent
    g. partial equation

16.2.P Decide whether each of the following equations is a redox equation. If it is, determine the oxidation numbers of each atom involved in the redox process (on both sides of the equation).
    a. $H_2S + HNO_3 \longrightarrow NO + S + H_2O$
    b. $HCl + Na_2Cr_2O_7 \longrightarrow NaCl + H_2O + CrCl_3 + Cl_2$
    c. $HCl + KOH \longrightarrow KCl + H_2O$
    d. $Cu + HNO_3 \longrightarrow Cu(NO_3)_2 + H_2O + NO_2$
    e. $NH_3 + O_2 \longrightarrow NO + H_2O$

16.3.Q For each of the redox equations in Problem 16.2, determine which substance is the reducing agent and which is the oxidizing agent.

16.4.Q Write the partial equations for the reactions in Problem 16.2, parts a, d, and e.

16.2 Balancing Redox Equations

Balance the following equations using the oxidation number method. (Make sure that you identify each oxidizing and reducing agent, that you know the oxidation numbers of each of these substances, and that you write balanced partial equations.)

16.5.P $?H_2S + ?HNO_3 \longrightarrow ?S + ?NO + ?H_2O$

16.6.P $?FeCl_3 + ?SnCl_2 \longrightarrow ?FeCl_2 + ?SnCl_4$

16.7.P $?Fe + ?O_2 \longrightarrow ?Fe_2O_3$

16.8.P $?KMnO_4 + ?HCl \longrightarrow ?KCl + ?H_2O + ?Cl_2 + ?MnCl_2$

16.9.P $?CuO + ?H_2 \longrightarrow ?Cu + ?H_2O$

16.10.P $?KClO_3 + ?Br_2 \longrightarrow ?KBrO_3 + ?Cl_2$

16.11.P $?HBr + ?H_2SO_4 \longrightarrow ?Br_2 + ?SO_2 + ?H_2O$

16.12.P $?K_2Cr_2O_7 + ?HCl \longrightarrow ?KCl + ?CrCl_3 + ?H_2O + ?Cl_2$

16.13.P $?N_2 + ?H_2 \longrightarrow ?NH_3$

16.14.P $?HClO_3 \longrightarrow ?HClO_4 + ?ClO_2 + ?H_2O$

(*Note:* The last equation is an example of a reaction in which the same element, chlorine, acts as both an oxidizing agent and a reducing agent.)

Self Test

Balance each equation using redox techniques. Then label the substance oxidized, the substance reduced, the oxidizing agent, and the reducing agent.

1. $?NH_3 + ?O_2 \longrightarrow ?NO + ?H_2O$
2. $?Cu + ?HNO_3 \longrightarrow ?Cu(NO_3)_2 + ?H_2O + ?NO_2$
3. $?Ag + ?Hg_2Cl_2 \rightleftharpoons ?AgCl + ?Hg$
4. $?CuO + ?H_2 \longrightarrow ?Cu + ?H_2O$
5. $?Na + ?H_2O \longrightarrow ?NaOH + ?H_2$
6. $?NaI + ?Cl_2 \longrightarrow ?NaCl + ?I_2$
7. $?NaMnO_4 + ?NaCl + ?H_2SO_4 \longrightarrow ?MnSO_4 + ?Na_2SO_4 + ?H_2O + ?Cl_2$

17
Carbon Chemistry

Objectives After completing this chapter, you should be able to:
1. Explain the scope of organic chemistry as opposed to inorganic chemistry.
2. Describe the structure of the carbon atom, both alone and in simple organic molecules.
3. Describe the structure of the major classes of organic molecules.
4. Describe the difference between the two allotropic forms of carbon.
5. Describe the major classes or families of carbon compounds: alkanes, alkenes, alkynes, alcohols, ethers, aldehydes, ketones, cyclic compounds, and carbohydrates.
6. Understand the difference between saturated and unsaturated compounds.
7. Understand the function of structural isomerism.
8. Use the IUPAC nomenclature of organic chemistry to recognize and name simple organic compounds from their formulas.
9. Write the formulas for simple organic compounds, given their names.
10. Describe the properties and uses of some of the more common organic compounds.
11. Describe some of the trends in properties of organic compounds as the molecular weight is increased within a given family of compounds.

17.1 Historical Introduction

Early Ideas About Organic and Inorganic Compounds

Since the beginning of chemistry, it has been recognized that all the known compounds can be divided roughly into two classes: organic compounds and inorganic compounds. Originally, organic compounds were obtained only from living things—from plants and animals. Inorganic compounds were obtained from the earth—from minerals.

This division into organic and inorganic matter stemmed from the ancient belief that living matter is intrinsically different from nonliving matter, being possessed of a "vital spirit" responsible for its life properties. This concept of **vitalism** persisted as a major force until well into the nineteenth century. The basic reason for the belief in vitalism was the fact that living matter has properties very hard to explain on a purely physical basis, such as the ability of a muscle to contract on command from the brain, the ability of an organism to reproduce itself, the ability of a person to feel emotion, and (most mysterious of all) the ability of a person to be aware of his or her own thoughts and perceptions. These abilities have still not been explained by science, but scientists are making progress in developing explanations and most scientists believe that ultimately an explanation is possible.

Back when little was known about biology, mystical explanations of life were commonly accepted. Organic compounds were considered to be different in nature from inorganic compounds, and it was believed that it was not possible to create organic compounds out of raw materials that had not been alive originally.

Many organic compounds were known at an early date. Some were obtained from plants: sugar, starch, vinegar, alcohol, etc. Others could be obtained in a pure state from animals, such as urea (obtained from urine) and glycerol (obtained as a byproduct of soap making). The fats and oils used in soap making are themselves mixtures of organic compounds obtained from animal tissue.

The structure of organic compounds was not understood at all until the end of the eighteenth century, when Lavoisier began to burn organic compounds and to weigh the combustion products. He found that these compounds were composed mainly of carbon, hydrogen, and oxygen. Later it was found that small quantities of nitrogen, sulfur, and phosphorus were present in some of the compounds.

An important breakthrough came in 1828, when Friedrich Wöhler created crystals of urea by heating ammonium cyanate. Here was an organic compound (urea) prepared from a compound (ammonium cyanate) that had never been part of a living body before. By proving that it was not necessary for organic compounds to come from living things, the experiment threw doubt on the theory of vitalism.

The controversy over vitalism continues to be fought to this very day. There are some who believe that the behavior of living matter can be explained only by the idea of a vital spirit. There are others who believe that the vital spirit concept explains nothing, and that living matter can best be explained by the same laws of nature that govern all other matter. The spirit of free inquiry is best preserved by the ongoing effort to understand and explain how living matter functions. Predictions about how far we can go in our understanding are premature at the present time, especially in view of the explosive increase in knowledge that has occurred during the past few decades.

Scientists are now at a point where they can determine the detailed structure of the extremely large molecules out of which living cells are built. We know what these molecules are made of and in what kinds of patterns the atoms are arranged. We can even synthesize (that is, make fron nonliving raw materials) an enormous number of organic compounds, many of which do not even exist in living matter.

But if organic compounds do not necessarily have anything to do with living matter, how do we define them? How do we tell the difference between organic and nonorganic compounds? We will take up these questions in the next section.

Current Definitions of Organic Compounds

Most simply, an **organic compound** is a compound of carbon together with other nonmetallic elements, such as hydrogen, oxygen, nitrogen, sulfur, chlorine, etc. There are also many organic compounds with metals in them; these are called **organometallic compounds**.

Among the organic compounds are those found in living organisms: carbohydrates, amino acids, proteins, lipids, fats, and many others. These compounds would have been considered organic under the old definition. In addition, there are completely man-made compounds that were previously unknown: plastics, dyes, explosives, drugs, artificial fibers such as nylon, and refrigerants such as freon.

Over a million different organic compounds are known, and it is said that this number increases by 5% every year. Organic molecules range from simple compounds of four or five atoms to complex molecules containing hundreds of thousands of atoms. One of the major recent advances in chemistry has been an increase in our ability to analyze the structure of the extremely large molecules that make up the basic substance of living tissue.

Most organic compounds contain hydrogen as well as carbon. There exists, in addition, a class of carbon compounds without hydrogen. This class includes carbon dioxide, carbon tetrachloride, and carbon disulfide. These molecules contain a carbon atom combined with oxygen, sulfur, or one of the halogens. Should these be considered to be organic compounds or inorganic compounds? The *Handbook of Chemistry and Physics* lists such substances in tables of both organic and inorganic compounds, so the dividing line is not clearcut.

The difference would appear to be functional—that is, it depends on the reaction being considered. When carbon dioxide combines with water to form carbonic acid, the reaction is thought of as inorganic:

$$CO_2 + H_2O \longrightarrow H_2CO_3$$

On the other hand, when carbonic acid (H_2CO_3) is produced from the reaction of carbonyl chloride ($COCl_2$) with water, we think of the reaction as organic:

$$COCl_2 + H_2O \longrightarrow H_2CO_3 + 2HCl$$

17.2 Properties of Carbon

Carbon Electron Structure

Carbon is usually thought of as a soft, black substance, as in graphite, soot, or coal. However, a diamond is a single crystal of pure carbon, is colorless, and is the hardest substance known. (The color of a diamond, if any, is due to small quantities of impurities.) Clearly, carbon can exist in more than one physical form, and the properties of one form are totally different from the properties of another. We will discuss these varieties of carbon in the next section, but first let us discuss the structure of the carbon atom itself.

A carbon atom has six orbital electrons arranged in two shells, as depicted in Figure 17-1. Recall from Chapter 5 that no more than two electrons are allowed in the innermost shell. The remaining four electrons reside in the outer shell, which is divided into two subshells. Two electrons occupy the s subshell, and the last two electrons make up the p subshell.

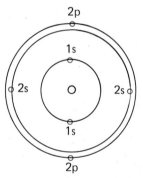

FIGURE 17-1 In the carbon atom there are six orbital electrons. Two electrons are located in the inner shell, and four valence electrons are located in the outer shell.

The shape of the electron wave function is represented in a diagram by a cloudlike figure called an orbital (see Figure 17-2). However, the orbitals of the electrons in a molecule are not the same as the orbitals in the free atom, because the carbon atom is bonded to other atoms in the molecule by the sharing of outer electrons (covalent bonds), which means that the electron orbitals must overlap the connected atoms.

For example, consider the structure of methane (CH_4). One way of representing the structure of this molecule is by means of a Lewis diagram, described in Section 7.4:

17.2 PROPERTIES OF CARBON

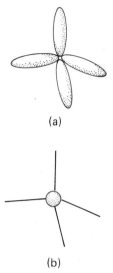

FIGURE 17-2 When a carbon atom is in a symmetrical molecule, its four valence electrons have their orbitals arranged so that there are equal angles between them, which puts them into a pyramidal (tetrahedral) pattern. Illustration (a) shows the orbital pattern. Illustration (b) shows the carbon atom at the center of the tetrahedron, with the bonds reaching out to the four corners. Each way of representing the carbon atom has its uses.

$$\begin{array}{c} H \\ H:\overset{..}{\underset{..}{C}}:H \\ H \end{array}$$

The diagram shows each of the four outer electrons from the carbon atom pairing up with the electron from a hydrogen atom to form a bond.

However, the Lewis diagram does not properly show how the electrons are arranged in space. According to the quantum model, each pair of electrons is represented by a probability function, or orbital, which surrounds the carbon and hydrogen atoms like a cloud. The probability function tells us the chance of locating an electron at any given point in that region of space.

What determines how the electrons arrange themselves in space? The molecule tries to organize itself so that it has the least amount of energy, taking up the shape that will result in the least amount of repulsion among the four pairs of electrons. It does this by placing the electrons so that they are as far from each other as they can get. Since each electron pair in the CH_4 molecules behaves in exactly the same way as every other electron pair, there is no reason for any pair to be farther away from the center of the molecule than any other pair. The form that provides maximum distance between the electrons is the pyramid (or tetrahedron) shown in Figure 17-3. (See Section 7.5 for further details.)

For this reason, symmetrical molecules such as CH_4 and CCl_4 are tetrahedral in shape. Many molecules that are well described by a tetrahedral representation were shown in Chapter 7.

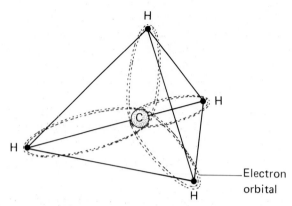

FIGURE 17-3 In a molecule of methane, the carbon atom is at the center and the four hydrogen atoms are at the corners of a tetrahedron. The orbitals of the shared electrons are like clouds surrounding both the carbon atom and the hydrogen atom, binding them together.

Because of the four outer electrons, the carbon atom almost always has an oxidation number of 4 in its compounds. That is, four electrons are available for bonding the carbon atom to other atoms.

One of the most important features of the carbon bond is the fact that one carbon atom can easily attach itself by covalent bonds to *another* carbon atom. This property is not possessed to as great an extent by any other atoms, except other valence-4 elements. In a metallic crystal—such as copper, for example—the atoms are joined by an attractive force resulting from the sharing of all the outer electrons in the crystal. The entire crystal can be thought of as a macromolecule.

Two carbon atoms, on the other hand, can be attached to each other with one set of paired electrons, and then the remainder of the available bonds can be satisfied with electrons from hydrogen atoms. The result, in the simplest case, is a compound that looks like this:

$$\begin{matrix} & H & H & \\ H & \!\!:\!\!\overset{..}{C}\!\!:\!\!\overset{..}{C}\!\!:\!\! & H \\ & H & H & \end{matrix}$$

This is a **saturated molecule**. That is, it has no tendency to join other molecules unless there is some particular reason for a reaction to take place. Therefore it does not form crystals under normal conditions.

The ability of carbon atoms to join each other in a chain—to *catenate*—is what allows molecules of carbon compounds to become so large and complex. The chain of carbon atoms acts like a backbone, holding the structure together.

Allotropes of Carbon

As mentioned in the last section, pure carbon is found in two different forms: one soft and black (graphite) and the other hard, transparent, and colorless (diamond). Two (or more) physical forms of the same element are called **allotropes** (or *allotropic forms*) of the element. In addition to carbon, a number of elements (including sulfur and phosphorus) are found with more than one allotropic form.

The two allotropes of carbon have many dissimilar properties in addition to color and hardness. Graphite is known to be a lubricant. That is, crystals of graphite slide easily over one another, so powdered graphite behaves like lubricating oil. (Spray containers of finely powdered graphite are sold for use in lubricating the interior mechanisms of locks. Graphite is preferable to oil for many uses because it does not dry out or harden.)

Powdered diamond, on the other hand, is used as an abrasive for grinding hard materials. Graphite is a moderately good electrical conductor, whereas diamond is an extremely good insulator. In short, the two allotropes behave quite differently.

All of the differences between the two forms of carbon arise from the different crystal structures. The forms of the two structures, in turn, are based on the properties of the electron orbitals discussed in the previous section.

In diamond the carbon atoms exist in their most symmetrical form, as shown in Figure 17-3. The four orbitals of the outer shell reach out in a tetrahedral pattern, linking each carbon atom to the next carbon atom in the crystal lattice. The result is a lattice of exceptional compactness and strength. (See Figure 17-4.)

FIGURE 17-4 In a diamond crystal, the tetrahedron is the basic building block of the structure. In this case there is a carbon atom at the center of the pyramid and at each of the four corners. Each corner atom is linked to another pyramid, forming a rigid structure throughout the entire crystal.

In the diamond lattice, all four of the outer electrons take part in the covalent bonds and are tightly bound to the atoms that share them. Since a solid needs free (unbound) electrons to be an electrical conductor, and since there are no free electrons in the diamond lattice, it follows that diamond is a nonconductor of electricity.

In graphite, only three of the electrons in the outer shell take part in the covalent bonds. These three electrons arrange themselves in a three-lobed pattern, all in one plane. As a result, the carbon atoms in graphite are arranged in a hexagonal formation, as in Figure 17-5. These six-atom rings are linked togther in a single plane, and other similar planes are found above and below. The linkage between planes is relatively loose, and therefore the planes can slide across each other. (Think of how easy it is to slide the cards in a deck of playing cards.) It is for this reason that graphite is so slippery.

The remaining outer-shell electrons have orbitals lying perpendicular to the

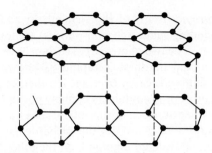

FIGURE 17-5 In graphite, the carbon atoms are arranged in flat hexagons, joined together in planes. Only three of the carbon bonds are needed to join the hexagons together. The fourth bond acts weakly in the direction perpendicular to the hexagons. As a result, the parallel planes are not joined together tightly and are free to slide over each other.

crystal plane, so they can be thought of as occupying the space between the planes. Since these electrons are relatively free, they are able to carry electric current through the crystalline structure.

Graphite (whose name comes from the Greek word meaning "to write") is the form of elemental carbon most commonly found in nature. Diamond is slightly unstable and will change to graphite if heated. However, under extremely high pressures (greater than 60,000 atm) and high temperatures (about 2000°C), graphite is converted into diamond. This process is responsible for the presence of diamonds in the earth's crust. Within the past few decades, methods have been developed for the manufacture of industrial-grade diamonds (for cutting and grinding).

Although the layers of graphite crystals slide easily over each other, making the graphite feel soft, the bonds between carbon atoms within the planes are quite strong. Long fibers of single graphite crystals (called "whiskers") can be made. These fibers have an extremely high tensile strength which makes them hard to break when they are pulled, and when embedded in plastics, they make a construction material that has the greatest strength-to-weight ratio of any material known. Such graphite fibers are used extensively in the aerospace industry for the manufacture of strong-but-light parts.

Diamond and graphite differ only in the amount of energy needed to maintain the crystal structure. Manufacturing diamond from graphite is an endothermic reaction. Thus, when you buy a diamond, you are paying a great deal for a small amount of energy.

17.3 Some Simple Carbon Compounds

Carbon-Hydrogen Compounds

The simplest compounds in organic chemistry are the **hydrocarbons**—compounds of carbon with hydrogen. What makes organic chemistry so complex is the fact that millions of different geometrical arrangements can be formed with only two kinds of atoms, carbon and hydrogen. Thus millions of different molecules can be created.

This feat is made possible by the property of **catenation**—the ability of two or more atoms to link together, allowing carbon atoms to form long chains. In addition, one or more side chains can branch out from the middle of a chain, so the number of possible arrangements of the carbon atoms is extremely large. As for the hydrogen atoms, they simply attach themselves to any unpaired electrons left over from the carbon-to-carbon links. (See Figure 17-6.)

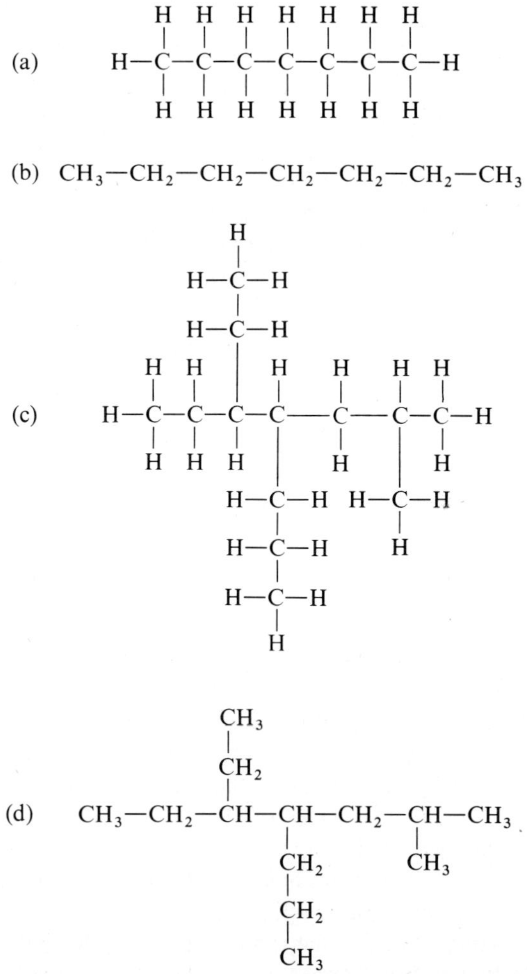

(a)
$$\begin{array}{c} \text{H H H H H H H} \\ \text{H}-\text{C}-\text{C}-\text{C}-\text{C}-\text{C}-\text{C}-\text{C}-\text{H} \\ \text{H H H H H H H} \end{array}$$

(b) $CH_3-CH_2-CH_2-CH_2-CH_2-CH_2-CH_3$

(c) [branched structural diagram with individual C and H atoms]

(d)
$$CH_3-CH_2-\underset{\underset{\underset{\underset{CH_3}{|}}{CH_2}}{\overset{\overset{\overset{CH_3}{|}}{CH_2}}{|}}}{CH}-\underset{\underset{CH_3}{|}}{CH}-CH_2-\underset{|}{CH}-CH_3$$

FIGURE 17-6 In organic compounds, the chains of carbon atoms can form straight lines, as in (a) and (b), or they can form branched patterns, as in (c) and (d). The structural diagrams can be drawn showing each individual carbon atom, or, to save space, they can be drawn showing CH_3 or CH_2 groups linked together. Parts (a) and (b) represent one compound, and (c) and (d) represent another compound.

Another complication is the fact that there are three kinds of carbon-to-carbon bonds: single bonds, double bonds, and triple bonds. Each type of bond has its own characteristics and produces a different kind of compound.

A **single bond** is a simple covalent bond formed by the sharing of a pair of electrons by two adjacent atoms. It may be diagrammed either with a single line representing the bond or with a pair of dots representing the electron pair:

$$\begin{array}{cc} H & H \\ | & | \\ H-C-C-H \\ | & | \\ H & H \end{array} \qquad \begin{array}{cc} H & H \\ H:\ddot{C}:\ddot{C}:H \\ H & H \end{array}$$

The three-dimensional model shown in Figure 17-7 gives a clearer concept of the molecule's spatial structure. In this model each of the carbon atoms is stationed at the center of its own tetrahedron, and the two tetrahedra are joined together at one of the corners to form the molecule. Each carbon atom has four electrons to share (that is, four bonds available to it), and in the above compound only one of these electrons is used to link the two carbon atoms together. Therefore there are three electrons (three bonds) left free on each carbon atom. (Notice that only one bond is drawn between adjacent atoms. It is as though the two atoms, like two people shaking hands, each contributed half a bond.)

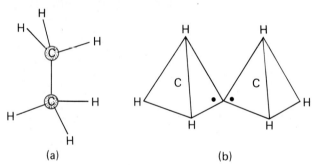

FIGURE 17-7 The ethane molecule, (a) represented by a skeletal structure and (b) represented as a pair of pyramids joined at one corner.

Many things can happen to these free electrons. The simplest possibility is that a hydrogen atom will attach itself to one of these carbon atoms by sharing its electron. There is also the possibility that another carbon atom may come along and join the chain, or that an element such as oxygen, nitrogen, or one of the halogens will join the chain.

A **double bond** exists where two carbon atoms each contribute two electrons to a covalent bond. Thus they share four electrons between them. The double bond is diagrammed either with a pair of lines or with four dots between the two carbon atoms:

$$\begin{array}{cc} H & H \\ | & | \\ C=C \\ | & | \\ H & H \end{array} \qquad \begin{array}{cc} H & H \\ \ddot{C}::\ddot{C} \\ H & H \end{array}$$

17.3 SOME SIMPLE CARBON COMPOUNDS

Since each carbon atom uses two electrons for carbon-carbon binding, there are only two electrons left on each atom for attaching other atoms. In three dimensions this compound may be pictured as a pair of tetrahedra joined at one edge, as in Figure 17-8(a). The paired electrons are at the two ends of this edge, and the hydrogen atoms are at the free corners of the tetrahedra.

The **triple bond** occurs when three electrons from each atom are shared. The diagram then shows either three lines or six electrons between the two carbons:

$$H-C\equiv C-H \qquad H:C:::C:H$$

Now each carbon atom has only one surplus electron left over to attach to another atom. In three dimensions we draw this molecule as a pair of tetrahedra joined at a common base. The one free corner of each tetrahedron holds one hydrogen atom, as in Figure 17-8(b).

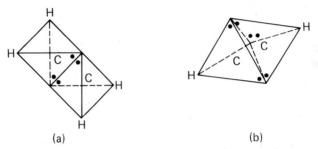

(a) (b)

FIGURE 17-8 (a) The tetrahedral structure of ethene. The double bond in this compound requires two pairs of shared electrons and is represented by joining two tetrahedra along one edge, so that they have two corners in common. (b) The tetrahedral structure of ethyne. The triple bond in this compound requires three pairs of shared electrons and is represented by joining two tetrahedra along an entire base, so that they have three corners in common.

Hydrocarbons containing nothing but single bonds are called **saturated hydrocarbons** because no more atoms can be connected to their molecules. All their bonds are satisfied. Hydrocarbons with double or triple bonds are called **unsaturated hydrocarbons** because it is possible to break one of the double or triple bonds and add more atoms to the molecule. Thus, the bond is unsatisfied, or unsaturated.

Let us now consider a few specific compounds that possess each of the three types of structures.

Single-Bond Compounds

The simplest organic compound is *methane* (CH_4). Its molecule has the tetrahedral structure described before, with the carbon atom in the middle and a hydrogen atom at each of the four corners. (See Figure 17-3.) Methane is a gas at normal atmospheric temperatures and is found naturally in a number of situations. It is popularly known as "marsh gas" because it is part of the gas that bubbles up from the bottoms of swamps, as a result of the decomposition of vegetable matter. Methane is found in petroleum and makes up the largest part of natural gas. It is highly flammable and when mixed with air is extremely explosive. (Methane is usually responsible for the explosions that take place periodically in coal mines.)

The boiling point of methane is $-164°C$. As a result, methane cooled to a temperature of $-164°C$ condenses into a liquid. Liquefied natural gas is mainly liquid methane.

If a hydrogen atom (both proton and electron) is removed from a molecule of CH_4, the resulting $\cdot CH_3$ group is called the *methyl group*. The methyl group is unstable because of the presence of an unpaired electron, represented by the dot $(\cdot)$ at the end of the $\cdot CH_3$ formula. (In older nomenclature such groups were called radicals. Although the modern practice is to give the term *radical* a more specialized meaning, the letter R is often used in formulas to represent groups of this nature.)

Two $\cdot CH_3$ groups join to form *ethane*, a compound with the formula C_2H_6. The formation of ethane may be visualized as follows:

$$CH_4 \longrightarrow \cdot CH_3 + \cdot H$$

$$\cdot CH_3 + \cdot CH_3 \longrightarrow CH_3{:}CH_3, \text{ or } C_2H_6$$

The structural formula of ethane is most simply represented by this diagram:

$$\begin{array}{c} \quad H \quad H \\ \quad | \quad \;\; | \\ H-C-C-H \\ \quad | \quad \;\; | \\ \quad H \quad H \end{array}$$

Each dash $(-)$ represents a pair of electrons $(:)$. As already shown in Figure 17-7, ethane may be represented three-dimensionally by a pair of tetrahedra joined at one corner. The remaining three corners of each tetrahedron represent electron pairs binding a total of six hydrogen atoms.

Ethane is a gas and, like methane, is found in petroleum. Its boiling point is $-88.6°C$, considerably higher than the boiling point of methane. Derivatives of methane and ethane are numerous and have important uses. (A **derivative** is a compound that is *derived*, or obtained, from another compound by a chemical reaction.) Ethanol (ethyl alcohol), for example, has the formula C_2H_5OH; it is obtained from ethane by replacing one of the H atoms with an OH group. (Note that this OH group is not the same as the OH^- hydroxyl ion.) The structural formula for ethanol is

$$\begin{array}{c} \quad H \quad H \\ \quad | \quad \;\; | \\ H-C-C-O-H \\ \quad | \quad \;\; | \\ \quad H \quad H \end{array}$$

For a three-dimensional structure of ethanol see Figure 17-9(a).

Methane and ethane have similar structures; the major difference is the additional carbon atom in ethane, together with its two accompanying hydrogen atoms. Therefore methane and ethane are considered to be members of the same family, or class, of compounds.

Another member of this class is obtained by removing a hydrogen atom from an ethane molecule and replacing it with a ·CH$_3$ group. The resulting compound is called *propane* (C$_3$H$_8$):

$$CH_3CH_2\cdot + \cdot CH_3 \longrightarrow CH_3CH_2CH_3$$

Its structural formula is

```
    H   H   H
    |   |   |
H — C — C — C — H
    |   |   |
    H   H   H
```

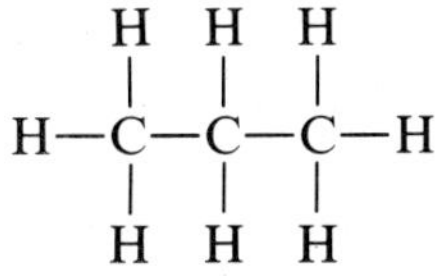

See Figure 17-9(b) for a three-dimensional structure of propane.

Propane is a gas bought in small tanks for heating, cooking, welding, etc. Its boiling point is −42°C.

To build the next higher compound in this class of compounds, we remove one of the hydrogen atoms and replace it with a methyl group, thus lengthening the carbon chain:

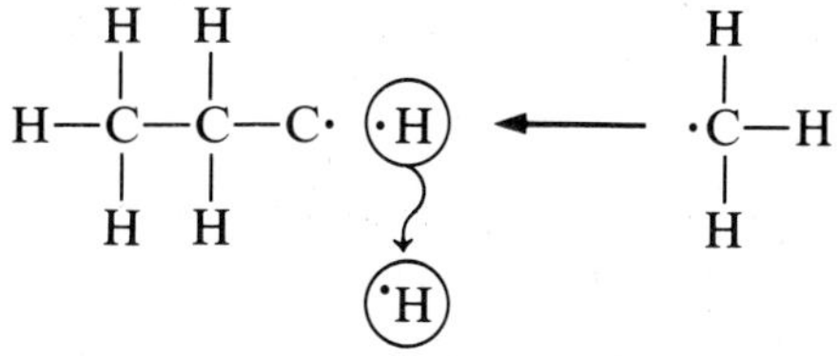

The unpaired electron (represented by the dot) in the methyl group allows the methyl group to attach itself to any other group with a similar unpaired electron. The compound formed by substituting a methyl group for a hydrogen atom in propane (as shown above) is the compound *butane*, with four carbon atoms. Its structural formula is

```
    H   H   H   H
    |   |   |   |
H — C — C — C — C — H
    |   |   |   |
    H   H   H   H
```

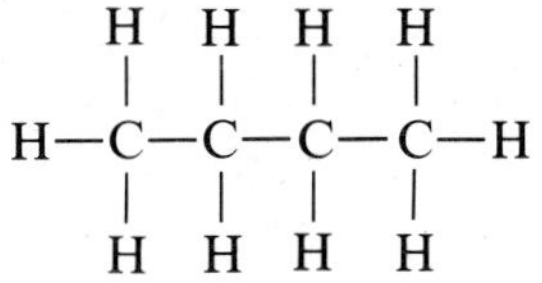

See Figure 17-9(c) for a three-dimensional structure of butane.

We can build longer and longer chains simply by replacing hydrogen atoms with methyl groups, one at a time, at the end of an existing chain. The class of compounds formed by this procedure is known as the **alkanes**, or *alkane hydrocarbons*. (The name comes from the German word for alcohol.) The old name for this class of compounds, *paraffin hydrocarbons*, orginated in the fact that paraffin—the waxy material used in making candles—is a member of this class (it contains over twenty carbon atoms).

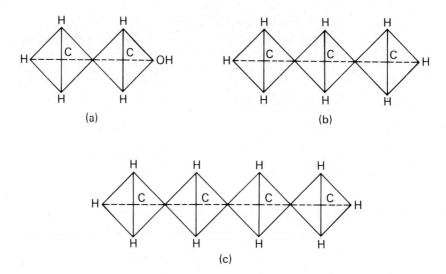

FIGURE 17-9 (a) Tetrahedral structure of ethanol. (b) Tetrahedral structure of propane. (c) Tetrahedral structure of butane.

An important insight into the properties of the alkane hydrocarbons can be obtained from Table 17-1, which lists the boiling points of some common hydrocarbons: the more carbon atoms there are in the chain, the higher the boiling point of the compound. This is a behavior typical of organic compounds.

One result of this trend is the fact that the lower members of the class (those with few carbon atoms) are gases at room temperatures. Compounds with more than six and less than eighteen carbon atoms are liquids under normal conditions, whereas compounds with greater numbers of carbon atoms are solids.

Another insight that can be derived from studying this table is that it is not

TABLE 17-1 The Alkanes

Common Name	Compound	Boiling Point (degrees Celsius)	Uses
natural gas	methane, CH_4	−164	heating
natural gas	ethane, C_2H_6	−88.6	
natural gas	propane, C_3H_8	−42.1	heating, welding
natural gas	butane, C_4H_{10}	−0.5	cigarette lighters
natural gas	pentane, C_5H_{12}	36.0	solvent
gasoline	hexane–nonane	69–200	fuel
kerosene	decane–hexadecane	200–300	fuel
lubricating oil	$C_{20}H_{42}$ up	300 up	lubrication
petroleum jelly	mixture	—[a]	lubrication
paraffin	mixture	—[b]	candles

[a] Semisolid.
[b] Solid.

necessary to memorize the properties of each of the thousands of organic compounds. Instead, it is more profitable to study the *relationships* between the various compounds. The classification schemes bring order out of apparent disorder. Once you become familiar with the general characteristics of a class of compound, and once you know how these characteristics vary as the molecular weight increases, you no longer need to deal with the confusion of independent compounds.

One further piece of information to be gleaned from Table 17-1 is that many members of the alkane hydrocarbons are familiar substances. Common petroleum products such as gasoline, kerosene, and petroleum jelly are essentially mixtures of these hydrocarbons.

Double-Bond Compounds

The simplest double-bond compound is *ethene*. Its formula can be written either as C_2H_4 or as CH_2CH_2, and its structural diagram is

$$\begin{array}{c} H \quad\quad H \\ \diagdown\quad\diagup \\ C = C \\ \diagup\quad\diagdown \\ H \quad\quad H \end{array}$$

See Figure 17-8(a) for the three-dimensional structure.

Another compound of the same class—*propene*—is obtained by substituting a methyl group for a hydrogen atom at the end of an ethene molecule. Its structure is

$$\begin{array}{c} H \quad\quad H \\ \diagdown\quad\diagup \\ C = C \quad\quad H \\ \diagup\quad\diagdown\quad\diagup \\ H \quad\quad C \\ \diagup\quad\diagdown \\ H \quad\quad H \end{array}$$

The three-dimensional structure of propene is shown in Figure 17-10(a).

These compounds belong to the class of **alkenes**. Members of this class are important in industry because of the relative ease with which one of the double

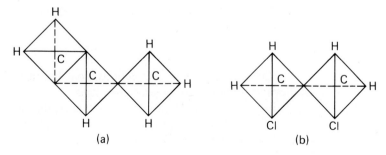

FIGURE 17-10 (a) Tetrahedral structure of propene. (b) Tetrahedral structure of dichloroethane.

bonds can be broken and replaced with a pair of atoms. For example, if ethene is treated with chlorine, a chlorine atom attaches itself to each of the two carbon atoms, forming the compound dichloroethane (known as a cleaning fluid under the common name ethylene chloride):

$$\begin{array}{c}H\\ \\ H\end{array}\!\!\!\diagdown\!\!\!C\!\!=\!\!C\!\!\!\diagup\!\!\!\begin{array}{c}H\\ \\ H\end{array} + Cl_2 \longrightarrow H-\underset{\underset{Cl}{|}}{\overset{\overset{H}{|}}{C}}-\underset{\underset{Cl}{|}}{\overset{\overset{H}{|}}{C}}-H$$

See Figure 17-10(b) for the three-dimensional structure of dichloroethane.

Because of the high chemical activity of these unsaturated hydrocarbons, the alkenes are useful as intermediates in the manufacture of many chemicals.

Triple-Bond Compounds

The simplest triple-bond compound is *ethyne* (C_2H_2), commonly known as acetylene. It has the structural formula

$$H-C\equiv C-H$$

See Figure 17-8(b) for the three-dimensional structure of ethyne.

Acetylene, a member of the class of **alkynes**, is well known as a flammable gas used in welding torches. It is also important in the chemical industry as the starting point for the manufacture of many other organic chemicals. The triple bond is highly reactive, and when one or two of the bonds are broken and the unpaired electrons are shared with other atoms or groups, a variety of new compounds are formed.

Another reason for the importance of acetylene is that it is readily manufactured from materials found in nature. One starting point is coal. When coal is heated to a high temperature in a closed vessel so that (in the absence of oxygen) it does not burn, volatile (gaseous) compounds are driven off and coke is left behind in the vessel. This process is known as **destructive distillation**. Coke (mostly carbon) is then heated with lime to form calcium carbide, and carbon monoxide is given off:

$$CaO + 3C \longrightarrow CaC_2 + CO$$

In the final step of the process, calcium carbide reacts with water to produce calcium hydroxide and gaseous acetylene:

$$CaC_2 + 2H_2O \longrightarrow Ca(OH)_2 + C_2H_2$$

Through this chain of reactions, coal becomes the starting point for the synthesis of thousands of organic chemicals.

Just as with the alkanes and the alkenes, the next member of the alkynes is obtained by removing a hydrogen atom from one end of the chain and replacing it with a ·CH_3 group. The compound *propyne* results:

$$H-C\equiv C\cdot + \cdot CH_3 \longrightarrow H-C\equiv C-CH_3$$

Carbon-Oxygen-Halogen Compounds

Carbon Monoxide and Carbon Dioxide

Carbon monoxide (CO) and carbon dioxide (CO_2) are both produced when carbon or carbon-containing compounds burn in oxygen. Normally the combustion of hydrocarbons such as methane results in the formation of CO_2 and H_2O:

$$CH_4 + 2O_2 \longrightarrow CO_2 + 2H_2O$$

When there is an excess of oxygen, CO_2 is the only oxide formed. However, when the oxygen is not sufficient for complete combustion, CO can result. CO is formed in automobile engines, in coal stoves, and in kerosene heaters under conditions that do not allow enough oxygen into the combustion chamber.

Carbon monoxide is a poisonous gas and must be avoided. It interferes with the normal function of the lungs, which is to transfer oxygen from the air into the red corpuscles of the bloodstream. Oxygen, of course, is needed to metabolize, or "burn," the parts of the food intake used as fuel, and CO_2 is given off as one of the waste products (again within the lungs). When CO is inhaled, it is absorbed into the blood, replacing the oxygen within the red blood cells. Those cells are thereby prevented from carrying oxygen to the parts of the body where it is needed.

Because of this property, CO is a poisonous gas and its inhalation is a common cause of death. For this reason automobiles should never be allowed to run inside a closed garage, and any kind of fuel-burning stove should be used only if there is adequate ventilation.

Both CO and CO_2 are colorless, odorless, and tasteless gases. The lack of odor adds to the danger. (Although CO_2 is not poisonous, people have suffocated in chambers filled with nothing but CO_2, because of the lack of oxygen.) Carbon monoxide condenses to the liquid form at a temperature of $-190°C$ (at a pressure of 1 atm). Carbon dioxide behaves in a more unusual manner. When its temperature is reduced to $-79°C$, it changes directly into a solid rather than into a liquid (at normal atmospheric pressure). This white solid, commonly known as *dry ice*, is used for refrigerating ice cream and other materials.

When warmed above $-79°C$, dry ice does not go through a liquid phase (at a pressure of 1 atm) but evaporates directly into gaseous CO_2. This solid-vapor transition is called *sublimation*. The white haze seen above a piece of dry ice, used for mist in cinema productions, is really water vapor from the atmosphere condensing in the cold air.

The behavior of CO_2 at high pressures is much different from its behavior at normal pressure. When compressed to about 59 atm (at room temperature), CO_2 becomes a liquid. Carbon dioxide purchased in a tank is generally in liquid form under high pressure. Opening the tank valve reduces the pressure inside the tank and allows the liquid to turn into CO_2 gas. When the gas escapes into the atmosphere, the sudden pressure reduction causes the gas temperature to drop to less than $-79°C$. This is a general principle of refrigeration: letting a gas expand against an outside pressure will cause a temperature drop. As a result the CO_2 condenses into white, powdery dry ice.

Carbon dioxide is important in life cycles. Animals inhale oxygen and exhale CO_2, the product of fuel combustion. Plants, on the other hand, breathe in CO_2 and, with the help of the energy from sunlight, convert it into carbohydrates and other substances necessary for living tissue. It is the absorption of CO_2 by plants that keeps the atmosphere in balance and maintains the concentration of CO_2 at about 0.03% of the total atmospheric composition.

However, furnaces and power plants that burn carbon-containing fuel pour more CO_2 into the atmosphere than the plant population can handle. The concentration of CO_2 in the atmosphere has been slowly increasing during the past century, according to measurements made over a long period of time. If we continue to rely on coal-burning power plants for energy into the twenty-first century, the concentration of CO_2 in the atmosphere will increase at an even greater rate than in the past.

The result of that increase of CO_2 is expected to be a general warming of the atmosphere, since CO_2 absorbs energy from sunlight. Although the increase in temperature may be only a few degrees, this effect can cause drastic climatic changes, including the melting of the polar ice caps. This "greenhouse effect" is considered to be a potentially serious problem and is the subject of a great deal of current research.

Carbon-Halogen Compounds

The halogens are important in organic chemistry because they are very active elements and therefore are able to displace hydrogen atoms from their positions in organic molecules. As a result the halogens easily form numerous compounds. Some of the compounds are useful in their own right, and some are used as intermediate steps in the synthesis of other compounds. In this section we will survey a few of the simpler carbon-halogen compounds.

A mixture of methane and chlorine reacts explosively when exposed to bright sunlight. In that reaction all the hydrogen atoms are stripped from the methane molecule, and bare carbon is left behind:

$$CH_4 + 2Cl_2 \longrightarrow C + 4HCl$$

However, if the gas mixture is exposed to a less intense light source, the chlorine can replace one hydrogen atom at a time in a gentler reaction. (The light stimulates the reaction by **dissociation** of the clorine molecule, separating the two chlorine atoms from each other so that they are free to react with the methane.) A series of reactions is possible, each reaction forming a molecule with different numbers of chlorine atoms. As the equations below show, one of the atoms from the dissociated chlorine molecule removes one of the hydrogen atoms from the organic molecule to form hydrogen chloride, while the other chlorine atom attaches itself to the carbon bond left free. The compounds produced are part of the general class of *halogenated hydrocarbons.*

Notice that in the equations below, two names are given for each compound. The first name is the older, more traditional name. The second name belongs to the modern system of nomenclature to be discussed in Section 17.4.

17.3 SOME SIMPLE CARBON COMPOUNDS

$$\underset{\text{methane}}{H-\underset{|}{\overset{|}{C}}-H} + Cl-Cl \longrightarrow \underset{\substack{\text{methyl chloride}\\\text{monochloromethane}}}{H-\underset{|}{\overset{|}{C}}-Cl} + H-Cl$$

$$\underset{}{H-\underset{|}{\overset{|}{C}}-Cl} + Cl-Cl \longrightarrow \underset{\substack{\text{methylene chloride}\\\text{dichloromethane}}}{H-\underset{|}{\overset{|}{C}}-Cl} + H-Cl$$

$$\underset{}{H-\underset{|}{\overset{|}{C}}-Cl} + Cl-Cl \longrightarrow \underset{\substack{\text{chloroform}\\\text{trichloromethane}}}{Cl-\underset{|}{\overset{|}{C}}-Cl} + H-Cl$$

$$\underset{}{Cl-\underset{|}{\overset{|}{C}}-Cl} + Cl-Cl \longrightarrow \underset{\substack{\text{carbon tetrachloride}\\\text{tetrachloromethane}}}{Cl-\underset{|}{\overset{|}{C}}-Cl} + H-Cl$$

The final result is a liquid commonly known as carbon tetrachloride, whose molecule consists of four chlorine atoms joined to a single carbon atom. This liquid does not easily burn and makes an excellent solvent for oils and greases. In the past it was much used in the home as a cleaning fluid and as a fire extinguisher. However, as will be described in the next section, CCl_4 may react to form phosgene—a highly poisonous gas—under certain conditions, and so its use in the home has been discontinued. Methylene chloride has replaced CCl_4 in many commercial cleaning solvents.

Chloroform was widely used in the past as a general anesthetic, but other safer anesthetics are now employed for the same purpose.

The melting and boiling points of the four methyl chlorides are compared in Table 17-2 to demonstrate how increasing the molecular weight of the compound tends to increase both of these temperatures. In other words, the volatility of the compound goes down as the molecular weight goes up. This table provides another example of the order that can be found by studying the relationships between compounds in the same class.

TABLE 17-2 Properties of Methyl Halides

Compound	Formula	Melting Point (degrees Celsius)	Boiling Point (degrees Celsius)
methane	CH_4	−182	−164
methyl chloride (monochloromethane)	CH_3Cl	−97.7	−24.2
methylene chloride (dichloromethane)	CH_2Cl_2	−95.1	40
chloroform (trichloromethane)	$CHCl_3$	−63.5	61.7
carbon tetrachloride (tetrachloromethane)	CCl_4	−23.0	76.5
Freon (dichlorodifluoromethane)	CCl_2F_2	−158	−29.8

Compounds similar to the four methyl chlorides are formed with the other halogens: fluorine, bromine, and iodine. Some of these are of considerable importance. One especially useful set of compounds is the *hybrid halides*—compounds containing both chlorine and fluorine atoms in the same molecule. An example of such a molecule is dichorodifluoromethane, commonly known as *Freon*. Freon is actually the trade name for a class of volatile hybrid halides with useful properties. These compounds are gases at normal temperatures and pressures, but they are easily liquefied under moderately high pressure. Since the evaporation of liquid Freon is highly endothermic, these compounds are widely used as refrigerants in air conditioners and refrigerators.

Carbon-Oxygen-Halogen Compounds

Numerous compounds contain both oxygen and one or more halogens bonded to carbon atoms. The simplest of these is *phosgene* (also known as carbonyl chloride), with the formula $COCl_2$. The molecular structure of phosgene is shown in the following diagram.

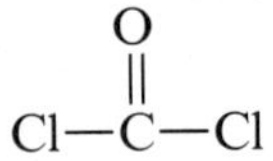

$$\underset{\text{Cl}-\text{C}-\text{Cl}}{\overset{\overset{\displaystyle O}{\|}}{}}$$

For a three-dimensional structure, see Figure 17-11. The C=O group is called the **carbonyl group** and is a significant component of many compounds.

When phosgene reacts with water, the Cl atoms are replaced by hydroxyl (OH) groups, and carbonic acid results:

$$\overset{\overset{\displaystyle O}{\|}}{\text{Cl}-\text{C}-\text{Cl}} + 2\,\text{HOH} \longrightarrow \underset{\text{carbonic acid}}{\overset{\overset{\displaystyle O}{\|}}{\text{HO}-\text{C}-\text{OH}}} + 2\,\text{HCl}$$

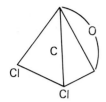

FIGURE 17-11
Tetrahedral structure of carbonyl chloride (phosgene)

Phosgene is an extremely poisonous gas; it was notorious as one of the gases used in chemical warfare during World War I. It can be a byproduct of a variety of chemical reactions, so good ventilation is important in handling compounds that can give rise to this gas. One early environmental concern was aroused by the reaction of carbon tetrachloride with steam to produce phosgene:

$$CCl_4 + H_2O \longrightarrow COCl_2 + 2HCl$$

Since CCl_4 was widely used as a cleaning fluid in establishments using steam irons, such a reaction was clearly a danger to health, and thus the use of carbon tetrachloride as a commercial cleaning fluid was eliminated. Similar health concerns resulted in the elimination of CCl_4 as the ingredient in fire extinguishers.

In general, the inhalation of halogenated hydrocarbons should be avoided as much as possible, both because they have an immediate toxicity and because many of them are known to cause cancer when ingested over a long period of time. (Yet there are those who inhale halogenated hydrocarbons recreationally by sniffing glue or other similar products, not appreciating the serious and permanent damage that may be sustained by the body's organs.)

17.4 Nomenclature of Organic Compounds

Like inorganic compounds, many organic compounds have three names: (1) the common, or trivial, name handed down from the beginnings of chemistry, (2) the more formal name developed by chemists over the past three centuries, and (3) the systematic name approved in 1957 by the International Union of Pure and Applied Chemistry (IUPAC). Nomenclature is exceedingly important in organic chemistry, because the compounds dealt with are often very complex and it is desirable for the name of the compound to represent, in an unambiguous way, the structure of the compound.

This section will present some of the rules for naming the simpler organic compounds. At the same time, you will learn more about the uses and properties of the various classes of compounds.

The Alkane Hydrocarbons

You have already seen that many of the hydrocarbons derived from petroleum have long, straight chains of carbon atoms. Other hydrocarbons have chains that branch off from each other. Still others are **cyclic compounds**, having ring-like forms. Molecules that are *not* cyclic are called **acyclic compounds**.

In this section we will look at the nomenclature for some of the unbranched, acyclic hydrocarbons: the alkanes. The names of these compounds have two parts: a prefix that depends on the number of carbon atoms in the chain and the suffix *-ane*. This suffix indicates that the named compound is saturated and acyclic.

TABLE 17-3 Summary of IUPAC Hydrocarbon Names

Carbon Root							
Number	Name	Alkane	(C_nH_{2n+2})	Alkene	(C_nH_{2n})	Alkyne	(C_nH_{2n-2})
1	meth-	methane	CH_4	—		—	
2	eth-	ethane	C_2H_6	ethene	C_2H_4	ethyne	C_2H_2
3	prop-	propane	C_3H_8	propene	C_3H_6	propyne	C_3H_4
4	but-	butane	C_4H_{10}	butene	C_4H_8	butyne	C_4H_6
5	pent-	pentane	C_5H_{12}	pentene	C_5H_{10}	pentyne	C_5H_8
6	hex-	hexane	C_6H_{14}	hexene	C_6H_{12}	hexyne	C_6H_{10}
7	hept-	heptane	C_7H_{16}	heptene	C_7H_{14}	heptyne	C_7H_{12}
8	oct-	octane	C_8H_{18}	octene	C_8H_{16}	octyne	C_8H_{14}
9	non-	nonane	C_9H_{20}	nonene	C_9H_{18}	nonyne	C_9H_{16}
10	dec-	decane	$C_{10}H_{22}$	decene	$C_{10}H_{20}$	decyne	$C_{10}H_{18}$

You have already encountered the first few alkanes: methane, ethane, propane, and butane. Table 17-3 gives the names and formulas of the first ten members of the alkane class (as well as the alkenes and alkynes). Starting with pentane, the prefixes are the well-known Greek prefixes representing the number of carbon atoms in the molecule, so these names are not difficult to remember.

In these compounds each of the carbon atoms (except for the two at the end) has two hydrogen atoms attached. Each of the two end carbon atoms has three hydrogen atoms attached. The structural formula for the alkanes can, therefore, be written in the form

$$CH_3CH_2 \cdots CH_2CH_3$$

or in the more condensed form

$$CH_3(CH_2)_xCH_3$$

The general formula for an alkane molecule is C_nH_{2n+2}, where n is any integer 1, 2, 3,.... Any acyclic molecule fitting this formula is an alkane.

You saw in Section 17.3 that the compound CH_3Cl, in which a chlorine atom has taken the place of one of the hydrogen atoms in the methane molecule, is called methyl chloride. The group $\cdot CH_3$ is called the *methyl group*. It is methane minus a hydrogen atom, and so has one free bonding site. (A bonding site is a location where another atom or group can be attached.)

In general, all alkane groups are named similarly. Ethane minus a hydrogen atom is the *ethyl group*; if it combines with a chlorine atom, ethyl chloride, C_2H_5Cl, is formed. These groups are called *alkyl groups*. When naming a particular alkyl group, replace the -ane suffix of the hydrocarbon with the suffix -yl. From hexane, for example, is derived the hexyl group, $\cdot C_6H_{13}$. It can attach itself to any number of other groups to form hexyl compounds (e.g., hexyl bromide, hexyl alcohol, etc.).

A well-known alkyl compound is tetra-ethyl lead, which consists of four ethyl radicals attached to a single lead atom: $Pb(C_2H_5)_4$. It has been added to gasoline to make it burn more evenly and reduce engine knock. However, because of pollution problems, use of tetra-ethyl lead is not permitted in newer automobiles. (The lead poisons the catalytic converters used to reduce nitrogen compounds in the emissions; in general, lead is an undesirable addition to the atmosphere.) Lead-free gasoline with other anti-knock additives is now manufactured for use in newer engines.

Unsaturated Hydrocarbons—Alkenes and Alkynes

The simplest double-bonded hydrocarbon has the structural formula $CH_2{=}CH_2$. The IUPAC rule for naming compounds with a single double bond is as follows: *Count the number of carbon atoms in the "backbone" of the molecule; the compound name has the same root as the alkane compound having the same number of carbon atoms, followed by the suffix -ene.* A molecule with two carbon atoms has a name that starts with *eth-*; therefore the compound $CH_2{=}CH_2$ is named *ethene*.

However, the IUPAC rules are not always followed. It has become customary in many cases to retain the older names—for example, ethylene instead of ethene. In a number of situations the common names for certain simple compounds are so familiar to chemists that to use the IUPAC nomenclature would cause some confusion. As a result, some compounds are known by more than one name.

The unsaturated hydrocarbon with three carbon atoms is *propene*: $CH_3{-}CH{=}CH_2$ (older name: propylene). Both propane and propene are gases under normal conditions. The introduction of the double bond has the effect of reducing the boiling point slightly: the boiling point of propane is $-42°C$, and the boiling point of propene is $-47°C$. This reduction of the boiling point is only a small effect—there are much larger differences between the chemical properties of propane and propene.

Notice that in propene it makes no difference whether the double bond goes on the left end or the right end. (It does not matter whether you draw the diagram from right to left or left to right.) However, the four-carbon compound *butene* has two possible structures that really are different. One structure has the double bond in the first position, at the end of the chain (see the formula below), and another structure places the double bond in the second position, in the middle of the chain. The two structures represent distinctly different compounds with different properties. To distinguish between these compounds, we add to the name of the compound a number indicating the first carbon in the chain to which the double bond is attached:

$$CH_2{=}CH{-}CH_2{-}CH_3 \qquad CH_3{-}CH{=}CH{-}CH_3$$
1-butene   2-butene

Always start counting from the end that gives the smallest number, even if it is necessary to count from right to left. For example, the compound

$$CH_3CH_2CH_2CH_2CH{=}CH_2$$

is named *1-hexene*.

Hydrocarbons with one double bond are called *alkenes* and have the general formula C_nH_{2n}. Many useful industrial chemicals and plastics are based on the alkenes, since they are highly reactive. One of the simplest derivatives of ethene is chloroethene: $CH_2\!=\!CHCl$. Under its older name, vinyl chloride, it is well known as an ingredient in the manufacture of plastics.

An important property of alkene molecules is their ability to fasten themselves together in a long chain—a **macromolecule**, containing up to millions of the simpler molecules. This process is known as *polymerization*. Most familiar plastics are **polymers** of simple alkene derivatives. Indeed, the entire plastics industry is based on the process of polymerization. (The word *polymer* comes from the Greek; *poly-merous* means "many parts.")

As we have seen, the simplest triple-bond compound is *acetylene*, $HC\!\equiv\!CH$. In the IUPAC system of nomenclature, the name of the compound is given a *-yne* ending, so the IUPAC name for acetylene is *ethyne*. However, many people still use the more common name, acetylene. The three-carbon compound $HC\!\equiv\!C\!-\!CH_3$ is called *propyne*; the four-carbon compound $HC\!\equiv\!C\!-\!CH_2\!-\!CH_3$ is *1-butyne*; and $CH_3C\!\equiv\!CCH_3$ is *2-butyne*.

As a class, compounds with a triple bond are called *alkynes* and have the general formula C_nH_{2n-2}. The alkynes undergo polymerization just like the alkenes and are highly useful in the chemical industry. Table 17-3 lists the first nine members of the alkyne family.

Isomers

Two or more compounds that have the same molecular formula but different structures are called **isomers**. (In Greek *iso-* means "equal" and *-mer* means "part.") Isomers play an important role in organic chemistry because the many ways of arranging a given number of atoms within a molecule add greatly to the variety of possible compounds.

Consider the compound butane (C_4H_{10}). So far we have thought of it only as a simple straight chain:

$$CH_3-CH_2-CH_2-CH_3$$

However, it is also possible to arrange the four carbon atoms (with associated hydrogens) into the following structure:

$$\begin{array}{c} CH_3-CH-CH_3 \\ | \\ CH_3 \end{array}$$

Both of these compounds are isomers of butane. The second is an example of a **branched chain** and is called *isobutane*. (*Note:* We do not think of one of these compounds as *the* compound and the other as the isomer. They are *both* isomers.)

In the IUPAC system, the above branched compound is named by finding the longest continuous carbon chain (three carbons) and identifying it as a saturated propane derivative with a methyl group attached to the second carbon atom. Accordingly, it receives the name *2-methylpropane*. Here is the formal rule for naming compounds like this: *Start by naming the longest chain according to the number of carbon atoms in that chain, and then add the prefix representing the side chain(s)*. So, for example, the following compound is called *2,3,5-trimethylhexane*:

$$CH_3-\underset{\underset{CH_3}{|}}{CH}-\underset{\underset{CH_3}{|}}{CH}-CH_2-\underset{\underset{CH_3}{|}}{CH}-CH_3$$

The numbers making up the prefix should be the lowest possible set of numbers. Therefore we number from left to right in this case. If we made the mistake of numbering from right to left, the numbers would be 2,4,5—which would not be the lowest possible set.

A similar situation arises when one or more halogen atoms are attached to a hydrocarbon. In the traditional notation, the compound

$$CH_3-CH_2-CH_2Cl$$

is called propyl chloride (C_3H_7Cl), whereas the compound

$$CH_3-\underset{\underset{Cl}{|}}{CH}-CH_3$$

is called *isopropyl chloride*. In the IUPAC notation, the first compound is called *1-chloropropane* and the second compound is called *2-chloropropane*, with the numbers identifying the carbon atoms to which the chlorine atoms are attached. Though these isomers contain the same number of carbon, hydrogen, and chlorine atoms, they are different compounds with different properties. They are both liquids, but the first compound has a boiling point of 47°C and the second boils at 36°C.

As you can see, the possibility of branched chains increases the number of organic compounds immeasurably. The greater the number of carbon atoms, the greater the number of ways of arranging those atoms.

Cyclic Compounds

In addition to straight and branched chains of carbon atoms, there are arrangements in which the carbon atoms form rings of various sizes. These molecules make up the class of cyclic compounds. Among the cyclic compounds are the most important of the organic compounds, including numerous dyes, perfumes, flavorings, drugs, and other useful chemicals, both natural and synthetic. In addition, a large number of the essential molecules that make up the living body (such as proteins and hormones) contain cyclic structures.

The following equation illustrates how a cyclic compound may be synthesized:

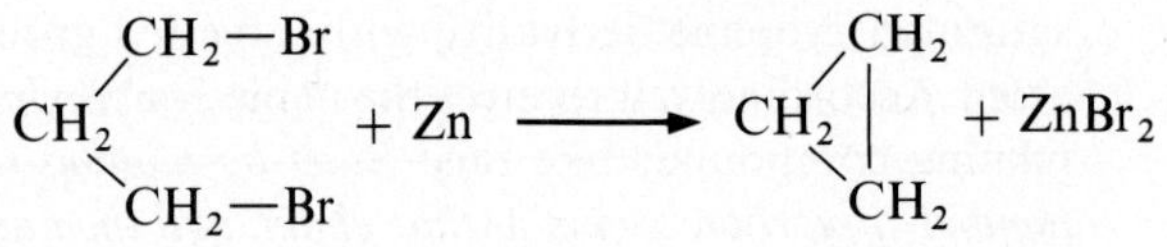

$$\begin{array}{c}\text{CH}_2\text{—Br}\\ \text{CH}_2\\ \text{CH}_2\text{—Br}\end{array} + \text{Zn} \longrightarrow \begin{array}{c}\text{CH}_2\\ \text{CH}_2 \;\;|\\ \text{CH}_2\end{array} + \text{ZnBr}_2$$

The left side of the equation shows 1,3-dibromopropane reacting with zinc. On the right side, the zinc has combined with the two bromine atoms to form zinc bromide, and the two ends of the propane molecule have closed in on themselves to form the cyclic compound *cyclopropane* (a gas used in operating rooms as a general anesthetic).

The molecular formula of cyclopropane is C_3H_6, the same as the formula for propene. However, the structural formula is completely different. In propene there are differences among the three carbon atoms (one is in the middle and two are on the end); also, two carbon atoms are next to a double bond, and the other carbon is not. In cyclopropane, by contrast, there are no differences among the three carbon atoms—the molecule is completely symmetrical. This and other important pieces of evidence lead us to the understanding that the molecule has the form of a ring—or, more precisely, a triangle.

Other simple cyclic compounds are cyclobutane, cyclopentane, and cyclohexane. These compounds tend to be similar in their properties to the alkane hydrocarbons, and they have the following structural formulas:

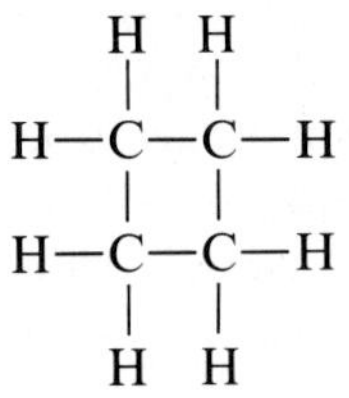

cyclobutane

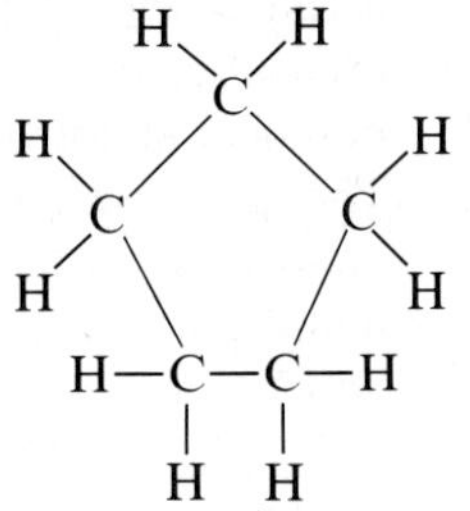

cyclopentane

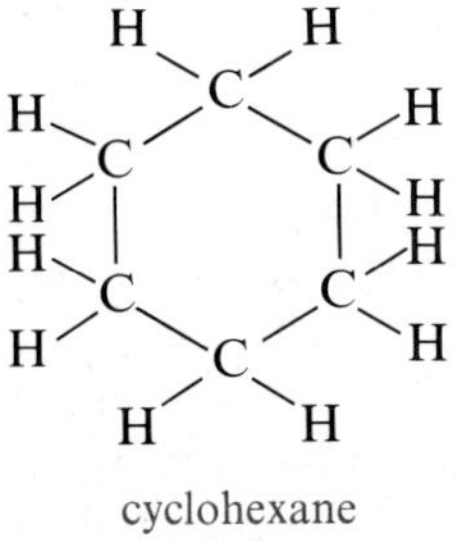

cyclohexane

From a historical point of view as well as from a chemical point of view, the most important cyclic compound is *benzene*. (Notice that the name of this compound has the ending *-ene*. In the past, a mixture of alkane hydrocarbons called *benzine* was commonly used as a solvent and as a fuel. The two names should not be confused.) Benzene is the simplest example of a class of compounds called **aromatic compounds**. The name arose because of the strong aroma emitted by most of these substances. Now, however, many compounds that do not have strong odors are classed as aromatic compounds because of their structure.

Benzene was the first compound whose structure was recognized as being a ring of carbon atoms. The discovery of a model for the benzene molecule was a great breakthrough in the development of organic chemistry. Credit for this discovery is given to the German chemist Friedrich Kekulé. Kekulé had been one of the leading workers in the field of organic chemistry for many years, and in 1861 he published an important textbook that, for the first time, advocated the use of structural formulas to describe organic molecules. It must be remembered that during that period of time there was still a great lack of understanding about atomic weights and molecular structure. Avogadro's hypothesis was only beginning to receive acceptance in 1860, and many people at that time still thought the atomic weight of oxygen was 8 instead of 16.

In 1865 Kekulé devoted a great deal of thought to the structure of benzene, a compound that chemists had been studying for many years. Its molecular formula was known to be C_6H_6, but it did not behave like any of the straight-chained hydrocarbons. The main difference was that each of the six carbon atoms was identical to all the others. It was impossible to say that one atom was on the end of the chain and another was in the middle. They all behaved in the same way. A completely new idea was needed to solve this puzzle. It was a problem that occupied the minds of many chemists. Some of them had indeed suggested cyclic structures, but none was taken seriously.

Kekulé described how the idea arose in his own mind. He had been working on his textbook without making any progress. One night, sitting by the fire, he fell asleep and began to dream of atoms and molecules forming themselves into long chains and writhing like snakes: "Suddenly one of the serpents caught its own tail and the ring thus formed whirled exasperatingly before my eyes. I woke as by lightning and spent the rest of the night working out the logical consequences of the hypothesis. If we learn to dream we shall perhaps discover truth. But let us beware of publishing our dreams until they have been tested by the waking consciousness."

Whether Kekulé conceived the cyclic idea on his own or whether he unconsciously remembered some of the previously published papers is not clear. Nevertheless, Kekulé's description of the benzene ring structure was influential in pointing other chemists in the right direction. The person who is able to propel a new idea into the mainstream of knowledge is often the one who receives credit for a discovery.

Kekulé's detailed account of the creative mind at work is striking, and the final sentence presents an important warning concerning the need to test hypotheses by observation before believing them too strongly.

The structural diagram for benzene is commonly written as

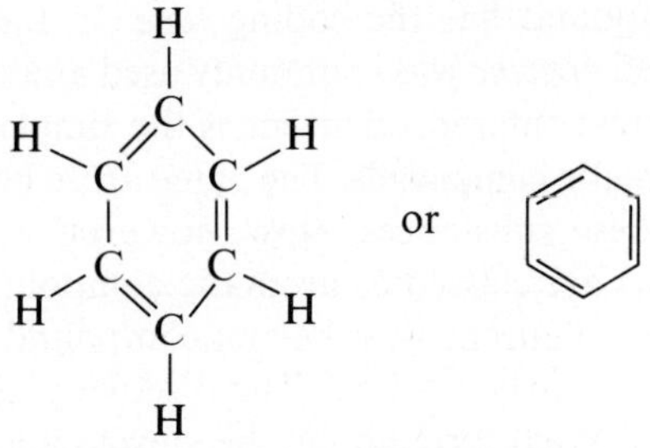

The atoms are often omitted from the diagram, since it is understood by chemists that a CH group is attached to each corner of the hexagon.

The double bonds provide a way of accounting for all the electrons. In the benzene molecule, each of the six carbon atoms is connected to two other carbon atoms plus a hydrogen atom. These bonds use up three of the four orbital electrons that each carbon atom has to share. There are then six unpaired electrons in the molecule. One way to account for these six electrons is to have adjacent carbon atoms share them. We do this by inserting three double bonds, as shown in the diagrams above.

This model is usually used to represent the benzene ring in structural diagrams. (Although the molecule actually has a hexagonal shape, it is called a ring for convenience.) One difficulty with this model is that the carbon atoms are still not completely identical. We should expect that replacing two of the hydrogen atoms with bromine atoms would yield two different compounds—one with the bromine atoms at the two ends of a single bond and one with the bromine atoms at the ends of a double bond, as shown below:

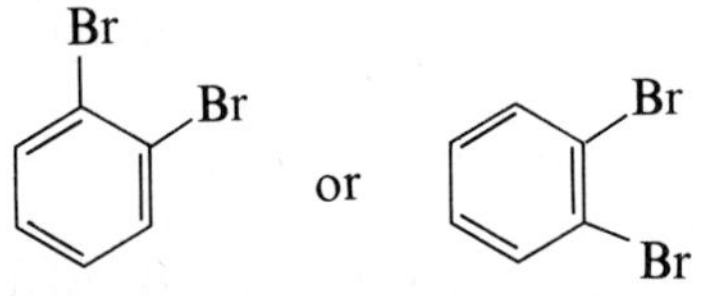

However, only one such compound is known. Modern quantum theory of molecular structure explains this phenomenon by envisioning the six electrons involved in the three double bonds as being shared equally among all six carbon atoms—as though they filled a doughnut-shaped cloud enclosing the entire molecule. (See Figure 17-12.) Accordingly, we often draw the structural formula for benzene in the following way:

This type of structure makes the benzene molecule completely symmetrical and explains why only one dibromo molecule exists. (The prefix *di-* refers to the two bromine atoms added to the basic molecule.)

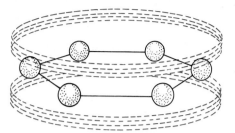

FIGURE 17-12 Quantum structure of the benzene ring. In the benzene ring, the three extra electrons form a cloud surrounding the entire ring, so we do not think of these electrons as belonging to any particular atom in the ring.

Derivatives of benzene have numerous uses. Replace one of the hydrogen atoms with an OH group and you get *phenol* (formerly known as carbolic acid).

First used in 1865 by Joseph Lister to kill bacteria during surgery, it is still widely employed as an antiseptic.

Replace one of the hydrogen atoms with an ethylene group and you get the compound *styrene*:

Polystyrene—a polymer of styrene—is well known as the light, foamy material used to make insulation and coffee cups.

It is also possible for two benzene rings to join together. *Napthalene* is a compound consisting of a double ring in which two of the carbon atoms are shared by the two rings:

Napthalene is the major ingredient in many moth-repellent products and also in the manufacture of dyes such as indigo.

The compounds formed from benzene and containing benzene rings make up the class of aromatic compounds, which has hundreds of thousands of members. Benzene and many of its derivatives are obtained by the destructive distillation of coal, described in Section 17.3. The volatile compounds (coal tar) driven off from the coal are collected and then distilled in such a way that the various compounds (with different boiling points) are separated. In this manner the numerous coal tar derivatives used in industry are obtained.

17.5 Other Common Classes of Compounds

Alcohols

The simplest of the alcohols is *methanol* (or methyl alcohol), a colorless liquid with a boiling point of 65°C. It is poisonous when ingested, so is not to be confused with ethanol (ethyl alcohol), the alcohol used in beverages. The structural formula of methanol is

$$\begin{array}{c} \text{H} \\ | \\ \text{H}-\text{C}-\text{OH} \\ | \\ \text{H} \end{array} \quad \text{or} \quad CH_3OH$$

We could write the formula as CH_4O, but it is important to keep the OH group separate and distinct in order to emphasize that this compound is an alcohol, according to the definition below.

The structural formula shows that methanol consists of an OH group attached to a methyl group. Methanol does not ionize in a water solution, so the OH group must not be thought of as a hydroxide ion. For this reason an alcohol is not to be confused with a base, even though both types of compounds have OH groups in their structures.

In general, an alcohol consists of a hydrocarbon group with an OH group covalently attached. In ethanol the OH group is attached to an ethyl group:

$$\begin{array}{c} \text{H} \quad \text{H} \\ | \quad | \\ \text{H}-\text{C}-\text{C}-\text{OH} \\ | \quad | \\ \text{H} \quad \text{H} \end{array} \quad \text{or} \quad C_2H_5OH$$

We can simplify the above definition by letting the symbol R represent any organic group—methyl, ethyl, propyl, etc. Then the formula for an alcohol can be written in the general form R—OH. (The use of R originated from the use of the term *radical* to mean group.)

In the older system, an alcohol is named by putting the name of the root group in front of the word *alcohol*. This is the origin of methyl alcohol, ethyl alcohol, propyl alcohol, and so on. In the IUPAC system, the suffix *-ol* is appended to the name of the corresponding hydrocarbon after dropping the final *-e*. Then methane yields *methanol*, ethane yields *ethanol*, propane yields *propanol*, etc.

The alcohols are among the best-known organic compounds. Methanol, used as a solvent and as a fuel, is popularly known as wood alcohol, since it originally was made by heating wood in a closed container and collecting the liquids driven off (the destructive distillation of wood). Methanol is very poisonous and causes blindness, so it must not be drunk in the erroneous belief that it is the same as ethanol. Methanol was the first antifreeze used in car radiators, but it fell out of favor because it had to be replaced periodically as a result of its low boiling point.

Ethanol, the active ingredient of wine, beer, and other alcoholic drinks, is the only alcohol that can be used as a beverage with relative safety. Even so, it must be consumed with care, for in large enough quantities it too is a poison. (A large number of organic compounds, if ingested or even inhaled over a long period of time, cause permanent damage to the liver. Ethanol shares this characteristic. In addition, ethanol sometimes has an effect more violent than expected, either because of an individual consumer's low tolerance or because of an interaction with various medications being taken at the same time.)

Isopropyl alcohol (2-propanol) has the following structural formula:

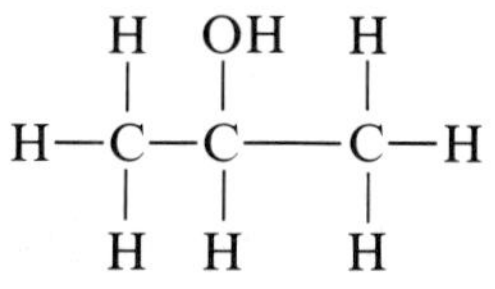

$$\begin{array}{c} \text{H} \quad \text{OH} \quad \text{H} \\ | \quad\;\; | \quad\;\; | \\ \text{H}-\text{C}-\text{C}-\text{C}-\text{H} \\ | \quad\;\; | \quad\;\; | \\ \text{H} \quad \text{H} \quad \text{H} \end{array}$$

This alcohol is the common rubbing alcohol sold in drug stores. Because isopropyl alcohol is not a potable substance (that is, it is not fit to drink), it is used externally in preference to ethanol. The alcohol control laws do not allow pure ethanol to be used in any product that is not labeled as a beverage or a medication to be sold under controlled conditions. Unlicensed ethanol must be **denatured**—that is, mixed with methanol or other poisonous substances to make it unsuitable for drinking.

It is possible to add more than one OH group to an alkane radical. Adding two OH groups gives the class of compounds called *glycols*. *Ethylene glycol*, shown below on the left, is popularly used as an antifreeze agent for car radiators, because when it is mixed with water the resulting liquid is noncorrosive, does not boil off easily, and has a freezing point lower than that of pure water. (See Chapter 13.) For these reasons it is the main ingredient in "permanent" antifreeze mixtures.

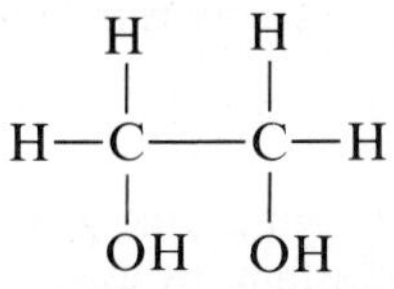

$$\begin{array}{cc} \begin{array}{c} \text{H} \quad \text{H} \\ | \quad\;\; | \\ \text{H}-\text{C}-\text{C}-\text{H} \\ | \quad\;\; | \\ \text{OH} \quad \text{OH} \\ \text{ethylene glycol} \end{array} & \begin{array}{c} \text{H} \quad \text{H} \quad \text{H} \\ | \quad\;\; | \quad\;\; | \\ \text{H}-\text{C}-\text{C}-\text{C}-\text{H} \\ | \quad\;\; | \quad\;\; | \\ \text{OH} \quad \text{OH} \quad \text{OH} \\ \text{glycerol} \end{array} \end{array}$$

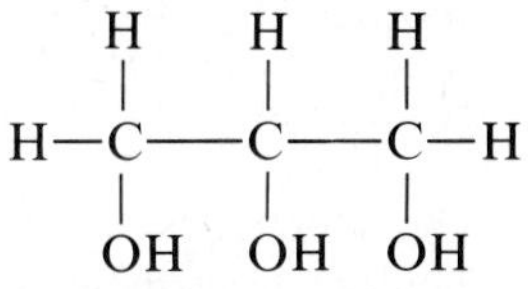

Glycerol, also known as glycerine (shown above on the right), is a propane derivative with one OH group attached to each of the three carbon atoms. Glycerol, originally known as a byproduct in the manufacture of soap, is a very sweet, viscous liquid used in the manufacture of foods, cosmetics, lotions, and the explosive nitroglycerine.

Ethers, Aldehydes, and Ketones

Ethers

An alcohol molecule can be thought of as a water molecule (H—OH) with one of the hydrogen atoms replaced by an organic group R (R—OH). In the same manner, we can think of an **ether** molecule as a water molecule with *both* hydrogen atoms replaced by organic groups (R—O—R′).

The simplest symmetrical ethers are dimethyl ether and diethyl ether:

```
    H   H              H H   H H
    |   |              | |   | |
H—C—O—C—H          H—C—C—O—C—C—H
    |   |              | |   | |
    H   H              H H   H H
  dimethyl ether         diethyl ether
```

It is also possible to have mixed ethers, such as ethyl methyl ether:

```
    H H   H
    | |   |
H—C—C—O—C—H
    | |   |
    H H   H
```

The ethers are very volatile and pungent liquids, and they are widely used in industry for dissolving fats and oils. (A *volatile* liquid is one that has a low boiling point.) The boiling point of diethyl ether—the most commonly used ether—is only 35°C (95°F), so this liquid will boil on a hot day simply from the natural heat of its surroundings. In comparison, the boiling point of ethyl alcohol is 79°C.

Diethyl ether was originally named diethyl oxide, before the adoption of the systematic organic nomenclature. From 1847 until fairly recently, this compound was the most commonly used general anesthetic. It had the disadvantage of being very explosive, however, especially when contaminated with minute quantities of impurities. As a result it could be set off in the operating room by an accidental spark, and extreme precautions had to be taken to prevent the accumulation of static electricity (which you can get by walking across a rug). Because of this hazard, ether has been replaced by less flammable compounds, such as cyclopropane.

Aldehydes

The aldehydes make up a populous group of compounds whose best-known member is *formaldehyde*, familiar to most biology students as the 37% solution frequently used for preserving biological specimens and commonly called *formalin*. Formaldehyde is also the starting ingredient for one of the first plastics, *bakelite*, invented in 1905 by L. H. Baekeland.

An aldehyde has the general structure

$$\text{R}-\overset{\overset{\displaystyle O}{\|}}{\text{C}}-\text{H}$$

In this formula, R represents (as before) any organic group, and the C=O group is the carbonyl group. This representation makes clear how to define an aldehyde: an **aldehyde** is a carbonyl group with an organic group and a hydrogen atom attached to the carbon atom.

Formaldehyde is the simplest of the aldehydes. In formaldehyde the group R is a single hydrogen atom, so the formula for formaldehyde is simply

$$H-\underset{\underset{\|}{O}}{C}-H$$

Another way of looking at the formaldehyde structure is as a methane molecule in which two of the hydrogen atoms have been replaced by a doubly bonded oxygen atom. Accordingly, in the systematic nomenclature this compound is called *methanal*. The name is obtained by first removing the final *-e* and then adding the suffix *-al* to the name of the corresponding hydrocarbon.

The next member of the aldehyde class is *acetaldehyde*, or, in the new system, ethanal. It has the formula CH_3-COH, with the methyl group in place of the generalized group R. As we will see when we discuss acids in the next section, the prefix *acet-* comes from the common compound *acetic acid*.

The aldehydes as a class are highly reactive and are the intermediate compounds from which a great number of other types of compounds can be manufactured.

Acetaldehyde is easily prepared from the raw material acetylene simply by adding water to the acetylene in the presence of sulfuric acid:

$$HC\equiv HC + HOH \xrightarrow{H_2SO_4} CH_3CH=O$$

You may recall that acetylene is made by reacting calcium carbide with water, and calcium carbide comes from heating a mixture of lime and coke (made from coal) in an oven. We thus trace the source of numerous organic compounds back to the coal dug out of the ground. The ease of preparation of acetaldehyde makes it the starting point for the manufacture of numerous plastics, dyes, and disinfectants.

Ketones

We saw that an aldehyde is a compound in which a carbonyl group is flanked on one side by an organic group and on the other side by a hydrogen atom. A **ketone** is a compound in which a carbonyl group has an organic group R on both sides. The general formula for a ketone is

$$R-\underset{\underset{\|}{O}}{C}-R'$$

The two groups R and R' may be identical or different.

The simplest ketone is *dimethyl ketone*, which goes by the common name of *acetone* because it was formerly made from acetic acid. The formula for this compound is

$$H-\underset{\underset{H}{|}}{\overset{\overset{H}{|}}{C}}-\overset{\overset{O}{\|}}{C}-\underset{\underset{H}{|}}{\overset{\overset{H}{|}}{C}}-H$$

Another way of envisioning this molecule is as a propane molecule (a three-carbon backbone) in which the central carbon atom is attached to =O instead of to two H atoms. Using this point of view, the IUPAC system of nomenclature calls this compound *2-propanone*; the *-e* at the end of propane is replaced by the suffix *-one*. The number 2 comes from the fact that the =O is attached to the second carbon atom in the chain. The advantage of using this nomenclature is that one can name a chain of any length without learning a new set of names.

The name acetone, however, is so familiar that it continues to be used even in technical literature. Acetone is made industrially by the action of bacteria on corn. It is a colorless, volatile liquid with a characteristic odor and a boiling point of 56°C. It has numerous uses as a solvent, in the manufacture of explosives, and as a paint and varnish remover. It is well known in the home as the chief ingredient of nail-polish remover.

Acids, Amines, and Carbohydrates

Organic Acids

Many familiar compounds fall into the class of organic acids. Vinegar is about 4% acetic acid, tartaric acid is found in baking powder, butyric acid is found in rancid butter, stearic acid is found in tallow and is used for making soap, palmitic acid is found in palm oil, and so on.

Organic acids differ from inorganic acids in their molecular structures. An inorganic acid consists simply of a hydrogen ion attached to a negative ion. An organic acid is defined as a compound in which the **carboxyl group** ·COOH is attached to an organic group to form the compound R—COOH. The following review of one method of preparing an organic acid will help you to understand the structure of the carboxyl group.

Start with an alcohol, R—OH. This alcohol can be oxidized to form an aldehyde simply by exposing the alcohol to the oxygen in the air over a long period of time. The reaction can be speeded up by the use of an oxidizing agent or a catalyst:

$$R-\underset{\underset{H}{|}}{\overset{\overset{H}{|}}{C}}-OH + \tfrac{1}{2}O_2 \longrightarrow R-\overset{\overset{O}{\|}}{C}-H + H_2O$$

alcohol   aldehyde

17.5 OTHER COMMON CLASSES OF COMPOUNDS

Two hydrogen atoms have been removed from the alcohol molecule, becoming oxidized to water. The oxygen atom remaining in the alcohol molecule forms a double bond with the carbon atom.

Further oxidation replaces the H atom at the end with an OH group, and an acid results:

$$R-\underset{\text{aldehyde}}{\overset{\overset{O}{\|}}{C}}-H + \tfrac{1}{2}O_2 \longrightarrow R-\underset{\text{acid}}{\overset{\overset{O}{\|}}{C}}-OH$$

The carboxyl group (COOH) is a combination of the carbonyl group (CO) and the OH group:

$$-\overset{\overset{O}{\|}}{C}-OH$$

The simplest organic acid is *formic acid*, in which the group R is nothing more than a hydrogen atom:

$$H-\overset{\overset{O}{\|}}{C}-OH$$

Formic acid is a colorless liquid with a pungent odor, irritating to the skin. It is emitted by ants and in 1670 was actually obtained by the distillation of ants. (The name is derived from *formicus*, the Latin word for ant.) Now it is synthesized in large quantities and is used in the manufacture of paper and textiles.

Putting a methyl group in the place of R gives *acetic acid*, perhaps the most familiar of the organic acids:

$$H-\underset{\underset{H}{|}}{\overset{\overset{H}{|}}{C}}-\overset{\overset{O}{\|}}{C}-OH$$

In dilute form acetic acid is the active ingredient of vinegar, giving it its characteristic sour flavor. Concentrated acetic acid, however, is a thick, oily liquid capable of causing serious skin burns. The oxidation of ethanol can result in the formation of acetic acid—one of the chief hazards of wine-making.

The organic acids are fairly weak as acids go, with a pH of about 5. Even so, they are stronger than carbonic acid. The organic acids react with bases to produce salts in the usual way. For example, acetic acid reacting with sodium hydroxide yields sodium acetate (CH_3COONa), in which the sodium atom replaces the hydrogen atom within the carboxyl group:

$$NaOH + CH_3COOH \longrightarrow CH_3COONa + H_2O$$

An alcohol and an organic acid can be combined to form an *ester*:

$$R-OH + HO-\overset{\overset{O}{\|}}{C}-R' \longrightarrow R-O-\overset{\overset{O}{\|}}{C}-R' + HOH$$

This reaction takes place when the mixture of acid and alcohol is heated with concentrated sulfuric acid. The sulfuric acid, having a strong affinity for water, removes a molecule of water from the two organic molecules so that they recombine to form an ester. When methyl alcohol, for example, is combined with acetic acid, the resulting ester is methyl acetate, known in the systematic nomenclature as acetic acid, methyl ester. Notice the similarity between the acid-alcohol and the acid-base reactions in the last two equations.

The esters tend to be oily liquids. Many of them have pleasant odors and flavors. They are responsible for the aromas and flavors of fruits and some flowers. In concentrated form, esters are frequently used as artificial flavoring materials. *Amyl acetate* (acetic acid, pentyl ester)—formed from amyl alcohol (pentanol) and acetic acid—is familiarly known as banana oil and is used by model airplane makers as a lacquer solvent. Esters are frequently added to perfumes and shampoos as aromatic ingredients, and they are added to candies as flavoring ingredients.

Amines

In the amines, an extremely important class of compounds, we encounter for the first time organic compounds containing nitrogen as well as carbon and hydrogen. You are already familiar with ammonia (NH_3) and the ammonium ion (NH_4^+). In organic chemistry the **amine group** is $\cdot NH_2$, and a compound consisting of an organic group R together with an amine group is called an *amine compound* ($R-NH_2$). For example, ethyl amine has the molecular structure

$$H-\underset{\underset{H}{|}}{\overset{\overset{H}{|}}{C}}-\underset{\underset{H}{|}}{\overset{\overset{H}{|}}{C}}-N\overset{H}{\underset{H}{\diagdown}}\diagup$$

Ethyl amine consists of ammonia with one of the hydrogens replaced by an ethyl group. Because of its structural similarity to ammonia, ethyl amine smells somewhat like ammonia and has many of the same properties.

The amines are important industrially, since they are used in the preparation of dyes, photographic developers, detergents, and drugs. In biology they have special significance because they take part in the formation of the amino acids, which are the building blocks from which the proteins are made.

An **amino acid** is a compound containing both an amine group and a carboxyl group. An example of an amino acid is *glycine*, with the structure

17.5 OTHER COMMON CLASSES OF COMPOUNDS

$$\begin{array}{c} \text{H} \quad \text{O} \\ | \quad \| \\ \text{H}-\text{C}-\text{C}-\text{OH} \\ | \\ \text{NH}_2 \end{array}$$

Glycine may be viewed as acetic acid in which one of the hydrogen atoms in the methyl group has been replaced by NH_2.

Although there are approximately 150 naturally occurring amino acids known, one group of 24 amino acids is especially important, for these compounds are the building blocks of living proteins. Proteins are large, complex molecules, containing many thousands of atoms. In recent years the analysis of protein structure has been simplified as a result of the realization that a protein molecule is simply a polymer of amino acids—a long chain composed of many amino acids, linked together like a bunch of sausages.

In the long protein chain, each amino acid is located in a particular position. The grouping of the amino acids forms a code, just as the grouping of letters into words and the position of the words in a sentence makes a meaningful message. If you consider the fact that the 26 letters of the alphabet can form hundreds of thousands of different words, you can see that combining the 24 amino acids in various patterns can result in hundreds of thousands of different proteins, each with a different set of properties and functions. These proteins are the backbone of living matter.

All but one of the 24 amino acids that make up proteins have a feature in common: in these molecules the amine group (NH_2) and the carboxyl group (COOH) are both attached to the same carbon atom. We see this feature in the compound *alanine*:

$$\begin{array}{c} \text{H} \quad \text{O} \\ | \quad \text{\textbackslash\textbackslash} \\ \text{CH}_3-\text{C}-\text{C}-\text{OH} \\ | \\ \text{NH}_2 \end{array}$$

When two amino acids join together in a chain, the link that holds them together is a special kind of bond called a **peptide bond**. The peptide bond is the C—N link within the group.

$$\begin{array}{c} \text{O} \\ \| \\ -\text{C}-\text{NH}- \\ \uparrow \\ \text{peptide bond} \end{array}$$

Consider a reaction in which the two amino acids described above—glycine and alanine—join to form a short chain called a *dipeptide*. The link involves removal of an OH from the ·COOH of the glycine and the H from the ·NH_2 of the alanine:

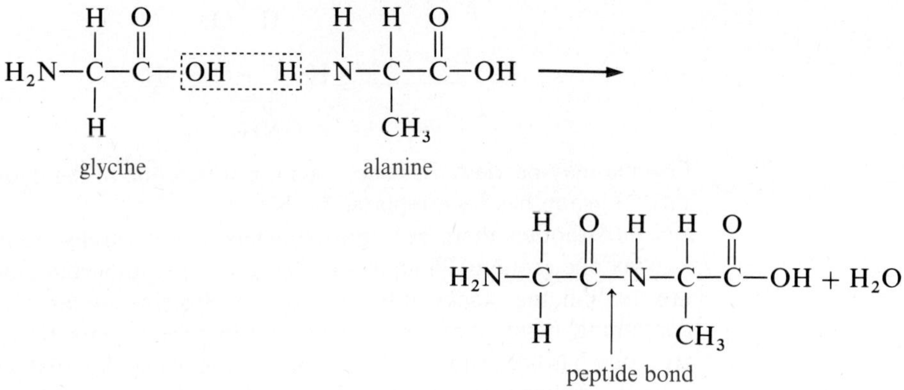

$$\underset{\text{glycine}}{H_2N-\underset{\underset{H}{|}}{\overset{\overset{H}{|}}{C}}-\overset{\overset{O}{\|}}{C}-\boxed{OH}} \quad \underset{\text{alanine}}{\overline{H}-N-\underset{\underset{CH_3}{|}}{\overset{\overset{H}{|}}{C}}-\overset{\overset{O}{\|}}{C}-OH} \longrightarrow$$

$$H_2N-\underset{\underset{H}{|}}{\overset{\overset{H}{|}}{C}}-\overset{\overset{O}{\|}}{C}-\overset{\overset{H}{|}}{N}-\underset{\underset{CH_3}{|}}{\overset{\overset{H}{|}}{C}}-\overset{\overset{O}{\|}}{C}-OH + H_2O$$

peptide bond

Notice that the dipeptide formed above has an ·NH₂ group at one end and a ·COOH group at the other end, just like the original amino acids. Thus it can continue to join with other amino acids or dipeptides to form longer and longer chains, until they reach a size where their complexity enables them to perform the functions of a living organism.

Carbohydrates Another class of compounds indispensable to life is the **carbohydrates**. This class contains many of our textiles, plastics, lacquers, paper, cellulose, explosives, and numerous types of sugars, which are the fuel on which our bodies run.

Originally the name *carbohydrate* meant a compound of carbon, hydrogen, and oxygen in which the hydrogen and oxygen have the same proportions that they do in water—two atoms of hydrogen to each atom of oxygen. However, as molecular structure became better known, this definition was found to be inadequate. Formaldehyde (CH_2O) is not a carbohydrate, nor is acetic acid ($C_2H_4O_2$). The definition of a carbohydrate depends on a particular molecular structure, and this structure is more complex than any of the structures we have considered so far. As we will see, carbohydrates contain aldehydes and ketones, as well as ·OH groups.

Two of the simplest carbohydrates are *glucose* and *fructose*, otherwise known as "grape sugar" and "fruit sugar," respectively. Glucose is particularly important in metabolism, since the carbohydrates in food are changed to glucose through the digestive process. It is the glucose that is actually "burned" to produce the body's energy. Glucose has the following structure:

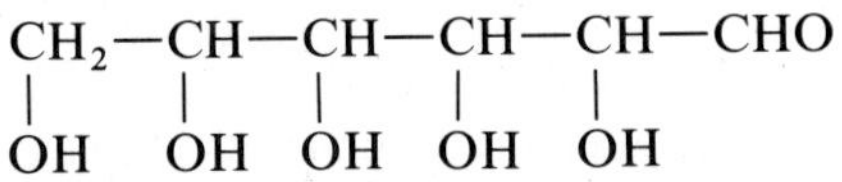

$$\underset{OH}{\overset{|}{CH_2}}-\underset{OH}{\overset{|}{CH}}-\underset{OH}{\overset{|}{CH}}-\underset{OH}{\overset{|}{CH}}-\underset{OH}{\overset{|}{CH}}-CHO$$

It consists of a carbon chain, with an OH group attached to each of five carbon atoms and with the aldehyde group (CHO) on the end. Therefore glucose is both an alcohol and an aldehyde.

Fructose is similar, with one important difference:

$$CH_2-CH-CH-CH-C-CH_2$$
$$||||\||$$
$$OHOHOHOHOOH$$

The five OH groups are the same as in glucose, but instead of the aldehyde group there is a carbonyl group (CO) in the second place from the right, making this molecule a ketone.

Sucrose (cane sugar) is a more complicated sugar, consisting of two simple sugar chains joined together:

```
         H—C────────┐              OH
         |          |              |
         H—C—OH     |          H—C—H
         |          |              |
         HO—C—H     O              C────┐
         |                         |    |
         H—C—OH                H—C—OH   |
         |                         |    O
         H—C────────┘          H—C—OH   |
         |                         |    |
         H—C—H                 H—C──────┘
         |                         |
         OH                    H—C—H
                                   |
                                   OH
```

Glucose and fructose, as single simple sugars, are called *monosaccharides*. Because sucrose is made of two simple sugars, it is a *disaccharide*. A molecule containing many simple sugar chains joined together is called a *polysaccharide*.

Cellulose and starches are polysaccharides with molecular weights in the hundreds of thousands. Cotton, wood pulp, and other plant products are made up largely of cellulose and are the major source of that substance for industrial use. In turn, cellulose is the raw material for the production of artificial fibers, explosives, and numerous plastics such as cellophane.

As you can see from the definitions of these compounds, carbohydrates must not be confused with hydrocarbons. Hydrocarbons are compounds of carbon and hydrogen alone, whereas carbohydrates contain oxygen, usually in the form of OH groups attached to each carbon atom. Their properties are much different, as are their sources. Hydrocarbons generally come from mineral sources (coal, petroleum, and natural gas), and carbohydrates, in their natural state, come from plant tissue. However, modern chemistry makes it possible to synthesize carbohydrates from the hydrocarbons. Thus, the original source of a compound—the way it is obtained—has nothing at all to do with its properties and uses.

Glossary

acyclic compound An organic compound that has a linear or branched structure; a noncyclic compound.

alcohol A class of organic compounds in which the hydroxyl (—OH) group is attached to an organic radical.

aldehyde A class of organic compounds in which an organic radical R and a hydrogen atom are attached to a carbonyl group:

$$R-\underset{\underset{H}{|}}{\overset{\overset{O}{\|}}{C}}-H$$

alkanes A class of acyclic organic compounds containing only carbon and hydrogen atoms in the ratio C_nH_{2n+2}; straight-chain, saturated hydrocarbons.

alkenes A class of acyclic organic compounds containing only carbon and hydrogen atoms in the ratio C_nH_{2n}; straight-chain compounds with a double bond.

alkynes A class of acyclic organic compounds containing only carbon and hydrogen atoms in the ratio C_nH_{2n-2}; straight-chain compounds with a triple bond.

allotropes Two or more crystalline or molecular structural forms of an element.

amine group The NH_2 group.

amino acid An organic molecule containing the amine group (NH_2) and the carboxyl group (COOH).

aromatic compound An organic compound containing one or more benzene rings.

branched chain compound An organic compound that has one or more groups of atoms attached to the backbone of the molecule.

carbohydrates Organic molecules containing carbon, hydrogen, and oxygen, with the hydrogen and oxygen in the ratio 2:1.

carbonyl group A group consisting of carbon and oxygen joined by a double bond (C═O).

carboxyl group The COOH group that indicates an organic acid.

catenation The ability of atoms to join in a chain.

cyclic compound An organic molecule with a carbon chain linked in a ring of three or more carbon atoms.

denatured Deliberately made impure to prevent illegal use. Ethyl alcohol is denatured by the addition of methanol.

derivative A compound created by a modification of another compound.

destructive distillation The process of heating a natural material such as coal or oil in the absence of oxygen so that the volatile products released can be collected.

dissociation The separation of molecules into their individual atoms or ions.

double bond A bond formed by the sharing of two pairs of electrons between two atoms.

ether An organic molecule of the form R—O—R′, where R and R′ are two organic radicals.

hydrocarbon A compound composed of hydrogen and carbon atoms.

isomers Two or more molecular arrangements of the same atoms.

ketone An organic molecule consisting of a carbonyl group between two organic radicals:

$$R-\overset{\overset{O}{\|}}{C}-R'$$

macromolecule A large molecule usually composed of many smaller molecules linked together.

organic compounds Compounds containing carbon.

organometallic compounds Organic molecules that contain one or more metal atoms.

peptide bond A connection between two amino acids.

polymer Large molecules consisting of many smaller molecules strung into a long chain.

saturated hydrocarbons Organic molecules containing only carbon and hydrogen, in which the carbon atoms are connected by single pairs of electrons (single bonds).

saturated molecule An organic molecule structured such that all the carbon-carbon bonds are composed of one pair of electrons.

single bond A covalent bond formed by the sharing of a single pair of electrons by two atoms.

triple bond A bond formed by the sharing of three pairs of electrons between two atoms.

unsaturated hydrocarbons Organic molecules containing carbon and hydrogen with one or more sets of double or triple bonds.

vitalism A theory that claims that living matter is fundamentally different from nonliving matter and that organic molecules can only come from living plants or animals.

Problems and Questions

17.1 Historical Introduction

17.1.Q What was the old definition of organic chemistry?

17.2.Q What was the old definition of inorganic chemistry?

17.3.Q What is the theory of vitalism?

17.4.Q What circumstances caused the change in the definition of organic chemistry?

17.5.Q a. What was the contribution of Lavoisier to the science of organic chemistry?
  b. What was the contribution of Friedrich Wöhler to the science of organic chemistry?

17.6.Q a. What are some of the organic compounds that were known prior to the development of formal chemistry?
  b. Which of these compounds are of animal origin and which are of vegetable origin?

17.7.Q What elements make up most organic compounds?

17.8.Q Name five organic compounds that you commonly buy at the drug store.

17.9.Q What is the modern definition of organic chemistry?

17.10.Q In what important way is the modern definition of organic chemistry different from the original definition?

17.11.Q a. What element in addition to carbon is most commonly found in organic compounds?
  b. What other elements are often found in organic compounds?

17.12.Q Name and identify a number of organic compounds that are not found in living organisms.

17.13.Q Name and identify a number of organic compounds found in living organisms.

17.2 Properties of Carbon

17.14.Q Describe the electron structure of the carbon atom.

17.15.Q Draw a Lewis diagram of the structure of CH_4.

17.16.Q Why is the CH_4 molecule shaped so that the four hydrogens are at the points of a tetrahedron rather than at the four corners of a square?

17.17.Q What do we mean when we say that a molecule is saturated?

17.18.Q What is an important difference between the covalent bond of the carbon atom and the ionic bond of the copper atom? How is this difference revealed in the structure of a copper crystal and that of the molecules of a carbon compound?

17.19.Q What do carbon atoms do when they catenate?

17.20.Q What feature of the carbon bond allows carbon atoms to catenate?

17.21.Q What is the practical result of catenation as far as the structure of organic molecules is concerned?

17.22.P Propose a three-dimensional molecular diagram for these compounds:
  a. CH_3CH_3 b. $CH_3CH_2CH_3$

17.23.Q What is an allotrope?

17.24.Q What are the two allotropes of carbon? How do they differ from each other with regard to the following properties?
  a. appearance
  b. crystalline structure
  c. hardness
  d. electrical conductivity

17.25.Q Give two practical uses for each of the allotropes of carbon.

17.26.Q Briefly describe how diamonds are made from graphite.

17.27.Q What is an important practical characteristic of single graphite crystal fibers? What are they used for?

17.3 Some Simple Carbon Compounds

17.28.Q What are the differences between saturated and unsaturated hydrocarbons?

17.29.Q How do we differentiate between an alkane, an alkene, and an alkyne from a structural point of view?

17.30.Q What do we mean when we say that an organic compound is a derivative?

17.31.Q a. Describe the methyl group.
   b. Does the methyl group exist by itself under normal conditions? Explain your answer.

17.32.P Show how C_5H_{12} can be obtained from C_4H_{10}.

17.33.Q What is the trend in the boiling points of the alkanes as the number of carbon atoms in the molecule is increased?

17.34.Q a. What is the common name of the alkane class of compounds?
   b. What characterizes the alkane class from a structural point of view?

17.35.Q Name some of the common members of the alkane class, and give their uses.

17.36.Q a. Why is acetylene an important compound?
   b. Describe how acetylene is derived from coal.

17.37.P Suggest the structure of the next larger molecule that is created by replacing an end hydrogen in each of the following with a methyl group:
   a. $CH_3CH_2CH_2CH_2CH_3$
   b. $CH_2{=}CHCH_2CH_3$ (There are two possible answers here.)
   c. $CH_3C{\equiv}CCH_3$

17.38.Q Why is it dangerous to sit in a parked car with the motor running and the windows closed?

17.39.Q What is meant by sublimation?

17.40.Q What process maintains a constant balance of CO_2 in the atmosphere?

17.41.Q What is the "greenhouse effect"? Why are scientists concerned about it?

17.42.P Using equations, show the conversion of CH_4 to CCl_4 by reaction with Cl_2.

17.43.Q Discuss the relationship between molecular weight and the ease of evaporation of organic molecules.

17.44.Q Why is CCl_4 no longer used as a cleaning fluid or a fire extinguisher?

17.45.Q What is the carbonyl group?

17.46.Q What is a halogenated hydrocarbon?

17.47.Q Chloroform ($CHCl_3$) reacts with oxygen in the presence of light in the following way:

$$2CHCl_3 + O_2 \xrightarrow{light} 2COCl_2 + 2HCl$$

   a. Why would it be dangerous to store an open bottle of chloroform on a shelf?
   b. What steps should be taken in the storage of chloroform to avoid danger?

17.48.Q a. What is Freon?
   b. What is the most important use of Freon?

17.4 Nomenclature of Organic Compounds

17.49.Q Which system of naming compounds is presently accepted by all chemists as the official system?

17.50.Q What is the chief purpose of a systematic way of naming compounds?

17.51.Q What characterizes the members of the alkane family from a structural point of view?

17.52.Q a. How many hydrogen atoms are attached to the end carbon atom of an alkane molecule?
   b. How many hydrogen atoms are attached to the middle carbon atoms of an alkane molecule?
   c. Write a general formula for any alkane molecule.

17.53.Q Name the first ten members of the alkane family, and write their formulas.

17.54.Q Remove one hydrogen atom from each of the molecules named in Question 17.53, and name the resulting group.

17.55.Q What is an acyclic hydrocarbon, and how does it differ from a cyclic hydrocarbon?

17.56.Q Which of the following molecules are members of the alkane family?
   a. $C_{20}H_{42}$
   b. $C_{16}H_{32}$
   c. $C_{100}H_{98}$
   d. $CH_3CH_2CH_2CH_3$
   e. C_6H_8O

17.57.Q What is the difference between an alkane, an alkene, and an alkyne?

17.58.Q Write the general formulas for the alkanes, the alkenes, and the alkynes.

17.59.Q Name the first nine members of the alkene family, and write their formulas.

17.60.Q Name the first nine members of the alkyne family, and write their formulas.

17.61.Q Write the structural formulas for 1-pentene, 2-pentene, and 3-pentene.

17.62.Q Name the compounds $CH_3C{\equiv}CCH_3$ and $CH{\equiv}CCH_2CH_3$.

17.63.Q What is the difference between $CH_2{=}CHCH_2CH_3$ and $CH_3CH_2CH{=}CH_2$?

17.64.Q What is the difference between $CH_3CH_2CH{=}CH_2$ and $CH_3CH{=}CHCH_3$?

17.65.Q What is polymerization? What practical use does it have?

17.66.Q To which class does each of the following compounds belong—alkane, alkene, or alkyne?
   a. $C_{10}H_{22}$ **b.** $C_{15}H_{28}$
   c. $C_{22}H_{44}$ **d.** C_9H_{16}
   e. $C_{27}H_{56}$

17.67.Q What are isomers?

17.68.Q Write the structural formulas for all the possible isomers of C_5H_{12}.

17.69.Q Give the IUPAC names for the following compounds:

a.
```
    H   H   H
    |   |   |
H — C — C — C — H
    |   |   |
    H   H   H
```

b.
```
    H   H   H
    |   |   |
H — C — C — C — H
    |   |   |
    H   |   H
        |
    H — C — H
        |
        H
```

c.
```
          H
          |
    H  H—C—H  H
    |   |    |
H — C — C — C — H
    |   |    |
    H  H—C—H  H
          |
          H
```

d.
```
              H
              |
          H — C — H
    H  H      |
    |  |      |
H — C— C — C — H
    |  |   |
    H  H   H
```

17.70.Q Name the following compounds using IUPAC rules:

a.
```
    H   H   H
    |   |   |
H — C — C — C — Cl
    |   |   |
    H   H   H
```

b.
```
    H   H   Cl
    |   |   |
H — C — C — C — H
    |   |   |
    H   H   H
```

c.
```
    H   Cl  H
    |   |   |
H — C — C — C — H
    |   |   |
    H   H   H
```

d.
```
    H   H   H
    |   |   |
H — C — C — C — H
    |   |   |
    H   H   Cl
```

e.
```
    Cl  H   H
    |   |   |
H — C — C — C — H
    |   |   |
    Cl  H   H
```

f.
```
    H   Cl  H
    |   |   |
H — C — C — C — H
    |   |   |
    Cl  H   H
```

g.
```
    H   H   H
    |   |   |
H — C — C — C — H
    |   |   |
    Cl  H   Cl
```

h.
```
    H   H   H
    |   |   |
H — C — C — C — H
    |   |   |
    Cl  Cl  Cl
```

17.71.Q What is a cyclic compound? How does it differ from an acyclic compound?

17.72.Q Draw the structural diagrams for cyclobutane, cyclopentane, and cyclohexane.

17.73.Q Describe how cyclopropane is made.

17.74.Q Expand the diagram below to show all the atoms present.

17.75.Q What was Kekulé's discovery? What was its historic significance?

17.76.Q Why was it necessary to postulate a cyclic structure to explain the properties of benzene?

17.77.Q Why do we draw the diagram for benzene in the following form?

17.78.Q What is an aromatic compound?

17.79.Q What is the relationship between plastic coffee cups and the benzene ring?

17.80.Q When was phenol first used as an antiseptic by Lister? Speculate on the conditions in operating rooms before the discovery of bacteria-killing methods. (*Note:* Do not confuse "antiseptic" with "anesthetic.")

17.81.Q What is one method of manufacturing benzene and its derivatives?

17.82.Q What is the difference between benzene and benzine?

17.5 Other Common Classes of Compounds

17.83.Q How do we recognize the class of compounds known as alcohols?

17.84.Q What is a glycol?

17.85.Q Name the following compounds:
   a. CH_3OH b. CH_3CH_2OH
   c. CH_2OHCH_2OH
   d. $CH_3(CH_2)_5CH_2OH$

17.86.Q Write the formulas for the following compounds:
   a. ethanol   b. glycerine
   c. 1-propanol   d. 2-propanol

17.87.Q Give one or more uses for each of the following compounds:
   a. methanol   b. ethanol
   c. isopropanol   d. ethylene glycol
   e. glycerine

17.88.Q What are the most important chemical differences between CH_3OH and NH_4OH?

17.89.Q a. Compare the boiling points of methane and methanol.
   b. Look up the boiling points of a number of alkanes and alcohols, and make a general rule about the effect on the boiling point when an ·OH group is added to an alkane molecule.

17.90.Q What is denatured alcohol?

17.91.Q Give the generalized structural formulas for the following:
   a. ethers
   b. aldehydes
   c. ketones

17.92.Q What do aldehydes and ketones have in common structurally?

17.93.Q Name the following compounds using IUPAC rules:
   a. CH_3-O-CH_3
   b. $CH_3-O-C_2H_5$
   c. CH_3CHO
   d. $HCHO$
   e. C_2H_5CHO
   f. $CH_3-CO-CH_3$
   g. $C_2H_5COC_2H_5$
   h. $CH_3COC_2H_5$

17.94.Q Write the structural formulas for the following compounds:
   a. acetone   b. diethyl ether
   c. ethanal   d. methyl propyl ether

17.95.Q a. When were anesthetics first used in operating rooms?
   b. Name two important contributions of organic chemistry to the practice of surgery since the mid-1800s.

17.96.Q Give at least two uses for each of the following compounds:
   a. acetone
   b. formaldehyde (methanal)
   c. diethyl ether

17.97.Q How do we identify an organic acid?

17.98.Q What is an organic amine?

17.99.Q What is an amino acid?

17.100.Q What is an ester?

17.101.Q How are esters synthesized?

17.102.Q Complete the following reactions:
  a. $CH_3COOH + C_2H_5OH \longrightarrow ? + ?$
  b. $C_2H_5COOH + CH_3CH_2OH \longrightarrow ? + ?$

17.103.Q What is the name of the bond that joins two amino acids together? What is the structure of this bond?

17.104.Q What is there about the combination of two amino acids that enables it to join with more amino acids to form a polymer?

17.105.Q What characterizes the structure of glucose?

17.106.Q How does the structure of fructose differ from that of glucose?

17.107.Q What is the importance of glucose in metabolism?

17.108.Q What is a simple sugar? How does it differ from a polysaccharide?

17.109.Q What are some of the uses of esters? What characteristic of esters makes them useful?

Self Test

1. Define or explain each of the following terms:
   a. organic chemistry
   b. allotrope
   c. saturated hydrocarbon
   d. unsaturated hydrocarbon
   e. alkane
   f. alkene
   g. alkyne
   h. halogenated hydrocarbon
   i. derivative
   j. cyclic compound
   k. polymerization
   l. ether
   m. aldehyde
   n. ketone
   o. alcohol
   p. organic acid
   q. amine
   r. peptide bond
   s. carbohydrate

2. Name the following compounds:
   a. $CH_3CH_2CH_3$
   b. CH_3CHCH_3
      $\quad\;\;|$
      $\;\;\;CH_3$
   c. $CH_3CH=CHCH_3$
   d. $CH_3C\equiv CCH_2CH_3$
   e. $CH_3CH_2CH_2OH$
   f. $CH_3CH_2CH_2Cl$
   g. $CH_3CHClCH_3$
   h. $CH_3CCH_2CH_3$
      $\quad\;\;||$
      $\quad\;\;O$
   i. CH_3CH_2CHO
   j. $CH_3CH_2-O-CH_2CH_3$

3. Write the formula for each of the following compounds:
   a. octane
   b. 3-chloropentane
   c. 1-butyne
   d. 2-butene
   e. diethyl ether
   f. methyl ethyl ketone
   g. acetone
   h. ethanol

4. Name the class (or family) to which each of the following compounds belongs:
   a. $C_{12}H_{26}$
   b. $C_{12}H_{24}$
   c. $C_{12}H_{22}$
   d. $CH_3CH_2CH_2-O-CH_3$
   e. $CH_3CH_2CH-CHO$
      $\qquad\quad\;\;|$
      $\qquad\quad CH_3$
   f. $NH_2CH_2CONHCHCOOH$
      $\qquad\qquad\qquad\qquad\;|$
      $\qquad\qquad\qquad\;\;CH_3$
   g. $CH_3CH_2CHCH_3$
      $\qquad\quad\;|$
      $\qquad\quad OH$
   h. $CH_2-CH-CH-CHO$
      $\;\;|\qquad|\quad\;\;|$
      $\;OH\;\;OH\;\;OH$
   i. $HCOOH$

18
Radioactivity

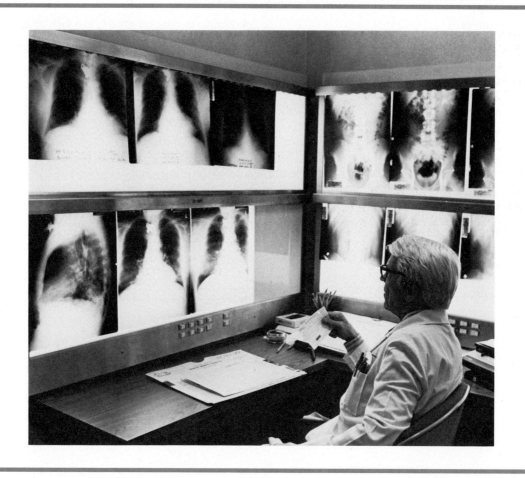

Objectives After completing this chapter, you should be able to:
1. Describe the various types of electromagnetic radiation.
2. Describe the discoveries of Roentgen and Becquerel.
3. Use the nomenclature of radiation and radioactivity correctly.
4. Distinguish among alpha particles, beta particles, and gamma rays.
5. Describe the basic reasons for the emission of radiation from atomic nuclei.
6. Write equations for nuclear reactions, including the atomic mass and the atomic number.
7. Describe the manufacture of artificial isotopes.
8. Describe the properties of alpha, beta, and gamma particles.
9. Describe qualitatively the shielding properties of various materials.
10. Describe several methods of detecting and measuring nuclear radiation.
11. Define qualitatively the units for measurement of radioactivity and radiation dose.
12. Make simple calculations using the concept of half-life.
13. Describe the major effects and uses of nuclear radiation in chemistry.
14. Discuss in a realistic manner the risks of nuclear radiation.

18.1 Radioactive Isotopes

Chapter 5 included an introduction to the discovery of radioactivity in which you saw how the study of radiation emitted by radioactive materials made possible an understanding of atomic structure. During the twentieth century, radioactivity has become a tool of great importance, and radioactive substances now serve useful functions in many branches of science.

Isotopes are different forms of the same element having different atomic masses (because of different numbers of neutrons inside the nucleus). Most elements have some stable isotopes and some unstable isotopes. The stable isotopes exist in nature, whereas the unstable ones must be manufactured in the laboratory (with the exception of a few elements such as radium and uranium, which have no stable isotopes but are found naturally). An unstable isotope approaches stability by emitting energy in the form of radiation, thus changing into another isotope. *Radioactivity* is the spontaneous emission of radiation from an isotope as it changes to another isotope.

It is the radiation that gives radioactive isotopes their importance, but this importance is a two-bladed sword. On the one hand, these substances have properties that make them extremely useful in such fields as chemistry and medicine. On the other hand, they carry with them dangers that have become widely publicized in connection with nuclear energy and nuclear weapons. These dangers are to many people more frightening than other hazards—partly because the hazards of radiation are invisible, and partly because radiation is unfamiliar and mysterious. For this reason all of us need to have factual knowledge of what radioactivity can and cannot do.

We begin with a short historical survey of basic concepts relating to radioactivity and radiation.

Historical Introduction

Before the last decade of the nineteenth century, only one kind of radiation was known to scientists: **electromagnetic radiation**. The light that enables us to see and the radiant heat that warms us are examples of electromagnetic radiation. Until the work of James Clerk Maxwell and Heinrich Hertz between 1870 and 1890, nothing was known about the nature of this radiation. As you saw in Section 5.4, Maxwell and Hertz proved that light consists of electromagnetic waves that travel through space at a speed of 300,000 km per second. Further work showed that the various types of radiation differ only in their wavelengths; **infrared radiation** has the longest wavelength, **ultraviolet radiation** has the shortest, and visible light is in the middle.

In 1895 a different kind of radiation was discovered in the laboratory of Wilhelm Roentgen in Würzburg, Germany. Roentgen had been experimenting with electric currents passing through vacuum tubes (cathode ray tubes). He found that when a high-voltage current was sent through his tube under certain conditions, a piece of cardboard standing near the tube would glow with a bright

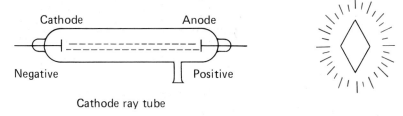

FIGURE 18-1 Roentgen's first x-ray tube was a cathode ray tube in which the electrons (accelerated by high voltage) struck the anode with great energy, producing x-rays. The x-rays caused a card covered with fluorescent paint to glow outside the tube.

green color. (See Figure 18-1.) This piece of cardboard had a thin coating of barium platinocyanide on it, because Roentgen had been studying the way this chemical glowed (fluoresced) when struck by the high-energy electrons in the cathode ray tube. He had not expected that the coating would glow even when the card was *outside* the tube.

Most astonishing was the fact that when a piece of opaque paper was placed between the vacuum tube and the coated screen, the glow persisted. Even a hand placed between the tube and the screen did not stop the fluorescence. However, the bones were not as transparent as the flesh of the hand to this new radiation and would cast shadows on the screen, as shown in Figure 18-2. Most dramatic of all, when the coated screen was replaced by a photographic plate, an image of the transparent hand was captured permanently on the plate.

Roentgen had discovered a kind of radiation able to pass through solid objects and containing enough energy to produce photographic images—something that no other kind of radiation could do. Because of the mysterious and unknown quality of this radiation, the rays coming from Roentgen's tube became known as **x-rays**. Even though we now know a great deal about them, the name *x-ray* persists (although in some places they are called Roentgen rays). Basically, x-rays are electromagnetic waves, just like visible light. However, the wavelengths of x-rays are very short—even shorter than those of ultraviolet light. The short wavelengths give x-rays their penetrating power and make it possible to photograph the interior of the human body.

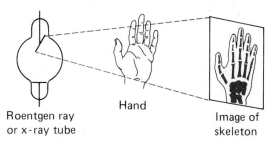

FIGURE 18-2 X-rays passing through a hand cast a shadow of the bones on a photographic plate and produce an image.

Roentgen's discovery triggered an event that opened up the entire topic of radioactivity and nuclear physics. It happened in Paris in 1896 in the laboratory of Antoine Henri Becquerel. Becquerel had been studying the properties of minerals that give off light on their own (phosphorescence) and also give off light when illuminated by ultraviolet radiation (fluorescence). Becquerel, curious to know whether any of these fluorescent minerals gave off x-rays as well as visible light, tried wrapping some photographic film in an opaque envelope and putting the envelope underneath a piece of fluorescent mineral that was exposed to sunlight.

He found that with most minerals nothing happened, but with one particular compound—potassium uranyl sulfate—the result was positive: the photographic film hidden beneath the mineral was found to be dark when it was developed. Becquerel's hypothesis appeared to have been verified.

Such a conclusion was premature, however. Becquerel soon found that the same darkening of the film took place even after the uranium ore and the film had been kept for a long time in a dark drawer. The sunlight had nothing to do with the experiment at all! Instead, the mineral was giving off radiation all by itself. Furthermore, when a key was put between the rock and the envelope, a shadow of the key was seen on the developed film. (See Figure 5-5.) The uranium mineral appeared to be continually and spontaneously giving off x-rays.

A great mystery was thus uncovered. The piece of uranium ore was giving off energy continuously, which seemed to violate the law of conservation of energy. The energy had to come from somewhere. Where did it come from?

It did not take long to discover that the element uranium was the one responsible for the emission of the radiation, and that any compound of uranium would behave in the same manner. However, this piece of information did not explain how the atoms of uranium could continue to emit radiant energy.

Becquerel's experiment was only the beginning of a long series of investigations carried on by many scientists. Among the most famous of these scientists was the Polish woman Marie Sklodovska Curie, who obtained her education and did her work in Paris. She was the first woman to obtain a doctorate in science and the first person to obtain two Nobel prizes in science. In 1903 Marie Curie, her husband Pierre Curie, and Becquerel were awarded the Nobel prize for their discoveries concerning the nature of the radiation being emitted from uranium. Then in 1911 Marie Curie received the Nobel Prize in chemistry for her discovery and isolation of two new radioactive elements: polonium and radium.

The discovery of these new elements proved that uranium was not the only element to emit radiation—there were entire families of elements sharing this property. In the decades following Marie Curie's discovery, our knowledge of radiation and radioactive isotopes increased greatly, as a large number of investigators leaped into this lively area of research.

The Basic Nomenclature of Radioactivity

Most naturally occurring elements do not emit radiation. However, during this century a large number of isotopes have been discovered that do emit radiation of one form or another. One of the main problems confronting atomic scientists early in the century was determining what causes some elements to emit radiation spontaneously.

Another important problem was determining the nature of the emitted radiation. In general, **radiation** is something that is emitted from a source (a light bulb, a radio antenna, a piece of uranium, etc.), travels through space with great speed, is generally invisible, and carries energy from the source to the receiver. However, since there are many different kinds of radiation, a complete definition of radiation requires a description of each type.

The first important feature of the radiation coming from elements such as uranium is the fact that it comes directly from the *nucleus* of each atom. It has nothing to do with the electrons orbiting the nucleus. For this reason, such radiation is called **nuclear radiation**, and the study of radioactivity was originally a part of nuclear physics. However, radioactivity is extensively used as a tool by chemists, and so the topic of *nuclear chemistry* has become an important branch of chemistry.

It is important to distinguish among the words *radioactive*, *radioactivity*, and *radiation*. A **radioactive material**, or element, is one that emits radiation. *Radioactivity* is the *process* of emitting *radiation*. In some newspaper accounts of nuclear accidents it is difficult to tell whether the writer is discussing the *release of radioactive material* or the *emission of radiation*, because the phrase "release of radiation" is often erroneously used in both senses. However, there is an enormous difference between uranium or plutonium leaking through a hole in a container and radiation passing through the solid, unbroken walls of a container. Those two events will have very different results.

We saw in Chapter 5 that there are three separate kinds of radiation emitted by uranium and similar elements. These three types of radiation are separated from one another when they pass between the poles of a strong magnet (see Figure 5-6); a photographic plate on the other side of the magnet will be darkened in three spots. This splitting of the radiation beam into three components occurs because positive and negative electric charges are deflected in opposite directions when they pass through a magnetic field, whereas neutral particles are not affected at all.

One of the three beams consists of particles with a positive electric charge—the alpha particles. Another is made of particles with a negative electric charge—the beta particles. The third beam has no electric charge at all and goes straight through the magnetic field without deflection; this uncharged beam consists of gamma rays. (These three rays were named by E. Rutherford in 1899 after the first three letters of the Greek alphabet.)

An **alpha particle** is a helium ion (He^{2+}) consisting of two protons and two neutrons bound so tightly together that they behave like a single particle. Thus, the alpha particle has an atomic mass of four and an atomic number of two, with two positive electric charges (see Figure 18-3).

The **beta particle** is nothing more than a single electron; beta rays are streams of electrons emitted from the nuclei of radioactive atoms. (Actually, each nucleus emits only one electron at a time, but we will discuss this later.)

FIGURE 18-3
An alpha particle consists of two neutrons and two protons. It is the same as the nucleus of a helium atom.

The **gamma ray** is identical to the x-ray discovered by Roentgen: it is an electromagnetic wave of very short wavelength. Even though gamma rays and x-rays are identical in nature, we call them gamma rays when they are emitted from a radioactive nucleus and x-rays when they are created artificially in a machine of some kind.

One of the interesting features of gamma rays (and indeed of all electromagnetic radiation) is that in certain circumstances they behave like waves and in other circumstances they behave like particles. These particles of radiation are the photons discussed in Section 5.4; their masses can be measured, and they can be counted just like any other kind of particle. Each gamma-ray photon possesses a certain amount of energy: the shorter the wavelength, the greater the energy of the photon. The notions of wave and particle become two aspects of the same thing in the concept of a photon as a wave packet, or bundle of waves. In this context, matter and energy are no longer two separate concepts.

The energy of alpha, beta, and gamma particles is usually given in units of electron-volts. We have used electron-volts to measure the energy levels of orbital electrons. There the energies involved are on the order of a few electron-volts. Alpha, beta, and gamma particles, on the other hand, generally have energies ranging from a few thousand to several million electron-volts. By artificial means we can accelerate these particles to energies of many hundreds of billions of electron volts.

Quantities of energy in radiation studies are often expressed in millions of electron-volts, or McV. Thus, we speak of radium emitting alpha particles with an energy of 4 MeV, together with gamma-ray photons with an energy of 0.19 MeV. Each of the radioactive isotopes in nature emits a unique set of particles, each with its own energy. We call this set of particles the **radiative spectrum** of the isotope; this unique spectrum enables us to identify small quantities of radioactive isotopes.

The Causes of Radioactivity

Beta Emission

The emission of alpha, beta, or gamma particles from an atomic nucleus is caused by events that take place within that nucleus. An atomic nucleus is like a tiny droplet composed of densely packed neutrons and protons in constant motion. The protons repel each other because they are all positively charged. Were it not for the neutrons, the nucleus would fly apart instantly.

The neutrons and their neighbouring protons attract each other by means of the **strong nuclear force**—a short-range force felt only when the neutrons and protons are practically touching each other. It is this strong nuclear force (sometimes likened to a "nuclear glue") that holds the atomic nucleus together.

Among the lighter elements, the number of neutrons in the nucleus is approximately equal to the number of protons. This equality is necessary for the nucleus to remain stable, since the neutron is not a stable particle. When left to itself, a neutron will separate into a proton and an electron (plus another particle with an extremely small mass and no electric charge, called an antineutrino). This process is known as the **beta decay** of the neutron:

$$\text{neutron} \longrightarrow \text{proton} + \text{electron} + \text{antineutrino}$$

$$n^0 \longrightarrow p^{+1} + e^{-1} + \text{antineutrino}$$

No matter how many neutrons you have at one instant in time, half of them will decay within about 12 minutes. The amount of time required for half the neutrons to decay is called the **half-life** of the neutron.

A neutron becomes stabilized when it is inside a nucleus with a proton. When the number of neutrons inside the nucleus is approximately equal to the number of protons, the nucleus can exist for a long time without change. However, if there are too many neutrons, the nucleus as a whole will be unstable. In order for the nucleus to stabilize itself, one of the neutrons must decay into a proton, and an electron and an antineutrino must be spit out. This is the cause of beta decay—the emission of a beta particle from a nucleus. (Note that it is the neutrons that decay, not the beta particles.) The antineutrinos emitted during the reaction have a negligible interaction with matter, so they escape into space without causing any other reactions.

For example, the isotope tritium (^{3}H—hydrogen with an atomic weight of 3) has one proton and two neutrons in the nucleus. This isotope is unstable. Sooner or later one of the neutrons must change into a proton, with an antineutrino and an electron (beta particle) emitted from the nucleus. The result will be a nucleus with two protons and one neutron, which is stable. The hydrogen isotope has changed into an isotope of helium (^{3}He) with an atomic weight of 3 instead of the more common weight of 4. The reaction is

$$^3\text{H} \longrightarrow {}^3\text{He} + e^- + \text{antineutrino} + 0.0186 \text{ MeV energy}$$

The energy on the right side of the equation represents the kinetic energy of the electron and the antineutrino emitted from the nucleus. (That is, the electron takes some of the energy, and the antineutrino takes the rest of the energy.) In addition, the atom must pick up another orbital electron from its surroundings, for with two nuclear protons it now needs two orbital electrons instead of one.

An interesting feature of such reactions is that it is completely impossible to predict exactly when a specific atom will react. You have absolutely no way of knowing when one particular tritium atom is going to decay into a helium atom. Some atoms decay immediately; others wait for a long time. You only know that (1) on the average a certain number will decay during each second of time and (2) at the end of 12 years half of the tritium atoms will be gone. That is, the half-life of tritium is 12 years.

This behavior is characteristic of radioactivity. Each beta particle is emitted by a different tritium atom. If you start with billions of tritium atoms, the result is a constant stream of beta particles coming out of the tritium sample—a stream that gets weaker and weaker as the number of tritium atoms decreases.

There are hundreds of radioactive isotopes known, each one with a different half-life. Indeed, the standard way of identifying an isotope is by measuring its half-life and the energy of the particles being emitted. The combination of the two gives a unique identification of the isotope.

Gamma Emission

Gamma-ray photons are often emitted together with beta particles. When a nucleus emits a beta particle, the nucleus remaining (the **daughter nucleus**) may be left in an excited state—that is, with more energy than normal. The law of conservation of energy tells us that this extra energy cannot simply disappear; it must be accounted for in some way. One way the nucleus can get rid of this energy is by giving off a gamma-ray photon. This is what happens in the majority of cases of beta decay. The process is similar to the way photons of visible light are emitted from atoms when their orbital electrons cascade down from upper energy levels to lower levels except that the energies involved are much greater.

To illustrate the gamma-ray photon emission energy, let us consider the isotope cesium-137 (^{137}Cs) which is a common beta and gamma source. It decays through the reaction

$$^{137}\text{Cs} \longrightarrow {}^{137}\text{Ba} + e^- + \text{antineutrino} + 1.18 \text{ MeV energy}$$

The 1.18 MeV is the *total* amount of energy that the cesium nucleus must get rid of. However, when we measure the energy of the electron (plus antineutrino), we find they together have 0.52 MeV of energy. Therefore we know that the barium atom has an excess of energy. From the law of conservation of energy we write

$$\text{total energy} = \text{beta and antineutrino energy} + \text{excess energy}$$

$$1.18 \text{ MeV} = 0.52 \text{ MeV} + \text{excess energy}$$

Therefore,

$$\text{excess energy} = 1.18 \text{ MeV} - 0.52 \text{ MeV}$$

$$= 0.66 \text{ MeV}$$

The nucleus has 0.66 MeV of energy to lose. It does this by emitting a gamma-ray photon with an energy of 0.66 MeV. Having done that, the barium atom settles down to its stable state.

Alpha Emission

The emission of alpha particles involves a situation quite different from that of beta and gamma emission. First of all, the elements that emit alpha particles are those with high atomic numbers—that is, those with many protons inside the nucleus. When there are many protons, the electrostatic repulsion between them is very strong. The presence of an equal number of neutrons is not enough to hold the nucleus together. There must be more neutrons than protons for the strong nuclear attraction to overcome the electrostatic repulsion between the protons. When there are enough neutrons, the balance between the attractions and repulsions is restored.

However, when there are more than about 84 protons inside the nucleus, the nucleus becomes unstable—the attractive force can never quite balance the repulsive force. In that situation the nucleus tends to break into smaller pieces.

There are a few cases in which the nucleus actually splits into two parts of roughly equal mass. This process, known as **nuclear fission**, is used to generate energy in nuclear power plants. However, in the majority of situations the nucleus attempts to cure its instability by emitting an alpha particle. When it does this, it loses two protons and two neutrons. Therefore its atomic number is reduced by 2 units and its atomic mass is reduced by 4 units.

Let us take as an example the first radioactive element to be studied: uranium-238 (^{238}U). Uranium has an atomic number of 92. When it emits an alpha particle, it goes through the following changes:

	Before (parent)	Change	After (daughter)
atomic number	92	-2	90
atomic mass	238	-4	234

The nucleus remaining behind (the daughter nucleus) has an atomic number of 90, and the element with an atomic number of 90 is thorium. Therefore the result of the decay of ^{238}U must be the isotope ^{234}Th.

$$^{238}_{92}U \longrightarrow {}^{234}_{90}Th + {}^{4}_{2}He$$

(Note that the number in the subscript is the *atomic number* of the isotope, and the number in the superscript is the *mass number*. We know that the second element on the right is helium, because helium is the element with an atomic number of 2.)

You may be wondering why, if uranium is constantly decaying, there is any left in the earth. After all, the earth is about 6 billion years old, and we assume that all the elements now in existence were also in existence at the time of the creation of the earth.

There is still uranium in the earth because ^{238}U has a very long half-life—about 4.5 billion years. This means that it took 4.5 billion years for half of the original supply of that isotope to disappear. Therefore, even though ^{238}U constantly decays, its rate of decay is so slow that there is still a large quantity of it left over from the original creation of the elements.

However, the decay of ^{238}U is not the end of the story. The ^{234}Th left over from the decay of the ^{238}U nucleus is by no means stable. The only reason we find this isotope in nature is because atoms of it are continually being created by the decay of ^{238}U, and we are able to catch them along the way before they disappear. Thorium-234 decays by beta emission, with a half-life of 24.5 days. The daughter nucleus resulting from the decay of ^{234}Th also undergoes beta decay, and it disappears even more rapidly than its parent. It has a half-life of only 1.14 minute.

$$^{234}_{90}Th \longrightarrow {}^{234}_{91}Pa + e^- \longrightarrow {}^{234}_{92}U + e^-$$

(*Note:* The atomic number rises from 90 to 91 and then on to 92 because a neutron in each nucleus converts to a proton. Therefore, during this reaction the number of positive charges in the nucleus increases.)

The above examples represent just the beginning of a long and complex chain of radioactive decays that begin with the ^{238}U atom. The decay series does not come to an end until it reaches the isotope lead-206, which is stable. In the course of this series of transformations, a total of 16 radioisotopes are formed, some with lifetimes of thousands of years and some with lifetimes of just a few minutes or seconds. (The term **radioisotope** is commonly used as shorthand for radioactive isotope.)

Most of the radioisotopes found in nature exist because they are created as part of this series or because they are part of another series beginning with the long-lived element thorium. For example, the well-known isotope radium-226, with a half-life of 1600 years, is found as part of the earth's crust because it is continually created as a result of the decay of uranium. Members of the uranium family of radioisotopes are all found in uranium ore, together with the final lead isotope.

Another radioisotope found in nature is potassium-40 (^{40}K), which has a half-life of 1.3 billion years. As a result of its long lifetime, some of its original atoms are still in existence. This potassium isotope is always found in small quantities (0.001% abundance) together with the stable isotopes ^{39}K and ^{41}K. Found within your body as part of its normal supply of potassium, it contributes to the natural radiation background that is within and about each of us at all times.

Artificial Radioisotopes

Many of the isotopes used in industry and in medicine are not found free in nature but are prepared by artificial means. These isotopes are produced by transmutation reactions—nuclear reactions that transform an isotope found in nature into another isotope.

For this purpose a number of reactions can be used, descriptions of which can be found in books on nuclear physics. However, here we will briefly describe the neutron-capture reaction, the reaction most commonly used in the production of radioactive isotopes.

Inside a nuclear reactor, free neutrons are copiously produced as a byproduct of uranium-235 fission, the source of the reactor energy. These neutrons are like the particles in a gas: they fly around in all directions and collide with anything that gets in their way.

When a neutron collides with the nucleus of an atom, the neutron may bounce off like a billiard ball or it may be absorbed—or *captured*—by the nucleus. Let us see what happens when a piece of cobalt is placed inside a nuclear reactor. Cobalt has only one stable isotope: ^{59}Co. A neutron encountering the nucleus of a ^{59}Co atom has a good chance of being captured by the nucleus. When that happens, the atom becomes an isotope of cobalt with an atomic mass one unit greater than that of the original isotope. The atomic number does not change.

$$^{59}_{27}\text{Co} + ^{1}_{0}\text{n} \longrightarrow ^{60}_{27}\text{Co}$$

Therefore, when a ^{59}Co nucleus absorbs a neutron, it becomes a ^{60}Co nucleus. Cobalt-60 is radioactive, with a half-life of about 5 years, and it emits both beta and gamma particles with a number of energies. The half-life of 5 years makes ^{60}Co a useful isotope. To manufacture it, you simply put a piece of cobalt into a nuclear reactor for 5 or 10 years and then remove it. By that time, it is intensely radioactive. The half-life of 5 years is long enough so that the cobalt can be used profitably for 10 years or so before its strength diminishes to such an extent that it is no longer valuable. Cobalt is commonly used for research requiring an intense source of gamma rays and for many types of industrial processes. (We will discuss these uses in more detail in a later section.)

Another isotope commonly used in chemical and biomedical research is tritium (^{3}H). Tritium is manufactured in a nuclear reactor from one of the isotopes of lithium according to the reaction

$$^6_3\text{Li} + ^1_0\text{n} \longrightarrow ^3_1\text{H} + ^4_2\text{He}$$

This is a fission reaction; the lithium nucleus absorbs a neutron, becomes unstable, and splits into two parts—one a nucleus of tritium and the other a nucleus of helium (an alpha particle). Tritium has a half-life of about 12 years and emits nothing but beta particles of fairly low energy (0.019 MeV). For this reason, as we will see, it is relatively safe to handle and may even be ingested by humans for medical purposes.

In this introduction to the nature of radioactivity, we have mentioned but a small fraction of the radioisotopes with known properties. A complete listing of all the known isotopes and the energies of the emitted radiation can be found in the *Handbook of Chemistry and Physics*. We now turn to a description of the radiation itself.

18.2 The Properties of Nuclear Radiation

Alpha Radiation

Each type of radiation has individual properties that determine how useful it is, how dangerous it is, what types of instruments to use to measure it, and so on.

All three types of radiation mentioned—alpha, beta, and gamma—have two important properties in common. First, they can all penetrate various thicknesses of matter. The penetrating power of each kind of radiation depends on a number of factors (which will be discussed shortly). Second, when one of any of these particles passes through a solid, a liquid, or a gas, it is able to knock one or more electrons away from the outer shells of the atoms encountered along the way, thus leaving a trail of positively ionized atoms and free electrons. It is this ionization that produces most of the chemical and biological effects of nuclear radiation and makes it possible to detect the radiation.

Alpha particles, being relatively massive as well as doubly charged, are considered to be strongly ionizing particles. Each alpha particle leaves a dense trail of ions in its wake. Since it takes a certain amount of energy to ionize an atom, the formation of many ions causes each alpha particle to lose all its kinetic energy in a short distance. Therefore alpha particles are "short range" particles; they travel only short distances through matter. It is as though each particle encountered a lot of friction along its path and was quickly brought to rest.

Even in air the alpha particles of uranium travel only a few centimeters. Many alpha particles are stopped by a sheet of paper or by a thin piece of aluminum foil. The **range of a particle** depends on the kind of material it is moving through as well as on the energy of the particle. In general, the higher the energy, the farther the particle will travel. Dense elements with a high atomic number (such as lead) stop alpha particles more readily than do light elements with a low atomic number (such as aluminum).

An alpha particle with an energy of 7 MeV, typical of those emitted from uranium or thorium, will pass through about 4.3 cm of air (at normal atmospheric pressure), but will be stopped by 0.026 mm of aluminum foil. For this reason it is very easy to contain or to shield emitters of alpha radiation. An ordinary metal box would be a safe container for an element that emitted nothing but alpha particles.

There are, however, two complications. First, most alpha emitters also emit gamma radiation. Gamma radiation is much more penetrating than alpha, and so requires more shielding. Second, none of the radioactive material must be allowed to get out into the air, so the container must be tightly sealed. The most dangerous aspect of alpha emitters is the fact that if the radioisotope itself is inhaled into the lungs, the alpha particles can penetrate directly into living tissue, with no shielding in the way. Even though they do not travel far, the tiny alpha particles can do a great deal of damage. For this reason it is very important to distinguish between (1) the effects of nuclear radiation and (2) the effects of ingesting radioactive materials directly into the body. Recent evidence suggests that one of the major causes of lung cancer is tiny particles of alpha emitters that are inhaled directly into the lungs with cigarette smoke.

Beta Particles

Beta particles, like alphas, possess the ability to ionize the atoms that lie along their paths as they travel through solids, liquids, or gases. Since beta particles are nothing but electrons, they are very light and create fewer ions per centimeter of path. Therefore they travel farther through matter than do alpha particles. Beta particles with an energy of 5 MeV will be stopped by 1 cm of aluminum; thus their range is several hundred times that of alpha particles with the same energy.

In addition, high-energy beta particles create x-rays as they pass through the aluminum, and these x-rays pass more easily through the material than do the original betas. Therefore it is necessary to use lead shielding to stop the x-rays. However, 5 MeV is an unusually high energy for beta radiation. Most isotopes give off beta rays of lower energy and so require less elaborate shielding.

Tritium, for example, emits beta particles with an energy of 0.019 MeV, and these are easily stopped by a piece of aluminum foil 0.004 mm thick. Beta particles from tritium outside the body will not even pass through the skin. For this reason tritium is much less harmful than most other radioisotopes and can be taken into the body for medical and research purposes.

Gamma Radiation

Like alpha and beta particles, gamma-ray photons ionize the materials they pass through. In general, gamma-ray photons ionize fewer atoms per centimeter of path than do the other two types of radiation. This means that a beam of gamma rays loses energy more slowly than does a beam of alpha and beta particles, and therefore it travels a greater distance through matter before all of its energy is gone.

There is another difference. A single alpha or beta particle produces numerous ionizations along its path, gradually losing energy until it comes to rest, but never actually disappearing. A gamma-ray photon, on the other hand, generally goes along without change until it hits an electron and knocks it out of an atom. The photon then suddenly disappears, having given all its energy to the electron. (The process is called the *photoelectric effect*.) As the photons in a beam vanish one by one, the beam gradually becomes weaker and weaker.

Since gamma-ray photons travel longer distances through matter, they are proportionately harder to shield. It takes about 1 cm of lead to remove half the photons from a beam of 1.1-MeV gamma rays (obtained from a cobalt-60 source). Another centimeter of lead would be needed to cut the gamma rays in half again. As a result, it requires 6.6 cm (2.6 in) of lead to reduce the strength of this gamma-ray beam to one one-hundredth of its original strength.

Lead is generally used for shielding gamma rays because it is a very dense element and has a high atomic number. Tungsten is an even better shield, but it is more expensive. (The reason we stress shielding in our discussion of properties of radiation is because one of the first things a chemist needs when handling

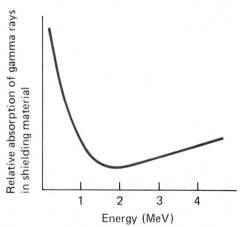

FIGURE 18-4 A graph showing the absorption of gamma rays in a material such as lead. Low-energy gamma rays are absorbed very strongly. There is minimal absorption of gamma rays with energies between 1 MeV and 2 MeV, and then the absorption increases as the energy of the gamma rays increases.

radioisotopes is a safe place to keep them, and this is where shielding enters the picture.)

The absorption of gamma rays in matter is a complex subject; at least three different processes are involved. Very low-energy gamma rays are absorbed most strongly and therefore have a shorter range in matter than do higher-energy gamma rays. As the energy of the gamma rays is increased, the amount of absorption in a given thickness of material becomes smaller and smaller. However, at an energy of about 1 MeV, a new absorption process starts coming into play, and the rate of absorption starts to increase again. (See Figure 18-4.) For this reason, 1-MeV gamma rays are the most penetrating.

Because gamma rays travel such long distances through matter, they are the most dangerous of the three types of radiation. A beam of gamma rays can pass through the entire human body (like a beam of x-rays) and produce ionizations throughout the body cells. The photons do not stop at the surface of the body, as do beta rays.

In Section 18.5 we will discuss some of the biochemical changes that result from the exposure of tissue to radiation. We will see that even though radiation has definite hazards, when carefully used it can have beneficial effects.

18.3 The Detection and Measurement of Radiation

Instrumentation

In order to deal intelligently with nuclear radiation, we must be able to detect its presence and measure its quantity. Generally speaking, the chemical and biological effects of radiation depend on the quantity of radiation entering the material of interest. For example, the effects of radiation on health are proportional to the amount of radiation absorbed by the body, so to do proper studies of these effects, it is necessary to measure quantities of radiation accurately.

Numerous types of instruments are available for these purposes. We list some of them here.

1. *Photographic film.* All ionizing radiation has the ability to darken the grains of silver in photographic film. After the film is developed, tracks of black-silver grains left by alpha and beta particles allow individual particles to be counted through a high-powered microscope. Gamma rays darken a film more uniformly, and the amount of darkening can be used to measure the quantity of gamma radiation that has passed through the film. This process is used in the film badges worn by workers in laboratories and industries where radiation is present. The badges are developed daily or weekly to see if any significant dose of radiation has been absorbed by the wearer.

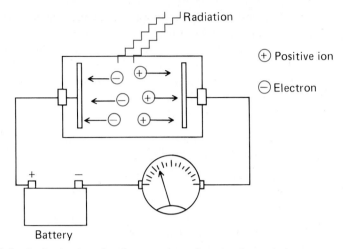

FIGURE 18-5 An ionization chamber consists of a pair of metal plates separated by the air in the chamber. A battery charges the plates, and when radiation goes through the chamber a small current passes between the plates. The amount of current shown on the meter is proportional to the amount of radiation passing through the chamber.

2. *Ionization chambers.* An ionization chamber has a small volume filled with air or other gas. On two sides of this chamber are a pair of metal plates connected to a battery or other source of electric charge as shown in Figure 18-5. If ionizing radiation passes through this chamber, it causes the air to conduct a small electric current from one plate to the other. This current can be measured to find the strength of the radiation. In a type of ionization chamber called a **dosimeter**, the total amount of electric charge passing between the plates in a given length of time is measured. This amount of electric charge is proportional to the total amount of radiation that has passed through the ionization chamber.

Dosimeters small enough that they can be clipped in a pocket like a pen are used to keep track of the radiation absorbed by a person on a day-to-day basis. They do the same job as film badges, and they can be read on the spot with a suitable instrument, whereas film badges have to be sent to the laboratory for developing.

3. *The Geiger counter.* Invented by Hans Geiger, the Geiger counter is similar to the ionization chamber but operates at a higher voltage. One electrode is shaped like a cylinder, and the other is a straight wire going down the center of the cylinder. When an alpha, beta, or gamma particle passes through the counter, it triggers a sudden discharge of current, like a spark. This spark is detected by an electronic circuit, and the number of such pulses is counted by another circuit. Thus, a Geiger counter gives an individual count of each particle that passes through it and so can be used to measure very small amounts of radiation.

4. *The scintillation counter.* The scintillation counter is based on the principle that a single crystal of sodium iodide gives off a short pulse, or **scintillation**, of light

(of a duration of less than a microsecond) when a gamma-ray photon is absorbed in the crystal. A sensitive light detector picks up this tiny burst of light and records the signal in an electronic counter. Single photons can be detected and their energies measured with a scintillation counter because the amount of light produced is proportional to the photon energy. (Photon energies cannot be measured with a Geiger counter.)

5. *Solid-state detectors.* Solid-state detectors are based on the principle that certain semiconductor crystals (such as specially treated germanium) produce short, very fast pulses of electricity when they absorb gamma-ray photons. These detectors can be very small, and they make very precise measurements of photon energies.

Each of the different types of radiation detectors has its own advantages and disadvantages, and each is best suited to certain types of jobs. The study of instrumentation is a specialty of its own.

In general we can divide radiation-measuring devices into three categories: (1) small monitoring devices, such as film badges and dosimeters, which a person can wear to measure the amount of radiation received; (2) portable, hand-held radiation monitors, which can be used for rough measurement of radiation in a laboratory, plant, or outdoor environment; and (3) complex and precise laboratory-grade equipment used for high-quality measurements.

In the next section we will consider the units used to measure radiation and radioactivity.

Units of Radiation

In measuring radiation, two kinds of quantities are of importance: the amount of radiation emitted by a radioactive sample and the amount of energy absorbed by an object receiving the radiation. In this sense, measuring radiation is like measuring the amount of light emitted by a light bulb—for some purposes it is important to know the brightness of the bulb itself, for other purposes it is more important to know the brightness of the light falling on a surface some distance away from the bulb.

The subject of radiation measurement is so new and is changing so rapidly that at present there is a confusing variety of units. In this chapter we shall mention only the units used most often by workers in nuclear radiation. The officially designated SI units are as yet little used but will be defined here for the sake of completeness.

Quantities of Radioactivity

The most common unit of radioactivity is the curie, named after Marie Curie, the discoverer of radium. Originally the curie was defined as the amount of radioactivity in 1 g of the pure element radium. Since a gram of radium represents a large quantity of radioactivity, radium is usually measured in milligrams (thousandths of a gram); a *millicurie* (*mCi*) is the amount of radioactivity in a milligram of radium. Even a millicurie of radium is too much to carry around in the bare hand with safety. However, a *microcurie* (μCi), a millionth of a curie, of radium in a sealed tube can be safely handled without extensive precautions.

After numerous radioisotopes were discovered, it was found convenient to redefine the curie so that instead of being based on the activity of radium it could be used to describe the strength of any radioactive material. In the new definition, the curie represents the amount of material in which a certain number of disintegrations takes place per second. A *disintegration* is the emission of an alpha or beta particle—that is, the transformation of one atom into another. (The gamma-ray emission is not counted separately from the beta, since both beta and gamma are emitted during a single transformation.)

Accordingly, the **curie (Ci)** is now defined as that quantity of any radioactive isotope undergoing 3.70×10^{10} disintegrations per second.

Notice that a curie is a quantity of *material*. It is not to be confused with a quantity of *radiation*.

We see from the definition that in a curie of any radioactive material, 3.70×10^{10} atoms are changing from one isotope to another during each second of time. Even in a microcurie, there are 3.70×10^4 (or 37,000) transformations occurring per second. This means that 37,000 alpha or beta particles are emitted every second from a single microcurie of material.

The SI unit of radioactivity is the **Becquerel (Bq)**, which is defined as the amount of material undergoing one disintegration per second. One curie equals 3.7×10^{10} Bq.

Notice that the modern definition of the curie has nothing to do with the number of grams of material. Although the curie was originally defined to be 1 g of radium, it may take more or less than 1 g of another isotope to emit 3.70×10^{10} particles per second. A small amount of a strongly radioactive isotope will emit a large amount of radiation.

The strength of the activity depends on the half-life. An isotope with a half-life of 1 year is decaying 10 times faster than an isotope with a half-life of 10 years. Therefore 1 mol of the 1-year isotope will have 10 times the activity of 1 mol of the 10-year isotope.

Measuring Radiation Dose

A **radiation dose** is the amount of radiant energy absorbed by each gram of the object receiving the radiation. This quantity depends on a number of factors: (1) the strength of the radiation source, (2) the distance between the source and the receiver, and (3) how strongly the receiver absorbs the radiation (some of the radiation may go all the way through the volume of interest instead of being absorbed). (See Figure 18-6.)

The original unit for measuring quantities of radiation was the *roentgen*, named after Wilhelm Roentgen, the discoverer of x-rays. The roentgen was designed specifically to measure x-rays. Later, when different kinds of radiation needed to be measured, scientists defined a slightly different unit called the **rem**, which stands for *roentgen equivalent mammal*. (The original meaning was "roentgen equivalent man," but the unisex version is taking its place.)

The rem is based on a number of factors. It primarily measures the amount of energy absorbed per gram of body tissue. But it also has a factor that accounts

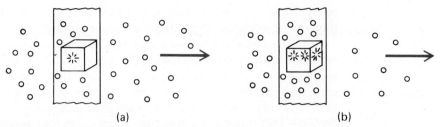

FIGURE 18-6 The energy absorbed in a cubic centimeter of matter from a beam of radiation depends on the absorption properties of the material. (a) If the matter is weakly absorbing, most of the radiation will go through any given cubic centimeter. (b) If the matter is strongly absorbing, more of the radiation will deposit its energy in the volume of interest.

for the biological effectiveness of the radiation, since some forms of radiation (such as neutrons) produce more damage in animal tissue than do x-rays or gamma rays. Therefore, using the rem as a unit of measurement allows us to compare the biological effects of different kinds of radiation. The rem is defined so that the same number of rems of different kinds of radiation produces the same biological effects.

By definition, 1 rem of x-rays or gamma rays delivers 10^{-5} J of energy per gram of body tissue as it ionizes the materials through which it passes. The SI unit of radiation dose is the **sievert (sv)**. One sievert of radiation delivers 1 J of energy per kilogram of body tissue, so 1 rem equals 0.01 sv.

One rem is a moderate amount of radiation. A person can absorb 1 rem of x-rays without great hazard. However, 100 rem is a large dose and should be accepted only when necessary, as for cancer therapy. At 500 rem the dose becomes lethal.

On the other end of the scale, small doses are measured in thousandths of a rem, called **millirems (mrem)**. The average person receives about 70 mrem of x-rays each year for diagnostic purposes. **Background radiation** comes to all of us from outer space, from the air, and from small amounts of radioactivity in the ground. It averages about 100 mrem per year for each person.

It is important to distinguish between **total dose** and **dose rate** (or radiation intensity). The rem measures the total dose, regardless of how much time it takes to receive that dose. Saying you received a dose of 10 rem is like saying you drank 10 gallons of beer. You have not specified whether you drank it in a day, a month, or a year.

The dose rate is described by the rem per unit time—for example, the number of rems per hour. If, for example, you receive 10 mrem per hour, then in 10 hours you will receive a total of 100 mrem. The effects of radiation on some biological and chemical processes depend on the dose rate as well as on the total dose.

News accounts of nuclear accidents are notorious for failing to specify whether they are describing total dose or dose rates. A dose rate of 10 mrem per hour measured outside a nuclear power plant is far different from a total dose of 10 mrem measured outside the plant during a 24-hour period.

18.4 Half-Life

The concept of half-life was first introduced in Section 18.1. Here we extend the discussion so as to clarify some details concerning the decay of radioactive substances. We can best do this by following a radioisotope through its lifetime.

Suppose we start out with 1 Ci of the radioactive isotope ^{60}Co, which has a half-life of 5.26 years. (By "start out" we mean that as the clock is started we measure out a 1-Ci sample of ^{60}Co.) In 1 curie of ^{60}Co, 3.7×10^{10} disintegrations take place every second, by definition of the curie. As each atom decays (changing to stable ^{60}Ni), fewer ^{60}Co atoms are left to decay.

The nature of radioactivity is such that a fixed fraction of the radioactive nuclei decay during a given period of time. In the present case, about 13% of the existing ^{60}Co atoms disintegrate each year. Therefore, as the number of radioactive nuclei present continues to diminish, the number of disintegrations per second keeps going down at the same rate.

By definition, the half-life of the isotope is the amount of time required for half of the atoms to undergo radioactive decay. This means that at the end of 5.26 years, half of the original number of ^{60}Co atoms remain, and only $\frac{1}{2} \times 3.7 \times 10^{10}$ disintegrations take place per second. In other words, *at the end of one half-life, half a curie of activity remains.*

The same thing happens during the next half-life. In that time half of the half-curie disintegrates, so only one-quarter ($\frac{1}{2} \times \frac{1}{2}$) of a curie remains. At the end of the next half-life, half of that quarter remains, so we are left with one-eighth of a curie. The process can best be described by drawing a graph (see Figure 18-7) showing the number of curies remaining at the end of a given period of time.

You can see that the chief feature of this curve is the gradual decay of the radioactivity. At any given time, the activity is half what it was 5.26 years before;

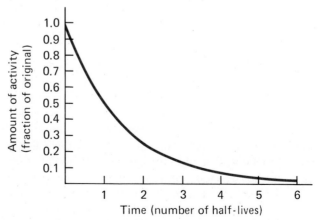

FIGURE 18-7 A curve showing the fraction of radioactive material remaining after a given number of half-lives. During each half-life, the amount of activity decreases by a factor of $\frac{1}{2}$.

and 5.26 years later, half the remaining activity will be gone. It is never possible to point to one particular moment when the radioactivity is all gone. The curve, theoretically, goes on forever. (However, if you wait for a really long time, there will come a moment when only one radioactive nucleus is left ... and then there will be none. At that point the theoretical equation no longer describes the real situation.)

In reality, after a long enough time the amount of radioactivity left becomes so small as to be negligible. If you start with 1 curie of ^{60}Co, the activity will be down to a microcurie (one-millionth of the original amount) in about 105 years (or 20 half-lives). You can check this number by multiplying $\frac{1}{2}$ times itself twenty times—you will get approximately one millionth.

The fact that you must wait many half-lives before the activity in a sample dies down to a *negligible* amount contributes to the problems arising in connection with the disposal of radioactive wastes from nuclear power plants. One of the chief isotopes among these numerous waste products is ^{137}Cs, which has a half-life of about 30 years. This does not sound like an enormous length of time. However, when a reactor has been in full operation for a year, its reactive core can contain extremely large amounts of radioactive substances. As much as a million curies of ^{137}Cs might be present. It would take 20 × 30, or 600, years for this quantity to decay down to 1 Ci, and another 600 years for it to decay down to 1 μCi, at which time it is no longer very dangerous.

18.5 Effects and Uses of Radiation

Chemical Uses

There are two major uses for radiation and radioisotopes in chemistry. Radiation can be used to trigger certain chemical reactions, and radioisotopes can be used as diagnostic tools to obtain information about the details of complex chemical reactions.

Reactions Caused by Radiation

As we have seen, the primary effect of the passage of radiation through matter is the production of ions and electrons: the ionization of atoms in the radiation path. There have been many efforts to put this effect to work as a trigger for chemical reactions. This effect can be especially useful in the case of organic compounds that do not normally ionize in solution. If large organic molecules are broken up into smaller fragments by the impact of nuclear radiation, then they may join together to make other molecules. Although this subject has been studied intensively in the laboratory, it has been found that as far as practical applications are concerned, radiation is an expensive way to make reactions happen.

18.5 EFFECTS AND USES OF RADIATION

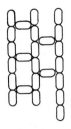

FIGURE 18-8
The molecular structure of plastics consists of long chains of short molecules linked together like sausages. Radiation can cause the chains to link together in the direction perpendicular to the chains (cross-linkage).

However, one type of reaction has been found useful in the making of plastics. There the aim is to join small molecules together into long chains, and then to join the chains together along the sides by *cross-linkages*, as shown in Figure 18-8. The cross-linkages strengthen the material. It has been found feasible to treat polyethylene with gamma radiation to make a plastic with a melting point higher than that of the starting material.

Some of the most important connections between radiation and chemistry are to be found in a number of undesirable effects that take place in materials exposed to large amounts of radiation, particularly in the vicinity of nuclear reactors. Organic materials are very susceptible to the effects of radiation. As a result, insulating materials used in nuclear power plants for electrical wiring, motors, and valves must be carefully chosen for their resistance to radiation. The insulation of an electric wire can be degraded dramatically by the presence of large amounts of gamma radiation. This is especially significant in safety-related electrical equipment, which must keep operating even *after* an accident that releases large amounts of radiation. Identification of radiation-resistant materials is an important new area of research.

Even metals are affected by certain types of radiation. Although gamma rays have little effect on metals, neutrons (found in large quantities in nuclear reactors) bombard the nuclei of the metal atoms like shotgun pellets. The neutrons can displace these atoms out of their normal places in the crystal lattice, and as a result the metal can become more brittle than normal. This effect is a matter of great concern because the containment vessel of a nuclear reactor cannot be allowed to become overly weakened by radiation.

Uses of Radiation as a Diagnostic Tool

A diagnostic tool is an instrument or a procedure that helps us to determine what is happening inside the thing being investigated. (The stethoscope and the thermometer are two of the physician's diagnostic tools.) A diagnostic tool that has become extremely important in chemistry and in biology during the past few decades is the radioisotope tracer method.

We use the tracer method to follow certain **tagged molecules** through all the chemical reactions in an animal (or plant) body. A molecule is tagged by attaching a radioactive isotope to it. For example, a molecule of sugar could be tagged either by replacing one of the carbon atoms with ^{14}C or by replacing one of the hydrogen atoms with tritium, ^3H. Then, as this molecule makes it way through the digestive system and through chains of reactions throughout the body, the radioactive atoms can be followed by means of sensitive radiation detectors. In that way we can determine the exact reactions through which the sugar molecule becomes converted into body carbohydrate and energy.

Carbon-14 is an important diagnostic tool for archeologists and historians, because it can be used to measure the ages of ancient objects with fairly good accuracy. Briefly, the method can be described as follows: ^{14}N makes up most of the earth's atmosphere. Cosmic rays (high-energy protons) coming to earth from outer space collide with the atoms of the upper atmosphere, producing a shower of free neutrons. These neutrons react with the nitrogen atoms in the air, causing some of them to change to ^{14}C according to the reaction

$$^{14}_{7}N + ^{1}_{0}n \longrightarrow ^{14}_{6}C + ^{1}_{1}H$$

The radioactive carbon atoms filter down to the lower parts of the atmosphere, eventually becoming part of the carbon dioxide in the air. They are then absorbed by the leaves of trees and other plants. As long as a plant is alive, the ratio of ^{14}C to ^{12}C in the plant remains fixed. As soon as the plant dies, however, there is no longer an intake of CO_2. Because of the radioactivity of ^{14}C, which has a half-life of 5730 years, the ratio of ^{14}C to ^{12}C begins to decrease.

With the use of radiation counters it is possible to measure the ratio of ^{14}C to ^{12}C in wood or other plant material. In this way it can be determined how many years ago a plant died. A newer method measures the number of ^{14}C atoms directly, instead of measuring the radiation emitted. This new method is so sensitive that the age of a piece of cloth can be determined from just a single thread. Radiocarbon dating has become an invaluable tool in dating archeological specimens up to 40,000 years old. If, however, the object is more than a few half-lives old, the amount of ^{14}C is so small that the method becomes unreliable. Other radioisotopes have been found that can provide dating for older materials. For example, the age of the earth itself has been determined by measuring the abundance of the various isotopes of lead found in nature, since ^{206}Pb is the end product of the uranium activity chain. Though estimates obtained by various methods may differ to some extent, there is complete agreement that the earth is at least several (5 or 6) billion years old.

Radiation in Medicine

There are many ways in which radiation is used in medicine. The role of x-rays in diagnosis is familiar. X-rays or gamma rays are also used for the treatment of cancer. Radioactive tracers are used extensively in the study of the biochemistry and physiology of the body—that is, the chemical reactions that make the body work.

In addition to the useful properties of radiation, there are harmful effects of which you should be aware. The risk/benefit ratio of every medical procedure must be considered before proceeding with it. The willingness to take a risk depends in part on the level of the potential benefits; if you have terminal cancer, you may be willing to take more risks than you would under other circumstances. When x-ray photographs are taken for medical reasons, we expect that the benefits outweigh the risks. The person being diagnosed or being given radiation therapy expects that the increase in longevity caused by the treatment will more than compensate for the loss due to the radiation effects.

Many early workers in x-rays and radium died before their time because they did not realize how harmful radiation can be. A notorious case during the 1930s involved women who painted luminous numbers on watch dials with a radioactive solution. They suffered radiation damage as a result of pointing the fine brushes with their lips! In this regard, radiation is similar to many other chemical hazards found in the workplace.

Because of these hazards, the study of the effects of radiation on animals and human beings is an important topic of research. It is a topic especially relevant to the controversy concerning the safety of nuclear power.

There are two major kinds of biological radiation effects: immediate and delayed. The effects depend largely on the size of the dose received. Little immediate effect is produced by 1 rem or less of radiation on the whole body. If it is received within a short time, 100 rem will cause mild radiation sickness and sometimes death. Five hundred rem will cause death 50% of the time, and 800 rem will kill almost all those exposed to it. (Notice that discussion of radiation risks immediately gets us into the topic of statistics. We cannot say definitely that an individual is going to die upon receipt of a 500-rem dose. We can only say that the probability of death in a short time is 50%.)

A more serious problem arises when alpha emitters such as uranium and plutonium are ingested directly, either by being swallowed or by being inhaled. Very minute quantities (a microgram or less) of such substances can cause serious health problems. First, these elements are chemically poisonous. Second, they lie in direct contact with sensitive tissue (such as lung tissue) and irradiate this tissue with highly ionizing alpha particles. There is evidence that some of the cancer-causing effects of cigarette smoking are the result of the inhalation of very small amounts of radon.

A separate problem occurs when radioisotopes fall from the air onto plants that are then eaten by animals, and so get into the food chain. Especially important in this connection are ^{90}Sr and ^{125}I, for strontium behaves chemically like calcium, and iodine becomes concentrated in the thyroid gland. Both of these isotopes may be present in fallout from nuclear explosions or in accidental emissions from nuclear power plants. (Fallout from nuclear testing was fairly widespread during the 1950s, but to date there has been no emission from a nuclear power plant producing serious, widespread effects to the public.)

The delayed effects of radiation, though difficult to observe, may be significant if large numbers of people are exposed. This is because radiation can stimulate biochemical processes that after a period of time (perhaps 20 years) may result in the growth of cancer cells. The probability that this will happen to a person is not great: a person has about 3 chances out of 10,000 of getting cancer from exposure to 1 rem of radiation. But if 10,000 people were all to receive 1 rem of radiation, then we would expect three cancer deaths (more or less) to result during the next 20 years. For this reason large populations should be exposed to as little radiation as possible.

However, it is impossible to eliminate all radiation exposure. Everybody is exposed to about 100 mrem (0.1 rem) of background radiation per year. This is the natural condition. It is believed that an additional 100 mrem per year from x-rays and other artificial sources of radiation produces relatively little harm. Indeed, those people living at high altitudes—such as in the Denver area—receive twice as much cosmic radiation as those living at sea level. The increased risk to health is small compared with other environmental factors, so no significant difference in life expectancy has been observed between the Denver population and the rest of the country.

Specifically, it has been calculated that increasing the annual radiation dose to every person in the population by 100 mrem would decrease mean life expectancy by about 500 hours (about 22 days), out of a total expectancy of 72 years. Eliminating cigarette smoking would *increase* the U.S. population's life expectancy by about 3 years—an effect about 50 times greater.

There has been a certain amount of concern over the possibility that exposure to radiation can produce genetic changes that will result in mutations in the offspring of the people exposed. Experiments with fruit flies would lead one to this conclusion. However, 40 years of research on the survivors of the Hiroshima and Nagasaki atomic blasts does not show evidence of genetic effects (over and above those that occur normally under all conditions). This somewhat surprising result remains a mystery and will be the subject of continuing study.

The fact that radiation is harmful to animal tissue is put to use in radiation therapy. Under some conditions cancer cells absorb x-rays or gamma rays more strongly than do healthy cells. Also, radiation can be aimed at a particular organ so that the undesirable cells receive more of a dose than the rest of the body. In that way the cancer cells are destroyed, with minimum harm to the body as a whole.

Glossary

alpha particle A particle emitted from the nucleus of a radioisotope consisting of two protons and two neutrons (equivalent to a helium atom without its two electrons).

background radiation Radiation present at a given location at all times, originating from outer space (cosmic rays) and from naturally occurring radioactive materials in the environment.

becquerel An SI unit of radioactivity, defined as the amount of radioactive material undergoing one disintegration per second.

beta decay A process in which an atomic nucleus emits a beta particle.

beta particle An electron emitted from an atomic nucleus as the result of the change of a neutron into a proton.

curie A unit of radioactivity, defined as the amount of an isotope that undergoes 3.7×10^{10} disintegrations per second.

daughter nucleus The nucleus remaining after a radioactive decay event.

dose rate The amount of radiation energy received by a person or object per unit time.

dosimeter A measuring device that records the quantity of radiation received by measuring the ionization caused by the radiation.

electromagnetic radiation Radiation consisting of electromagnetic energy (characterized as either waves or photons). It includes infrared, visible light, ultraviolet, x-rays, and gamma rays.

gamma ray High-energy electromagnetic radiation emitted by a nucleus. Gamma rays are the same as x-rays, but from a different source.

half-life The time it takes for half of the atoms in a radioactive sample to decay.

infrared radiation Electromagnetic radiation with wavelengths longer than those of visible light and shorter than those of radio waves.

millirem (mrem) One-thousandth of a rem.

nuclear fission The splitting of a nucleus into two pieces having approximately equal mass.

nuclear radiation Energetic particles (including gamma-ray photons) emitted from the nucleus of an atom.

radiation Any kind of energy capable of traveling through a vacuum at a very high speed, including electromagnetic waves and a variety of particles. See also **nuclear radiation**.

radiation dose The amount of radiation energy absorbed per unit volume of material.

radiative spectrum The set of waves or particles, each with a unique energy, radiated from a given source.

radioactive material A material that exhibits radioactivity.

radioisotope An isotope that emits nuclear radiation, as opposed to a stable isotope that is not radioactive.

range of a particle The distance that a nuclear particle can penetrate into a target material.

rem Abbreviation for roentgen equivalent mammal; a unit of radiation measurement defined as the amount of radiation that produces the same biological effect as 10^{-5} J of x-ray energy absorbed in 1 g of body tissue.

scintillation A short pulse of light emitted when ionizing particles are absorbed by materials such as sodium iodide.

sievert (sv) A unit of radiation dose defined as the amount of radiation that produces the same biological effect as 1 J of x-ray energy absorbed in 1 kg of body tissue; equal to 100 rem.

strong nuclear force The force that binds neutrons and protons together inside an atomic nucleus.

tagged molecules Molecules in which one or more of the atoms is radioactive.

total dose The total amount of radiation energy received by a person or object over a specified period of time.

ultraviolet radiation Electromagnetic radiation with wavelengths shorter than those of visible light and longer than those of x-rays.

x-rays Electromagnetic radiation with wavelengths shorter than 10^{-8} m. X-rays are basically the same as gamma rays but are produced by different sources.

Problems and Questions

18.1 Radioactive Isotopes

18.1.Q a. What is radioactivity?
b. What two events take place in radioactivity?

18.2.Q Briefly outline the important work of the following persons:
a. Maxwell and Hertz
b. Roentgen
c. Becquerel
d. Pierre and Marie Curie

18.3.Q What kind of radiation is visible light?

18.4.Q How do the various colors of visible light differ from one another?

18.5.Q a. How does infrared radiation differ from red light?
b. How does ultraviolet light differ from violet light?

18.6.Q a. How are x-rays similar to visible light?
b. How are x-rays different from visible light?
c. What important characteristics of x-rays makes them useful in medicine?

18.7.Q How did the discovery of x-rays lead to the discovery of radioactivity?

18.8.Q Define the following:
a. alpha ray (particle)
b. beta ray (particle)
c. gamma ray
d. isotope
e. photon

18.9.Q What is the difference between radioactivity and radiation?

18.10.Q Describe how to separate the three kinds of radiation coming from uranium.

18.11.Q What part of the atom does uranium radiation come from?

18.12.Q a. What unit of energy is generally used in describing nuclear radiation?
b. Compare the energy of the visible-light photons given off by the outer electrons of an atom with the energy of the gamma-ray photons emitted by the nucleus of a typical radioactive isotope.

18.13.Q Describe the model that allows us to think of a gamma-ray photon as a wave and as a particle at the same time.

18.14.Q Define the following:
   a. strong nuclear force
   b. half-life
   c. beta decay
   d. tritium
   e. daughter nucleus
   f. nuclear fission

18.15.Q a. Describe the origin of beta rays.
   b. Describe the origin of gamma rays.
   c. Describe the origin of alpha rays.

18.16.Q What is the major difference between the cause of beta emission and that of alpha emission?

18.17.Q Explain why uranium exists on earth today, even though it is radioactive.

18.18.Q Explain why radium exists on earth today, even though its half-life is only 1600 years.

18.19.Q a. What is one source of the background radiation that we find around us continually?
   b. Can this background radiation be avoided?

18.20.Q Name the two long-lived radioactive elements that are the origin of most of the naturally occuring radioisotopes in the earth's crust.

18.21.Q What is nuclear fission? What is the main function of this reaction?

18.22.Q What is an artificial radioisotope?

18.23.Q What is transmutation?

18.24.Q Describe the process of neutron capture.

18.25.Q a. Describe how radioactive ^{60}Co is prepared.
   b. What nuclear reaction is used in this process?

18.26.P Complete the following equations:
   a. $^{59}_{27}Co + ? \longrightarrow {}^{60}_{27}Co$
   b. $^{6}_{3}Li + {}^{1}_{0}n \longrightarrow {}^{3}_{1}H + ?$
   c. $^{238}_{92}U \longrightarrow ? + {}^{4}_{2}He$
   d. $^{234}_{90}Th \longrightarrow ? + e^{-}$
   e. $? \longrightarrow {}^{3}_{2}He + e^{-}$
   f. $^{1}_{1}H + {}^{1}_{0}n \longrightarrow ?$
   g. $^{3}_{1}H \longrightarrow ? + e^{-} + $ antineutrino

18.2 The Properties of Nuclear Radiation

18.27.Q a. What properties do alpha, beta, and gamma radiation have in common?
   b. In what ways do they differ?

18.28.Q What specific properties are attributed to alpha particles?

18.29.Q What is meant by the range of an alpha particle?

18.30.Q Which has a greater range—a low-energy or a high-energy alpha particle?

18.31.Q How does the range of an alpha particle compare with that of a beta particle for a given energy?

18.32.Q How is the range of an alpha particle related to its ionizing power?

18.33.Q a. What is the energy of a typical uranium alpha particle?
   b. What is the range of this particle in air?
   c. What is its range in aluminum?

18.34.Q What is the most serious biological danger of alpha radiation?

18.35.Q A 5-MeV beta particle is stopped by 1 cm of aluminum. Why is aluminum of this thickness not a sufficient shield for the beta radiation? What kind of shield must be used?

18.36.Q What characteristic of tritium allows it to be ingested safely for medical purposes?

18.37.Q The hydrogen in sugar ($C_{12}H_{22}O_{11}$) can be replaced by radioactive tritium. Suggest possible biological experiments utilizing this sugar that might prove interesting.

18.38.Q The following is a well-known reaction:

$$CH_3\underset{\underset{O}{\|}}{C}-O-CH_3 + HOH \longrightarrow CH_3\underset{\underset{O}{\|}}{C}-OH + CH_3OH$$

For many years chemists could not determine the exact mechanism for this reaction. Was the —CH$_3$ group replaced by one of the H atoms from the HOH, or was the —O—CH$_3$ group replaced by an —OH group from the HOH? Suggest how the question could be resolved by using a radioactive isotope of oxygen.

18.39.Q Describe two ways in which the absorption of gamma rays in matter is different from the absorption of alpha or beta rays.

18.40.Q Why is heavy shielding required for gamma radiation?

18.41.Q Which material makes the best shield for gamma rays—aluminum, lead, or tungsten?

18.42.Q What thickness of lead is required to cut in half the intensity of a gamma-ray beam with an energy of about 1.1 MeV?

18.43.Q What thickness of lead is required to reduce the intensity of 1.1-MeV gamma rays to 0.01 times the original intensity?

18.44.Q Gamma rays of what energy require the greatest thickness of lead shielding for a given reduction of intensity?

18.45.Q What characteristic of gamma rays is responsible for their great hazard?

18.46.Q Civil defense films suggest that school children hide under desks or move into hallways to protect themselves against radiation. In a situation where radioactive substances became mixed with air, would such maneuvers have any useful effect?

18.3 The Detection and Measurement of Radiation

18.47.Q Why are film badges worn by people working in places where they might be exposed to radiation?

18.48.Q What advantage does the dosimeter have over a film badge?

18.49.Q What would be the simplest kind of instrument to place in various locations around a nuclear power plant in order to record the amount of radiation received outside the perimeter of the plant?

18.50.Q What kind of instrument would you use to measure the number of gamma-ray photons as well as their energy?

18.51.Q What kind of instrument would you use to measure very small quantities of alpha or beta rays?

18.52.Q What is an important difference between a Geiger counter and a scintillation counter?

18.53.Q Beautiful glazes for pottery are produced from uranium ores. However, the production of such glazes has been discontinued because of the radioactivity in the glaze. Also, a green glass called "vaseline glass" receives its special green color from uranium compounds. How would you test pottery and glassware to see if it contained uranium?

18.54.Q Describe the operation of an ionization chamber.

18.55.Q Describe the operation of a Geiger counter.

18.56.Q a. In what way is a Geiger counter similar to an ionization chamber?
   b. In what way is a Geiger counter different from an ionization chamber?

18.57.Q What is meant by radiation dose?

18.58.Q a. Give the original definition of the curie.
   b. Give the modern definition of the curie.
   c. Why was it found useful to change the definition of the curie?

18.59.Q a. Does a gram of radium have the same amount of radioactivity as a gram of uranium? Explain your answer.
   b. Isotope A has a half-life of 3000 years. Isotope B has a half-life of 10,000,000 years. Which sample would be most radioactive—1 mol of isotope A or 1 mol of isotope B? Explain your answer.

18.60.P An isotope is found to disintegrate at a rate of 7.4×10^6 disintegrations per second. What is the activity of this isotope in millicuries?

18.61.Q Define the rem.

18.62.Q Does the curie measure the same kind of thing as the rem? If not, explain the difference.

18.63.Q Why is it important to be careful about the difference between total dose and dose rate in dealing with radiation?

18.64.Q a. Is 100 mrem considered a large amount of radiation?
   b. Is 100 rem considered a large amount of radiation?

18.65.Q What amount of radiation is lethal?

18.66.Q a. How much background radiation do you receive every year?
   b. Compare this amount with the lethal amount.

18.4 Half-Life

18.67.P The half-life of $^{234}_{90}$Th is 24 days. If you start with 60 g of this isotope, how much will be left after
   a. 24 days?
   b. 48 days?
   c. 72 days?

18.68.P The half-life of $^{238}_{92}$U is 4.5×10^9 years. If you start with 6×10^{24} atoms of this isotope, how many atoms will be left after 4.5×10^9 years?

18.69.P A lab technician stores 2.0 g of $^{210}_{83}$Bi in a cabinet on January 1, 1983. The half-life of this isotope is 5.0 days. How many grams of $^{210}_{83}$Bi will be in the bottle 365 days later?

18.70.P The isotope $^{214}_{82}$Pb has a half-life of 27 minutes. If 10 g of the isotope is shipped by a supplier to a laboratory and the trip takes 72 hours, how much $^{214}_{82}$Pb will arrive at the lab? What do you think of the practicality of this shipping procedure?

18.71.P A lab receives 10 g of $^{214}_{83}$Bi, whose half-life is 20 minutes. The trip from the supplier to the lab took 4 hours. How much of the isotope was originally shipped from the supplier?

18.72.P The half-life of radon is 3.8 days.
   a. How long does it take for 50% of the sample to decay?
   b. How long does it take for 75% of the original sample to decay?

18.73.P A glass block containing 10,000 curies of ^{137}Cs is buried in a salt mine. Three hundred years later, this mine is inspected. How many curies of ^{137}Cs are found? (*Note:* Encapsulating high-level radioactive wastes in a solid material such as glass and burying the container deeply in the earth is a favored way of disposing of such wastes.)

18.5 Effects and Uses of Radiation

18.74.Q a. Do gamma rays have a permanent effect on metals? Describe the effects, if any.
   b. Do neutrons have a permanent effect on metals? Describe the effects, if any.

18.75.Q What is the immediate primary effect of the passage of gamma rays through organic materials?

18.76.Q Describe how some plastics can be strengthened by gamma radiation.

18.77.Q Describe the effect of gamma radiation on the electrical resistance of organic insulators.

18.78.Q Name three areas of research that use radioactive isotopes.

18.79.Q How is radioactivity used as a diagnostic tool in chemistry and in biology?

18.80.Q A researcher wants to know if fructose in the diet is used in the brain cells and, if so, how long it takes for ingested fructose to reach the brain. Suggest an experiment that could be used to follow the travels of fructose from the digestive system to the brain.

18.81.Q A pipe seems to be leaking badly, but it is 2000 ft long and is buried underground. Suggest a possible way to use radioactive materials to detect the location of the leak. Think about safety problems in deciding on a choice of method.

18.82.Q An art dealer suspects he has a Rembrandt painting and wants evidence to show that it is not a modern forgery. Suggest a possible test, using radioactivity, that could be carried out to determine the age of the painting.

18.83.Q Explain what is meant by a risk/benefit calculation.

18.84.Q What are the two major roles of radiation in medicine?

18.85.Q Explain what happens when you use your lips to put a point on a fine brush that has been dipped in a radium solution.

18.86.Q What are common sources of background radiation?

18.87.Q What are the two major types of biological effects of radiation?

18.88.Q a. How many rem of radiation will cause death in half the cases of radiation exposure?
   b. How many rem of radiation will cause death in almost all the cases of radiation exposure?

18.89.Q Describe the importance of the radioisotopes ^{90}Sr and ^{125}I in the food chain.

18.90.Q What is the most important delayed effect of radiation?

18.91.Q Suggest a way to arrange one or more x-ray beams so that undesirable cancer cells get a lethal dose while the rest of the body receives a relatively safe dose.

18.92.P Estimate how many cancer deaths would be induced in a population of 100,000 by a radiation dose of 10 mrem. Compare this number with the normal cancer death rate of approximately 170 deaths per 100,000 population.

Self Test

1. Define or explain each of the following terms:
    a. alpha particle
    b. beta particle
    c. gamma ray
    d. photon
    e. isotope
    f. strong nuclear force
    g. half-life
    h. daughter nucleus

2. Complete the following reactions:
    a. $^{238}_{92}U \longrightarrow\ ^{234}_{90}Th + ?$
    b. $^{234}_{90}Th \longrightarrow\ ^{234}_{91}Pa + ?$
    c. $^{59}_{27}Co + ? \longrightarrow\ ^{60}_{27}Co$
    d. $^{27}_{13}Al + ? \longrightarrow\ ^{30}_{15}P + ^{1}_{0}n$

3. A geologist finds a sample of a mineral that has a high content of ^{206}Pb.
    a. What other elements should be present in that sample?
    b. How should the geologist test the sample for those suspected elements?

4. Explain the different between dose rate and total dose.

5. Some commercial smoke detectors make use of the reaction

$$^{241}_{95}Am \longrightarrow\ ^{237}_{93}Np + ?$$

   a. Complete the reaction.
   b. The detector measures the presence of ions in the air. What might cause the occurrence of such ions?
   c. It is estimated that the maximum exposure possible from the detector is 0.5 mrem per year. Considering the presence of the radioactive materials and the radiation emitted, determine whether this detector poses any appreciable hazards.

6. A TV reporter is heard to say that 100 mrem of radiation was discharged from a power plant.
    a. What facts are missing from this report? (*Hint:* At least two major facts are missing.)
    b. Later it turns out that the report referred to a coal-burning power plant and not to a nuclear power plant. How could you explain the release of radiation from a coal-burning plant? (*Note:* Under normal operating conditions, coal-burning plants emit as much radiation as nuclear plants because coal contains minute amounts of uranium and thorium.)

7. The half-life of ^{14}C is about 5730 years. If a tree was cut down in the year 1, what percent of the original ^{14}C in the tree is now present? (*Hint:* Since we have not studied the equation for radioactive decay, a simple way of solving this problem is to pick the answer off the graph of activity versus time in Figure 18-7.)

8. Iodine tends to concentrate in the thyroid gland. Iodine-131, a beta and gamma emitter with a half-life of 8 days, has been extensively used to diagnose malfunctions of the thyroid gland. (A gamma-ray camera creates an image of the thyroid gland from the gamma rays emitted by the iodine-131 located within the gland itself.)
    a. On the day of treatment, there are 2.5 microcuries of ^{131}I in the thyroid of a patient. Approximately how many days will it take for the activity to decrease to 0.1 microcurie?
    b. What is one possible adverse side-effect of this procedure?

9. A patient is fed a dose of a radioactive isotope. Suggest a possible course of clinical testing that might be carried out on the patient for the purpose of diagnosis.

10. a. List five common sources of radiation found in the environment and in society.
    b. Which of these sources can be controlled and reduced?
    c. Is it possible to live in an environment completely free of radiation?

Answers to Problems and Questions

Note: All answers have been rounded off to the appropriate number of significant digits. See Section 2.8 for details.

Chapter 1

1.1 **a.** 2.3×10^1 **c.** 9.3×10^7 **e.** 3.4×10^{-2} **g.** 1.0×10^{-4} **1.2** **a.** 5600 **c.** 0.002783 **e.** 0.000 000 000 001 593 2 **g.** 602 200 000 000 000 000 000
1.3 **a.** 10^{12} **c.** 10^{10} **e.** 10^{45} **1.4** **a.** 10^{-5} **c.** 10^{-18} **e.** 10^{-113}
1.5 **a.** 10^{-2} **c.** 10^{-2} **e.** 10^{-11} **1.6** **a.** 1.2×10^9 **c.** 5.3×10^{-15} **e.** 1.1×10^3
1.7 **a.** 10^4 **c.** 10^3 **e.** 10^{-2} **1.8** **a.** 10^2 **c.** 10^{-8} **e.** 10^8 **1.9** **a.** 10^{21} **c.** 10^8 **e.** 10^4 **1.10** **a.** 5.0×10^{11} **c.** 2.0×10^{11} **e.** 1.97×10^{23} **1.12** **a.** 6 **c.** 6.02×10^{23} **1.13** **a.** 3 **c.** 10^{14} **1.14** **a.** 1.0×10^{15} **c.** 2.00×10^{18}
1.15 **a.** 1.79×10^7 **c.** 1.619×10^5 **e.** 7.37×10^{-8} **1.16** **a.** 5.59×10^3 **c.** -5.23×10^{21} **e.** -6.77×10^3 **1.17** **a.** 7.53×10^{10} **c.** 4.5×10^5 **e.** 4.75×10^1
1.18 **a.** 10^8 **c.** 10^{-12} **e.** 10^{-3} **1.19** **a.** 4×10^8 **c.** 2×10^{-4} **e.** 3.6×10^{-5}
1.20 **a.** 10^{-6} in^2 **1.21** 10 ft **1.23** 20 in **1.24** **a.** $b = c - a$ **c.** $d = \dfrac{x-b}{c-1}$
e. $x = \dfrac{6}{y}$ **g.** $b = \dfrac{24c - ma}{n}$ **1.25** **a.** $R = \dfrac{PV}{nT}$ **1.26** **a.** $x = \dfrac{y-b}{m}$
1.27 **a.** $q = \dfrac{z - (a - 21)p}{b - 14}$

Chapter 2

2.1 2400 oz **2.3** 1 gal at $7.50 **2.5** $A = 93$ oz, $B = 111$ oz, $C = 296$ oz
2.7 4482 lb **2.9** 1.2 pages per day **2.17** **a.** 10^8 atoms **c.** 10^{16} atoms
2.19 5×10^6 meter sticks **2.21** **a.** 10^{-8} cm **c.** 10^{-4} μm **2.23** 10^{24} cubes
2.25 2.64×10^4 L; 26.4 m^3 **2.27** 250 g **2.29** **a.** 132 lb **c.** 22 lb
2.30 **a.** 1.4 L **c.** 1.4 kg **2.31** **a.** 10^6 g **c.** 10^{12} μg **2.33** **a.** 5.526×10^{-3} kg **c.** 5.526×10^6 μg **2.35** **a.** 750 cm^3 **2.36** 3.2 kg **2.38** 3.15×10^5 L
2.40 **a.** 1346 msec **2.42** **b.** 3.8×10^4 hr **2.43** 53 hr **2.45** **a.** 7×10^{-3} sec **c.** 7×10^3 μsec **2.46** **b.** 10 msec **2.47** **a.** 98.4 ft **c.** 91.4 cm; 914 mm
2.48 **a.** 61 in^3 **c.** 16.4 L **2.50** 121 g **2.52** 1.35×10^{-2} g; 4.76×10^{-4} oz
2.54 0.795 qt **2.56** 3.94×10^{-11} in **2.58** 8.00×10^5 cm^3

2.61 $V(Al)/V(Pt) = 21.45/2.7 = 7.94$ **2.66 a.** 1.55 L **2.72** 35°C **2.74** 10800°F **2.78** 37°C **2.80** 2.8°C **2.82** −269°C **2.84** 310 K **2.87** 144 K **2.91** 1235 K **2.92 a.** 24.674 cm³ **b.** 0.046 cm³ **c.** 0.19% **2.95 a.** 5 weighings **b.** 5 weighings **2.97 a.** 2 **c.** 3 **2.98 a.** 16,000 **c.** 5,100,000 **2.99 a.** 68,000 **c.** 4.5×10^6 **2.100 a.** 15.67 **c.** 599.2

Chapter 3

3.53 3.14×10^7 J **3.55 a.** 2.55×10^4 cal **d.** 1.07×10^5 J **3.57 a.** 92.4 Btu **b.** 23.3 kcal **c.** 9.75×10^4 J **3.62 a.** 1.05×10^6 J **c.** 9.25×10^5 kcal **3.66** 0.085 kcal/kg·°C **3.68** oxygen **3.73** 1.30×10^8 J **3.75 a.** 4.77×10^8 J **c.** 31.6 kg

Chapter 4

4.12 a. %C = 27.3%, %O = 72.7% **c.** %C = 27.3% **4.14 a.** 1168 kg **4.17 a.** %H = 2.0% **4.24 a.** N_2O, N_2O_3

Chapter 5

5.35 a. 3.14×10^{14} **5.42** 388.7 g **5.50** 108.83 **5.52 b.** 24.31 **c.** 12

Chapter 6

6.8 atomic volume of C = 6.0 cm³/mol
atomic volume of Fe = 7.11 cm³/mol

Chapter 7

7.9 a. 36 electrons, 35 protons, 45 neutrons **c.** 2 electrons, 1 proton, 0 neutrons **e.** 18 electrons, 20 protons, 20 neutrons

Chapter 8

8.26 a. N: −3; H: +1 **c.** K: +1; Cl: −1 **e.** H: +1; O: −2 **g.** H: +1; O: −1 **i.** S: +4; O: −2 **8.31 a.** Na_2CrO_4 **c.** $(NH_4)_2CO_3$ **e.** Rb_2SO_4 **g.** SO_3 **i.** H_3PO_4 **8.39 a.** sulfuric acid **c.** sodium phosphate **e.** nitrous acid **8.45 a.** lye; sodium hydroxide **c.** spirit of hartshorn; ammonium hydroxide **e.** milk of lime; calcium hydroxide **g.** ferrous hydroxide; iron(II) hydroxide **8.51 a.** sodium carbonate **c.** carbonic acid **8.53 a.** potassium sulfate **c.** cobalt(II) chloride **e.** iron(II) bromide **g.** potassium hydrogen sulfate

Chapter 9

9.9 32 cm³ of NH_3 **9.11** 7.5 L of N_2; 22.5 L of F_2 **9.16** 1 mol, or 6.02×10^{23} molecules **9.18 a.** 6.02×10^{23} electrons **9.20 a.** 3.01×10^{23} atoms of Mg **9.24** 500 km **9.28** 200 g **9.32** 1.20×10^{24} molecules **9.35 a.** 2 mol of MgO **c.** 0.5 mol of K_2O **9.36 a.** 1 mol of O_2, or 6.02×10^{23} molecules **c.** 0.43 mol, or 2.6×10^{23} molecules **9.39 a.** 22.9 g **c.** 74.6 g **e.** 78.1 g **9.41 a.** 394.5 g **c.** 761.6 g **e.** 547.9 g **9.43** 0.83 g of H_2 **9.45 a.** 760 g of H_2 **b.** 150 g of NH_3 **9.51 a.** 1 mol **9.53** 19.2 L **9.59 a.** 11.2 L

Chapter 10

10.5 a. 0.5 mol of HCl; 0.5 mol of NaOH b. 0.5 mol of NaCl; 0.5 mol of H_2O
c. 29.2 g of NaCl; 9.0 g of H_2O **10.9** a. 1 mol b. 2 mol c. 1.70 g of H_2
10.23 a. $2HCl + Zn \rightarrow ZnCl_2 + H_2$ c. $CuSO_4 + Fe \rightarrow FeSO_4 + Cu$
e. $2NaI + Cl_2 \rightarrow 2NaCl + I_2$ **10.29** a. $NaOH + HCl \rightarrow NaCl + H_2O$
c. $KCl + AgNO_3 \rightarrow AgCl + KNO_3$ e. $CaCO_3 + 2HCl \rightarrow CaCl_2 + H_2CO_3$
10.30 a. $H_2SO_4 + 2NaOH \rightarrow Na_2SO_4 + 2H_2O$ c. $Mg + 2HCl \rightarrow MgCl_2 + H_2$
e. $CO + Fe_3O_4 \rightarrow CO_2 + 3FeO$ **10.31** a. 6 mol c. 3 mol e. 6 mol g. 1.5 mol
i. 4.5 mol

Chapter 11

11.2 a. 2:1 c. 2:1   **11.3** a. 0.75 mol of Al_2O_3
c. 1.5 mol of KCl; 2.25 mol of O_2 **11.5** a. 1.35 mol of Na_2SO_4; 2.7 mol of H_2O
c. 1.35 mol of NH_4Cl **11.7** a. 1.55 g of KCl b. 0.99 g of O_2
11.9 a. 36.2 g of $Mg(OH)_2$ b. 11.2 g of H_2O **11.11** a. 256 g of $MgSO_4$
b. 42.8 mol of H_2O **11.13** a. 0.765 mol of $AgNO_3$ b. 82.5 g of Ag c. yes
11.15 a. 29.5 g of HCl b. 0.101 mol of $MnCl_2$ c. 0.253 mol of Cl_2
11.17 a. 0.128 mol of C_6H_6 b. 0.384 mol of H_2O **11.19** 82.3 g of $BaCrO_4$
11.21 214.1 g of ZnO   **11.23** 61.8 g of N_2
11.25 a. 24.7% K; 34.8% Mn; 40.5% O c. 2.0% H; 32.7% S; 65.3% O
11.27 26.3 g of sodium hypochlorite; 473.7 g of water   **11.29** a. 0.80% folic acid
b. 40.0% niacin c. 59.2% other   **11.31** 23.6 mL of alcohol   **11.33** 0.2% K
11.35 a. RDA = 50 g of protein b. 1.40 kg; 3.125 lb

Chapter 12

12.24 a. lower
b. (1) 730.5 mmHg (2) 730.5 torr (3) 0.9612 atm (4) 97,376 N/m^2 (5) 973.8 mb
12.26 a. 0.1 mm Hg b. 0.13 mb c. 1.3×10^{-4} atm d. 13 Pa   **12.36** 6600 mL
12.38 65.0 torr   **12.40** 2770 kPa   **12.42** 2.1 atm   **12.51** 3.2 L
12.53 576°C   **12.55** 1150 L   **12.57** no   **12.59** b. 2.7 L   **12.73** 0.683 L
12.75 0.013 torr   **12.77** 3400 L   **12.88** 56 L   **12.90** a. -64°C
b. 7050 torr   **12.92** 9.8 g   **12.104** 1.32 atm, 1000 torr
12.106 1.0 L (*Hint*: Total volume at STP equals tank volume plus raft volume.)
12.108 a. 150 kPa b. 225 kPa

Chapter 13

13.2 0.245 g/mL   **13.4** 3.4×10^{-6} g   **13.6** 3570 g   **13.8** a. 96.5 g
b. 567.1 g c. 193 g   **13.10** a. $CaCl_2$ b. 166.7 mL c. 60.0 mL   **13.21** a. 1.0 M
c. 1.4 M e. 2.0 M **13.23** a. 1000 mL c. 17 mL e. 6.7×10^{-2} mL
13.25 a. 0.79 M b. 0.79 N **13.27** a. 0.5 N b. 16 g   **13.29** 8 mL
13.31 It raises the boiling point by 1.1°C and lowers the freezing point by 4.05°C.
13.34 It raises the boiling point by 7.8°C and lowers the freezing point by 16.1°C.

Chapter 14

14.22 4.03×10^{19} ions/cm^3 **14.24** 5.15×10^{22} ions/L   **14.26** 1.8×10^{-7}%
14.30 a. 6.7×10^{-4} mol/L b. 1.34%   **14.36** 1×10^{-10} mol/L
14.38 1×10^{-5} mol/L   **14.40** 1×10^{-13} mol/L   **14.45** a. pH = 4
14.46 a. 1×10^{-6} mol/L   **14.48** a. 1×10^{-5} mol/L c. 1×10^{-1} mol/L

14.51 a. 1×10^{-1} mol/L; 1×10^{-2} mol/L b. ratio = 10:1 **14.57** pH = 7
14.59 pH = 4.5 **14.61** 3.2×10^{-8} mol/L
14.65 2.0×10^{-3} mol/L; new pH = 2.4 **14.73** pH = 10.4
14.75 5.0×10^{-8} mol/L
14.77 $[OH^-] = 7.9 \times 10^{-13}$ mol/L; $[H^+] = 1.3 \times 10^{-2}$ mol/L
14.79 1.5×10^{-5} mol/L **14.82** $[H^+]$ is increased by a factor of 10.

Chapter 15

15.26 8.8×10^{-10} **15.28** 0.015 mol/L **15.30** a. 3.5×10^{-3} mol/L b. 2.5
15.32 $[H^+]$ in $HClO_3$ is 1.5×10^3 times greater than in HNO_2.
15.45 Combustion of hydrogen to form H_2O yields 15.9 kJ/g.

Chapter 16

16.2 a. $H^{+1}, S^{-2}, N^{+5}, O^{-2} \rightarrow H^{+1}, S^0, N^{+2}, O^{-2}$ c. not redox
e. $N^{-3}, H^{+1}, O^0 \rightarrow N^{+3}, H^{+1}, O^{-2}$
16.5 $3H_2S + 2HNO_3 \rightarrow 3S + 2NO + 4H_2O$; oxidizing agent = N, reducing agent = S
16.7 $4Fe + 3O_2 \rightarrow 2Fe_2O_3$; oxidizing agent = O, reducing agent = Fe
16.9 $CuO + H_2 \rightarrow Cu + H_2O$; oxidizing agent = Cu, reducing agent = H
16.11 $2HBr + H_2SO_4 \rightarrow Br_2 + SO_2 + 2H_2O$; oxidizing agent = S, reducing agent = Br
16.13 $N_2 + 3H_2 \rightarrow 2NH_3$; oxidizing agent = H, reducing agent = N

Chapter 17

17.22

a. [structure of cyclopropane-like fused rings with two C atoms] b. [structure with three C atoms]

17.32 [structural diagram: H–C–C–C–C with H's, plus C–H, yielding H–C–C–C–C–C–H pentane, with H₂ released]

17.37 a. $CH_3CH_2CH_2CH_2CH_2CH_3$
b. $CH_2=CHCH_2CH_2CH_3$
or $CH_3CH=CHCH_2CH_3$
c. $CH_3C\equiv CCH_2CH_3$

17.42 $2CH_4 + Cl_2 \longrightarrow 2CH_3Cl + H_2$
$2CH_3Cl + Cl_2 \longrightarrow 2CH_2Cl_2 + H_2$
$2CH_2Cl_2 + Cl_2 \longrightarrow 2CHCl_3 + H_2$
$2CHCl_3 + Cl_2 \longrightarrow 2CCl_4 + H_2$

Chapter 18

18.26 a. ^1_0n c. $^{234}_{90}Th$ e. ^3_1H g. ^3_2He **18.60** 0.2 mCi
18.67 a. 30 g c. 7.5 g **18.68** 3×10^{24} atoms
18.70 160 half-lives equal 8 periods of 20 half-lives each. Each period gives a reduction in activity of 10^{-6}. Therefore, essentially nothing is left at the end of the delivery time.
18.72 a. 3.8 days b. 7.6 days

Answers to Self Tests

Note: All answers have been rounded off to the appropriate number of significant digits. See Section 2.8 for details.

Chapter 1
1. **a.** 6.23×10^{11} **b.** 2.36×10^{-10} **c.** 5×10^{-7} **d.** $1.234\,567\,89 \times 10^8$ **e.** 5.40×10^{-3} 2. **a.** 7 910 000 **b.** 0.000 85 **c.** 60 200 000 000 **d.** 0.000 019 8 **e.** 9 064 000 000 000 000   3. **a.** 2.1×10^{10} **b.** 7.5×10^{-1} **c.** 5.6×10^{62} **d.** 1.2×10^1 **e.** 4.4×10^3 4. **a.** 2.6×10^2 **b.** 6.4×10^{11} **c.** 2.2×10^{59} **d.** 1.0×10^{-4} **e.** 4.2×10^{13} 5. **a.** 3.84×10^{-2} **b.** -7.54×10^3 **c.** 5.645×10^2 **d.** 2.1×10^{-8} **e.** -6.02×10^{23} 6. 3.14×10^8 in^3
7. 2.7×10^5 in^3 8. 30 ft   9. 2×10^3 in   10. **a.** $y = 128$ **b.** $x = 6.5$ **c.** no **d.** no **e.** Let $b = 0$ 11. $d = -2.4$ 12. **a.** $x = 124.9$ **b.** yes

Chapter 2
1. $972.50   2. **a.** 16.19 kg **b.** 35.7 lb   3. **a.** 3.7×10^8 capsules/yr **b.** cost = 2.5¢/bottle   4. 12.14 kg of chlorine   5. **a.** 2000 cm **b.** 65.6 ft
6. **a.** 4.7×10^7 mm **b.** 29.2 mi   7. **a.** 4×10^3 micron **b.** 0.16 in
8. **a.** 8.9×10^5 cm **b.** 2.9×10^4 ft   9. **a.** 25 mm^2 **b.** 840 cm^2 **c.** 2.3×10^3 cm^2 **d.** 0.9 km^2 = 9×10^5 m^2 10. **a.** 1728 cm^3 **b.** 4.2×10^{-18} cm^3
11. **a.** 3.0×10^7 cm^3, 30 m^3
b. Assume that the box is 40 cm long and 20 cm high. The room has 3 rows of boxes, each 4 m high and 10 m long. The room width is at least 2.75 m. (Other solutions are possible.)
12. **a.** 1380 kg **b.** 3036 lb   13. **a.** 352 lb **b.** 58.7 lb (See Section 2.4.) **c.** 160 kg **d.** 160 kg **e.** by pushing against it and feeling its inertia   14. 2.88×10^{10} μsec
15. 1.69 kg   16. **a.** 196 m^3; 1.96×10^5 L **b.** 1.8×10^5 kg   17. 254 L; 0.254 m^3
18. **a.** no **b.** no; need more than two points to prove linearity
19. 0°C; 32°F; 273 K   20. -269°C; -452°F   21. **a.** 1% **b.** 2% **c.** Precision is better because random error is less than systematic error.
22. **a.** air currents, vibration, dirt, friction, variations in temperature
b. See chapter glossary. **c.** no, because "noise" is always present
23. A: 3.3% error; B: 0.0048% error   24. **a.** 300 **b.** 6.74×10^{11} **c.** 1.19×10^7 **d.** 0.010 **e.** 2.3×10^6

Chapter 3

1. **a.** physical—melting wax
    chemical—burning wax and wick
    b. physical—change from liquid to gas phase
    c. physical—changes only in the pressure, temperature, and volume
    d. physical—salt can be recovered from solution by evaporation
    e. chemical—splitting of crude oil molecules into simpler molecules of refined oil
2. **a.** Fuel molecules combine with oxygen molecules, increasing the kinetic energy of gaseous waste molecules. Therefore, the exhaust gas is hot.
    b. The rapidly moving gas molecules collide with the piston head, forcing it to move out and thereby expand the cylinder volume.
3. The molecules of the compounds collide with each other, causing a rearrangement of the atoms within the molecules. Sometimes there is an exchange of atoms between the reacting molecules; sometimes two molecules combine to form a third molecule.
4. **a.** During boiling, some of the molecules move too rapidly to stick together. They then escape from the liquid and form a gas.
    b. During melting, the molecules move too rapidly to be bound into a rigid, solid form (the kinetic energy is greater than the potential energy of the binding force), but they are not moving fast enough to be completely free of each other. Therefore, they slide over each other, loosely bound in a liquid phase.
5. The ball is given kinetic energy when it is thrown up. As it rises, its kinetic energy is converted into potential energy until, at the top of the trajectory, all the energy is potential. As the ball falls down, the potential energy changes back into kinetic energy.
6. (1) Mass and energy are equivalent, so the mass of an object can be changed into energy; conversely, the kinetic energy of a moving object increases the object's mass. (2) No signal or object can travel faster than the speed of light.
7. In a closed system, the total amount of energy is constant. Therefore, no kind of reaction can create energy from nothing.
8. When the hydrogen combines with oxygen, the resulting water molecule contains less energy than do the separate hydrogen and oxygen molecules. The excess energy is given off in the form of heat and light.
9. **a.** 7.2×10^4 kcal; 3.0×10^8 J **b.** 1440 kg
10. **a.** 6.05×10^8 J **b.** 12.7 kg
11. The hydrogen and oxygen molecules must get past a potential barrier in order to react. At room temperature they do not have enough energy to get past this potential barrier.
12. **a.** 9.54×10^7 J **b.** 415 kg
13. The molecules within the object move less rapidly.
14. Quantities of heat are measured by observing the change of temperature of a known mass of a known substance in a calorimeter.

Chapter 4

1. **a.** The philosopher's stone was believed to be a stone whose touch would turn base metals into gold. (Needless to say, such a stone has never been found.)
    b. The search for transmutations led from alchemy to the modern study of chemistry.
2. During the Middle Ages, Latin was the language used by scientists for scholarly purposes.
3. **a.** The mass of a closed system cannot change.
    b. A pure compound is always composed of the same elements, which are always present in the same proportions by weight.

c. If two elements can form more than one compound, the ratio of the weight of the first element to the weight of the second element is always a simple integer ratio.
4. No change; the chamber is a closed system, so conservation of mass applies.
5. **a.** helium **b.** manganese **c.** phosphorus **d.** potassium **e.** silver **f.** gold **g.** mercury **h.** zinc **i.** fluorine **j.** chromium **k.** tungsten **l.** uranium
6. **a.** H **b.** Na **c.** Ca **d.** Cl **e.** Fe **f.** Co **g.** Ra **h.** Ba **i.** O **j.** N **k.** Ne **l.** C
7. **a.** CH_4: 1 carbon, 4 hydrogen; **b.** N_2O: 2 nitrogen, 1 oxygen; **c.** $CoCl_2$: 1 cobalt, 2 chlorine; **d.** $CaBr_2$: 1 calcium, 2 bromine; **e.** NH_4Cl: 1 nitrogen, 4 hydrogen, 1 chlorine; **f.** NaF: 1 sodium, 1 fluorine; **g.** FeI_2: 1 iron, 2 iodine; **h.** H_2SO_4: 2 hydrogen, 1 sulfur, 4 oxygen; **i.** K_3PO_3: 3 potassium, 1 phosphorus, 3 oxygen; **j.** H_2O: 2 hydrogen, 1 oxygen; **k.** XeF_4: 1 xenon, 4 fluorine
8. law of definite proportions
9. **a.** law of definite proportions **b.** law of multiple proportions
10. **a.** 1.5 g of oxygen for every gram of sulfur
    b. mass of sulfur atom = 2 × mass of oxygen atom
11. According to law of multiple proportions, CO_3 is a reasonable compound.
12. Qualitative analysis determines which elements are present within a substance. Quantitative analysis determines the amount of each element present in a substance.

Chapter 5

1. A force is an interaction capable of changing the state of motion of an object; a push or a pull.
2. gravitational force, electromagnetic force, and nuclear force
3. induction coil and vacuum pump
4. the electron
5. The Rutherford scattering experiment showed that alpha particles colliding with an atom scatter at large angles.
6. **a.** They repel each other (electrostatic repulsion).
   b. They repel each other (electrostatic repulsion).
   c. They attract each other (electrostatic attraction).
   d. There is no electrostatic attraction or repulsion. There is a small magnetic force and a negligible gravitational attraction.
7. The force varies inversely as the square of the distance between them, so the closer together they are, the stronger the repulsion between them. However, when they get very close, they are pulled together by the nuclear interaction (attraction).
8. The atomic mass of a single isotope is the sum of the masses of the neutrons and protons. The atomic number is the number of protons; the number of neutrons must be known to find the atomic mass. The atomic mass of natural silver is the average atomic mass of all the stable silver isotopes.
9. $$\text{percentage mass} = \frac{\text{electron mass} \times 100}{\text{atomic mass}}$$
$$= \frac{\frac{1}{1837} \times 47}{108} = 0.024\%$$

10. $$\frac{(4889 \times 63.93) + (2781 \times 65.93) + (411 \times 66.93) + (1857 \times 67.92) + (62 \times 69.93)}{10000}$$
$$= 65.39$$

11. The photoelectric effect is evidence: a photon striking a material surface knocks out an electron from the surface. The energy of the electron depends on the photon energy, which in turn is inversely proportional to the wavelength of the light. A beam of light consists of numerous photons, each with a specific amount of energy.

12. Protons very close together attract each other with the strong nuclear force. The presence of neutrons in the nucleus provides enough nuclear binding to overcome the electrostatic repulsion between the protons.

13. **a.** (1) In the hydrogen atom, there is a single electron in a circular orbit. (2) Only certain orbits are allowed; the radii and energies of the allowed orbits are calculated by a mathematical procedure developed by Bohr. (3) The electron is normally in its lowest orbit, or energy level. (4) If the electron is placed in a higher orbit, it will quickly drop down to the lowest orbit, or ground state. (5) In dropping down from one level to another, the electron emits a photon of light whose wavelength is inversely proportional to the energy lost by the electron.

   b. (1) The orbital electron is described by a wave packet. (2) The wavelength of the packet is related to the electron's energy. (3) An integral number of wavelengths fit around the circumference of the orbit; this requirement allows the electron energy to be calculated. The rest of the assumptions are the same as those that apply to the Bohr model.

   c. The Bohr model says nothing about a wave packet and it makes an arbitrary assumption about the energy of allowed orbits.

15. The Schroedinger model explains why only certain orbits are allowed and why the ground state is the lowest orbit possible. It also explains why the electron cannot fall into the nucleus.

16. **a.** A quantum number is an integer or fraction that appears in equations describing the properties of electrons and other fundamental particles.

    b. four

    c. The quantum number n determines the allowed electron energies.

17. **a.** two

    b. helium (and all heavier elements)

    c. second shell

    d. lithium

18. An electron configuration is a description of the number of electrons in each shell and subshell of a given atom.

19. **a.** It is an electron configuration having two electrons in the first ($1s$) subshell and two electrons in the second ($2s$) subshell.

    b. beryllium

20. If a model is not able to predict a specific observable occurrence, then it is not possible to verify the model by experiment and/or observation. If therefore makes no difference whether or not you believe in the model, as far as physical events are concerned.

21. The Schroedinger model predicts: (1) the wavelengths of light emitted by excited hydrogen atoms, (2) the energy needed to ionize a hydrogen atom, (3) the effects of a magnetic field on a hydrogen atom, and (4) the structure of a hydrogen molecule. Also, with modifications this model makes the same predictions for more complex atoms.

Chapter 6
1. a. fluorine—pale-yellow corrosive gas
    b. chlorine—pale-green poisonous gas
    c. bromine—red-brown liquid
    d. iodine—purple solid
    e. astatine—unstable, radioactive solid; very rare
2. atomic radius and volume, ionization energy, oxidation number
3. a. Weigh 12 g of carbon, or, since we know 6.0 cm^3 is the volume of 1 gram-atomic mass, measure out 6.0 cm^3.
    b. A gram-atomic mass of any element contains the same number of atoms. Therefore, 35.5 g of chlorine has the same number of atoms as 12 g of carbon.
4. a. When the known atoms were arranged in order of atomic mass and lined up in families with similar properties, there were three spaces belonging to no known elements. The spaces occurred simply because the elements having atomic masses of 45, 68, and 70, together with the required properties, had not yet been discovered.
    b. gallium, scandium, and germanium
5. a. atomic number
    b. the number of protons in the nucleus
    c. Henry Moseley
    d. Bombard the element with high-speed electrons, and measure the wavelength of the x-rays emitted by the element.
6. a. 1 b. 2 c. 1 d. 8 e. 2 f. 7 g. 8 h. 1 i. 7 j. 4
7. Metals have outer electrons that are easily separated from the atom; that is, the atom is easily ionized. Therefore, a piece of solid metal has many free electrons floating inside it. In a nonmetal, the outer electrons are not easily separated from the atom, so there are very few free electrons found within the substance. In a semimetal the ease of ionization is intermediate, resulting in some, but not many, free electrons.
8. the Pauli exclusion principle
9. The electron goes into a state that has the least amount of energy of all the available states.
10. a. two electrons in the first shell ($n = 1$), plus one electron in the s subshell of the second shell ($n = 2$)
    b. two electrons in the first shell, plus two electrons in the s subshell of the second shell ($n = 2$), plus four electrons in the p subshell of the second shell
    c. two electrons in the s subshell of the first shell, plus two electrons in the s subshell of the second shell, plus six electrons in the p subshell of the second shell (making a total of eight electrons in the second shell)
11. lithium, oxygen, and neon
12. There are more metals than nonmetals because there are more elements with one or two electrons in the outer shell. This happens because once there are two electrons in the outer shell, additional electrons tend to be added to one of the lower shells, leaving two electrons still in the outer shell.
13. Silicon is used in the manufacture of glass and construction materials (concrete) and in the manufacture of semiconductors, which are used in transistors.
14. a. It requires about 1 eV of energy to release an outer electron from an atom in a crystal of pure silicon or germanium. At room temperature, there are enough free electrons in the crystal to make it a weak electrical conductor.
    b. In a carbon (diamond) crystal, the energy required to free electrons from the atoms is fairly large, so that at room temperature there are very few free electrons in the

crystal. Therefore, the electrical conductivity is very low; the crystal is a nonconductor.

15. Cesium has one outer electron with a fairly small ionization energy. Since that electron is easily transferred from the cesium atom, this element is very active. Helium, on the other hand, has eight outer electrons, which is an especially stable configuration. Therefore, helium does not attract or give up electrons and is an inert element.

Chapter 7

1. a.–h. See chapter glossary for definitions.

2. In covalent bonds, the valence electrons are shared equally between two atoms. In polar covalent bonds, the shared electrons spend more time near one of the atoms than the other. An ionic bond is an extreme case of a polar covalent bond; we may think of the outer electron of the one atom being separated from that atom and captured by the other atom.

3. When one atom is more electronegative than another, it has a stronger attraction for electrons. In that case the shared electrons remain closer to the more electronegative atom than to the other.

4. In the hydrogen molecule, two hydrogen atoms attract the shared electrons equally. Therefore, the molecule is symmetrical and unpolarized. In HCl the chlorine atom attracts the shared electrons more strongly than does the hydrogen atom, causing the chlorine end of the molecule to be more negative than the other end.

5. If the substance is a liquid, determine how good an electrical conductor it is. (Introduce a pair of electrodes connected to a battery and measure how much current flows.) If it is a solid, either dissolve it in water (if possible) or melt it, and perform the same test. If it is a good conductor, the substance has ionic bonding.

6. Hydrogen has just one orbital electron, which may be transferred to (or shared by) a nonmetal atom.

7. a. Na, 1+; I, 1−   **b.** Na, 1+; O, 1−   **c.** Mg, 2+; H, 1−
   d. Fe, 3+; Br, 1−   **e.** C, 4+; O, 2−

8. a. H:S̈:
       H

   b. H:F̈:

   c. H:N̈:H
       H

   d. :C̈l:N̈:C̈l
           :C̈l:

   e. H:Ö:
       H

   f.   H
         H:C̈:H
             H

9. An atom tends to be in the most stable state when it has eight electrons in its outer shell (except for atoms with atomic numbers of less than 5, which reach such stability with two outer-shell electrons).

10.

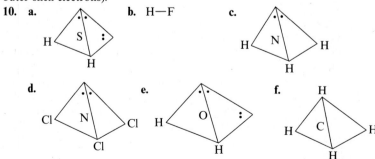

ANSWERS TO SELF TESTS 567

11. The four pairs of shared electrons repel each other, spreading out into a four-lobed figure that can be represented by a tetrahedron. (A tetrahedron has four corners.)

Chapter 8

1. **a.** CO_2 **b.** $KHSO_4$ **c.** $K_2Cr_2O_7$ **d.** NH_4Cl **e.** $Fe_2(SO_3)_3$ **f.** $Al_2(SO_4)_3$
   g. $LiNO_3$ **h.** $Mg(BrO_3)_2$ **i.** HNO_3 **j.** $Pb(OH)_3$
2. **a.** hydrogen sulfate (sulfuric acid in aqueous solution) **b.** calcium hydroxide
   c. magnesium phosphate **d.** sodium iodide
   e. potassium chlorite **f.** hydrofluoric acid **g.** ammonia
   h. calcium carbonate **i.** strontium oxide **j.** acetic acid

Chapter 9

1. **a.** 110.99 g **b.** 180.16 g **c.** 98.00 g
2. **a.** 3.0 mol **b.** 0.522 mol **c.** 434 mol **d.** 0.628 mol
3. **a.** 142 g **b.** 467 g **c.** 8.7×10^5 g **d.** 7.0×10^5 g
4. **a.** 219 g **b.** 37.2 g of KCl; 24.0 g of O_2
5. **a.** 22.2 L **b.** 71.7 g/mol
6. **a.** 1.43 g/L **b.** 32 g/mol
7. **a.** 1.0 L **b.** 1.2 L
8. **a.** 0.29 mol **b.** 8.1 g
9. 7.0 L
10. 0.716 g/L

Chapter 10

1. The iron gradually disappears, and a red-brown precipitate of copper appears. The blue solution changes to a pale yellow.
2. Zinc is higher in the activity series than hydrogen and therefore has less affinity for electrons than hydrogen does. As a result, the zinc atoms give up their outer electrons to hydrogen ions. The hydrogen ions become hydrogen atoms and escape from the solution, leaving Zn^+ ions in the solution. In this manner, the zinc replaces the hydrogen in the hydrochloric acid solution.
3. **a.** double displacement reaction **b.** 0.5 mol **c.** 30.2 g of H_2SO_4
4. **a.** synthesis, decomposition, displacement, double displacement
   b. synthesis: $2H_2 + O_2 \rightarrow 2H_2O$
   decomposition: $2KClO_3 \rightarrow 2KCl + 3O_2$
   displacement: $Mg + 2HCl \rightarrow MgCl_2 + H_2$
   double displacement: $NaCl + AgNO_3 \rightarrow NaNO_3 + AgCl\downarrow$
5. A chemical equation does not mean that the left side of the equation "equals" the right side. It means that the substances on the left side "yield" the substances on the right side.
6. Bubble the unknown gas through water, and add a solution of calcium hydroxide (lime water). If the gas contains carbon dioxide, a white precipitate of calcium carbonate is obtained, as a result of these reactions:

$$H_2O + CO_2 \longrightarrow H_2CO_3$$
$$H_2CO_3 + Ca(OH)_2 \longrightarrow CaCO_3\downarrow + 2H_2O$$

7. Precipitation causes one of the products of a reaction to be removed from the reacting system. The reaction then proceeds in the direction that produces more of the depleted product.

8. a. synthesis: $2H_2 + CO \rightarrow CH_3OH$
   b. synthesis: $3H_2 + N_2 \rightarrow 2NH_3$
   c. displacement: $Fe_3O_4 + 4H_2 \rightarrow 3Fe + 4H_2O$
   d. displacement: $B_2H_6 + 6H_2O \rightarrow 2H_3BO_3 + 6H_2$
   e. double displacement: $H_2SO_4 + 2NaOH \rightarrow Na_2SO_4 + 2H_2O$
   f. displacement: $C_3H_8 + 5O_2 \rightarrow 3CO_2 + 4H_2O$
   g. displacement: $Mg + 2HCl \rightarrow MgCl_2 + H_2$
   h. displacement: $2AgNO_3 + Cu \rightarrow Cu(NO_3)_2 + 2Ag$
   i. synthesis: $CO + Fe_3O_4 \rightarrow CO_2 + 3FeO$
   j. displacement: $6NaOH + 2Al \rightarrow 3H_2 + 2Na_3AlO_3$
   k. double displacement: $2HCl + Na_2CO_3 \rightarrow CO_2 + H_2O + 2NaCl$
   l. displacement: $6CO_2 + 6H_2O \rightarrow C_6H_{12}O_6 + 6O_2$

Chapter 11

1. a. $Cl_2 + 2NaBr \rightarrow 2NaCl + Br_2$
      NaCl: 4.64 mol; 271 g
      Br_2: 2.32 mol; 371 g
   b. $TiCl_4 + 2H_2O \rightarrow TiO_2 + 4HCl$
      TiO_2: 0.296 mol; 23.6 g
      HCl: 1.18 mol; 43.0 g
   c. $CaO + SO_3 \rightarrow CaSO_4$
      $CaSO_4$: 44.00 mol; 5990 g
   d. $2H_2O_2 \rightarrow 2H_2O + O_2$
      H_2O: 8.09×10^{-4} mol; 0.0146 g
      O_2: 4.04×10^{-4} mol; 0.0129 g
   e. $C_6H_{12}O_6 + 6O_2 \rightarrow 6CO_2 + 6H_2O$
      CO_2: 5.0 mol; 220 g
      H_2O: 5.0 mol; 90 g
2. a. 0.077 mol b. 1.1 g
3. a. 0.38 g b. 0.0063 mol
4. a. 600 g b. 1.46×10^4 g
5. 4.6 g of HF
6. 2.73×10^4 g, or 27.3 kg
7. a. 16.7% b. 33.3% c. 50%
8. a. 10.6% b. 81.1% c. 3.5% d. 1.3% e. 3.5%
9. a. 0.0018 mL b. approximately 15 mL
10. a. 40 mL b. 10 mL

Chapter 12

1. a. The pressure increases.
   b. The volume increases.
   c. As the temperature increases, the gas molecules move faster and therefore apply more force to the walls of the container. If the container walls are fixed, the pressure goes up. If the walls can move (as in a cylinder with a piston), the gas expands.
2. a. The pressure doubles.
   b. The volume doubles.
   c. The Kelvin temperature must be halved.

ANSWERS TO SELF TESTS 569

    d. The temperature cannot change by itself; the gas must be cooled from the outside. That is, energy must be removed from the system if the product PV is to remain unchanged as the number of moles of gas (n) increases.

3. 1.80×10^5 Pa

4. The Kelvin temperature must be doubled, an increase of 293 K (from 293 K to 586 K). The Centigrade temperature increases by 293° from 20°C to 313°C.

5. 0.72 atm

6. 777 torr

7. **a.** Each time a molecule hits the container wall, it imparts a certain amount of impulse to the wall. The faster the molecule, the greater the impulse. Also, the faster the molecules move, the more often they strike the walls. Therefore, the pressure is proportional to the square of the molecular velocity.

    b. The gas temperature is proportional to the average kinetic energy of the molecules, which in turn is proportional to the square of the molecular velocity. Therefore, using the result of part a, the gas pressure is directly proportional to the temperature.

8. **a.** Lowering the temperature reduces the molecular velocity. Since the molecules move more slowly, it takes more time for them to travel from one container wall to another. As a result, there are fewer collisions per second.

    b. The pressure is reduced by increasing the volume, decreasing the temperature, *or* decreasing the number of molecules in the container. Any of these reduces the number of collisions per second.

    c. Increasing the volume decreases the number of collisions per second, because it takes a longer time for each molecule to bounce from one wall to another.

    d. Adding more molecules increases the number of collisions per second, because there are more molecules in the container to undergo collisions.

9. Ideal gas molecules have no volume and do not interact with each other, except at the instant of collision. Real gas molecules have definite volumes and interact with each other at a distance. For this reason, real gases do not obey the ideal gas law exactly, especially near their condensation points. In fact, the condensation of a gas into a liquid is the result of interactions among the molecules.

10. An ideal gas has zero volume and pressure at a temperature of 0 K (absolute zero). A real gas condenses and then freezes when cooled toward that temperature, and therefore does not lose all its volume.

Chapter 13

1. See chapter glossary for definitions.

2. Both ethanol and water have polar covalent molecules. The dipole-dipole interaction creates an attraction between ethanol molecules and water molecules, causing the ethanol to remain intimately mixed with the water.

3. When sodium chloride, an ionic solid, mixes with water, the attraction of each sodium ion to a chloride ion is overcome by the attraction of the surrounding polar water molecules. The sodium and chloride ions separate, and each is surrounded by a cluster of water molecules. The negative (oxygen) end of each water molecule points toward the positive sodium ion, and the positive end of the water molecule (hydrogens) points toward the negative chloride ion.

4. **a.** 0.27 M **b.** 0.27 N

5. **a.** 0.14 M **b.** 0.28 N

6. **a.** 75 mL   **b.** 3.75×10^{-3} mol of each reactant
7. **a.** 21.9 g   **b.** 1.37 N   **c.** 2.20 N
8. **a.** The freezing point is at $-3.6°C$.   **b.** The boiling point is at 101°C.

Chapter 14

1. See chapter glossary for definitions.
2. The acid neutralizes the base (that is, the OH^- ions react with the H^+ ions, forming water) to remove the bitter alkali taste and thereby improve the flavor of the product.
3.

	Acid	Base	Acid	Base
a.	HCl	Cl^-	H_3O^+	H_2O
b.	HCH_3CO_2	$CH_3CO_2^-$	H_3O^+	H_2O
c.	H_2SO_4	HSO_4^-	$H_3SO_4^+$	H_2SO_4

4. **a.** 7.0 **b.** 4.7 **c.** 13.5
5.

	$[H^+]$	$[OH^-]$
a.	1.6×10^{-3}	6.3×10^{-12}
b.	1.3×10^{-1}	7.9×10^{-14}
c.	7.9×10^{-11}	1.3×10^{-4}

6. **a.** 1.0×10^{-9} mol of dissociated molecules per mole of solute
   b. 1.6×10^{-5} mol of dissociated molecules per mole of solute
   c. 0.5 mol of dissociated molecules per mole of solute
7. The addition of acetic acid to reach a pH of 6.1 increases the hydrogen ion concentration from its neutral value of 10^{-7} mol/L to 7.9×10^{-7} mol/L. The hydroxide ion concentration decreases to 1.25×10^{-8} mol/L.

Chapter 15

1. See chapter glossary for definitions.
2. **a.** $r = k[HCOOH][NH_4OH]$ **b.** $r = k[HCl][KOH]$
   c. $r = k[AgNO_3][NaCl]$ **d.** $r = k[N_2][H_2]^3$
3. When the temperature increases, the molecules taking part in a reaction move more rapidly, and there are a greater number of molecules with kinetic energies greater than the activation energy of the reaction. Therefore, there are a greater number of effective collisions resulting in reaction.
4. Since the rate of developing increases with temperature, the higher the temperature, the shorter will be the developing time.
5. The rough rule of thumb for reaction rates is that the rate doubles for every increase in temperature of 10°C. Therefore, the rate at 10°C is double the rate at 0°C. At 50°C the rate is 32 times the rate at 0°C.
6. **a.** $K_i = \dfrac{[H_3O^+][CH_3COO^-]}{[CH_3COOH]}$ **b.** $K_i = \dfrac{[NH_4^+][OH^-]}{[NH_3]}$

7. **a.** $[OH^-] = 8.1 \times 10^{-5}$ mol/L   **b.** $[OH^-] = 1.6 \times 10^{-3}$ mol/L
   c. $[NH_4^+]$ decreases by a factor of 20, and $[OH^-]$ increases by a factor of 20.
   d. If a base, such as NaOH, is added to the solution, the hydroxide ion concentration is increased. The reaction rate from right to left then increases, decreasing the concentration of NH_4^+.

ANSWERS TO SELF TESTS   571

8. Given the reaction

$$\text{solid}_1 + \text{gas} \rightleftharpoons \text{solid}_2$$

increasing the gas pressure drives the reaction to the right, thereby decreasing the amount of gas (and thus reducing the pressure) and increasing the amount of solid$_2$.

9. **a.** 2.86×10^4 kJ **b.** 1600 g of O$_2$
10. **a.** 5.36×10^4 kJ **b.** 3800 g of F$_2$
    c. The hydrogen-oxygen reaction yields 15.9 kJ per gram of fuel; the hydrogen-fluorine reaction yields 13.4 kJ per gram of fuel. Therefore, the hydrogen-oxygen reaction is preferable.
11. According to Le Chatelier's principle, a reaction goes in the direction that will reduce the stress applied to the system. Increasing the temperature of the system is an applied stress. The equilibrium state will shift so as to reduce the increase of temperature. If the left-to-right reaction is exothermic, the reaction will tend to go in the opposite direction, which is endothermic. Therefore, the equilibrium shifts toward the left.
12. Since dissolving salt in water is endothermic, the reaction tends to reduce the temperature of the system. Therefore, the ice-salt-water mixture has a lower temperature than the simple ice-water mixture.

Chapter 16
1. **a.** $4NH_3 + 5O_2 \rightarrow 4NO + 6H_2O$
   b. N^{-3} is oxidized to N^{+2}. O^0 is reduced to O^{-2}.
   c. O^0 is the oxidizing agent. N^{-3} is the reducing agent.
2. **a.** $Cu + 4HNO_3 \rightarrow Cu(NO_3)_2 + 2H_2O + 2NO_2$
   b. Cu^0 is oxidized to Cu^{+2}. N^{+5} is reduced to N^{+4}.
   c. N^{+5} is the oxidizing agent. Cu^0 is the reducing agent.
3. **a.** $2Ag + Hg_2Cl_2 \rightarrow 2AgCl + 2Hg$
   b. Ag^0 is oxidized to Ag^{+1}. Hg^{+1} is reduced to Hg^0.
   c. Hg^{+1} is the oxidizing agent. Ag^0 is the reducing agent.
4. **a.** $CuO + H_2 \rightarrow Cu + H_2O$
   b. H^0 is oxidized to H^{+1}. Cu^{+2} is reduced to Cu^0.
   c. Cu^{+2} is the oxidizing agent. H^0 is the reducing agent.
5. **a.** $2Na + 2H_2O \rightarrow 2NaOH + H_2$
   b. Na^0 is oxidized to Na^{+1}. H^{+1} is reduced to H^0.
   c. H^{+1} is the oxidizing agent. Na^0 is the reducing agent.
6. **a.** $2NaI + Cl_2 \rightarrow 2NaCl + I_2$
   b. I^{-1} is oxidized to I^0. Cl^0 is reduced to Cl^{-1}.
   c. Cl^0 is the oxidizing agent. I^{-1} is the reducing agent.
7. **a.** $2NaMnO_4 + 10NaCl + 8H_2SO_4 \rightarrow 2MnSO_4 + 6Na_2SO_4 + 8H_2O + 5Cl_2$
   b. Cl^{-1} is oxidized to Cl^0. Mn^{+7} is reduced to Mn^{+2}.
   c. Mn^{+7} is the oxidizing agent. Cl^{-1} is the reducing agent.

Chapter 17
1. See chapter glossary for definitions.
2. **a.** propane **b.** 2-methylpropane (isobutane) **c.** 2-butene **d.** 2-pentyne
   e. propanol **f.** 1-chloropropane **g.** 2-chloropropane
   h. 2-butanone (ethyl methyl ketone) **i.** propanal **j.** diethyl ether

3. **a.** $CH_3CH_2CH_2CH_2CH_2CH_2CH_2CH_3$ (C_8H_{18}) **b.** $CH_3CH_2CHClCH_2CH_3$
   c. $HC\equiv CCH_2CH_3$ **d.** $CH_3CH=CHCH_3$
   e. $CH_3CH_2OCH_2CH_3$ ($C_2H_5OC_2H_5$) **f.** $CH_3COCH_2CH_3$ ($CH_3COC_2H_5$)
   g. CH_3COCH_3 **h.** CH_3CH_2OH (C_2H_5OH)
4. **a.** alkane **b.** alkene **c.** alkyne **d.** ether **e.** aldehyde
   f. amino acid (dipeptide) **g.** alcohol **h.** carbohydrate **i.** acid

Chapter 18

1. See chapter glossary for definitions.
2. **a.** $^{238}_{92}U \rightarrow {}^{234}_{90}Th + {}^{4}_{2}He$ **b.** $^{234}_{90}Th \rightarrow {}^{234}_{91}Pa + e^-$
   c. $^{59}_{27}Co + {}^{1}_{0}n \rightarrow {}^{60}_{27}Co$ **d.** $^{27}_{13}Al + {}^{4}_{2}He \rightarrow {}^{30}_{15}P + {}^{1}_{0}n$
3. **a.** Uranium, protactinium, radium, and other members of the uranium radioactivity chain should be found.
   b. A Geiger counter will detect the presence of radioactive isotopes. A scintillation counter can be used to measure the energy of characteristic gamma rays and thus identify the isotopes present.
4. The total dose is a measure of the total amount of radiant energy absorbed in a given amount of time. The dose rate is the amount of energy absorbed per unit time (per hour, per day, etc.).
5. **a.** $^{241}_{95}Am \rightarrow {}^{237}_{93}Np + {}^{4}_{2}He$
   b. Ions may be created by emitted alpha particles colliding with air molecules or smoke particles.
   c. Compared with the background radiation normally encountered, 0.5 mrem is an insignificant amount of radiation. The radioactive material itself is not harmful as long as it is kept sealed in its container.
6. **a.** The report does not indicate the period of time over which this dose was recorded; therefore, the dose rate is not known. Second, the report does not specify whether the radiation came directly from the plant in the form of gamma rays or from radioactive gases released into the atmosphere.
   b. Since coal contains small amounts of uranium and thorium, the gases and soot emitted from the power plant through the smokestacks will contain particles of radioactive isotopes.
7. 80%
8. **a.** 4.5 half-lives; 36 days
   b. There could be possible damage to cells from radiation and a slight increase of cancer risk (but small compared with that associated with normal background radiation).
9. The isotope can be injected in the form of a compound that tends to concentrate in a particular organ; the emitted gamma radiation can be used to form an image of that organ with a gamma-ray camera. (See Problem 8.) Metabolism in the body can be studied by measuring the rate of excretion of certain compounds tagged with radioactive isotopes.
10. **a.** Five sources of radiation are (1) cosmic rays, (2) radioactive isotopes in the earth's crust, (3) diagnostic and therapeutic x-rays, (4) radioactive isotopes used for diagnosis and treatment, and (5) nuclear reactors.
    b. Diagnostic and therapeutic x-rays and isotopes can be reduced to a minimum by proper techniques. Nuclear reactors can be shielded to reduce emitted radiation to an acceptable level.
    c. Background radiation is always present. It can be reduced by living in a deep cave, but then the radioactivity in the earth's crust is still present.

Index

Abscissa, 59, 73
Absolute pressure, 339, 371
Accuracy, 66, 73
Acetic acid, 452, 513
Acetone, 512
Acetylene (*see* Ethyne)
Acid anhydride, 291, 302
Acid strength, 414, 429
Acid-base pair, 412, 429
Acidity, degree of, 419, 429
Acids, 247–249, 255, 408
   Arrhenius, 408, 429
   Bronsted-Lowry, 411, 429
   Lewis, 413, 429
   strong, 408, 415, 429
   weak, 408, 415, 429
Actinide, 185, 205
Activity series, 293, 302
Addition, 19, 33
Affinity, electron, 213–214, 230
Alanine, 515
Alchemy, 124, 132, 186
Alcohols, 508–509, 518
Aldehyde, 511, 518
Alkali metal, 180, 195, 205
Alkaline earth metal, 197, 205
Alkalinity, degree of, 420, 429
Alkanes, 491, 499–501, 518
   boiling points, 492
Alkenes, 493, 500–502, 518
Alkyl groups, 500
Alkynes, 494, 500–502, 518
Allotropic forms, 200, 205, 484, 518
Alloy, 95, 323, 382, 402
Alpha emission, 534
Alpha particle, 143, 531, 537, 550
Aluminum, 198, 292
Amine compound, 514–516

Amine group, 514, 518
Amino acid, 514–516, 518
Ammonia, 239, 449, 462
Analysis, 1, 125
Angstrom, 147, 173
Ångström, Anders Jonas, 147
Anion, 217, 230
Anode, 141, 217, 230
Antimony, 199
Antomons, Guillaume, 350
Aqua regia, 201
Area, units of, 45
Arrhenius, Svante August, 408
Atmosphere, 340, 371
Atom, 85
Atomic mass, 152, 173, 181, 268
Atomic mass unit (amu), 153, 173
Atomic model, 2, 4, 142, 162
Atomic number, 152, 157, 173, 183
Atomic radius, 201
Atomic structure, 190–193
Atomic volume, 181, 182
Atomic weight, 128, 153
Autoionization, 414, 429
Avogadro, Amedeo, 263
Avogadro's hypothesis, 264–266, 280
Avogadro's number, 268–271, 280

Bakelite, 510
Balmer series, 165
Bar, 341, 371
Barometer, 335, 344, 371
Bases, 249, 250, 255, 409
   Arrhenius, 409, 429
   Bronsted-Lowry, 411, 429
   Lewis, 413, 429

   strong, 415, 429
   weak, 415, 429
Becquerel (unit of radiation), 543, 550
Becquerel, Antoine, 530
Benzene, 504–507
Benzene ring, structure, 507
Berthollet, Claude Louis, 125
Berzelius, Jons Jacob, 446
Beta decay (*see* Beta emission)
Beta emission, 533, 550
Beta particle, 143, 531, 550, 538
Binary acids, 247, 248
Binary compound, 242, 255
Biochemistry, 2
Biology, molecular, 2
Bohr, Niels, 162
Bohr model, 163, 166, 187
Boiling point, 62, 397, 402
Boltzmann, Ludwig, 367
Bond strength, 221, 230
Bonding, 214, 230
Bonds
   covalent, 215, 218, 230
   double, 221, 230, 488, 518
   hydrogen, 387, 402
   ionic, 215–218, 230, 388
   polar-covalent, 222, 230, 388
   single, 221, 230, 488, 518
   triple, 221, 230, 489, 519
Boyle, Robert, 124, 344
Boyle's law, 344–349, 371
Branched chain (*see* Compounds, branched chain)
British thermal unit, 102, 114
Bromine, 180
Bronsted, Johannes, 410
Buret, 395–396, 402
Butane, 491

Calcium, 197
Calibration, 62, 66, 73
Calorie
  gram, 102, 114
  kilogram, 102, 114
Calorimeter, 102
Cannizzaro, Stanislao, 265
Carbohydrate, 516–517, 518
Carbon, 482–486
  electron structure, 482
Carbon dioxide, 495
  solid, 370, 371, 495
Carbon monoxide, 495
Carbonyl group, 498, 518
Carboxyl group, 512, 518
Catalysis, 446, 463
Catalyst, 446, 463
Catenation, 487, 518
Cathode, 141, 217, 230
Cathode rays, 141
Cation, 217, 230
Celsius scale, 61, 63, 73
Chadwick, James, 145
Changes
  chemical, 84, 114
  physical, 86
Charge number, 240, 255, 471, 476
Charles, Jacques Alexander, 350
Charles's law, 349–356, 371
Chemistry, uses of, 1, 2
Chemists, types of, 1
Chlorine, 180, 201
Cobalt-60, 536, 545
Collision rate
  effective, 442, 463
  total, 442, 463
Collisions
  effective, 442, 463
  elastic, 367, 371, 437
  inelastic, 437, 463
  molecular, 436–441
Colloidal suspension, 383, 402
Combining volumes, law of, 263, 280
Combustion, 110, 114
  heat of, 111, 114
Completion of reaction, 449, 463
Compounds, 87, 93, 114
  acyclic, 499, 518
  aromatic, 505–507, 518
  branched chain, 502, 518
  carbon-halogen, 496
  covalent, 253

cyclic, 499, 503–507, 518
  double-bond (see Alkenes)
  organic, 87, 200, 480–519
  triple-bond (see Alkynes)
Concentration, 383, 402
Condensation, 91, 114
Condensation point, 352, 371
Conservation of energy, law of, 3, 8, 98, 101, 114, 534
Conservation of mass, law of, 101, 114
Constant, 27
Constant composition, law of (see Definite proportions, law of)
Conversion factor, 40, 52, 73, 287
Crookes, Sir William, 140
Cryogenics, 354, 357, 371
Crystal, 85, 86, 114, 217
Crystalline material (see Materials, crystalline)
Curie (unit of radiation), 542, 550
Curie, Marie, 143, 197, 530
Cyclopropane, 504

Dalton, John, 85, 128, 142, 152, 262, 364
Dalton's law (see Partial pressures, law of)
Davy, Sir Humphry, 196, 446
de Broglie, Louis, 166
Definite proportions, law of, 125, 132
Democritus, 84
Denatured alcohol, 509, 518
Denial, laws of, 4
Density, 53, 73, 332, 371
  relative (see Specific gravity)
Derivative, 490, 518
Deuterium, 156, 173
Diamond, 485
Diatomic molecule, 219, 230
Diffusion, 266, 280
Dipeptide, 515
Dipole, electric, 222, 230
Dipole-dipole interaction, 222, 230, 387
Dissociation, 496, 518
  electrolytic, 408, 410, 429
Distillation, destructive, 494, 507, 508, 518
Division, 17, 32
Dose, total radiation, 544, 551
Dose rate, 544, 550
Dosimeter, 541, 550

Dry ice (see Carbon dioxide, solid)
Ductility, 195, 205

Einstein, Albert, 100, 161, 166
Electric charge, 138, 141, 268
Electrolysis, 196, 205, 217, 267
Electrolytes, 408, 429
Electron, 141, 173, 268, 532–533
Electron configuration, 170, 188–190
Electron transfer, 410
Electron-dot model (see Models, Lewis)
Electronegativity, 214, 230
Electron-volt (eV), 164, 173
Elements, 93, 114, 122, 123
  symbols for, 124
  transition (see Transition element)
Energy, 3, 95
  activation, 441, 463
  average, 443, 463
  chemical potential, 99, 114
  conservation of (see Conservation of energy, law of)
  internal, 460, 463
  ionization, 203
  kinetic, 96, 114, 367, 368, 441, 463
  potential, 97, 115
  thermal, 98, 102, 105, 115
Energy distribution curve, 443, 463
Energy level, 164, 166, 168, 173
Enthalpy, 461, 463
Equations
  algebraic, 24, 31, 33
  balanced, 297–302
  chemical, 124, 237, 255, 286–305
  linear, 27, 30, 31
  van der Waals, 369, 372
Equilibrium, 450, 463
  and concentration, 456–457
  and pressure, 457–459
  and stress, 455
  and temperature, 462
Equilibrium constant, 450, 463
Errors, 66, 73
  random, 67, 73
  systematic, 67, 73
Error analysis, 66, 71
Ester, 514
Ethane, 490
Ethanol, 490, 508

Ethene, 493, 501
Ether, 510, 518
Ethyne, 494, 502, 511
Evaporation, 91, 114
Excited state, 164, 173
Exponent, 12, 31

Fahrenheit scale, 61, 62, 73
Family of elements, 87, 180, 185, 188–193, 205
Ferrous chloride, 288
Fluid, 333
Fluorescence, 140, 173
Fluorine, 180, 201
Footpound, 96, 114
Force, 96, 138, 173
   electromagnetic, 138, 139, 212, 230
   gravitational, 138
   inverse-square, 139, 173
   nuclear, 146, 173, 532, 551
   van der Waals, 219, 220, 390
Formaldehyde, 510
Formation, heat of, 112, 114
Formic acid, 513
Formula, 238, 244–246
   molecular, 238, 255
   structural, 238, 255
Francium, 197
Freezing point, 352, 371, 397
Freon, 498
Fructose, 516
Fusion, 297, 302
   heat of, 111, 114

Galilei, Galileo, 336
Gamma emission, 534
Gamma radiation, 143, 532, 539, 550
Gas law
  combined, 358–360, 371
  ideal, 362, 371
  universal, 5, 360–364, 372
Gases, 87, 90, 114, 332, 371
  general properties, 333, 371
  ideal, 362, 369, 371
  noble, 93
  particular properties, 332, 371
  real, 369, 372
Gauge pressure, 339, 371
Gay-Lussac, Joseph Louis, 262, 350
Gay-Lussac's law, 350

Geiger counter, 541
Geissler, Heinrich, 140
Germanium, 199
Glucose, 516
Glycerol, 509
Glycine, 515
Glycols, 509
Gram atomic mass, 182, 205, 268
Gram molecular mass, 270, 280
Graph, 28, 31, 58
Graphite, 485
Gravitational force (*see* Force, gravitational)
Gravity, specific (*see* Specific gravity)
Ground state, 164, 173
Group (*see* Family of elements)
Guericke, Otto von, 344

Half-life, 533, 545–546, 550
Halogens, 180, 201, 205
Handbook of Chemistry & Physics, 40, 153, 157, 461, 481, 537
Heat (*see* Energy, thermal)
Heat capacity, 108, 114
Hertz, Heinrich, 159, 528
Heterogeneous, 94, 114
Homogeneous, 93, 114
Hydration, heat of, 389, 402
Hydrocarbons, 486–494, 518
  halogenated, 496–499
  paraffin (*see* Alkanes)
  saturated, 489, 518
  unsaturated, 489, 501–502, 519
Hydrochloric acid, 201, 286–289
Hydrogen, 219, 294
Hydrogen ion concentration, 420
Hydrogen peroxide, 292
Hydrolysis, 296, 302
Hydronium ion, 410, 429
  concentration, 420, 429
Hydroxide, 249, 255
Hydroxide ion concentration, 420, 429
Hypothesis, 3, 8, 262

Indicator, 395, 402
Inert substance (*see* Substance, inert)
Inertia, 49, 73
Inhibitors, 448, 463
Insulators, 200, 205
International System, 45, 53, 73

International Union of Pure and Applied Chemistry, 124, 236, 255, 499
Inverse-square force, 139, 173
Iodine, 180
Ion, 155, 173, 214, 230
  polyatomic, 238–243, 255
Ion concentration, 416, 429
Ion product constant, 421, 429
Ionic charge, 218, 239
Ionic compound, 216, 239
Ionization, 155, 173, 408
  degree of, 416, 429
Iron chloride (*see* Ferrous chloride)
Isomers, 502, 518
Isotopes, 157, 173, 528
  carbon, 158
  hydrogen, 156
  mercury, 158
  neon, 156
  separation of, 159
IUPAC (*see* International Union of Pure and Applied Chemistry)

Joule, 96, 105, 114
Joule, James Prescott, 98, 367

Kekulé, Friedrich, 505
Kelvin scale, 63, 65, 73, 352–354
Ketone, 511, 518
Kilogram, 50
Kilomole, 274
Kilopascal, 340, 371
Kinetic theory, 105, 114, 366–369, 371

Lanthanide, 185, 189, 205
Lattice, crystal, 88, 114, 216, 230
Lavoisier, Antoine, 124, 480
Laws
  of denial, 4
  of nature, 3, 8, 84
  of permission, 4
Le Chatelier, Henri Louis, 455
Le Chatelier's principle, 455, 462, 463
Length, units of, 45
Lewis, G. N., 226, 412
Limestone, 292
Linear relationship, 60, 73, 351, 371
Liquid, 87, 89, 115
Liquid crystal, 92, 115
Litmus test, 414, 429

Logarithms, 424–426, 429
Loschmidt, Joseph, 266
Loschmidt's number, 266, 280
Lowry, Thomas, 410

Macromolecule, 502, 518
Magdeburg hemispheres, 347
Magnetite, 294, 302
Malleability, 195, 205
Manometer, 339, 371
Mariotte, Edmé, 349
Mariotte's law, 349
Mass, 48, 49, 73, 100, 115
    conservation of (*see* Conservation of mass, law of)
    standard, 50
    units of, 48
Mass action, law of, 450, 463
Mass number, 157, 173
Mass ratio, 309–312, 325
Mass spectrometer, 150, 173
Mass-mass problems, 315–321, 325
Materials
    amorphous, 88, 114
    crystalline, 88, 114
Matter, forms of, 84
Matter wave, 166, 173
Maxwell, James Clerk, 138, 159, 367, 528
Mayer, Julius Robert, 367
Melting point, 62, 90, 115
Mendeleev, Dmitri, 181
Metals, 193, 205, 409
    alkali (*see* Alkali metal)
    alkaline earth (*see* Alkaline earth metal)
Metastable molecule, 292, 302
Meter, 45
Methanal, 511
Methane, 482, 489
Methanol, 508
Methyl halides, boiling points of, 498
Metric system, 45, 73
Meyer, Julius Lothar, 181–183
Millibar, 341, 371
Millikan, Robert A., 141, 268
Millirem, 544, 551
Mixtures, 94
Models
    Arrhenius, 408
    Bronsted-Lowry, 410
    Lewis, 226, 230, 412
    in science, 171
Molar mass, 271, 280
Molar volume, 277, 280
Molarity, 390–392, 402
Mole, 269–273, 280
Mole ratio, 308, 309, 325
Molecular mass, 270, 272, 280, 347
Molecular solids, 220
Molecules, 85, 115, 262, 266
    saturated, 484, 518
    tagged, 547, 551
Mole-mass problems, 312–314, 325
Motion, perpetual, 3
Multiple proportions, law of, 130, 132
Multiplication, 14, 32

Nanometer, 46
Naphthalene, 507
Nature, laws of, 3, 8, 84
Neon, 155
Neutralization, 390, 402, 413
Neutron, 145, 173, 532–533
Newton (unit of force), 96, 340, 371
Niobium, 197, 357
Nitrogen, 200
Nitroglycerine, 292
Noble gases, 185, 186, 201, 205
Nonelectrolyte, 408, 429
Nonmetals, 180, 199, 205, 409
Normality, 392–396, 402
Notation, scientific, 12, 14, 31, 32
Nuclear fission, 535, 537, 551
Nuclear force (*see* Force, nuclear)
Nucleus, 145, 173, 532
    daughter, 534, 550

Observations, 3
Octet rule, 226, 230
Orbital, 171, 173
Ordinate, 59, 73
Organic acids, 512–514
Organic compounds (*see* Compounds, organic)
Organo-metallic compounds, 481
Osmosis, 382, 402
Oxidation, 180, 205, 224, 470, 476
Oxidation number, 212, 224–225, 230, 240–243, 470
Oxidizing agent, 225, 230, 471, 476
Oxo-acid, 247, 248, 255
Oxygen, 200, 291
Ozone, 295, 302, 447

Parameters, 28
Partial equation, 471, 476
Partial pressures, law of, 364–366, 371
Particles, 2, 141
Pascal (unit of pressure), 340, 371
Pascal, Blaise, 337
Pauli exclusion principle, 191, 205
Peptide bond, 515, 518
Percent composition, 321–325
Period, 185, 205
Periodic properties, 181, 186
Periodic Table, 180, 205
Permission, laws of, 4
Peroxide, 242, 255
Perpetual motion machine, 3
pH, 422–428, 429
Phenolphthalein, 395
Phenomenological description, 286, 302
Phosgene, 498
Phosphorus, 200
Photoelectric effect, 161, 173, 539
Photon, 161, 173, 187, 532
Planck, Max, 166
Plasma, 91, 115, 155
Polar molecule, 222
Polarization, 222–223
Polymers, 502, 518
Potassium chloride, 87
Potassium-40, 536
Power-of-ten system, 12
Power(s), 22, 31, 32
    negative, 13
    zeroth, 18
Precipitation, 295, 302
Precision, 66, 73
Prefixes, 45, 46
Pressure, 333, 368, 372
    units of, 339–344
Priestley, Joseph, 124, 291
Probability density, 168, 173
Products, 122, 132
Propane, 491
Propanol, 509
Propene, 493
Properties
    chemical, 86, 87, 114, 152
    colligative, 397–401, 402
    intrinsic, 53, 73

INDEX

Properties (cont.)
  physical, 86, 115
Proportion, direct, 29, 60, 73
Proteins, 515
Proton, 142, 173, 532–533
Proton transfer, 411, 429
Proust, Joseph Louis, 125

Qualitative analysis, 125, 132
Qualitative chemistry, 286, 302
Quantitative analysis, 125, 132
Quantitative chemistry, 286, 302
Quantum, 161, 173
Quantum mechanics, 166, 173
Quantum number, 163, 168, 173, 190

Radiation, 531, 551
  background, 544, 550
  electromagnetic, 159, 528–530, 550
  infrared, 528, 550
  measurement of, 540–544
  nuclear, 531, 551
  ultraviolet, 528, 551
  units of, 542–544
  uses of, 546–548
Radiation dose, 543, 551
Radical, 238, 255, 490, 508
Radioactive material, 531, 551
Radioactivity, 143, 173, 528, 530–539
Radioisotope, 536, 551
Range, particle, 538, 551
Rankine scale, 63, 65, 73
Rare earth element (*see* Lanthanide)
Rate, 42, 73
Rate equation, 445
Reactants, 122, 132
Reaction
  chemical, 436
  decomposition, 291, 302
  displacement, 292, 302
  double-displacement, 295–297, 302
  endothermic, 110, 114, 291, 302, 461
  exothermic, 110, 114, 291, 302, 460
  nuclear, 535, 536–537
  oxidizing (*see* Oxidation)
  reversible, 294, 302
  synthesis, 1, 290, 302

Reaction probabilities, 436, 441
Reaction rate, 441, 445, 463
Reactive substance (*see* Substance, reactive)
Redox reactions, 470–476
Reducing agent, 224, 230, 471
Reduction, 470, 476
Relativity, principle of, 4, 100, 115
Rem (unit of radiation dose), 543, 551
Research, 1, 8
Richter, J. B., 308
Roentgen, Wilhelm, 143, 528
Roots, 22, 31, 33, 246, 255
Rounding off, 69
Rumford, Count, 98
Rutherford, Ernest, 142, 145

Salts, 247, 250–253, 255, 410
Schroedinger, Erwin, 166
Schroedinger model, 166, 171
Scintillation counter, 541, 551
Selenium, 200
Semiconductors, 198, 205
Semimetals, 198, 205
Shell, electron, 169, 173, 188–193
Shielding, radiation, 539
Sievert, 544, 551
Significant figures, 69
Silicon, 198
Silver chloride, 296
Sodium, 180, 197
Sodium chloride, 86, 180, 214, 225
Solid, 87, 88, 115
Solubility, 95, 115, 385, 402
  temperature dependence of, 385
Solute, 94, 115
  ionic, 388
  nonpolar, 390
  polar covalent, 388
Solutions, 94, 382–401, 402
  concentrated, 383, 402
  dilute, 383, 402
  electrolytic, 408, 429
  heat of, 111, 114, 389, 402
  saturated, 95, 385, 402
  supersaturated, 386, 402
  types of, 382
Solvation, heat of, 389, 402
Solvent, 94, 115
Specific gravity, 56, 73
Specific heat, 106, 115
Spectrum, 160, 162, 174
  radiative, 532, 551

Standard atmospheric pressure, 335, 340, 372
Standard temperature and pressure, 277, 280, 340
State ground (*see* Ground state)
  intermediate, 437, 463
States of matter, 87, 88
Stoichiometry, 308–325
STP (*see* Standard temperature and pressure)
Structure, molecular, 228
Styrene, 507
Sublimation, 180, 205, 398, 402, 495
Substance
  inert, 87, 114
  pure, 92, 115
  reactive, 87, 115
Subtraction, 19, 33
Sucrose, 517
Sulfur, 89, 200
Sulfuric acid, 293
Superconductivity, 198, 205, 357, 372
Superoxide, 242, 255
Surface tension, 387, 402
Synthesis (*see* Reaction, synthesis)

Technical name, 236, 255
Temperature, 61, 102, 105, 115, 333, 367
Ternary acids, 248
Ternary compound, 242, 255
Tetrahedron, 228, 230
Theory, 4
Thomson, J. J., 141, 142, 154, 155
Time, units of, 51
Titration, 395, 402
Torr, 341, 372
Torricelli, Evangelist, 336, 344
Transition elements, 185, 197, 205
Transmutation, 123, 132, 186
Transuranium elements, 185
Trends in properties, 201
Tritium, 157, 174, 533, 537
Trivial name, 236, 255
Tungsten, 198

Unit factor method, 39, 73
Units, 39, 45, 73
Uranium, 530, 535

Vacuum, 335, 341, 372

Valence, 212, 230
van't Hoff's law, 462
Vapor, 332, 372
Vapor pressure, 398, 402
Variable, 27, 31, 58
Vitalism, 480, 519

Volume, units of, 47

Water, ionization of, 418
Wave mechanics, 166, 174
Wave packet, 148, 161, 166, 174
Weight, 49, 73

Wohler, Friedrich, 480
Work, 96, 115

X-rays, 529, 551

Zero, absolute, 65, 73, 353, 357, 371

Atomic and Molar Masses of the Elements[a]

ELEMENT	SYMBOL	ATOMIC NUMBER	ATOMIC MASS (u) MOLAR MASS (g/mol)	ELEMENT	SYMBOL	ATOMIC NUMBER	ATOMIC MASS (u) MOLAR MASS (g/mol)
Actinium	Ac	89	227.0278[b]	Erbium	Er	68	167.26
Aluminium	Al	13	26.981 54	Europium	Eu	63	151.96
Americium	Am	95	(243)[c]	Fermium	Fm	100	(257)[c]
Antimony	Sb	51	121.75	Fluorine	F	9	18 998 403
Argon	Ar	18	39.948	Francium	Fr	87	(223)[c]
Arsenic	As	33	74.9216	Gadolinium	Gd	64	157.25
Astatine	At	85	(210)[c]	Gallium	Ga	31	69.72
Barium	Ba	56	137.33	Germanium	Ge	32	72.59
Berkelium	Bk	97	(247)[c]	Gold	Au	79	196.9665
Beryllium	Be	4	9.012 18	Hafnium	Hf	72	178.49
Bismuth	Bi	83	208.9804	Helium	He	2	4.002 60
Boron	B	5	10.81	Holmium	Ho	67	164.9304
Bromine	Br	35	79.904	Hydrogen	H	1	1.007 94
Cadmium	Cd	48	112.41	Indium	In	49	114.82
Calcium	Ca	20	40.08	Iodine	I	53	126.9045
Californium	Cf	98	(249)[c]	Iridium	Ir	77	192.22
Carbon	C	6	12.011	Iron	Fe	26	55.847
Cerium	Ce	58	140.12	Krypton	Kr	36	83.80
Cesium	Cs	55	132.9054	Lanthanum	La	57	138.9055
Chlorine	Cl	17	35.453	Lawrencium	Lr	103	(260)[c]
Chromium	Cr	24	51.996	Lead	Pb	82	207.2
Cobalt	Co	27	58.9332	Lithium	Li	3	6.941
Copper	Cu	29	63.546	Lutetium	Lu	71	174.967
Curium	Cm	96	(247)[c]	Magnesium	Mg	12	24.305
Dysprosium	Dy	66	162.50	Manganese	Mn	25	54.9380
Einsteinium	Es	99	(252)[c]	Mendelevium	Md	101	(258)[c]

[a] The atomic masses of many elements are not invariant but depend on the origin and treatment of the material; the values given here apply to elements as they exist naturally on the earth and to certain artificial elements.

[b] For these radioactive elements the mass given is that for the longest lived isotope.

[c] Atomic masses for these radioactive elements cannot be quoted precisely without knowledge of the origin of the elements; the value given is the atomic mass number of the isotope of that element of longest known half-life.